# Recent Advances in Polyphenol Research

**Recent Advances in Polyphenol Research**

A series for researchers and graduate students whose work is related to plant phenolics and polyphenols, as well as for individuals representing governments and industries with interest in this field. Each volume in this biennial series focuses on several important research topics in plant phenols and polyphenols, including chemistry, biosynthesis, metabolic engineering, ecology, physiology, food, nutrition, and health.

**Volume 4 Editors:**

Annalisa Romani, Vincenzo Lattanzio, and Stéphane Quideau

**Series Editor-in-Chief:**

Stéphane Quideau (University of Bordeaux, France)

## Dedication

This fourth volume of *Recent Advances in Polyphenol Ressearch* is dedicated to the memory of Edwin Haslam, Professor of Physical-Organic Chemistry at the University of Sheffield, UK, who peacefully passed away at his home in Exeter on October 3, 2013, aged 81. Professor Haslam had been a long-standing and faithful member of Groupe Polyphénols, and was for many of us a model, a helpful mentor, a great colleague, and a friend. His pioneering and outstanding contributions to the field of plant polyphenols were, still are, and will continue to be a great source of knowledge and inspiration.

# Acknowledgments

The editors wish to thank all of the members of the Groupe Polyphénols Board Committee (2010–2012) for their guidance and assistance throughout this project.

***Groupe Polyphénols Board 2010–2012***

Dr Véronique Cheynier
Dr Catherine Chèze
Prof. Gilles Comte
Prof. Olivier Dangles
Dr Kevin Davies
Prof. Victor de Freitas
Prof. Ann E. Hagerman
Dr Johanna Lampe
Prof. Vincenzo Lattanzio
Dr Virginie Leplanquais
Dr Stephan Martens
Prof. Stéphane Quideau
Prof. Anna-Lisa Romani
Prof. Celestino Santos-Buelga
Prof. Kristiina Wähälä
Prof. Kumi Yoshida

# Contents

# Contributors


**Nickolas A. Anderson,** Department of Biochemistry, Purdue University, West Lafayette, IN, USA

**Renato Bruni,** The $\varphi^2$ Laboratory of Phytochemicals in Physiology, LS9 Bioactives and Health Interlab Group, Department of Food Sciences, University of Parma, Parma, Italy

**Luca Calani,** The $\varphi^2$ Laboratory of Phytochemicals in Physiology, LS9 Bioactives and Health Interlab Group, Department of Food Sciences, University of Parma, Parma, Italy

**Ilaria Campesi,** Laboratory of Sex-Gender Medicine, National Institute of Biostructures and Biosystems, Osilo, Italy

**Tak Hang Chan,** Department of Chemistry, McGill University, Montreal, QC, Canada

**Clint Chapple,** Department of Biochemistry, Purdue University, West Lafayette, IN, USA

**Di Chen,** Department of Oncology and Barbara Ann Karmanos Cancer Institute, School of Medicine, Wayne State University, Detroit, MI, USA

**C. Peter Constabel,** Centre for Forest Biology, Department of Biology, University of Victoria, Victoria, BC, Canada

**Margherita Dall'Asta,** The $\varphi^2$ Laboratory of Phytochemicals in Physiology, LS9 Bioactives and Health Interlab Group, Department of Food Sciences, University of Parma, Parma, Italy

**Daniele Del Rio,** The $\varphi^2$ Laboratory of Phytochemicals in Physiology, LS9 Bioactives and Health Interlab Group, Department of Food Sciences, University of Parma, Parma, Italy

**Richard A. Dixon,** Department of Biological Sciences, University of North Texas, Denton, TX, USA

**Q. Ping Dou,** Departments of Oncology, Pharmacology, and Pathology and Barbara Ann Karmanos Cancer Institute, School of Medicine, Wayne State University, Detroit, MI, USA

**Rebecca L. Edwards,** Institute of Food Research, Norwich Research Park, Norwich, UK

**Flavia Franconi,** Department of Biomedical Science, Centre of Excellence for Biotechnology Development and Biodiversity, University of Sassari, Sassari, Italy

**Lina Gallego-Giraldo,** Department of Biological Sciences, University of North Texas, Denton, TX, USA

**Clarissa Gerhauser,** Epigenomics and Cancer Risk Factors, German Cancer Research Center (DKFZ), Heidelberg, Germany

**Fathima R. Kona,** Department of Oncology and Barbara Ann Karmanos Cancer Institute, School of Medicine, Wayne State University, Detroit, MI, USA

**Paul A Kroon,** Polyphenols and Health Group, Institute of Food Research, Norwich Research Park, Norwich, UK

**Maria Marino,** Department of Biology, University Roma Tre, Rome, Italy

**Ken Ohmori,** Department of Chemistry, Tokyo Institute of Technology, Tokyo, Japan

**Fernando Pina,** REQUIMTE, Department of Chemistry, Faculty of Science and Technology, New University of Lisbon, Lisbon, Portugal

**M. S. Srinivasa Reddy,** Forage Genetics International, West Salem, WI, USA

**Annalisa Romani,** PHYTOLAB, Department of Statistics, Informatics, Applications, University of Florence, Florence, Italy

**Kazuki Saito,** RIKEN Center for Sustainable Resource Science, Yokohama, Japan, and Graduate School of Pharmaceutical Sciences, Chiba University, Chiba, Japan

**Juha-Pekka Salminen,** Laboratory of Organic Chemistry and Chemical Biology, Department of Chemistry, University of Turku, Turku, Finland

**Min Shen,** Department of Pharmacology, Department of Oncology and Barbara Ann Karmanos Cancer Institute, School of Medicine, Wayne State University, Detroit, MI, USA

**Keisuke Suzuki,** Department of Chemistry, Tokyo Institute of Technology, Tokyo, Japan

**Vincent Walker,** Centre for Forest Biology, Department of Biology, University of Victoria, Victoria, BC, Canada

**Keiko Yonekura-Sakakibara,** RIKEN Center for Sustainable Resource Science, Yokohama, Japan

**Kazuko Yoshida,** Centre for Forest Biology, Department of Biology, University of Victoria, Victoria, BC, Canada

# Preface

During the last 10 years there has been increasing interest in the study of plant polyphenols and their innumerable roles in a variety of very different contexts. Plant polyphenols are secondary metabolites and constitute one of the most common and widespread groups of substances in plants. Their structural diversity is likely the result of plant adaptive responses to natural selection.

Polyphenols express a large and diverse range of beneficial effects in plants and in humans consuming plant-derived food and beverages. For example, polyphenols are well known for their antioxidation activity, hormone-like behavior, and role as natural neurotransmitters, among many other biological activities. They also provide antimicrobial activity for the plant's own defense against invading pathogens.

The diversity of structures and activities of plant polyphenolic compounds has resulted in the emergence of numerous investigations in various and often interdisciplinary research areas, encompassing scientific domains as diverse as chemistry, biochemistry, biotechnology, ecology, physiology, nutrition and food chemistry, pharmacy and medicine, cosmetics, and textile technology, as well as in quality and environment controls and assessments.

It is thus the aim of the International Conference on Polyphenols, which is a biennial event that is organized under the auspices of Groupe Polyphénols, to provide scientists across disciplines with a forum for sharing new findings and for exchanging views and ideas on polyphenol research at large.

For the first time in its history, in 2012 the 26th International Conference on Polyphenols was organized in Florence, Italy. The interest in polyphenol science at the University of Florence involves many departments, including Pharmaceutical Sciences, Chemistry, Plant Sciences and Ecology, Food Science, and Medicine, as well as The Multidisciplinary Centre of Research on Food Sciences (CeRA – MCRFS) and the laboratory of Commodity Sciences and Quality Control, Environment Assessments and Certification. In these fields, particular attention has been dedicated to functional-food, nutraceutical, and cosmeceutical discoveries and applications.

At the 26th International Conference on Polyphenols, five different main topics were selected for the scientific program:

(1) *Phenols and Polyphenols Chemistry*: Covering (i) isolation and structural elucidation, and (ii) synthesis, reactivity, and physical-chemical properties.
(2) *Biosynthesis, Genetics, and Metabolic Engineering*: Dealing with biosynthesis and genetic manipulation.

(3) *Roles in Plants and Ecosystems*: Covering phenolic functions in plants and correlation with biotic and abiotic stresses.
(4) *Health and Nutrition*: Focusing on polyphenol metabolism and bioavailability, as well as cancer prevention and perspectives on gender-dependent human health effects.
(5) *Polyphenols and Drug Discovery*: Including new findings on sources of isolated and standardized polyphenolic fractions and novel epigenetic polyphenol mechanisms.

More than 400 scientists from 42 countries attended the conference in July 2012, with nearly 400 paper contributions, comprising 52 oral communications and 327 poster presentations (Fig. P.1).

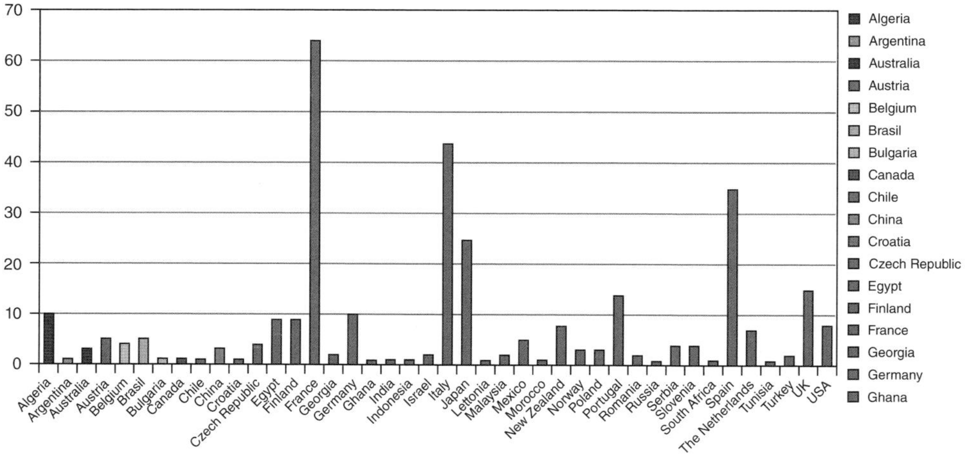


**Fig. P.1** Contributions to the 26th International Conference on Polyphenols (number of papers presented) by country.

The success of this 26th edition of the International Conference on Polyphenols would not have been possible without the support of both public and private sponsors. The Scientific and Technological Pole and the Social Pole of the University of Florence, PIN of Prato, the National Council for Research, and several private-company sponsors (Agilent Technologies, BioTech Power, Indena, ISR Ecoindustria, Domus Olea, Force A, Biokyma, PhenoFarm, Dermaresia, Silva Team, Bioscen Future) are gratefully acknowledged.

All of the lectures, oral communications, and ensuing discussions and debates were broadcast live on RadioSpin, the University of Florence webradio, and through Ustation (the Italian university radio stations network), on the other connected university radios of the network. These radiophonic conference proceedings are available in podcasts on the RadioSpin Web site: www.radiospin.it.

Annalisa Romani
Vincenzo Lattanzio
Stéphane Quideau

Chapter 1

# Monolignol Biosynthesis and its Genetic Manipulation: The Good, the Bad, and the Ugly

Richard A. Dixon[1], M.S. Srinivasa Reddy[2], and Lina Gallego-Giraldo[1]
[1]*Department of Biological Sciences, University of North Texas, Denton, TX, USA*
[2]*Forage Genetics International, West Salem, WI, USA*

**Abstract:** Economic and environmental factors favor the adoption of lignocellulosic bioenergy crops for production of liquid transportation fuels. However, lignocellulosic biomass is recalcitrant to saccharification (sugar release from cell walls), and this is, at least in part, due to the presence of the phenylpropanoid-derived cell-wall polymer lignin. A large body of evidence exists documenting the impacts of lignin modification in plants. This technology can lead to improved forage quality and enhanced processing properties for trees (paper pulping) and lignocellulosic energy crops. We here provide a comprehensive review of the literature on lignin modification in plants. The pathway has been targeted through down-regulation of the expression of the enzymes of the monolignol pathway and down-regulation or over-expression of the transcription factors that control lignin biosynthesis and/or programs of secondary cell-wall development. Targeting lignin modification at some steps in the monolignol pathway can result in impairment of plant growth and development, often associated with the triggering of endogenous host-defense mechanisms. Recent studies suggest that it may be possible to decouple negative growth impacts from lignin reduction.

**Keywords:** monolignol biosynthesis, genetic modification, transcription factor, gene silencing, saccharification

*Recent Advances in Polyphenol Research, Volume 4*, First Edition.
Edited by Annalisa Romani, Vincenzo Lattanzio, and Stéphane Quideau.

## 1.1 Introduction

Lignin is a major component of plant secondary cell walls, and the second most abundant plant polymer on the planet. It constitutes about 15–35% of the dry mass of vascular plants (Adler, 1977). Considerable attention has been given over the past several years to the reduction of lignin content in model plant species, forages, trees, and dedicated bioenergy feedstocks. This is because forage digestibility, paper pulping, and liquid fuel production from biomass through fermentation are all affected by recalcitrance of lignocellulose, primarily due to the presence of lignin, which blocks access to the sugar-rich cell-wall polysaccharides cellulose and hemicellulose for enzymes and microorganisms (Pilate *et al.*, 2002; Reddy *et al.*, 2005; Chen & Dixon, 2007).

Much is now known of the biosynthesis of lignin and its control at the transcriptional level. This informs the targets that have been selected for genetic modification of lignin content and composition in transgenic plants. Which gene is down- or up-regulated has a considerable effect on lignin content and composition. Equally, lignin modification can have profound impacts on plant growth and development, ranging from good through bad to "downright ugly," but these impacts are again strongly target-dependent. Understanding the mechanisms that can impact plant growth – which equate to agronomic performance – in crop species "improved" through lignin modification is critical for economic advancement of the forage and biofuels industries. Although still poorly understood, these mechanisms may also throw light on basic plant developmental and defense processes.

## 1.2 Function and distribution of lignin in plants

Lignin is an aromatic heteropolymer derived primarily from three hydroxycinnamyl alcohols: 4-coumaryl alcohol, coniferyl alcohol, and sinapyl alcohol, which give rise, respectively, to the 4-hydroxyphenyl (H), guaiacyl (G), and syringyl (S) subunits of lignin (Freudenberg & Neish, 1968; Ralph *et al.*, 2004). G units are mono-methoxylated, S units are di-methoxylated, and H units are not methoxylated (Fig. 1.1). These monomers are linked through oxidative coupling catalyzed by both peroxidases and laccases (Boudet *et al.*, 1995). Unlike cellulose and other polymers that have labile linkages (e.g. glycosidic or peptide) between their building blocks, the units of lignin are linked by strong ether and carbon–carbon bonds (Sarkanen, 1971). Lignin is present in the secondarily thickened cell walls of plants, where it is critical to cell-wall structural integrity and gives strength to stems (Chabannes *et al.*, 2001b; Jones *et al.*, 2001). Lignin also imparts hydrophobicity to vascular elements for water transport. The lignin content of the mature internodes of stems of alfalfa (*Medicago sativa*), the world's major forage legume and a target of much of the work to be described in this article, is about 17% of the dry weight (Guo *et al.*, 2001a).

Lignin composition varies among major phyla of vascular plants (Boerjan *et al.*, 2003). Dicotyledonous and monocotyledonous angiosperm lignins contain G and S units as the two major monomer species, with low levels of H units. Monocotyledonous lignins have more H units than dicotyledonous lignins (Baucher *et al.*, 1998), but care must be taken not to attribute other components to H units, as often happens (Boerjan *et al.*, 2003). Fern

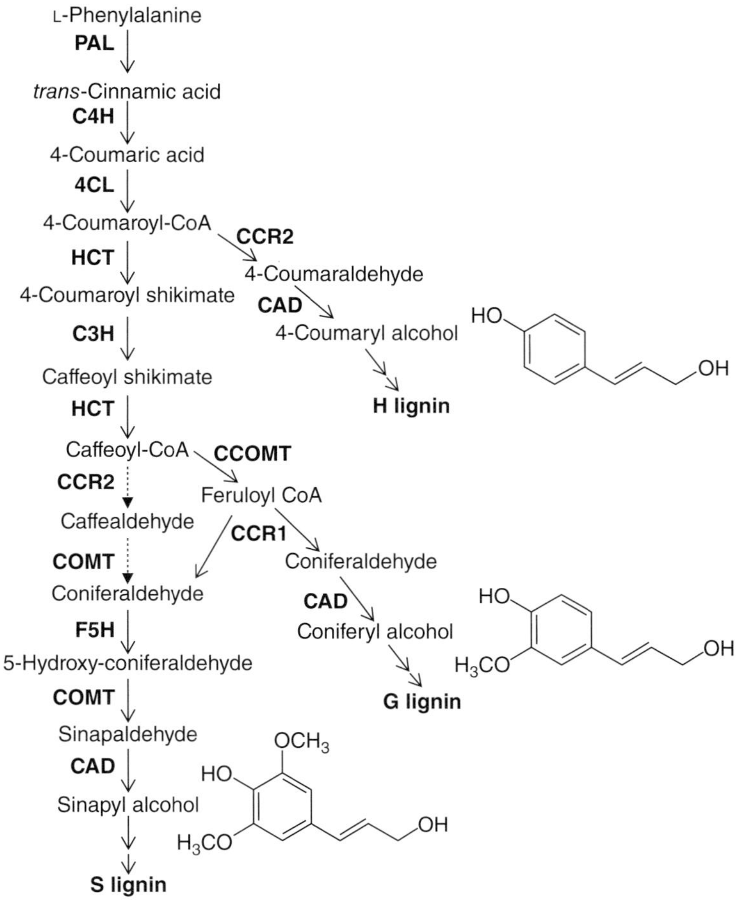


**Fig. 1.1** Scheme for monolignol biosynthesis in dicotyledonous angiosperms, including revisions encompassing the different biochemical activities of cinnamoyl-CoA reductase (CCR) forms in *Medicago truncatula* (Zhou *et al.*, 2010). See text for enzyme abbreviations.

and gymnosperm lignins have primarily G units and low levels of H units, but S units have been found in cuplet fern, yew plum pine, sandarac-cypress, and a few genera in the Gnetophyta (Weng *et al.*, 2008b). Some lower plants, like *Selaginella moellendorfii* (Weng *et al.*, 2008a,b) and *Marchantia polymorpha*, have both G and S units in their lignins (Espineira *et al.*, 2011), despite predating hardwoods/dicots and even softwoods. The apparent presence of H, G, and S units in the lignin from the seaweed *Calliarthron cheilosporioides* (Martone *et al.*, 2009) may indicate convergent evolution of lignin.

The presence of each methoxyl group on a monolignol unit results in one less reactive site, and therefore fewer available potential coupling combinations during polymerization. Thus, S lignin is more linear and less crosslinked than G/S lignin, and provides a strong yet flexible polymer that is especially advantageous to herbaceous angiosperms (Bonavitz & Chapple, 2010). A correlation has been shown between the degradability of the cell walls in forages and the amount of G lignin, as lignin rich in G units is more highly condensed, making it less amenable to degradation (Jung & Deetz, 1993). Thus, transgenic poplar plants with lignin rich in G units are, like softwoods, more difficult to pulp because of their more condensed lignin (Lapierre *et al.*, 1999).

Lignin content increases with progressive maturity of stems; this relationship has been studied in detail in alfalfa (Jung *et al.*, 1997; Chen *et al.*, 2006), ryegrass (Tu *et al.*, 2010), tall fescue (Buxton & Redfearn, 1997; Chen *et al.*, 2002), and switchgrass (Mann *et al.*, 2009; Shen *et al.*, 2009). Decreasing the lignin content increases the digestibility of alfalfa for ruminant animals (Baucher *et al.*, 1999; Guo *et al.*, 2001a,b; Reddy *et al.*, 2005) and improves processing efficiency for the production of liquid biofuels through saccharification and fermentation (Chen & Dixon, 2007). Lignin composition has also been linked with reduced cell-wall digestibility (Jung & Deetz, 1993). However, the importance of lignin composition for digestibility has been questioned based on the results of studies with synthetic lignins, which show lignin composition *per se* to have no effect (Grabber *et al.*, 1997).

Plants have primary and secondary cell walls, which differ in both function and composition. Primary walls allow cells to expand and divide, while providing mechanical strength. Once cell growth stops, a much thicker secondary cell wall is deposited in some specialized cell types. These include vessels and fibers in the stem, sclereid cells, endodermal tissue of roots, some cells of anthers and pods important for dehiscence (Zhong & Ye, 2009), and seed coats (Marles *et al.*, 2008; Chen *et al.*, 2012). Generally, secondary cell walls consist of three layers, named S1 (outer), S2 (middle), and S3 (inner). Lignin deposition starts at the cell corners in the region of the middle lamella and the primary wall when S1 formation has started. Most of the lignin is deposited in the S2 layer and impregnates the cellulose and hemicelluloses there (Donaldson, 2001; Boerjan *et al.*, 2003). Based on UV microscopy, the density of lignin is higher in the middle lamella and primary walls than in the secondary walls of secondarily thickened cells, but the secondary walls have more lignin content as they constitute the largest proportion of the total cell wall (Fergus *et al.*, 1969). Usually H units are deposited first during cell-wall formation, followed by G units and then S units (Terashima *et al.*, 1993, 1998; Donaldson, 2001). However, S units have been identified in lignin from corn coleoptiles, indicating that S lignin deposition may also start early in development (Musel *et al.*, 1997). H lignin is believed to determine the shape of the cells by acting as a matrix for deposition of G and S units (Terashima *et al.*, 1998). Vascular cells without H units may be free to expand and assume a round shape. In general, a higher amount of G units is present in vessels than in fibers, which are rich in S units (Saka & Goring, 1985).

There is considerable variation in lignin content and composition not only between different plant species but also in different tissues of the same plant, between various developmental stages of the plant, and in response to environmental conditions (Terashima *et al.*, 1993; Musel *et al.*, 1997; Vermerris & Boon, 2001; Donaldson, 2002; Chen *et al.*, 2006). For example, the S/G ratio increases to alter the cell-wall mechanical properties in poplar plants grown under simulated wind influence compared to plants grown in non-windy conditions (Koehler & Telewski, 2006).

Two examples, one for a bioenergy crop (switchgrass, *Panicum virgatum*), the other for a forage crop (alfalfa), are given to demonstrate the extent of variation in lignin encountered in wild-type, non-genetically-modified plant biomass. The lignin contents and compositions of switchgrass cv. Alamo grown in the field, in greenhouses, and in growth chambers were compared using different techniques (Mann *et al.*, 2009). Lignin content was not different in leaves from different parts of the tiller, but stem tissues had increasing lignin content from the top to the bottom of the tiller. Younger stem tissue from the field had slightly higher lignin content compared to the greenhouse- and growth chamber-grown plants. The S/G ratio of leaf and stem tissues varied between different environments (Mann *et al.*, 2009). Similar observations have been made in tall fescue (Chen *et al.*, 2002) and perennial ryegrass (Tu *et al.*, 2010), where lignin content increases moderately during the stem-elongation stage and then dramatically on progression from the elongation to the reproductive stage.

In a study with alfalfa, the lignin content of young internodes (internodes 1–2) was 93 mg/g CWR (cell-wall residue), increasing towards a value of 250 mg/g CWR in the mature eighth internode (Chen *et al.*, 2006). This was accompanied by an increase in S/G ratio from 0.087 to 0.640. The lignin contents (thioacidolysis yields) and lignin monomer compositions of individual cell types (vascular elements, phloem fibers, and vascular parenchyma) from the fifth internode were quite different (Nakashima *et al.*, 2008). Thioacidolysis yields were higher in vascular cells than in parenchyma and fiber cells (430, 267, and 76 μmol/g dry weight, respectively), with fiber and parenchyma cells enriched in S lignin units with an S/G ratio of 0.60 and 0.72 respectively, and vascular cells enriched in G lignin units with an S/G ratio of 0.17 (Nakashima *et al.*, 2008). The H/total lignin ratios were 0.06, <0.01, and 0.03, respectively, in vascular elements, fiber, and parenchyma cells from the fifth internodes of greenhouse-grown alfalfa.

Coherent anti-Stokes Raman scattering microscopy (CARS) has been used to determine the spatial distribution of lignin across secondary cell walls from the stems of alfalfa plants (Zeng *et al.*, 2010). At the tissue level, CARS intensity decreased in the order fiber > xylem > epidermis > phloem > parenchyma. In general, the CARS signal at the cellular level was highest in the cell corner compared to the compound middle lamella (middle lamella and primary walls from adjacent cells), and the signal in the middle lamella was higher than that in the secondary walls (Zeng *et al.*, 2010).

## 1.3 Targets for modification of lignin biosynthesis

Fig. 1.1 shows a current view of the pathways for monolignol biosynthesis, and Fig. 1.2 outlines the transcriptional control mechanisms that regulate lignin deposition during secondary cell-wall formation. Both biosynthetic enzymes and transcription factors (TFs) have been targeted to reduce (or occasionally increase) lignin levels. In the following sections we briefly describe the gene targets and then describe the effects of their down-regulation on lignin content and composition and plant phenotype in model systems, forages, and industrial pulp or bioenergy species.

### *1.3.1 Gene targets 1. Biosynthetic enzymes*

The reader is referred to Fig. 1.1, which illustrates the positions of the various enzymes in the monolignol pathway.

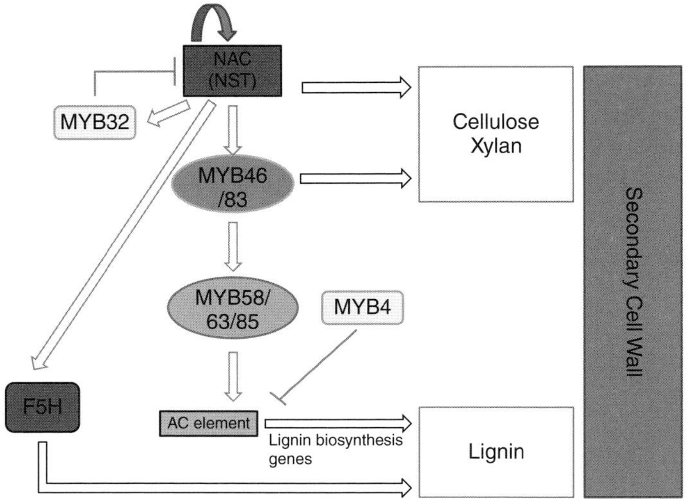


**Fig. 1.2** Transcriptional controls for the biosynthesis of monolignols in the context of secondary cell-wall biosynthesis (based on Arabidopsis and *Medicago truncatula*). Both NAC and MYB genes can activate the entire secondary cell-wall biosynthesis pathway. In *M. truncatula*, F5H (ferulate 5-hydroxylase) is regulated by the NAC master switch (which is also under autoregulatory control), whereas other lignin genes are regulated by MYB58/63/85 through AC elements in their promoters. MYB 4 is a lignin/phenylpropanoid pathway repressor.

#### *1.3.1.1 L-phenylalanine ammonia-lyase (PAL)*

PAL has been characterized biochemically from many plant species since its discovery in 1961 (Koukol & Conn, 1961). *PAL* genes were first cloned from French bean (*Phaseolus vulgaris*) (Edwards *et al.*, 1985). The enzyme is tetrameric, and contains an unusual methylidene imidazolone residue at the active site that is formed post-translationally (Calabrese *et al.*, 2004). PAL is usually encoded by a multigene family, and it is possible that expression of different members can lead to formation of heterotetramers, the functional significance of which is not clear (Reichert *et al.*, 2009).

#### *1.3.1.2 Cinnamate 4-hydroxylase (C4H)*

C4H catalyzes 4-hydroxylation of cinnamic acid to 4-coumaric acid (Russell & Conn, 1967; Russell, 1971). C4H is the most abundant plant cytochrome P450 enzyme. The cloning of the *C4H* gene was described almost simultaneously from alfalfa (Fahrendorf & Dixon, 1993), artichoke (Teutsch *et al.*, 1993), and mung bean (Mizutani *et al.*, 1993).

#### *1.3.1.3 4-coumarate: coenzyme-A ligase (4CL)*

4CL has been studied extensively since the 1970s (Hahlbrock & Grisebach, 1970; Knobloch & Hahlbrock 1975, 1977). The enzyme converts hydroxycinnamic acids, preferably 4-coumaric acid, to the corresponding coenzyme-A thioesters. 4CL exists as a multigene family in those species studied to date (Hu *et al.*, 1998; Ehlting *et al.*, 1999; Lindermayr *et al.*, 2002; Dixon & Reddy, 2003; Xu *et al.*, 2009). Only four of the eleven putative 4CLs in Arabidopsis appear to encode catalytically active 4CL enzymes, and the knock-out mutant of At4CL5 does not show any changes in lignin content or monomer composition (Costa *et al.*, 2005), suggesting functional redundancy.

### *1.3.1.4 Enzymes of the coumaroyl shikimate shunt*

Even though the biochemical formation of 4-coumaroyl shikimate and 4-coumaroyl quinate had been known for many years (Rhodes & Wooltorton, 1976; Ulbrich & Zenk, 1980), it was not originally appreciated that these reactions might be involved in lignin biosynthesis. Discovery of the *Arabidopsis thaliana* cytochrome P450-dependent monooxygenase enzyme CYP98A3 (4-coumaroyl shikimate 3′-hydroxylase, C3′H), which hydroxylates the shikimate and quinate esters of 4-coumarate, prompted a revision of the monolignol pathway with the suggestion of a new route for 3-hydroxylation of the 4-hydroxyphenyl moiety (Schoch *et al.*, 2001; Franke *et al.*, 2002a). Soon after, tobacco hydroxycinnamoyl-CoA: shikimate hydroxycinnamoyl transferase (HCT) was characterized (Hoffmann *et al.*, 2003). The monolignol pathway was then revised to involve HCT utilizing 4-coumaroyl-CoA as the acyl donor and shikimate or quinate as the acceptor, followed by 3-hydroxylation of the 4-coumaroyl moiety by C3′H, leading to a caffeoyl ester and its subsequent conversion to caffeoyl-CoA by HCT acting in the reverse direction (Fig. 1.1). Identification of a separate hydroxycinnamoyl-CoA: quinate hydroxycinnamoyl transferase (HQT) involved in the synthesis of chlorogenic acid (caffeoyl quinate) (Niggeweg *et al.*, 2004) suggested that the 4-coumaroyl ester of shikimate was likely the preferred intermediate in lignin biosynthesis. In tomato, HQT down-regulation or over-expression did not change lignin content but led, respectively, to a decrease or an increase in chlorogenic acid content (Niggeweg *et al.*, 2004).

There is good genetic evidence for the operation of the "shikimate shunt" in lignin biosynthesis in several plant species (Franke *et al.*, 2002a; Hoffmann *et al*, 2004; Reddy *et al.*, 2005; Shadle *et al.*, 2007; Wagner *et al.*, 2007; Coleman *et al.*, 2008a). However, it is still not clear whether this pathway operates universally (e.g. in monocots). Other enzyme systems are known to exist for the conversion of a coumaroyl moiety to a caffeoyl moiety (e.g. Kneusel *et al.*, 1989), although most have yet to be analyzed beyond the level of protein biochemistry.

### *1.3.1.5 Caffeoyl-CoA 3-O-methyltransferase (CCoAOMT)*

CCoAOMT is an S-adenosyl L-methionine and divalent cation-dependent *O*-methyltransferase that preferentially converts caffeoyl-CoA to feruloyl-CoA (Kuhnl *et al.*, 1989; Ye *et al.*, 1994; Inoue *et al.*, 1998; Parvathi *et al.*, 2001). Demonstration of the involvement of CCoAOMT in lignin biosynthesis first came from studies on xylogenesis in Zinnia (Ye *et al.*, 1994), and this resulted in the first major revision of the monolignol pathway. Previously, caffeic acid 3-*O*-methyltransferase (COMT) was believed to be involved in methylation at both the C3 and C5 positions of monolignols (Finkle & Nelson, 1963; Davin & Lewis, 1992). The alfalfa CCoAOMT crystal structure has been obtained, and the enzyme forms a homodimer in solution, although the dimer is not necessary for substrate recognition and transmethylation as the substrate and cofactor interact with the monomer (Ferrer *et al.*, 2005).

#### *1.3.1.6 Ferulate 5-hydroxylase (F5H)*

F5H is the third cytochrome P450-dependent monooxygenase enzyme in the monolignol pathway. The *F5H* gene was cloned from the *fah1* mutant of Arabidopsis using a forward-genetics approach (Chapple *et al.*, 1992). The F5H enzyme has a higher affinity for coniferaldehyde and coniferyl alcohol compared to ferulate and is therefore more correctly referred to as coniferaldehyde 5-hydroxylase or Cald5H (Humphreys *et al.*, 1999; Osakabe *et al.*, 1999). This discovery led to a reappraisal of the monolignol pathway that no longer supported involvement of ferulate and sinapate in lignin biosynthesis.

#### *1.3.1.7 Caffeic acid 3-*O*-methyltransferase (COMT)*

COMT has been studied for many years (Finkle & Nelson, 1963). However, the common name of the enzyme appears to be a misnomer; caffeic acid may not be a substrate for COMT during monolignol biosynthesis, as COMT from many species, including Arabidopsis, aspen, and alfalfa, has a significantly higher affinity for 5-hydroxyconiferaldehyde than for caffeic acid (Li *et al.*, 2000; Parvathi *et al.*, 2001). In Arabidopsis, *O*-methylation of 5-hydroxyconiferyl alcohol is inhibited in the presence of 5-hydroxyconiferaldehyde such that, when both substrates are present, AtCOMT preferentially catalyzes *O*-methylation of 5-hydroxyconiferaldehyde (Nakatsubo *et al.*, 2008). Alfalfa COMT can efficiently methylate caffealdehyde and caffeyl alcohol (Parvathi *et al.*, 2001). COMT from tall fescue (Chen *et al.*, 2004) and wheat (Ma & Xu, 2008) efficiently utilizes both caffealdehyde and 5-hydroxyconiferaldehyde. Further studies are needed to determine unequivocally the preferred routes for monolignol *O*-methylation *in vivo*.

#### *1.3.1.8 Cinnamoyl-CoA reductase*

Cinnamoyl-CoA reductases (CCRs) are involved in the reduction of hydroxycinnamoyl-CoA thioesters to the corresponding aldehydes, and have been studied for many years (Gross *et al.*, 1973). There are two well-characterized CCRs in Arabidopsis: AtCCR1 is five times more efficient with feruloyl-CoA and sinapoyl-CoA than is AtCCR2, and is involved in developmentally regulated lignification, whereas AtCCR2 is expressed in response to pathogen infection and hence may be involved in disease resistance (Lauvergeat *et al.*, 2001). Feruloyl-CoA and caffeoyl-CoA are the most and least preferred substrates, respectively, for CCRs from Arabidopsis (Patten *et al.*, 2005). Feruloyl-CoA and sinapoyl-CoA are the preferred substrates for *M. truncatula* MtCCR1, whereas caffeoyl-CoA and 4-coumaroyl-CoA are the preferred substrates for MtCCR2 (Zhou *et al.*, 2010). MtCCR2 may be involved in a route to lignin biosynthesis whereby caffeoyl-CoA is converted to caffealdehyde, which is then 3-*O*-methylated to coniferaldehyde by COMT (Zhou *et al.*, 2010) (Fig. 1.1), a pathway previously suggested to occur in Arabidopsis (Do *et al.*, 2007).

#### *1.3.1.9 Cinnamyl alcohol dehydrogenase (CAD)*

CAD is a zinc-dependent enzyme that catalyzes the reduction of hydroxy-cinnamaldehydes to their corresponding alcohols (Mansell *et al.*, 1974). Coniferaldehyde and sinapaldehyde are preferred substrates for CAD in tall fescue (Chen *et al.*, 2003).

### *1.3.2 Gene targets 2. Transcription factors*

The genes involved in cellulose, xylan, and lignin synthesis are coordinately expressed during secondary wall biosynthesis. In the past decade, many of the TFs involved in the biosynthesis of lignin and the other secondary cell-wall polymers have been identified and characterized (Fig. 1.2). In Arabidopsis, several closely related NAC TFs regulate secondary wall biosynthesis; these are NST1 (NAC secondary wall-thickening promoting factor 1), NST2/NST3/SND1 (secondary wall-associated NAC domain protein 1), VND6 (vascular-related NAC-domain 6), and VND7 (Mitsuda *et al.*, 2005, 2007; Zhong *et al.*, 2006; Yamaguchi *et al.*, 2008). Although two *NST* genes act redundantly to control lignification in the interfascicular tissues in the stem of Arabidopsis (Mitsuda *et al.*, 2005, 2007), a single *NST* gene was cloned from the model legume *Medicago truncatula* by forward-genetic screening and shown to function as a master switch for secondary wall synthesis in various tissues (Zhao *et al.*, 2010a). MYB-family TFs are important for the biosynthesis of other secondary wall components, and act downstream of the *NST* genes (Zhong *et al.*, 2007; Ko *et al.*, 2009; McCarthy *et al.*, 2009; Zhou *et al.*, 2009) (Fig. 1.2). Among these, MYB46 is a direct target of NST3/SND1 *in vitro* and *in vivo*, and may be the master switch that turns on expression of genes for the biosynthesis of cellulose, xylan, and lignin (Zhong *et al.*, 2007; Ko *et al.*, 2009). Detailed promoter and electrophoretic mobility shift-assay analyses have revealed that AC-rich elements corresponding to MYB TF binding motifs in the promoters of many monolignol pathway genes are necessary for the coordinated activation of these genes (Lacombe *et al.*, 2000; Patzlaff *et al.*, 2003; Zhou *et al.*, 2009). Interestingly, F5H is not regulated by a "lignin MYB" gene in *M. truncatula*, but directly by the upstream NAC TF (Zhao *et al.*, 2010b). Finally, MYB and WRKY TFs can act as transcriptional repressors of lignification (Wang *et al.*, 2010; Shen *et al.*, 2012). Further details of the TFs that regulate secondary wall formation can be found in the following review articles: Demura & Fukuda (2007), Zhong & Ye (2007), Demura & Ye (2010), and Zhong *et al.* (2010).

## 1.4 Impacts of lignin modification through targeting of the monolignol biosynthetic pathway

In this section we review both the biochemical and the broader plant-growth phenotypes that result from down-regulation of monolignol biosynthesis at each of the enzymatic steps in the pathway. A similar analysis of the impacts of targeting the TFs that affect the monolignol pathway is given in Section 1.5. Clearly, reducing lignin levels in transgenic plants is facile. However, depending on the step targeted, the results can be good, bad, or ugly!

### *1.4.1 L-phenylalanine ammonia-lyase (PAL)*

The lignin content of PAL down-regulated alfalfa total stem tissue (internodes 1–8) was 97 mg/g of total CWR, compared to ~150 mg/g CWR for control alfalfa, and varied from 72 mg/g CWR in internodes 1–2 to 127 mg/g CWR in the eighth internode, indicating that the lignin content was reduced in all developmental stages of the stem (Chen *et al.*, 2006). Analysis of the cell-specific effects of PAL down-regulation in the fifth internode, using laser-capture microdissection to collect specific cell types, showed an increase in S/G ratio in vascular elements and fibers but not in parenchyma (Nakashima *et al.*, 2008). Cell-wall autofluorescence almost completely disappeared in the stem sections of these plants, except in a few vascular cells.

Down-regulation of PAL in transgenic tobacco leads to a decrease in lignin content (Elkind *et al.*, 1990; Bate *et al.*, 1994; Sewalt *et al.*, 1997a; Korth *et al.*, 2001), altered leaf shape and texture, stunted growth, reduced pollen viability, and altered flower morphology and pigmentation. The leaves have 10-fold reduced levels of the hydroxycinnamate ester chlorogenic acid and 70% reduced salicylic acid content compared to controls (Pallas *et al.*, 1996). The plants are more susceptible to the fungal pathogen *Cercospora nicotianae* (Maher *et al.*, 1994), and to viral pathogens (Felton *et al.*, 1999).

*Arabidopsis thaliana* has four PAL genes. Single mutants in *pal1* or *pal2*, *pal1 pal2* double mutants, and *pal1 pal2 pal3* triple mutants show normal morphology with a slight to significant reduction in lignin content (Rohde *et al.*, 2004; Huang *et al.*, 2010). In contrast, *pal1 pal2 pal4* triple mutants and *pal1 pal2 pal3 pal4* quadruple mutants have a dwarf phenotype and are sterile (Huang *et al.*, 2010). The *pal1 pal2* double mutants have lower levels of feruloyl malate esters and increased phenylalanine content, lack three kaempferol glycosides, and are sensitive to UV light, but are more drought-resistant than wild-type (Rohde *et al.*, 2004; Huang *et al.*, 2010). The quadruple mutant has lower salicylic acid levels and is susceptible to the bacterial pathogen *Pseudomonas syringae*.

In *Salvia miltiorrhiza*, PAL down-regulation leads to a reduction in the contents of lignin and the water-soluble phenolics rosmarinic acid and salvianolic acid B. This is accompanied by stunted growth, altered leaf morphology, and delayed root formation (Song & Wang, 2011).

Because of the extensive negative pleiotropic effects of PAL down-regulation on phenylpropanoid metabolism and growth, this enzyme is probably not a suitable target for reducing the recalcitrance of cell-wall material in forage or bioenergy crops.

### *1.4.2 Cinnamate 4-hydroxylase (C4H)*

C4H down-regulation in alfalfa leads to reduced lignin content in stem tissue, accompanied by a decreased S/G ratio, but has less of an effect on soluble phenolic compounds than is observed in PAL down-regulated plants (Reddy *et al.*, 2005; Chen *et al.*, 2006). Based on laser-capture microdissection analysis, the S/G ratio is decreased in vascular and fiber cells but is similar to control values in parenchyma cells (Nakashima *et al.*, 2008). Stem sections of these plants show a few distorted vascular cells, and the sections have very low lignin autofluorescence, much like PAL down-regulated alfalfa. Vascular cells are smaller than in wild-type, and reduced in number (Fig. 1.3a). The saccharification efficiency of C4H down-regulated alfalfa is higher than that of wild-type plants, but lower than that of HCT down-regulated alfalfa (Chen & Dixon, 2007). C4H down-regulated alfalfa forage is approximately 15% more digestible in the rumens of steers than is forage from wild-type plants (Reddy *et al.*, 2005).

Down-regulation of C4H activity in tobacco results in a reduced lignin content with decreased S/G ratio, and in decreased levels of chlorogenic acid and soluble esters of caffeic acid in leaf tissue (Sewalt *et al.*, 1997a; Blount *et al.*, 2000; Blee *et al.*, 2001). C4H mis-sense mutants of Arabidopsis show decreased lignin content and decreased levels of several phenolic compounds, as well as pleiotropic effects including male sterility (Schilmeller *et al.*, 2009). Sense suppression of C4H in transgenic tomato results in reduced lignin content with increased S/G ratio (Millar *et al.*, 2007), in contrast to the decreased S/G ratio in tobacco and alfalfa. Down-regulation of C4H in all these species results in a dwarf phenotype.

### *1.4.3 4-coumarate: coenzyme-A ligase (4CL)*

Down-regulation of 4CL activity results in decreased lignin content in alfalfa, Arabidopsis, tobacco, aspen (*Populus tremuloides*), and hybrid white poplar (*Populus tremula* X *Populus alba*). This is accompanied by a slight increase in the S/G ratio and increased salicylic acid levels in alfalfa, a greater increase in S/G ratio in Arabidopsis, an unchanged S/G ratio in aspen, and decreased S/G ratio in hybrid white poplar.

Autofluorescence of stem sections of 4CL down-regulated alfalfa confirms an overall reduction in lignin content, while Mäule staining shows lower amounts of S lignin in vascular cells and pith rays (Fig. 1.3a). The lignin thioacidolysis yield is most noticeably decreased in vascular and fiber cells (Nakashima *et al.*, 2008). These lines also have reduced mean areas of individual vascular cells and decreased numbers of vascular cells compared to wild-type. The biomass yield of 4CL down-regulated alfalfa at early bud stage is decreased by around 37% (Nakashima *et al.*, 2008).

Down-regulation of 4CL in tobacco decreases the lignin content and leads to plants with a dwarf phenotype and/or brown stems. In these plants, the amount of 4-hydroxybenzaldehyde increases, levels of vanillin and syringaldehyde are reduced, and there are changes in several other phenolic compounds (Kajita *et al.*, 1996, 1997).

The reduction in lignin level in Arabidopsis following down-regulation of 4CL is achieved by a decrease in G units but not S units, leading to a higher S/G ratio, and the plants appear phenotypically normal (Lee *et al.*, 1997).

Transgenic aspen trees with suppressed Pt4CL1 expression have an up to 45% reduction in lignin content with no significant change in S/G ratio. The reduction in lignin content has been proposed to be compensated for by an up to 15% increase in cellulose content, accompanied by increased or unchanged biomass (Hu *et al.*, 1999; Li *et al.*, 2003). It is debatable as to whether the plants are synthesizing additional cellulose or the effect is simply one of compensation. Aspen trees with 4CL1 down-regulation and F5H over-expression show an additive effect, with an up to 52% reduction in lignin, 64% higher S/G ratio, and 30% more cellulose (Li *et al.*, 2003).

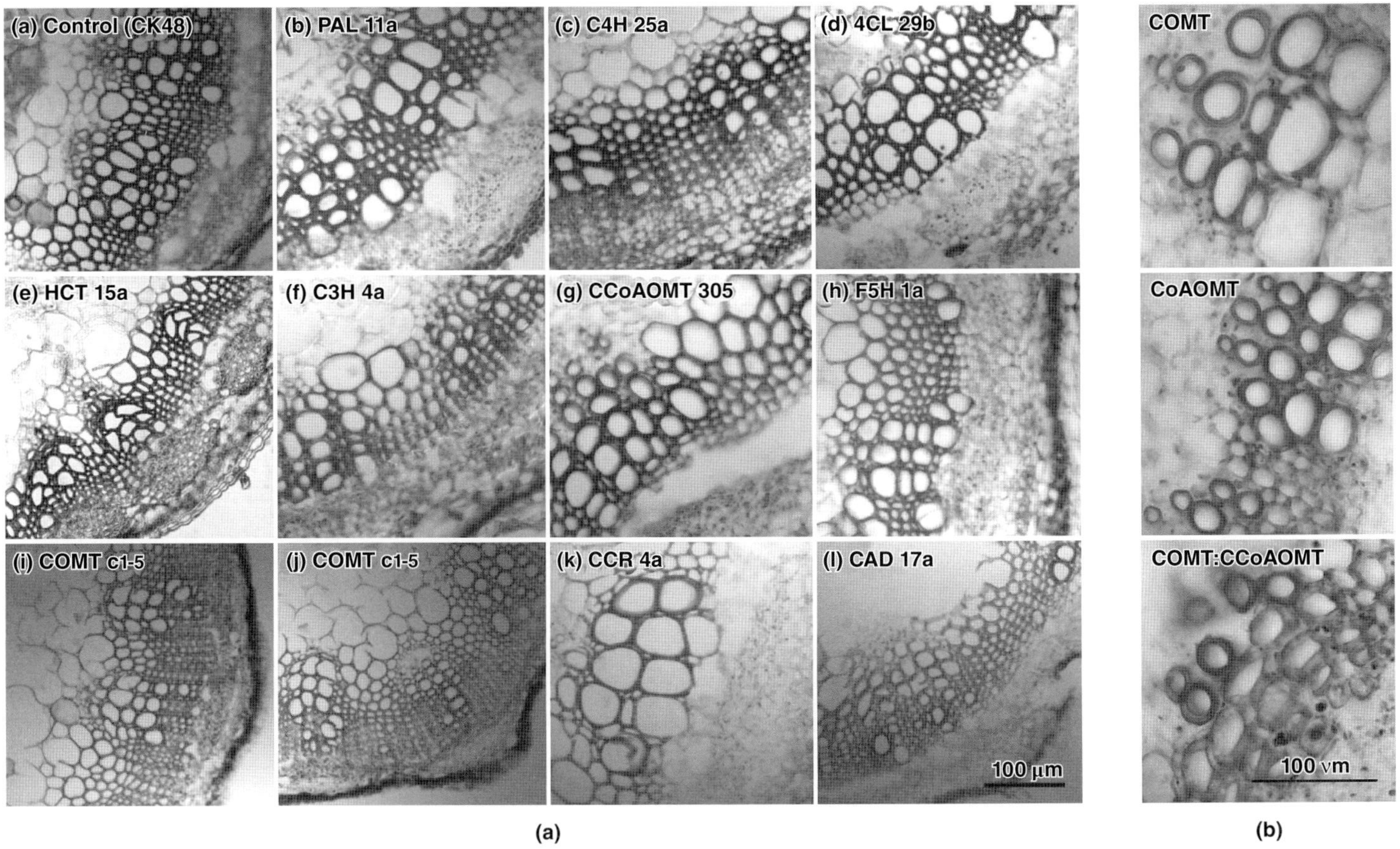


**Fig. 1.3** Cellular impacts of the targeting of monolignol biosynthesis in alfalfa through RNAi-mediated silencing of specific enzymatic steps. (a) Micrographs showing cross-sections through the fifth internodes of transgenic alfalfa plants down-regulated independently at each of the enzymatic steps shown. (b) As in (a), but showing a comparison of single down-regulation of COMT or CCoAOMT with the result of down-regulating both genes simultaneously. (Nakashima *et al.*, 2008. Reproduced with permission of John Wiley & Sons.) For color details, please see Plate 1.3.

Transgenic field-grown hybrid white poplar trees down-regulated in 4CL1 have decreased lignin content with decreased S/G ratio (Voelker *et al.*, 2010). However, unlike aspen (Hu *et al.*, 1999; Li *et al.*, 2003), the white poplar with greatest suppression of 4CL shows reduced biomass, with brown stems similar to those seen in tobacco (Kajita *et al.*, 1996) and alfalfa (Nakashima *et al.*, 2008). The brown wood is enriched in phenolics such as naringenin, dihydrokaempferol, and their corresponding glucosides. However, the saccharification rates of these trees do not correlate well with reduced lignin content.

### 1.4.4 Hydroxycinnamoyl-CoA: shikimate hydroxycinnamoyl transferase (HCT)

Silencing of HCT causes a strong reduction of lignin content and a large increase of H units in alfalfa, Arabidopsis, *Nicotiana benthamiana*, and *Pinus radiata*. This is accompanied by a dwarf phenotype in alfalfa, Arabidopsis, and *N. benthamiana*.

In alfalfa, silencing of HCT results in an approximately 50% reduction in lignin content, accompanied by a decrease in S and G units and a corresponding increase in H units from trace amounts to almost 50% of the total lignin, as determined by thioacidolysis. The H unit-rich lignin is of low molecular weight and more extractable than the bulk lignin (Ziebell *et al.*, 2010). The thioacidolysis yield of HCT down-regulated alfalfa is more severely reduced in parenchyma and fiber cells compared to vascular cells. Stem sections show very low lignin autofluorescence in vascular cells, which are very small (<50% of the size of the cells in the control), reduced in number, and severely distorted (Fig. 1.3a). The pith rays show a characteristic pattern, with S lignin localized to the middle layers of the cell walls (Nakashima *et al.*, 2008). Levels of wall-bound vanillin and ferulic acid are decreased, whereas levels of 4-hydroxybenzaldehyde and 4-coumaric acid are increased (Chen *et al.*, 2006).

Alfalfa HCT down-regulated transgenics exhibit an up to 166% increase in enzymatic saccharification efficiency (Chen & Dixon, 2007) and an up to 20% increase in *in vitro* dry-matter digestibility using rumen fluid (Shadle *et al.*, 2007). However, the most strongly down-regulated lines are dwarf with a bushy phenotype (Fig. 1.4a), and exhibit reduced photosynthetic and transpiration rates, and delayed flowering in some lines (Chen *et al.*, 2006; Shadle *et al.*, 2007; Gallego-Giraldo *et al.*, 2011b). Surprisingly, these plants also show increased drought and fungal tolerance (Gallego-Giraldo *et al.*, 2011b).

Transgenic suppression of HCT in *P. radiata* tracheary elements results in a 42% reduction in lignin content, an increased proportion of H units (from trace amounts to 31%), and a decrease in the proportion of G units (from 99.6% in controls to 69% in transgenics) (Wagner *et al.*, 2007). NMR analysis of lignin from HCT down-regulated *P. radiata* and alfalfa confirms the increase in H units and some corresponding interunit linkage ratio changes in the polymer structure (Wagner *et al.*, 2007; Pu *et al.*, 2009).

Virus-induced gene silencing of HCT in *N. benthamiana* results in dwarf plants with a 15% reduction in lignin content, increased H units from trace amounts to 8% of total lignin units, and a 9% decrease in S units, without any change in G units (Hoffmann *et al.*, 2004). The corresponding increase in saccharification efficiency is similar to that seen in HCT down-regulated alfalfa.

**Fig. 1.4** Growth phenotypes of alfalfa and *M. truncatula* plants as a result of down-regulation (silencing or insertional mutagenesis) of monolignol biosynthetic enzymes or regulatory TFs. In each case, a corresponding wild-type control grown in parallel is included for comparison. (a,b) RNAi-mediated down-regulation of HCT and C3′H, respectively, in alfalfa. (c–e) Transposon insertion mutants in CCR1, CCoAOMT, and CAD, respectively, in *M. truncatula*. (f,g) Lignin autofluorescence in cross-sections of the sixth internodes of wild-type *M. truncatula* and a line harboring a transposon insertion in the NST1 NAC TF gene. (h) Growth phenotypes of the plants in (f,g) showing lodging in the *nst1* mutant. (i,j) Lignin autofluorescence in cross-sections of the sixth internodes of wild-type *M. truncatula* and a line harboring a transposon insertion in the WRKY-12 TF gene (Wang *et al.*, 2010. Reproduced with permission of National Academy of Sciences of the United States of America). (k) Growth phenotypes of the plants in (i,j). For color details, please see Plate 1.4.

In Arabidopsis, silencing of HCT results in dwarf plants with green/purple coloration, reduced lignin content with increased amount of H units (from trace to 85%), and increased flavonoid content (Hoffmann *et al.*, 2004; Besseau *et al.*, 2007). The HCT-deficient plants have small vascular cells, many of which are collapsed (Besseau *et al.*, 2007).

## 1.4.5 4-coumaroyl shikimate 3′-hydroxylase (C3′H)

Silencing of C3′H expression in alfalfa (Fig. 1.4b), Arabidopsis, and hybrid poplar (*Populus grandidentata* X *Populus alba*) results in all cases in dwarf plants with distorted xylem elements, reduced lignin content, and a massive increase in the proportion of H lignin units. These plants also have increased flavonoid content.

Down-regulation of C3′H in alfalfa results in a greater than 30% reduction in lignin content, an increase in the H/(total lignin) ratio from 0.03 to 0.48, and a small increase in S/G ratio (Reddy *et al.*, 2005; Chen *et al.*, 2006). Stem sections of C3′H down-regulated alfalfa show very low lignin autofluorescence, except in a few vascular cells that have low amounts of S lignin, and some of the vessels are collapsed (Nakashima *et al.*, 2008) (Fig. 1.3a). Patten *et al.* (2007) reported that C3′H-silenced alfalfa plants produce more xylem than control lines, but Nakashima *et al.*, (2008) found fewer and smaller vessel elements compared to controls; however, this study was restricted to the fifth internode. Laser-capture microdissection studies indicated that lignin thioacidolysis yield was considerably reduced in vascular, fiber, and parenchyma cells (Nakashima *et al.*, 2008). NMR studies confirmed the increased proportion of H units in the lignin, and revealed a low molecular weight for the H-rich lignin, higher levels of phenylcoumarans and resinols, and doubling of the proportion of – and an increased variability in – dibenzodioxocin structures (Ralph *et al.*, 2006; Pu *et al.*, 2009; Ziebell *et al.*, 2010). There was an increase in the content of wall-bound 4-hydroxybenzaldehyde and 4-coumaric acid and a decrease in vanillin and ferulic acid in these plants (Chen *et al.*, 2006). Salicylic acid content was also increased (Lee *et al.*, 2011; Gallego-Giraldo *et al.*, 2011b), and there was a corresponding increase in the expression of pathogenesis-related (PR) genes, which are inducible by salicylate.

The saccharification efficiency of C3′H down-regulated alfalfa is higher than that of wild-type but slightly lower than that of HCT down-regulated alfalfa (Chen & Dixon, 2007). The *in situ* digestibility of C3′H silenced alfalfa measured in fistulated steers increased by more than 18% (Reddy *et al.*, 2005); the separation in digestibility between C3′H silenced and control alfalfa was measurable within 12 hours of incubation in the rumen, and the difference increased towards 72 hours of incubation (Reddy *et al.*, 2005).

C3′H down-regulation in hybrid poplar leads to dwarf plants with >50% reduction in lignin content, with H units increased from trace amounts in controls to ~20% of the total, accompanied by a reduction in G units but little change in S units (Coleman *et al.*, 2008a,b). As with HCT down-regulated alfalfa, these plants have collapsed xylem, and reduced photosynthesis and stomatal conductance.

C3′H mutants of Arabidopsis are dwarf, and have more than 60% reduction in lignin content, with approximately 95% of the lignin being made of H units (Franke *et al.*, 2002a,b; Abdulrazzak *et al.*, 2006). The plants have collapsed vessel elements, with smaller cells similar to C3′H down-regulated alfalfa, and flavonoid content is increased (Abdulrazzak *et al.*, 2006). There is high expression of genes related to response to biotic and abiotic stress, in particular those mediated by jasmonate and abscisic acid. Sinapoylated cyanidin glucosides accumulate in C3′H gene-silenced Arabidopsis, but very low levels of sinapate esters are seen in the true null mutants (Abdulrazzak *et al.*, 2006). The cell walls of C3′H mutants are approximately 80% more digestible by polysaccharide hydrolases than are those of wild-type (Franke *et al.*, 2002b).

### *1.4.6 Caffeoyl CoA 3-O-methyltransferase (CCoAOMT)*

CCoAOMT down-regulated alfalfa plants appear phenotypically normal, although knock-out mutants in *Medicago truncatula* have reduced biomass yield (Fig. 1.4d). The lignin has an increased proportion of $\beta$–5-linked dimers of G units (Guo *et al.*, 2001a; Chen *et al.*, 2006). The approximately 50% reduction in G units with little change in S units suggests that CCoAOMT may not be essential for S lignin synthesis in alfalfa, and recent studies in *M. truncatula* have led to the hypothesis that there may be a shunt around the CCoAOMT reaction involving the activity of CCR with caffeoyl-CoA (Zhou *et al.*, 2010); this leads to the formation of caffealdehyde, a preferred substrate for Medicago COMT (Parvathi *et al.*, 2001). This model is supported by mathematical modeling of flux partitioning into lignin in alfalfa lines down-regulated at different steps in the monolignol pathway (Lee *et al.*, 2011).

The large change in S/G ratio in CCoAOMT down-regulated alfalfa is primarily in the vascular elements (Fig. 1.3a), but the thioacidolysis yields are much lower in parenchyma and fiber cells compared to vascular elements (Nakashima *et al.*, 2008). S lignin deposition in pith rays shows a distinctive pattern, with S lignin confined to the middle layer of the cell wall (Nakashima *et al.*, 2008). The mean area of the vascular cells is smaller than that of the wild-type (Nakashima *et al.*, 2008). CCoAOMT down-regulated alfalfa accumulates caffeic acid $\beta$-D-glucoside and salicylic acid (Guo *et al.*, 2001a; Chen *et al.*, 2006; Lee *et al.*, 2011). The CCoAOMT down-regulated alfalfa plants have increased saccharification efficiency (Chen & Dixon, 2007) and a 6% increase in forage digestibility (Guo *et al.*, 2001b; Reddy *et al.*, 2005) compared to wild-type.

CCoAOMT down-regulation in tobacco results in an up to 15% reduction in lignin content with preferentially more reduction in G lignin, resulting in an increased S/G ratio (Zhong *et al.*, 1998; Pincon *et al.*, 2001a). Reduction in plant growth was reported in one of the two studies.

CCoAOMT down-regulation in poplar results in a 12–40% reduction in lignin content, decreases in both G and S units, but an overall 11% increase in S/G ratio, with no change in plant growth (Meyermans *et al.*, 2000; Zhong *et al.*, 2000). Levels of 4-hydroxybenzoate and glucosides of caffeic acid, vanillic acid, and sinapic acid are increased (Meyermans *et al.*, 2000). Cross-sections of CCoAOMT down-regulated poplar stems have red (Meyermans *et al.*, 2000) or orange (Zhong *et al.*, 2000) coloration, with a few slightly deformed vessels.

Flax lignin is unusual in having a high proportion of G units; down-regulation of CCoAOMT in flax reduces the lignin content by 8–18%, with a reduction in the proportion of G units. The plants are phenotypically normal, except that the thickness of the vascular cells is reduced (Day *et al.*, 2009).

Arabidopsis CCoAOMT knock-out mutants exhibit an approximately 26% reduction in lignin content, accompanied by a 56% higher S/G ratio due to a 16% reduction in the level of G units. The stems are slightly shorter than in wild-type under short-day growth conditions (Do *et al.*, 2007). FTIR microscopy showed that the reduction in lignin content was more severe in xylem fibers with only G lignin than in fiber cells with both S and G lignins. The fact that the mutant has a significant amount of lignin with G units despite the complete loss of CCoAOMT enzyme activity suggests the presence of an alternative pathway for 3-*O*-methylation. Arabidopsis double mutants generated by crossing *CCoAOMT* and *COMT* mutants (Goujon *et al.*, 2003a) are compromised in development and have very low lignin content, composed mostly of H units. The flavonoid isorhamnetin is not detectable in the mutants (consistent with roles for COMT and CCoAOMT in methylation of flavonoids) and sinapoylmalate is barely detectable (Do *et al.*, 2007). Simultaneous strong down-regulation of both CCoAOMT and COMT in alfalfa likewise results in plants that are compromised in development, with aberrantly thickened vascular cell walls (Fig. 1.3b) (Nakashima *et al.*, 2008).

Down-regulation of CCoAOMT in tracheary element cultures of *Pinus radiata* results in a 5–20% reduction of lignin content, with reduction in G units, leading to a 10-fold increase in H/G ratio. The lignin of these transgenic cultures has a unique catechyl (C) unit signature in addition to the H and G units that are normally present (Wagner *et al.*, 2011). NMR analysis revealed small amounts of caffeyl alcohol as the new monolignol in these transgenic tracheary elements. The incorporation of caffeyl alcohol into the lignin polymer reflects the incorporation of 5-hydroxyconiferyl alcohol in COMT down-regulated alfalfa and other angiosperms (Marita *et al.*, 2003a). However, CCoAOMT down-regulation in alfalfa and other angiosperms does not result in the incorporation of C units into lignin. Recently, a lignin composed entirely of caffeyl alcohol units has been discovered in the seed coats of vanilla orchid and several species of cactus (Chen *et al.*, 2012).

### *1.4.7 Ferulate 5-hydroxylase (F5H)*

The *fah1* mutant of Arabidopsis is devoid of F5H activity, fails to accumulate sinapoyl malate (a major leaf metabolite in wild-type Arabidopsis), and contains lignin that is almost completely made up of G units, as in gymnosperms (Chapple *et al.*, 1992; Meyer *et al.*, 1998; Marita *et al.*, 1999). The lignin of the mutant has enhanced levels of phenylcoumaran and dibenzodioxocin structures, and over-expression of the wild-type copy of F5H in the *fah1* background results in restoration of the wild-type lignin profile and the disappearance of these structures (Marita *et al.*, 1999).

Over-expression of F5H under control of the *C4H* promoter results in lignin composed of close to 100% S units, mostly linked through $\beta$-aryl ether linkages (Marita *et al.*, 1999). Benzodioxane structures are also detected in F5H over-expressing Arabidopsis, probably due to increased flux into 5-hydroxyconiferyl alcohol that is not sufficiently rapidly methylated by the Arabidopsis COMT (Ralph *et al.*, 2001b). Similarly, over-expression of F5H from *Selaginella moellendorffii* (SmF5H) in the *fah1* mutant results in lignin with up to 80% S lignin (Weng *et al.*, 2008b). Neither over-expression nor down-regulation of F5H alters the overall growth phenotype of Arabidopsis. F5H suppression does not affect the saccharification efficiency or digestibility of Arabidopsis, but over-expression of F5H increases saccharification efficiency, most noticeably after treatment with hot water prior to enzyme hydrolysis (Li *et al.*, 2010a).

Down-regulation of F5H in alfalfa results in only a slight reduction in S units, with no significant change in lignin content and no effect on plant growth (Reddy *et al.*, 2005; Chen *et al.*, 2006). It is hard to detect the effects of F5H down-regulation based on autofluorescence of stem sections, but Mäule staining showed a reduction in S lignin content (Nakashima *et al.*, 2008) (Fig. 1.3a). The plants had smaller vascular cells than controls, but the number of vascular cells per cluster was normal. Thioacidolysis showed no reduction in the S lignin content of vascular cells despite the fact that the *PAL* promoter driving the F5H transgene is most strongly expressed in this tissue type. However, there was a reduction in S units in fibers and vascular parenchyma cells (Nakashima *et al.*, 2008). The F5H-suppressed alfalfa showed neither improved digestibility nor increased fermentable sugar yield (Reddy *et al.*, 2005; Chen & Dixon, 2007).

Over-expression of sweetgum F5H under the control of the *Populus trichocarpa 4CL1* promoter in transgenic aspen results in accelerated lignification of stem xylem cells, without a change in lignin content, and a 2.5-fold increase in S/G ratio compared to controls, with S units constituting approximately 80% of the lignin units (Li *et al.*, 2003). Similarly, over-expression of F5H in hybrid poplar results in lignin comprising up to 97.5% S units, leading to a more linear and less polymerized lignin (Stewart *et al.*, 2009). The benzodioxane structures detected in F5H over-expressing Arabidopsis (Ralph *et al.*, 2001b) are barely detectable in F5H over-expressing aspen (Li *et al.*, 2003) and hybrid poplar, suggesting that COMT is not limiting in these plants (Stewart *et al.*, 2009). In hybrid poplar, *C4H* promoter-driven F5H over-expression results in lower-molecular-weight lignin, a twofold decrease in 4-hydroxybenzoate incorporation, decreased levels of phenylcoumarans, and increased levels of resinols and spirodienones (Stewart *et al.*, 2009). Pulping efficiency is increased in F5H over-expressing poplar (Huntley *et al.*, 2003), possibly as a result of the lower molecular weight of the lignin, as well as the high syringyl (and hence, high $\beta$-ether) content (Stewart *et al.*, 2009).

### *1.4.8 Caffeic acid* O*-methyltransferase (COMT)*

COMT down-regulation has been analyzed in many plant species. In alfalfa (Fig. 1.3a), Arabidopsis, aspen (*Populus tremuloides*), corn, perennial ryegrass, poplar (*Populus tremula* x *alba*), silver birch, *Stylosanthes humilis,* switchgrass, tall fescue, and tobacco, COMT down-regulation causes a large reduction in S units, leading to a lower S/G ratio and highlighting the involvement of COMT in S lignin biosynthesis. COMT down-regulation is usually accompanied by the appearance of 5-hydroxyguaiacyl units in lignin, due to incorporation of 5-hydroxy-coniferyl alcohol with unique benzodioxane structures (Jouanin *et al.*, 2000; Ralph *et al.*, 2001a,b; Marita *et al.*, 2003a; Morreel *et al.*, 2004; Lu *et al.*, 2010). An increase in biphenyl (5–5) and phenylcoumaran ($\beta$–5) lignin linkages has also been detected in COMT-deficient plants. Decreased lignin content is observed in most studies. Cell-wall digestibility is increased in alfalfa (Guo *et al.*, 2001b; Reddy *et al.*, 2005), perennial ryegrass, *Stylosanthes humilis*, switchgrass, tall fescue, and tobacco; saccharification efficiency has been shown to be increased in alfalfa (Chen & Dixon, 2007) and sorghum (Sattler *et al.*, 2010); and pulping efficiency is decreased in poplar as a result of COMT down-regulation (Lapierre *et al.*, 1999). Analysis of early- and late-maturity COMT down-regulated alfalfa revealed a mean ethanol conversion efficiency increase of 8% compared to wild-type (Dien *et al.*, 2011).

In stem cross-sections of COMT down-regulated alfalfa, lignin accumulates unevenly between the pith rays, with lignin autofluorescence being stronger in some portions than in others (Nakashima *et al.*, 2008). The portions with strong autofluorescence have larger vascular cells with a normal number of cells per cluster, whereas the portions with less autofluorescence have normal-sized cells with a lower number of cells per cluster. This imbalance across the stem sections may be the reason for the increased lodging compared to controls observed in the field (Nakashima *et al.*, 2008). There is greater reduction in thioacidolysis lignin content in fiber cells, and to some extent in parenchyma cells, but an increase in lignin content of vascular cells in the fifth internode of COMT down-regulated alfalfa compared to controls (Nakashima *et al.*, 2008). However, it must be remembered that thioacidolysis yields are deceptive here, as 5-hydroxyguaiacyl monomers are released from benzodioxane structures significantly less efficiently than are the normal S and G monomers from $\beta$-ether units.

In the forage legume *Stylosanthes humilis*, histochemical analysis reveals that COMT down-regulation leads to no apparent change in lignin content but a drastic reduction in S lignin levels (Rae *et al.*, 2001). The *in vitro* digestibility in rumen fluid and acid pepsin is increased by 16% compared to controls.

The corn *bm3* mutant has a brown coloration of the leaf midrib in internodes at the flowering stage. The Klason lignin content is reduced by ~20%, with reduced S lignin content, the appearance of 5-hydroxy guaiacyl units, and elevated levels of ferulic acid monomers and dimers (Marita *et al.*, 2003b). *bm3* mutant and transgenic COMT down-regulated corn show similar (30–40%) decreases in wall-esterified 4-coumaric acid content (Piquemal *et al.*, 2002; Marita *et al.*, 2003b). The S/G ratio in *bm3* is decreased by more than twofold to 0.06, and NMR analysis reveals the presence of benzodioxane units in the isolated lignin. The *bm3* mutant has 15% higher cell-wall degradability by cellulase treatment (Marita *et al.*, 2003b), whereas in COMT down-regulated corn the digestibility of stems and leaves is increased by 7 and 2%, respectively (Piquemal *et al.*, 2002). Commercial *bm3* corn has been developed; it makes up about 5% of the silage market in Canada (Zandbergen, 2006). It has lower yield compared to controls, but improves milk production (Oba & Allen, 1999).

Sorghum *bmr12* is a knock-out mutant of COMT with 13% reduced acid detergent lignin content, decreased S lignin content, and increased *in vitro* dry-matter digestibility (Sattler *et al.*, 2010). Conversion of cellulose to ethanol in sorghum *bmr12* biomass after acid pretreatment is increased by 21% compared to controls (Dien *et al.*, 2009).

COMT down-regulation in switchgrass results in plants with normal growth and development (Fu *et al.*, 2011a); the same has been observed in perennial ryegrass (Tu *et al.*, 2010) and tall fescue (Chen *et al.*, 2004). The digestibility of switchgrass, perennial ryegrass, and tall fescue biomass increased by 9, 8.6, and 10.8% respectively compared to their respective controls. In switchgrass, the overall ethanol yield after saccharification and fermentation was increased by 38% compared to controls (Fu *et al.*, 2011a).

COMT suppression results in the formation of a red–brown woody stem with no change in lignin content in aspen (Tsai *et al.*, 1998) and brownish coloration at the basal internode of the stem in switchgrass (Fu *et al.*, 2011b). Suppression of COMT in poplar was shown in one study to reduce lignin content, as determined by the Klason and acetyl bromide methods (Jouanin *et al.*, 2000), but no change was reported in another using the same analytical methods (Van Doorsselaere *et al.*, 1995). The pulping efficiency of COMT-suppressed poplar plants with reduced S lignin content is reduced drastically (Lapierre *et al.*, 1999; Pilate *et al.*, 2002). This is presumably because, despite there being $\beta$-ethers, benzodioxanes are inefficiently cleaved under alkaline pulping conditions, unlike the syringyl counterparts they replace (Ralph *et al.*, 2004).

Field trials of COMT down-regulated transgenic poplar plants showed that growth indicators and interactions with insects were normal (Pilate *et al.*, 2002). The down-regulation of COMT in silver birch led to no change in lignin content and did not affect the feeding preferences of insect herbivores or interactions with the ectomycorhhizal fungus *Paxillas involutus* (Tiimonen *et al.*, 2005; Sutela *et al.*, 2009).

The *Atcomt1* mutant of Arabidopsis totally lacks S units and has a slight increase in lignin content (Goujon *et al.*, 2003a). COMT down-regulation in tobacco has been reported to either lower the lignin content (Ni *et al.*, 1994) or result in no change (Dwivedi *et al.*, 1994; Atanassova *et al.*, 1995; Zhong *et al.*, 1998). The cell-wall digestibility of COMT down-regulated tobacco is increased by up to 48% (Sewalt *et al.*, 1997b).

### *1.4.9 Cinnamoyl-CoA reductase (CCR)*

CCR-suppressed alfalfa, Arabidopsis, *Medicago truncatula*, Norway spruce, poplar (*Populus tremula* x *alba*), perennial ryegrass, tobacco, and tomato have all been reported with various degrees of reduction in lignin content, and with severe down-regulation leading to a dwarf phenotype (Fig. 1.4c) with collapsed vessels.

In alfalfa, CCR down-regulation results in reduced biomass, 30% reduction in the lignin content of mature stem internodes, and an increase in S/G ratio (Jackson *et al.*, 2008). Total lignin autofluorescence is low in stem sections, with significantly weaker S lignin staining than in controls and with the tangential walls staining unevenly, indicating the presence of pits between vascular elements (Nakashima *et al.*, 2008). The diameter of the vascular cells of these plants is significantly increased, and the vessels are rounder in shape than those of the wild-type (Fig. 1.3a). The thioacidolysis lignin yield is reduced drastically in vascular, fiber, and parenchyma cells, with H lignin almost completely absent in vascular cells (Nakashima *et al.*, 2008). This may explain the unusual cell shape, as H lignin has been proposed to determine the shape of vascular cells by acting as a matrix for the deposition of G and S units (Terashima *et al.*, 1998). *In vitro* dry-matter digestibility increases by 5–10% in these plants, and the saccharification efficiency is increased by up to 50% (Jackson *et al.*, 2008).

Several Arabidopsis *AtCCR1* knock-out mutants have been characterized. An *irx4* mutant (Jones *et al.*, 2001) and *ccr1s* and *ccr1g* (Mir Derikvand *et al.*, 2008) exhibit roughly 25–50% reductions in lignin content, similar to transgenic *CCR1* down-regulated Arabidopsis (Goujon *et al.*, 2003b). *CCR1* mutants of Arabidopsis show delayed senescence compared to wild-type (Patten *et al.*, 2005; Laskar *et al.*, 2006; Mir Derikvand *et al.*, 2008) and dwarfness with altered architecture of the cell walls where the S2 sublayer is affected (Jones *et al.*, 2001; Goujon *et al.*, 2003b; Ruel *et al.*, 2009). The S/G ratio is decreased in interfascicular fibers but increased in vascular bundles (Ruel *et al.*, 2009), and the overall S/G ratios are decreased at the 6-week-old stage but comparable to wild-type at the 9-week-old stage (Laskar *et al.*, 2006). Increased accumulation of ferulic acid is seen in the walls of *ccr1* mutants (Mir Derikvand *et al.*, 2008), and there are also changes in soluble phenolics, including an increase in feruloyl malate. *CCR1* down-regulation has a positive impact on the cell-wall digestibility of Arabidopsis (Goujon *et al.*, 2003b). Simultaneous suppression of CCR and CAD in Arabidopsis leads to a severe dwarf phenotype, red coloration of xylem, accumulation of soluble phenolics (sinapoyl malate, feruloyl malate, and flavanol glucosides), and male sterility due to the absence of lignified thickening in the endothecium, resulting in indehiscence of anthers (Thévenin *et al.*, 2011).

Transgenic tobacco plants with strong down-regulation of CCR activity leading to a 50% reduction in lignin exhibit a dwarf phenotype, abnormal leaf development, and orange–brown xylem, probably due to the presence of phenolics such as ferulate and sinapate (Piquemal *et al.*, 1998; Chabannes *et al.*, 2001a). The S/G ratio is increased, with lignin deposition affected in the inner S2 and S3 layers of the fibers and vessels (Piquemal *et al.*, 1998; Chabannes *et al.*, 2001a). The lignin of CCR down-regulated tobacco has an increased content of tyramine ferulate, an unusual component of tobacco cell walls (Ralph *et al.*, 1998). Inclusion of ferulic acid, as a monomer, into lignin provides a new mechanism for producing branch points in the polymer (Ralph *et al.*, 2008). Transcriptomic and metabolomic analyses of CCR down-regulated tobacco suggest increased photo-oxidative stress and photorespiration (Dauwe *et al.*, 2007). Surprisingly, and in sharp contrast to the situation described in Arabidopsis, crossing of CCR- and CAD-down-regulated tobacco results in progeny that look normal and have an increased S/G ratio and slightly elevated tyramine ferulate levels when compared to plants down-regulated in CAD or CCR alone (Chabannes *et al.*, 2001a).

CCR down-regulated tomato plants show a severe dwarf phenotype with 38% reduction in Klason lignin content, brown–orange coloration of the xylem, and slightly collapsed xylem vessels (van der Rest *et al.*, 2006). These plants have higher contents of chlorogenic acid and rutin, and accumulate N-caffeoyl putrescine and kaempferol rutinoside, which are not seen in the wild-type.

CCR1 down-regulated perennial ryegrass shows an up to 37% reduction in lignin content, little change in S/G ratio, and increased accumulation of ferulic, sinapic, and *O*-caffeoyl quinic acids and flavonol glucosides. Field-grown transgenics exhibited a normal growth phenotype and up to 13.6% improved digestibility (Tu *et al.*, 2010).

In poplar, down-regulation of CCR is associated with reduced growth, orange–brown coloration of the outer xylem, concentric sublayers in the S2 layer of fibers and the S2 and S3 layers of vessels in the orange–brown area, reduced lignin and hemicellulose content, lower S/G ratio, and increased accumulation of ferulate esters (Leplé *et al.*, 2007). Like in CCoAOMT down-regulated poplar plants (Meyermans *et al.*, 2000), glucosides of vanillic and sinapic acids accumulate in CCR down-regulated poplar, but caffeic acid glucosides are absent. Down-regulation of CCR in Norway spruce results in plants with reduced stem diameter, 8% reduced lignin content, and 34% reduced H lignin compared to the control (Wadenbäck *et al.*, 2008). CCR down-regulated tobacco, poplar, and Norway spruce show improved pulping efficiency, but the plants are compromised due to the growth defects (O'Connell *et al.*, 2002; Leplé *et al.*, 2007; Wadenbäck *et al.*, 2008).

CCR1 down-regulated alfalfa lines have a reduced lignin content and an increased S/G ratio, but plants do not show a colored xylem phenotype (Jackson *et al.*, 2008). Some lines exhibit reduced biomass production. *M. truncatula CCR1* completely (and *CCR2* partially) complements the *irx4 CCR* mutation in Arabidopsis (Zhou *et al.*, 2010). *M. truncatula CCR1* knock-out mutants exhibit a severely dwarf phenotype (Fig. 1.4c), with highly reduced lignin content, whereas *MtCCR2* mutants have only a moderate reduction in lignin content and appear phenotypically normal (Zhou *et al.*, 2010).

### 1.4.10 Cinnamyl alcohol dehydrogenase (CAD)

CAD down-regulation has been studied in transgenic alfalfa, tobacco, poplar (*Populus tremula* x *alba*), flax, tall fescue, and switchgrass, and CAD mutants of Arabidopsis, *Medicago truncatula*, pine, corn, rice, and sorghum have been identified. In general, CAD-suppressed plants do not show a dwarf phenotype, have modest changes in lignin content, and have a red–brown coloration of the xylem in their stems.

Alfalfa and *M. truncatula* plants with reduced CAD activity show decreased biomass yields in some lines (Fig. 1.4e), red pigmentation in vascular tissue due to accumulation of monolignol precursor hydroxycinnamaldehyde units, an up to 10% reduction in lignin content, and a decreased S/G ratio (Baucher *et al.*, 1999; Jackson *et al.*, 2008). The mean area of individual vascular cells is reduced in comparison to controls (Nakashima *et al.*, 2008) (Fig. 1.3a). The thioacidolysis yield from lignin is reduced by more than 70% in vascular, fiber, and parenchyma cells. Sinapaldehyde and coniferaldehyde markers (Lapierre *et al.*, 2004) have been detected using GC/MS in CAD-down-regulated alfalfa during thioacidolysis, but they could not be quantified due to the low amount (Jackson *et al.*, 2008); incorporated aldehydes as their 8-*O*-4-ethers cleave more poorly than their *β*-*O*-4-ether counterparts in normal units. CAD down-regulation had little impact on saccharification efficiency, which was improved marginally only in a few lines, but the *in vitro* dry-matter digestibility of CAD-down-regulated alfalfa was increased by up to 16% compared to controls (Baucher *et al.*, 1999; Jackson *et al.*, 2008).

The *gold hull and internode (gh)* mutants of rice exhibit reddish-brown pigmentation in the hull and internode and turn golden yellow at maturation. Using map-based cloning, the *gh2* mutation was identified as being in a *CAD* gene (Zhang *et al.*, 2006). The lignin content was reduced by 5–6% compared to wild-type, and H, G, and S units were reduced in similar proportion (Zhang *et al.*, 2006). The *brown midrib6* mutant of sorghum is a null CAD allele (Saballos *et al.*, 2008, 2009; Sattler *et al.*, 2009) with a reduction in H, G, and especially S units, leading to a reduced S/G ratio (Sattler *et al.*, 2009). Cows fed with *bmr6* silage exhibit a 10% increase in milk production (Oliver *et al.*, 2004).

The *bm1* mutation of corn has very low CAD activity (Halpin *et al.*, 1998) and increased levels of cinnamaldehydes (Halpin *et al.*, 1998; Marita *et al.*, 2003b). Lignin content is reduced by 20%, with no change in S/G ratio. The *bm1* gene may encode a TF that controls monolignol genes and suppresses CAD (Guillaumie *et al.*, 2007).

CAD down-regulation in tall fescue leads to ~15% decreased lignin content, with reduction in G and S but not H units, and a decreased S/G ratio (Chen *et al.*, 2003). The *in vitro* dry-matter digestibility is increased by 7.2–9.5% compared to controls (Chen *et al.*, 2003). Similarly, CAD down-regulation in switchgrass results in a 14–22% reduction in lignin content and a decreased S/G ratio, no changes in wall-bound phenolics, an increase in soluble chlorogenic acid content, and an up to 20% increased digestibility and 15–35% increased saccharification efficiency (Fu *et al.*, 2011b). Another study shows a more modest (2–11%) increase in saccharification efficiency in CAD-down-regulated switchgrass (Saathoff *et al.*, 2011). CAD-deficient sorghum also shows improved saccharification efficiency (Dien *et al.*, 2009).

The *Atcad-c* and *Atcad-d* mutants of Arabidopsis have reduced CAD activities, but only *Atcad-d* shows a (slight) reduction in lignin content, a reduction of S units, and incorporation of sinapaldehyde in lignin (Sibout *et al.*, 2003). According to Kim *et al.* (2004), CAD mutants of Arabidopsis attain wild-type lignin levels and composition through maturity. The Arabidopsis *cad-c cad-d* double mutant has reduced lignin content, with incorporation of coniferaldehyde and sinapaldehyde into lignin, in both xylem vessels and fibers, leading to limp floral stems (Sibout *et al.*, 2005). The Arabidopsis *cad-c cad-d ccr1* triple mutant exhibits a 50% reduction in lignin content, accompanied by severely reduced growth and male sterility due to failure of anther dehiscence (Thévenin *et al.*, 2011).

Down-regulation of CAD in poplar is associated with red xylem but no change in overall lignin amount or S/G ratio (Baucher *et al.*, 1996). Pulping efficiency is higher in such plants (Baucher *et al.*, 1996; Lapierre *et al.*, 1999; Pilate *et al.*, 2002). Levels of thioacidolysis-derived indene compounds that originate from the incorporation of sinapaldehyde into lignins through 8–*O*–4-cross-coupling increase as a function of CAD suppression in poplar (Lapierre *et al.*, 2004). The interaction of insects with field-grown transgenic poplar trees with reduced CAD activity is normal, and no changes are detected in the microbial communities growing below these trees (Pilate *et al.*, 2002).

The CAD mutant of loblolly pine has a reduced lignin content and slightly bent stems (MacKay *et al.*, 1997). In addition to mildly elevated vanillin and coniferaldehyde levels in the lignin, dihydroconiferyl alcohol is present in high amounts (Ralph *et al.*, 1997); its derivation remains unclear.

Down-regulation of CAD in tobacco does not alter the lignin content (Stewart *et al.*, 1997), although it leads to higher levels of hydroxycinnamaldehyde and hydroxybenzaldehyde (mainly syringaldehyde) incorporation (Kim *et al.*, 2003) and improved pulping efficiency (O'Connell *et al.*, 2002). Transcripts of genes related to light harvesting/signaling and cell-wall expansion are elevated in CAD-down-regulated tobacco (Dauwe *et al.*, 2007).

In flax, decreased CAD activity results in reduced lignin content and increased levels of ferulic, 4-coumaric, caffeic, and chlorogenic acids (Wróbel-Kwiatkowska *et al.*, 2007). Many of the transformants show dramatic growth reduction, but some plants look phenotypically normal. These CAD-down-regulated plants also show improved mechanical properties but reduced resistance to *Fusarium oxysporium* (Wróbel-Kwiatkowska *et al.*, 2007).

## 1.5 Impacts of lignin modification through targeting of TFs

Lignin deposition during plant secondary cell-wall formation is controlled by a network of tissue-specific TFs (Umezawa, 2009; Zhao & Dixon, 2011). This network coordinates monolignol biosynthesis into the broader program of secondary cell-wall biosynthesis, and therefore consists of both master switches, which control multiple features of secondary wall formation, and lignin pathway-specific downstream TFs (Fig. 1.2). There are clear distinctions between the down-regulation of lignin biosynthesis through the targeting of biosynthetic enzymes versus regulatory TFs. In the former case, the monolignol pathway may remain functional up to the step that has been down-regulated, resulting in metabolic spillover. In the latter, metabolic spillover may not occur as the pathway has not been activated at any stage. It might therefore be predicted that targeting TFs will have less deleterious effects on plant growth and development than the targeting of biosynthetic enzymes. This appears to be at least partly correct, as can be seen in the following examples.

### 1.5.1 NAC master switches

Secondary wall-containing cell types in stems provide mechanical strength to support a plant's weight and facilitate the transport of water and nutrients. In reproductive organs, secondary cell walls at the valve margin of the siliques and in the endothecium of the anthers are required for seed shedding and pollen release, respectively. In the anther endothecium of *Arabidopsis thaliana*, secondary wall thickening is controlled non-redundantly by two NAC TFs – NST1 and NST2 – and knock-out of both is required to prevent anther dehiscence (Mitsuda *et al.*, 2005). Likewise, double knock-out of *NST1* and another NAC TF, *NST3/SND1*, is necessary for the striking phenotype of loss of secondary cell-wall development in interfascicular and vascular fibers of the stem. In contrast to the requirement for redundant NAC TFs in Arabidopsis, *M. truncatula* possesses a single fiber-specific secondary wall NAC TF, MtNST1, which orchestrates multiple features of cell-wall development in different tissues (Zhao *et al.*, 2010a). *Tnt1* retrotransposon insertions in the *MtNST1* gene result not only in loss of lignification in the interfascicular region of the stem (Fig. 1.4f,g) and prevention of anther dehiscence but also in alterations in the cell walls of stomatal guard cells (loss of fluorescence likely due to the loss of wall-bound ferulate esters). The saccharification efficiency of *MtNST1* transposon insertion mutants is increased by up to 23% following acid pretreatment and by 18% with no pretreatment. Loss of function of *MtNST1* has a relatively minor impact on overall plant growth; stem length is reduced by around 25%, but average leaf area is increased by around 14%. The reduction in stem length is mainly the result of a reduction of internode length in the most mature internodes. Flowering time and stem diameter are not affected in the mutant lines, but lodging is increased (Zhao *et al.*, 2010a) (Fig. 1.4h).

### 1.5.2 MYB repressors of monolignol biosynthesis

MYB transcriptional repressors of phenylpropanoid biosynthesis were first described in *Antirrhinum majus* (Tamagnone *et al.*, 1998). A switchgrass ortholog, PvMYB4, binds to the AC-rich elements of lignin biosynthetic genes both *in vitro* and in a yeast transcription system (Shen *et al.*, 2012). Ectopic over-expression of PvMYB4 in transgenic switchgrass results in the down-regulation of most monolignol pathway genes, reduced lignin content and ester-linked 4-coumarate to ferulate ratio, reduced plant stature but increased tillering, and an approximately threefold increase in sugar-release efficiency from CWRs without pretreatment (Shen *et al.*, 2012).

### 1.5.3 WRKY repressors of lignification in pith

During a forward-genetic screen for *Medicago truncatula* mutants with reduced lignin autofluorescence, some plants were observed to possess ectopic lignification throughout the central pith tissues of the stem (Fig. 1.4i,j) (Wang *et al.*, 2010). This phenotype was caused by the presence of a *Tnt1* retrotransposon insertion in a gene encoding a WRKY-family TF. *T-DNA* insertional knock-out of the orthologous *WRKY-12* gene in Arabidopsis led to the same phenotype, and this could be complemented by the Medicago *WRKY* gene. In addition to increased lignification, *WRKY* knock-out plants possess highly thickened pith cell walls with increased levels of cellulose and hemicellulose, such that the overall biomass density of the stems is increased by up 50% (Wang *et al.*, 2010). Remarkably, this does not appear to negatively affect overall biomass in either Arabidopsis or Medicago (Fig. 1.4k).

## 1.6 Monolignol pathway modification and plant growth

Generally, down-regulation of lignin biosynthesis at the level of CCoAOMT, COMT, F5H, and CAD does not result in severe negative effects on plant growth. This can be explained by the possibility of redundancy at the 3-*O*-methylation step (COMT/CCoAOMT double knock-down lines do have a dwarf phenotype), replacement of S units by 5-OH-G units in F5H knock-outs, or incorporation of aldehydes into the lignin of CAD-deficient plants. Excluding PAL and C4H, deficiency of which has major pleiotropic effects, this leaves the question of why targeting 4CL, HCT, C3H, and CCR leads to dwarf plants. Several hypotheses have been proposed, of two types. In the first type, metabolic spillover from monolignol biosynthesis into flavonoids (levels of which are clearly elevated in C3′H- and HCT-down-regulated alfalfa and Arabidopsis) indirectly leads to growth inhibition through processes such as inhibition of auxin transport (Besseau *et al.*, 2007), or else growth is affected by other metabolites that might interfere with hormone action. In the second type, it is not the metabolic spillover itself that is the primary reason for impaired growth, but rather the "incorrectly" lignified cell walls, which the plant senses (Li *et al.*, 2010b), possibly through the release of oligosaccharide signal molecules (Gallego-Giraldo *et al.*, 2011b). Another obvious question is whether the mechanisms of growth inhibition are the same for plants down-regulated in 4CL, C3′H, HCT, or CCR.

These points are well illustrated by attempts to understand the growth phenotype of HCT down-regulated Arabidopsis and alfalfa. These plants are dwarf, and silencing *HCT* in Arabidopsis leads to accumulation of flavonoids and inhibition of auxin transport (Besseau *et al.*, 2007). Restoring auxin transport by reducing flavonoid content has been reported to overcome the dwarf phenotype while maintaining the reduced lignin phenotype (Besseau *et al.*, 2007), although these results could not be reproduced in a more recent study (Li *et al.*, 2010b). Furthermore, HCT down-regulated alfalfa plants also have increased levels of flavonoids, but auxin transport appears normal (Gallego-Giraldo *et al.*, 2011b). The dwarf phenotype of HCT-down-regulated Arabidopsis plants is overcome when lignin accumulation is restored by expression of a *Selaginella* enzyme that bypasses the HCT reactions (Li *et al.*, 2010b). This suggests that structural effects in partially lignified secondary cell walls are in some way linked to the growth defects. These could operate in one of two ways: either a defective cell-wall structure leads to physiological stress effects associated, for example, with the leakiness of conducting vessels or (perhaps) incorrectly lignified cell walls release signal molecules that actively perturb growth.

Evidence for the latter model comes from studies with HCT down-regulated alfalfa. Microarray analysis indicates that these plants massively over-express a number of PR genes, as though they were under microbial attack (Gallego-Giraldo *et al.*, 2011b), and cold-water treatment of cell walls from the transgenic plants, but not control plants, releases pectic polysaccharides, which induce the same set of genes when added to cell-suspension cultures (Gallego-Giraldo *et al.*, 2011b). Furthermore, the plants have elevated levels of the defense hormone salicylic acid, known to be required for induction of PR proteins (Gallego-Giraldo *et al.*, 2011b). At this time it is not possible to be sure whether the elevated salicylate levels occur as a result of the action of wall-released pectic elicitors or through spillover from the lignin pathway (salicylate is biosynthetically related to lignin via the shikimic acid pathway, and can also be formed from cinnamate). The interesting observation that levels of salicylate are inversely proportional to lignin levels and growth in a series of transgenic alfalfa plants in which lignin biosynthesis has been perturbed at different biosynthetic steps (Lee *et al.*, 2011; Gallego-Giraldo *et al.*, 2011a) is essentially consistent with either hypothesis.

Reducing salicylate levels by genetically blocking its formation or causing its removal restores growth in HCT down-regulated Arabidopsis, although the plants maintain reduced lignin levels (Gallego-Giraldo *et al.*, 2011a). This important observation indicates that it is theoretically possible to engineer plants with highly reduced cell-wall recalcitrance without negatively impacting growth.

HCT-down-regulated alfalfa and Arabidopsis lack responsiveness to gibberellic acid, and this responsiveness is restored upon genetic removal of salicylate in plants that still retain the low-lignin phenotype (Gallego-Giraldo *et al.*, 2011a). These data place salicylate as a central component in growth-signaling pathways that either sense flux into the monolignol pathway or respond to secondary cell-wall integrity. However, salicylate is clearly not central to the orchestration of growth and genetic responses in all lignin down-regulated plants. For example, although salicylate levels are increased in CCR-down-regulated alfalfa, microarray analysis of CCR-down-regulated tobacco showed activation of a set of stress-responsive genes that did not include the PR protein genes induced in alfalfa and Arabidopsis; rather, these genes were associated with oxidative stress, perhaps arising as a result of increased photorespiration (Dauwe *et al.*, 2007).

## 1.7 Conclusions: it isn't all that bad!

In summary, lignin modification is a well-supported approach to reducing cell-wall recalcitrance in higher plants. Some genes, such as CCoAOMT and COMT, can be targeted to give significant improvements in saccharification efficiency or forage digestibility without producing significant effects on agronomic properties. Targeting other genes such as HCT can lead to much better improvements in saccharification efficiency, but at the cost of significantly reduced biomass yield. The negative growth impacts can, however, be overcome by additional breeding or engineering, for example to remove excess salicylate. This latter manipulation is probably not viable for field-grown plants, as it may compromise disease resistance. The fact that low lignin and poor growth can be uncoupled, however, gives hope that metabolic engineering strategies can be developed that will allow for improved growth of very low-lignin plants. An alternative approach is to rewire the transcriptional control networks to target lignin modification to specific cell types, and to replace the reduced lignin with additional polysaccharide biosynthesis; this has recently been demonstrated in Arabidopsis (Yang *et al.*, 2013). Finally, although it is beyond the scope of the present article, there is increasing evidence that a reduction in lignin levels results, paradoxically, in enhanced resistance against microbes (Funnell-Harris *et al.*, 2010; Gallego-Giraldo *et al.*, 2011b). Clearly, the bad and the ugly outcomes of lignin modification may yet prove to have good and attractive sides.

## References

Abdulrazzak, N., Pollet, B., Ehlting, J., *et al.* (2006) A *coumaroyl-ester-3-hydroxylase* insertion mutant reveals the existence of nonredundant *meta*-hydroxylation pathways and essential roles for phenolic precursors in cell expansion and plant growth. *Plant Physiology*, 140, 30–48.

Adler, E. (1977) Lignin chemistry – past, present and future. *Wood Science and Technology*, 11, 169–218.

Atanassova, R., Favet, N., Martz, F., *et al.* (1995) Altered lignin composition in transgenic tobacco expressing *O*-methyltransferase sequences in sense and antisense orientation. *Plant Journal*, 8, 465–477.

Bate, N.J., Orr, J., Ni, W., *et al.* (1994) Quantitative relationship between phenylalanine ammonia-lyase levels and phenylpropanoid accumulation in transgenic tobacco identifies a rate-determining step in natural product synthesis. *Proceedings of the National Academy of Sciences USA*, 91, 7608–7612.

Baucher, M., Chabbert, B., Pilate, G., *et al.* (1996) Red xylem and higher lignin extractability by down-regulating a cinnamyl alcohol dehydrogenase in poplar. *Plant Physiology*, 112, 1479–1490.

Baucher, M., Monties, B., Van Montagu, M. & Boerjan, W. (1998) Biosynthesis and genetic engineering of lignin. *Critical Reviews in Plant Science*, 17, 125–197.

Baucher, M., Bernard-Vailhé, M.A., Chabbert, B., *et al.* (1999) Down-regulation of cinnamyl alcohol dehydrogenase in transgenic alfalfa (*Medicago sativa* L.) and the effect on lignin composition and digestibility. *Plant Molecular Biology*, 39, 437–447.

Besseau, S., Hoffmann, L., Geoffroy, P., *et al.* (2007) Flavonoid accumulation in *Arabidopsis* repressed in lignin synthesis affects auxin transport and plant growth. *Plant Cell*, 19, 148–162.

Blee, K., Choi, J.W., O'Connell, A.P., *et al.* (2001) Antisense and sense expression of cDNA coding for CYP73A15, a class II cinnamate 4-hydroxylase, leads to a delayed and reduced production of lignin in tobacco. *Phytochemistry*, 57, 1159–1166.

Blount, J.W., Korth, K.L., Masoud, S.A., *et al.* (2000) Altering expression of cinnamic acid 4-hydroxylase in transgenic plants provides evidence for a feedback loop at the entry point into the phenylpropanoid pathway. *Plant Physiology*, 122, 107–116.

Boerjan, W., Ralph, J. & Baucher, M. (2003) Lignin biosynthesis. *Annual Review of Plant Biology*, 54, 519–546.

Bonavitz, N.D. & Chapple, C. (2010) The genetics of lignin biosynthesis: connecting genotype to phenotype. *Annual Review of Genetics*, 44, 337–363.

Boudet, A.M., Lapierre, C. & Grima-Pettenati, J. (1995) Biochemistry and molecular biology of lignification. *New Phytologist*, 129, 203–236.

Buxton, D.R. & Redfearn, D.D. (1997) Plant limitations to fiber digestion and utilization. *Journal of Nutrition*, 127, 814S–818S.

Calabrese, J.C., Jordan, D.B., Boodhoo, A., *et al.* (2004) Crystal structure of phenylalanine ammonia-lyase: multiple helix dipoles implicated in catalysis. *Biochemistry*, 43, 11403–11416.

Chabannes, M., Barakate, A., Lapierre, C., *et al.* (2001a) Strong decrease in lignin content without significant alteration of plant development is induced by simultaneous down-regulation of cinnamyl alcohol dehydrogenase (CAD) in tobacco plants. *Plant Journal*, 28, 257–270.

Chabannes, M., Ruel, K., Yoshinaga, A., *et al.* (2001b) In situ analysis of lignins in transgenic tobacco reveals a differential impact of individual transformations on the spatial patterns of lignin deposition at the cellular and subcellular levels. *Plant Journal*, 26, 271–282.

Chapple, C.C., Vogt, T., Ellis, B.E. & Somerville, C.R. (1992) An Arabidopsis mutant defective in the general phenylpropanoid pathway. *Plant Cell*, 4, 1413–1424.

Chen, F. & Dixon, R.A. (2007) Lignin modification improves fermentable sugar yields for biofuel production. *Nature Biotechnology*, 25, 759–761.

Chen, L., Auh, C.-K., Chen, F., *et al.* (2002) Lignin deposition and associated changes in anatomy, enzyme activity, gene expression, and ruminal degradability in stems of tall fescue at different developmental stages. *Journal of Agricultural and Food Chemistry*, 50, 5558–5565.

Chen, L., Auh, C.K., Dowling, P., *et al.* (2003) Improved forage digestibility of tall fescue (*Festuca arundinacea*) by transgenic down-regulation of cinnamyl alcohol dehydrogenase. *Plant Biotechnology Journal*, 1, 437–449.

Chen, L., Auh, C.-K., Dowling, P., *et al.* (2004) Transgenic down-regulation of caffeic acid *O*-methyltransferase (COMT) led to improved digestibility in tall fescue (*Festuca arundinacea*). *Functional Plant Biology*, 31, 235–245.

Chen, F., Reddy, M.S.S., Temple, S., *et al.* (2006) Multi-site genetic modulation of monolignol biosynthesis suggests new routes for formation of syringyl lignin and wall-bound ferulic acid in alfalfa (*Medicago sativa* L.). *Plant Journal*, 48, 113–124.

Chen, F., Tobimatsu, Y., Havkin-Frenkel, D., *et al.* (2012) A polymer of caffeyl alcohol in plant seeds. *Proceedings of the National Academy of Sciences USA*, 109, 1772–1777.

Coleman, H.D., Park, J.Y., Nair, R., *et al.* (2008a) RNAi-mediated suppression of *p*-coumaroyl-CoA 3′-hydroxylase in hybrid poplar impacts lignin deposition and soluble secondary metabolism. *Proceedings of the National Academy of Sciences USA*, 105, 4501–4506.

Coleman, H.D., Samuels, A.L., Guy, R.D. & Mansfield, S.D. (2008b) Perturbed lignification impacts tree growth in hybrid poplar-A function of sink strength, vascular integrity, and photosynthetic assimilation. *Plant Physiology*, 148, 1229–1237.

Costa, M.A., Bedgar, D.L., Moinuddin, S.G., *et al.* (2005) Characterization *in vitro* and *in vivo* of the putative multigene 4-coumarate:CoA ligase network in Arabidopsis: syringyl lignin and sinapate/sinapyl alcohol derivative formation. *Phytochemistry*, 66, 2072–2091.

Dauwe, R., Morreel, K., Goeminne, G., *et al.* (2007) Molecular phenotyping of lignin-modified tobacco reveals associated changes in cell-wall metabolism, primary metabolism and photorespiration. *Plant Journal*, 52, 263–285.

Davin, L.B. & Lewis, N.G. (1992) Phenylpropanoid metabolism: biosynthesis of monolignols, lignans and neolignans, lignins and suberins. In: *Phenolic Metabolism in Plants* (eds H.A. Stafford & R.K. Ibrahim), pp. 325–375. Plenum Press, New York.

Day, A., Neutelings, G., Nolin, F., *et al.* (2009) Caffeoyl coenzyme A *O*-methyltransferase down-regulation is associated with modifications in lignin and cell-wall architecture in flax secondary xylem. *Plant Physiology and Biochemistry*, 47, 9–19.

Demura, T. & Fukuda, H. (2007) Transcriptional regulation in wood formation. *Trends in Plant Science*, 12, 64–70.

Demura, T. & Ye, Z.H. (2010) Regulation of plant biomass production. *Current Opinion in Plant Biology*, 13, 299–304.

Dien, B.S., Sarath, G., Pedersen, J.F., *et al.* (2009) Improved sugar conversion and ethanol yield for forage sorghum (*Sorghum bicolor* L. Moench) lines with reduced lignin contents. *BioEnergy Research*, 2, 153–164.

Dien, B.S., Miller, D.J., Hector, R.E., *et al.* (2011) Enhancing alfalfa conversion efficiencies for sugar recovery and ethanol production by altering lignin composition. *Bioresource Technology*, 102, 6479–6486.
Dixon, R.A. & Reddy, M.S.S. (2003) Biosynthesis of monolignols. Genomic and reverse genetic approaches. *Phytochemical Reviews*, 2, 289–306.
Do, C.-T., Pollet, B., Thevenin, J., *et al.* (2007) Both caffeoyl coenzyme A 3-*O*-methyltransferase 1 and caffeic acid *O*-methyltransferase 1 are involved in redundant functions for lignin, flavonoids and sinapoyl malate biosynthesis in Arabidopsis. *Planta*, 226, 1117–1129.
Donaldson, L.A. (2001) Lignification and lignin topochemistry-an ultrastructural view. *Phytochemistry*, 57, 859–873.
Donaldson, L.A. (2002) Abnormal lignin distribution in wood from severely drought stressed *Pinus radiata* trees. *IAWA Journal*, 23, 161–178.
Dwivedi, U.N., Campbell, W.H., Yu, J., *et al.* (1994) Modification of lignin biosynthesis in transgenic *Nicotiana* through expression of an antisense *O*-methyltransferase gene from *Populus*. *Plant Molecular Biology*, 26, 61–71.
Edwards, K., Cramer, C.L., Bolwell, G.P., *et al.* (1985) Rapid transient induction of phenylalanine ammonia-lyase mRNA in elicitor-treated bean cells. *Proceedings of the National Academy of Sciences USA*, 82, 6731–6735.
Ehlting, J., Buttner, D., Wang, Q., *et al.* (1999) Three 4-coumarate: coenzyme A ligases in *Arabidopsis thaliana* represent two evolutionary classes in angiosperms. *Plant Journal*, 19, 9–20.
Elkind, Y., Edwards, R., Mavandad, M., *et al.* (1990) Abnormal plant development and down-regulation of phenylpropanoid biosynthesis in transgenic tobacco containing a heterologous phenylalanine ammonia-lyase gene. *Proceedings of the National Academy of Sciences USA*, 87, 9057–9061.
Espineira, J.M., Novo Uzal, E., Gomez Ros, L.V., *et al.* (2011) Distribution of lignin monomers and the evolution of lignification among lower plants. *Plant Biology*, 13, 59–68.
Fahrendorf, T. & Dixon, R.A. (1993) Stress responses in alfalfa (*Medicago sativa* L.). XVIII: molecular cloning and expression of the elicitor-inducible cinnamic acid 4–hydroxylase cytochrome P450. *Archives of Biochemistry and Biophysics*, 305, 509–515.
Felton, G.W., Korth, K.L., Bi, J.L., *et al.* (1999) Inverse relationship between systemic resistance of plants to microorganisms and to insect herbivory. *Current Biology*, 9, 317–320.
Fergus, B.J., Procter, A.R., Scott, J.A.N. & Goring, D.A.I. (1969) The distribution of lignin in spruce-wood as determined by ultraviolet microscopy. *Wood Science and Technology*, 3, 117–138.
Ferrer, J.-L., Zubieta, C., Dixon, R.A. & Noel, J.P. (2005) Crystal structures of alfalfa caffeoyl Coenzyme A 3-*O*-methyltransferase. *Plant Physiology*, 137, 1009–1017.
Finkle, B.J. & Nelson, R.F. (1963) Enzyme reactions with phenolic compounds: a *meta-O*-methyltransferase in plants. *Biochimica et Biophysica Acta*, 78, 747–749.
Franke, R., Humphreys, J.M., Hemm, M.R., *et al.* (2002a) The Arabidopsis REF8 gene encodes the 3-hydroxylase of phenylpropanoid metabolism. *Plant Journal*, 30, 33–45.
Franke, R., Hemm, M.R., Denault, J.W., *et al.* (2002b) Changes in secondary metabolism and deposition of an unusual lignin in the ref8 mutant of Arabidopsis. *Plant Journal*, 30, 47–59.
Freudenberg, K. & Neish, A.C. (1968) *Constitution and Biosynthesis of Lignin*. Springer-Verlag, Berlin, Heidelberg, New York.
Fu, C., Mielenz, J.R., Xiao, X., *et al.* (2011a) Genetic manipulation of lignin reduces recalcitrance and improves ethanol production from switchgrass. *Proceedings of the National Academy of Sciences USA*, 108, 3803–3808.
Fu, C.-X., Xiao, X., Xi, Y., *et al.* (2011b) Downregulation of cinnamyl alcohol dehydrogenase (CAD) leads to improved saccharification efficiency in switchgrass. *BioEnergy Research*, 4, 153–164.
Funnell-Harris, D.L., Pedersen, J.F. & Sattler, S.E. (2010) Alteration in lignin biosynthesis restricts growth of *Fusarium* spp. in brown midrib sorghum. *Phytopathology*, 100, 671–681.
Gallego-Giraldo, L., Escamilla-Trevino, L., Jackson, L.A. & Dixon, R.A. (2011a) Salicylic acid mediates the reduced growth of lignin down-regulated plants. *Proceedings of the National Academy of Sciences USA*, 108, 20814–20819.

Gallego-Giraldo, L., Jikumaru, Y., Kamiya, Y., *et al.* (2011b) Selective lignin downregulation leads to constitutive defense response expression in alfalfa (*Medicago sativa* L.) *New Phytologist*, 190, 627–639.
Goujon, T., Sibout, R., Pollet, B., *et al.* (2003a) A new *Arabidopsis thaliana* mutant deficient in the expression of *O*-methyltransferase impacts lignins and sinapoyl esters. *Plant Molecular Biology*, 51, 973–989.
Goujon, T., Ferret, V., Mila, I., *et al.* (2003b) Down-regulation of the *AtCCR1* gene in *Arabidopsis thaliana*: effects on phenotype, lignins and cell wall degradability. *Planta*, 217, 218–228.
Grabber, J.H., Ralph, J., Hatfield, R.D. & Quideau, S. (1997) *p*-Hydroxyphenyl, guaiacyl, and syringyl lignins have similar inhibitory effects on wall degradability. *Journal of Agricultural and Food Chemistry*, 45, 2530–2532.
Gross, G.G., Stockigt, J., Mansell, R.L. & Zenk, M.H. (1973) Three novel enzymes involved in reduction of ferulic acid to coniferyl alcohol in higher plants: ferulate:CoA ligase, feruloyl-CoA reductase and coniferyl alcohol reductase. *FEBS Letters*, 31, 283–286.
Guillaumie, S., Pichon, M., Martinant, J.P., *et al.* (2007) Differential expression of phenylpropanoid and related genes in *brown-midrib bm1*, *bm2*, *bm3*, and *bm4* young near-isogenic maize plants. *Planta*, 226, 235–250.
Guo, D., Chen, F., Inoue, K., *et al.* (2001a) Down-regulation of caffeic acid 3-*O*-methyltransferase and caffeoyl CoA 3-*O*-methyltransferase in transgenic alfalfa. Impacts on lignin structure and implications for the biosynthesis of G and S lignin. *Plant Cell*, 13, 73–88.
Guo, D., Chen, F., Wheeler, J.S., *et al.* (2001b) Improvement of in-rumen digestibility of alfalfa forage by genetic manipulation of lignin *O*-methyltransferases. *Transgenic Research*, 10, 457–464.
Hahlbrock, K. & Grisebach, H. (1970) Formation of coenzyme A esters of cinnamic acids with an enzyme preparation from cell suspension cultures of parsley. *FEBS Letters*, 11, 62–64.
Halpin, C., Holt, K., Chojecki, J., *et al.* (1998) *Brown-midrib* maize (*bm1*) – a mutation affecting the cinnamyl alcohol dehydrogenase gene. *Plant Journal*, 14, 545–553.
Hoffmann, L., Maury, S., Martz, F., *et al.* (2003) Purification, cloning, and properties of an acyltransferase controlling shikimate ester intermediates in phenylpropanoid metabolism. *Journal of Biological Chemistry*, 278, 95–103.
Hoffmann, L., Besseau, S., Geoffroy, P., *et al.* (2004) Silencing of hydroxycinnamoyl-coenzyme A shikimate/quinate hydroxycinnamoyltransferase affects phenylpropanoid biosynthesis. *Plant Cell*, 16, 1446–1465.
Hu, W.-J., Kawaoka, A., Tsai, C.-J., *et al.* (1998) Comparmentalized expression of two structurally and functionally distinct 4-coumarate ligase genes in aspen (*Populus tremuloides*). *Proceedings of the National Academy of Sciences USA*, 95, 5407–5412.
Hu, W.-J., Harding, S.A., Lung, J., *et al.* (1999) Repression of lignin biosynthesis promotes cellulose accumulation and growth in transgenic trees. *Nature Biotechnology*, 17, 808–812.
Huang, J., Gu, M., Lai, Z., *et al.* (2010) Functional analysis of the Arabidopsis *PAL* gene family in plant growth, development, and response to environmental stress. *Plant Physiology*, 153, 1526–1538.
Humphreys, J.M., Hemm, M.R. & Chapple, C. (1999) Ferulate 5-hydroxylase from Arabidopsis is a multifunctional cytochrome P450-dependent monooxygenase catalyzing parallel hydroxylations in phenylpropanoid metabolism. *Proceedings of the National Academy of Sciences USA*, 96, 10045–10050.
Huntley, S.K., Ellis, D., Gilbert, M., *et al.* (2003) Significant increases in pulping efficiency in C4H-F5H-transformed poplars: improved chemical savings and reduced environmental toxins. *Journal of Agricultural and Food Chemistry*, 51, 6178–6183.
Inoue, K., Sewalt, V.J.H., Balance, G.M., *et al.* (1998) Developmental expression and substrate specificities of alfalfa caffeic acid 3-*O*-methyltransferase and caffeoyl Coenzyme A 3-*O*-methyltransferase in relation to lignification. *Plant Physiology*, 117, 761–770.
Jackson, L.A., Shadle, G.L., Zhou, R., *et al.* (2008) Improving saccharification efficiency of alfalfa stems through modification of the terminal stages of monolignol biosynthesis. *Bioenergy Research*, 1, 180–192.

Jones, L., Ennos, A.R. & Turner, S.R. (2001) Cloning and characterization of *irregular xylem4* (*irx4*): a severely lignin-deficient mutant of Arabidopsis. *Plant Journal*, 26, 205–216.

Jouanin, L., Goujon, T., deNadai, V., *et al.* (2000) Lignification in transgenic poplars with extremely reduced caffeic acid *O*-methyltransferase activity. *Plant Physiology*, 123, 1363–1373.

Jung, H.G. & Deetz, D.A. (1993) Cell wall lignification and degradability. In: *Forage Cell Wall Structure and Digestibility* (eds H.G. Jung, D.R. Buxton, R.D. Hatfield & J. Ralph), pp. 315–346. ASA-CSSA-SSSA, Madison, WI.

Jung, H.G., Sheaffer, C.C., Barnes, D.K. & Halgerson, J.L. (1997) Forage quality variation in the US alfalfa core collection. *Crop Science*, 37, 1361–1366.

Kajita, S., Katayama, Y. & Omori, S. (1996) Alterations in the biosynthesis of lignin in transgenic plants with chimeric genes for 4-coumarate:CoA ligase. *Plant and Cell Physiology*, 37, 957–965.

Kajita, S., Hishiyama, S., Tomimura, Y., *et al.* (1997) Structural characterization of modified lignin in transgenic tobacco plants in which the activity of 4-coumarate: coenzyme A ligase is depressed. *Plant Physiology*, 114, 871–879.

Kim, H., Ralph, J., Lu, F., *et al.* (2003) NMR analysis of lignins in CAD-deficient plants. Part 1. Incorporation of hydroxycinnamaldehydes and hydroxybenzaldehydes into lignins. *Organic and Biomolecular Chemistry*, 1, 268–281.

Kim, S.J., Kim, M.R., Bedgar, D.L., *et al.* (2004) Functional reclassification of the putative cinnamyl alcohol dehydrogenase multigene family in *Arabidopsis*. *Proceedings of the National Academy of Sciences USA*, 101, 1455–1460.

Kneusel, R.E., Matern, U. & Nicolay, K. (1989) Formation of *trans*-caffeoyl-CoA from *trans*-4-coumaroyl-CoA by $Zn^{2+}$-dependent enzymes in cultured plant cells and its activation by an elicitor-induced pH shift. *Archives of Biochemistry and Biophysics*, 269, 455–462.

Knobloch, K.-H. & Hahlbrock, K. (1975) Isoenzymes of *p*-coumarate: CoA ligase from cell suspension cultures of *Glycine max*. *European Journal of Biochemistry*, 52, 311–320.

Knobloch, K.-H. & Hahlbrock, K. (1977) 4-coumarate: CoA ligase from cell suspension cultures of *Petroselinum hortense* Hoffm. Partial purification, substrate specificity, and further properties. *Archives of Biochemistry and Biophysics*, 184, 237–248.

Ko, J.H., Kim, W.C. & Han, K.H. (2009) Ectopic expression of MYB46 identifies transcriptional regulatory genes involved in secondary wall biosynthesis in Arabidopsis. *Plant Journal*, 60, 649–665.

Koehler, L. & Telewski, F.W. (2006) Biomechanics and transgenic wood. *American Journal of Botany*, 93, 1433–1438.

Korth, K.L., Blount, J.W., Chen, F., *et al.* (2001) Changes in phenylpropanoid metabolites associated with homology-dependent silencing of phenylalanine ammonia-lyase and its somatic reversion in tobacco. *Physiologia Plantarum*, 111, 137–143.

Koukol, J. & Conn, E.E. (1961) The metabolism of aromatic compounds in higher plants. IV. Purification and properties of the phenylalanine deaminase of *Hordeum vulgare*. *Journal of Biological Chemistry*, 236, 2692–2698.

Kuhnl, T., Koch, U., Heller, W. & Wellmann, E. (1989) Elicitor induced S-adenosyl-L.methionine: caffeoyl-CoA 3-O-methyltransferase from carrot cell suspension cultures. *Plant Science*, 60, 21–25.

Lacombe, E., Van Doorsselaere, J., Boerjan, W., *et al.* (2000) Characterization of *cis*-elements required for vascular expression of the cinnamoyl CoA reductase gene and for protein-DNA complex formation. *Plant Journal*, 23, 663–676.

Lapierre, C., Pollet, B., Petit-Conil, M., *et al.* (1999) Structural alterations of lignins in transgenic poplars with depressed cinnamyl alcohol dehydrogenase or caffeic acid *O*-methyltransferase activity have opposite impact on the efficiency of industrial Kraft pulping. *Plant Physiology*, 119, 153–163.

Lapierre, C., Pilate, G., Pollet, B., *et al.* (2004) Signatures of cinnamyl alcohol dehydrogenase deficiency in poplar lignins. *Phytochemistry*, 65, 313–321.

Laskar, D.D., Jourdes, M., Patten, A.M., *et al.* (2006) The Arabidopsis cinnamoyl CoA reductase *irx4* mutant has a delayed but coherent (normal) program of lignification. *Plant Journal*, 48, 674–686.

Lauvergeat, V., Lacomme, C., Lacombe, E., *et al.* (2001) Two cinnamoyl-CoA reductase (CCR) genes from *Arabidopsis thaliana* are differentially expressed during development and in response to infection with pathogenic bacteria. *Phytochemistry*, 57, 1187–1195.

Lee, D., Meyer, K., Chapple, C. & Douglas, C.J. (1997) Antisense suppression of 4-coumarate: coenzyme A ligase activity in Arabidopsis leads to altered lignin subunit composition. *Plant Cell*, 9, 1985–1998.

Lee, Y., Chen, F., Dixon, R.A. & Voit, E.O. (2011) Integrative analysis of transgenic alfalfa (*Medicago sativa* L.) suggests new metabolic control mechanisms for monolignol biosynthesis. *PloS Computational Biology,* May 2011, 7 (5) e1002047.

Leplé, J.C., Dauwe, R., Morreel, K., *et al.* (2007) Downregulation of cinnamoyl-coenzyme A reductase in poplar: multiple-level phenotyping reveals effects on cell wall polymer metabolism and structure. *Plant Cell*, 19, 3669–3691.

Li, L., Popko, J.L., Umezawa, T. & Chiang, V.L. (2000) 5-hydroxyconiferyl aldehyde modulates enzymatic methylation for syringyl monolignol formation, a new view of monolignol biosynthesis in angiosperms. *Journal of Biological Chemistry*, 275, 6537–6545.

Li, L., Zhou, Y., Cheng, X., *et al.* (2003) Combinatorial modification of multiple lignin traits in trees though multigene cotransformation. *Proceedings of the National Academy of Sciences USA*, 100, 4939–4944.

Li, X., Ximenes, E., Kim, Y., *et al.* (2010a) Lignin monomer composition affects *Arabidopsis* cell-wall degradability after liquid hot water pretreatment. *Biotechnology for Biofuels*, 3, 27.

Li, X., Bonawitz, N.D., Weng, J.K. & Chapple, C. (2010b) The growth reduction associated with repressed lignin biosynthesis in *Arabidopsis thaliana* is independent of flavonoids. *Plant Cell*, 22, 1620–1632.

Lindermayr, C., Mollers, B., Fliegmann, J., *et al.* (2002) Divergent members of a soybean (*Glycine max* L.) 4-coumarate:coenzyme A ligase gene family. *European Journal of Biochemistry*, 269, 1304–1315.

Lu, F., Marita, J.M., Lapierre, C., *et al.* (2010) Sequencing around 5–hydroxyconiferyl alcohol-derived units in caffeic acid *O*-methyltransferase-deficient poplar lignins. *Plant Physiology*, 153, 569–579.

Ma, Q.H. & Xu, Y. (2008) Characterization of a caffeic acid 3-*O*-methyltransferase from wheat and its function in lignin biosynthesis. *Biochimie*, 90, 515–524.

MacKay, J.J., O'Malley, D.M., Presnell, T., *et al.* (1997) Inheritance, gene expression, and lignin characterization in a mutant pine deficient in cinnamyl alcohol dehydrogenase. *Proceedings of the National Academy of Sciences USA*, 94, 8255–8260.

Maher, E.A., Bate, N.J., Ni, W., *et al.* (1994) Increased disease susceptibility of transgenic tobacco plants with suppressed levels of preformed phenylpropanoid products. *Proceedings of the National Academy of Sciences USA*, 96, 7802–7806.

Mann, D.J.G., Labbé, N., Sykes, R., *et al.* (2009) Rapid assessment of lignin content and structure in switchgrass (*Panicum virgatum* L.) grown under different environmental conditions. *BioEnergy Research*, 2, 246–256.

Mansell, R.L., Gross, G.G., Stockigt, J., *et al.* (1974) Purification and properties of cinnamyl alcohol-dehydrogenase from higher plants involved in lignin biosynthesis. *Phytochemistry*, 13, 2427–2435.

Marita, J.M., Ralph, J., Hatfield, R.D. & Chapple, C. (1999) NMR characterization of lignin in Arabidopsis altered in the activity of ferulate 5-hydroxylase. *Proceedings of the National Academy of Sciences USA*, 96, 12328–12332.

Marita, J.M., Ralph, J., Hatfield, R.D., *et al.* (2003a) Structural and compositional modifications in lignin of transgenic alfalfa down-regulated in caffeic acid 3-*O*-methyltransferase and caffeoyl coenzyme A 3-*O*-methyltransferase. *Phytochemistry*, 62, 53–65.

Marita, J.M., Vermerris, W., Ralph, J. & Hatfield, R.D. (2003b) Variations in the cell wall composition of maize brown midrib mutants. *Journal of Agricultural and Food Chemistry*, 51, 1313–1321.

Marles, M.A., Vandenberg, A. & Bett, K.E. (2008) Polyphenol oxidase activity and differential accumulation of polyphenolics in seed coats of pinto bean (*Phaseolus vulgaris* L.) characterize postharvest color changes. *Journal of Agricultural and Food Chemistry*, 56, 7049–7056.

Martone, P.T., Estevez, J.M., Lu, F., *et al.* (2009) Discovery of lignin in seaweed reveals convergent evolution of cell wall architecture. *Current Biology*, 19, 169–175.

McCarthy, R.L., Zhong, R. & Ye, Z.H. (2009) MYB83 is a direct target of SND1 and acts redundantly with MYB46 in the regulation of secondary cell wall biosynthesis in Arabidopsis. *Plant and Cell Physiology*, 50, 1950–1964.

Meyer, K., Shirley, A.M., Cusumano, J.C., *et al.* (1998) Lignin monomer composition is determined by the expression of a cytochrome P450-dependent monooygenase in *Arabidopsis*. *Proceedings of the National Academy of Sciences USA*, 95, 6619–6623.

Meyermans, H., Morreel, K., Lapierre, C., *et al.* (2000) Modifications in lignin and accumulation of phenolic glucosides in poplar xylem upon down-regulation of caffeoyl-coenzyme A O-methyltransferase, an enzyme involved in lignin biosynthesis. *Journal of Biological Chemistry*, 275, 36899–36909.

Millar, D.J., Long, M., Donovan, G., *et al.* (2007) Introduction of sense constructs of cinnamate 4-hydroxylase in transgenic tomato plants shows opposite effects on flux into stem lignin and fruit flavonoids. *Phytochemistry*, 68, 1497–1509.

Mir Derikvand, M., Sierra, J.B., Ruel, K., *et al.* (2008) Redirection of the phenylpropanoid pathway to feruloyl malate in Arabidopsis mutants deficient for cinnamoyl-CoA reductase 1. *Planta*, 227, 943–956.

Mitsuda, N., Seki, M., Shinozaki, K. & Ohme-Takagi, M. (2005) The NAC transcription factors NST1 and NST2 of Arabidopsis regulate secondary wall thickenings and are required for anther dehiscence. *Plant Cell*, 17, 2993–3006.

Mitsuda, N., Iwase, A., Yamamoto, H., *et al.* (2007) NAC transcription factors, NST1 and NST3, are key regulators of the formation of secondary walls in woody tissues of Arabidopsis. *Plant Cell*, 19, 270–280.

Mizutani, M., Ward, E., DiMaio, J., *et al.* (1993) Molecular cloning and sequencing of a cDNA encoding mung bean cytochrome P450 (P450C4H) possessing cinnamate 4-hydroxylase activity. *Biochemical and Biophysical Research Communications*, 190, 875–880.

Morreel, K., Ralph, J., Lu, F., *et al.* (2004) Phenolic profiling of caffeic acid *O*-methyltransferase-deficient poplar reveals novel benzodioxane oligolignols. *Plant Physiology*, 136, 4023–4036.

Musel, G., Schindler, T., Bergfeld, R., *et al.* (1997) Structure and distribution of lignin in primary and secondary cell walls of maize coleoptiles analysed by chemical and immunological probes. *Planta*, 201, 146–159.

Nakashima, J., Chen, F., Jackson, L., *et al.* (2008) Multi-site genetic modification of monolignol biosynthesis in alfalfa (*Medicago sativa*): effects on lignin composition in specific cell types. *New Phytologist*, 179, 738–750.

Nakatsubo, T., Kitamura, Y., Sakakibara, N., *et al.* (2008) At5g54160 gene encodes *Arabidopsis thaliana* hydroxyconiferaldehyde *O*-methyltransferase. *Journal of Wood Science*, 54, 312–317.

Ni, W., Paiva, N.L. & Dixon, R.A. (1994) Reduced lignin in transgenic plants containing a caffeic acid *O*-methyltransferase antisense gene. *Transgenic Research*, 3, 120–126.

Niggeweg, R., Michael, A.J. & Martin, C. (2004) Engineering plants with increased levels of the antioxidant chlorogenic acid. *Nature Biotechnology*, 22, 746–754.

Oba, M. & Allen, M.S. (1999) Effects of brown midrib 3 mutation in corn silage on dry matter intake and productivity of high yielding dairy cows, *Journal of Dairy Science*, 82, 135–142.

O'Connell, A., Holt, K., Piquemal, J., *et al.* (2002) Improved paper pulp from plants with suppressed cinnamoyl-CoA reductase or cinnamyl alcohol dehydrogenase. *Transgenic Research*, 11, 495–503.

Oliver, A.L., Grant, R.J., Pedersen, J.F. & O'Rear, J. (2004) Comparison of *brown midrib-6* and *-18* forage sorghum with conventional sorghum and corn silage in diets of lactating dairy cows. *Journal of Dairy Science*, 87, 637–644.

Osakabe, K., Tsao, C.C., Li, L., *et al.* (1999) Coniferyl aldehyde 5–hydroxylation and methylation direct syringyl lignin biosynthesis in angiosperms. *Proceedings of the National Academy of Sciences USA*, 96, 8955–8960.

Pallas, J.A., Paiva, N.L., Lamb, C. & Dixon, R.A. (1996) Tobacco plants epigenetically suppressed in phenylalanine ammonia-lyase expression do not develop systemic acquired resistance in response to infection by tobacco mosaic virus. *Plant Journal*, 10, 281–293.

Parvathi, K., Chen, F., Guo, D., *et al.* (2001) Substrate preferences of *O*-methyltransferases in alfalfa suggest new pathways for 3-*O*-methylation of monolignols. *Plant Journal*, 25, 193–202.

Patten, A.M., Cardenas, C.L., Cochrane, F.C., *et al.* (2005) Reassessment of effects on lignification and vascular development in the *irx4 Arabidopsis* mutant. *Phytochemistry*, 66, 2092–2107.

Patten, A.M., Jourdes, M., Brown, E.E., *et al.* (2007) Reaction tissue formation and stem tensile modulus properties in wild-type and p-coumarate-3-hydroxylase downregulated lines of alfalfa, *Medicago sativa* (*Fabaceae*). *American Journal of Botany*, 94, 912–925.

Patzlaff, A., McInnis, S., Courtenay, A., *et al.* (2003) Characterisation of a pine MYB that regulates lignification. *Plant Journal*, 36, 743–754.

Pilate, G., Guiney, E., Holt, K., *et al.* (2002) Field and pulping performances of transgenic trees with altered lignification. *Nature Biotechnology*, 20, 607–612.

Pincon, G., Maury, S., Hoffmann, L., *et al.* (2001) Repression of *O*-methyltransferase genes in transgenic tobacco affects lignin synthesis and plant growth. *Phytochemistry*, 57, 1167–1176.

Piquemal, J., Lapierre, C., Myton, K., *et al.* (1998) Down-regulation in cinnamoyl-CoA reductase induces significant changes of lignin profiles in transgenic tobacco plants. *Plant Journal*, 13, 71–83.

Piquemal, J., Chamayou, S., Nadaud, I., *et al.* (2002) Down-regulation of caffeic acid *O*-methyltransferase in maize revisited using a transgenic approach. *Plant Physiology*, 130, 1675–1685.

Pu, Y., Chen, F., Ziebell, A., *et al.* (2009) NMR characterization of C3H and HCT downregulated alfalfa lignin. *BioEnergy Research*, 2, 198–208.

Rae, A.L., Manners, J.M., Jones, R.J., *et al.* (2001) Antisense suppression of the lignin biosynthetic enzyme, caffeate *O*-methyltransferase, improves *in vitro* digestibility of the tropical pasture legume, *Stylosanthes humilis*. *Australian Journal of Plant Physiology*, 28, 289–297.

Ralph, J., MacKay, J.J., Hatfield, R.D., *et al.* (1997) Abnormal lignin in a loblolly pine mutant. *Science*, 277, 235–239.

Ralph, J., Hatfield, R.D., Piquemal, J., *et al.* (1998) NMR characterization of altered lignins extracted from tobacco plants down-regulated for lignification enzymes cinnamyl alcohol dehydrogenase and cinnamoyl-CoA reductase. *Proceedings of the National Academy of Sciences USA*, 95, 12803–12808.

Ralph, J., Lapierre, C., Marita, J.M., *et al.* (2001a) Elucidation of new structures in lignins of CAD- and COMT-deficient plants by NMR. *Phytochemistry*, 57, 993–1003.

Ralph, J., Lapierre, C., Lu, F., *et al.* (2001b) NMR evidence of benzodioxane structures resulting from incorporation of 5-hydroxyconiferyl alcohol into lignins of *O*-methyltransferase-deficient plants. *Journal of Agricultural and Food Chemistry*, 49, 86–91.

Ralph, J., Lundquist, K., Brunow, G., *et al.* (2004) Lignins: natural polymers from oxidative coupling of 4-hydroxyphenyl-propanoids. *Phytochemical Reviews*, 3, 29–60.

Ralph, J., Akiyama, T., Kim, H., *et al.* (2006) Effects of coumarate 3-hydroxylase downregulation on lignin structure. *Journal of Biological Chemistry*, 281, 8843–8853.

Ralph, J., Kim, H., Lu, F., *et al.* (2008) Identification of the structure and origin of a thioacidolysis marker compound for ferulic acid incorporation into angiosperm lignins (and an indicator for cinnamoyl CoA reductase deficiency). *Plant Journal*, 53, 368–379.

Reddy, M.S.S., Chen, F., Shadle, G., *et al.* (2005) Targeted down-regulation of cytochrome P450 enzymes for forage quality improvement in alfalfa (*Medicago sativa* L.). *Proceedings of the National Academy of Sciences*, 102, 16573–16578.

Reichert, A.I., He, X.Z. & Dixon, R.A. (2009) Phenylalanine ammonia-lyase (PAL) from tobacco (*Nicotiana tabacum*): characterization of the four tobacco PAL genes and active heterotetrameric enzymes. *Biochemical Journal*, 424, 233–242.

Rhodes, M.J.C. & Wooltorton, L.S.C. (1976) The enzymic conversion of hydroxycinnamic acids to the p-coumaroyl quinic and chlorogenic acids in tomato fruits. *Phytochemistry*, 15, 947–951.

Rohde, A., Morreel, K., Ralph, J., *et al.* (2004) Molecular phenotyping of the *pal1* and *pal2* mutants of *Arabidopsis thaliana* reveals far-reaching consequences on phenylpropanoid, amino acid, and carbohydrate metabolism. *Plant Cell*, 16, 2749–2771.

Ruel, K., Berrio-Sierra, J., Mir Derikvand, M., *et al.* (2009) Impact of *CCR1* silencing on the assembly of lignified secondary walls in *Arabidopsis thaliana*. *New Phytologist*, 184, 99–113.

Russell, D.W. (1971) The metabolism of aromatic compounds in higher plants. X. Properties of the cinnamic acid 4-hydroxylase of pea seedlings and some aspects of its metabolic and developmental control. *Journal of Biological Chemistry*, 246, 3870–3878.

Russell, D.W. & Conn, E.E. (1967) The cinnamic acid 4-hydroxylase of pea seedlings. *Archives of Biochemistry and Biophysics*, 122, 256–258.

Saathoff, A.J., Sarath, G., Chow, E.K., *et al.* (2011) Downregulation of cinnamyl-alcohol dehydrogenase in switchgrass by RNA silencing results in enhanced glucose release after cellulase treatment. *PLoS One*, Jan 27, 6(1):e16416.

Saballos, A., Vermerris, W., Rivera, L. & Ejeta, G. (2008) Allelic association, chemical characterization and saccharification properties of *brown midrib* mutants of sorghum (*Sorghum bicolor* (l.) Moench). *BioEnergy Research*, 1, 193–204.

Saballos, A., Ejeta, G., Sanchez, E., *et al.* (2009) A genomewide analysis of the cinnamyl alcohol dehydrogenase family in sorghum [*Sorghum bicolor* (L.) Moench] identifies SbCAD2 as the brown midrib6 gene. *Genetics*, 181, 783–795.

Saka, S. & Goring, D.A.I. (1985) Localization of lignins in wood cell walls. In: *Biosynthesis and Biodegradation of Wood Componenets* (ed. T Higuchi), pp. 51–62. Academic, Orlando, FL.

Sarkanen, K.V. & Ludwig, C.H. (1971) *Lignins: Occurrence, Formation, Structure and Reactions*. Wiley-Interscience, New York.

Sattler, S.E., Saathoff, A.J., Haas, E.J., *et al.* (2009) A nonsense mutation in a cinamyl alcohol dehydrogenase gene is responsible for sorghum *brown midrib6* phenotype. *Plant Physiology*, 150, 584–595.

Sattler, S.E., Funnell-Harris, D.L. & Pederson, J.F. (2010) Efficacy of singular and stacked brown midrib 6 and 12 in the modification of lignocellulose and grain chemistry. *Journal of Agricultural and Food Chemistry*, 58, 3611–3616.

Schilmeller, A.L., Stout, J., Weng, J.K., *et al.* (2009) Mutations in the cinnamate 4-hydroxylase gene impact metabolism, growth and development in Arabidopsis. *Plant Journal*, 60, 771–782.

Schoch, G., Goepfert, S., Morant, M., *et al.* (2001) CYP98A3 from *Arabidopsis thaliana* is a 3′-hydroxylase of phenolic esters, a missing link in the phenylpropanoid pathway. *Journal of Biological Chemistry*, 276, 36566–36574.

Sewalt, V.J.H., Ni, W., Blount, J.W., *et al.* (1997a) Reduced lignin content and altered lignin composition in transgenic tobacco down-regulated in expression of L-phenylalanine ammonia-lyase or cinnamate 4-hydroxylase. *Plant Physiology*, 115, 41–50.

Sewalt, V.J., Ni, W., Jung, H.G. & Dixon, R.A. (1997b) Lignin impact on fiber degradation: increased enzymatic digestibility of genetically engineered tobacco (*Nicotiana tabacum*) stems reduced in lignin content. *Journal of Agricultural and Food Chemistry*, 45, 1977–1983.

Shadle, G., Chen, F., Srinivasa Reddy, M.S., *et al.* (2007) Down-regulation of hydroxycinnamoyl CoA: shikimate hydroxycinnamoyl transferase in transgenic alfalfa affects lignification, development and forage quality. *Phytochemistry*, 68, 1521–1529.

Shen, H., Fu, C., Xiao, X., *et al.* (2009) Developmental control of lignification in stems of lowland switchgrass variety Alamo and the effects on saccharification efficiency. *BioEnergy Research*, 2, 233–245.

Shen, H., He, X., Poovaiah, C.R., *et al.* (2012) Functional characterization of the switchgrass (*Panicum virgatum*) R2R3–MYB transcription factor *PvMYB4* for improvement of lignocellulosic feedstocks. *New Phytologist*, 193, 121–136.

Sibout, R., Eudes, A., Pollet, B., *et al.* (2003) Expression pattern of two paralogs encoding cinnamyl alcohol dehydrogenases in Arabidopsis. Isolation and characterization of the corresponding mutants. *Plant Physiology*, 132, 848–860.

Sibout, R., Eudes, A., Mouille, G., *et al.* (2005) *CINNAMYL ALCOHOL DEHYDROGENASE-C* and *-D* are the primary genes involved in lignin biosynthesis in the floral stem of Arabidopsis. *Plant Cell*, 17, 2059–2076.

Song, J. & Wang, Z. (2011) RNAi-mediated suppression of the phenylalanine ammonia-lyase gene in *Salvia miltirrhiza* causes abnormal phenotypes and a reduction in rosmarinic acid biosynthesis. *Journal of Plant Research*, 124, 183–192.

Stewart, D., Yahiaoui, N., McDougall, G.J., *et al.* (1997) Fourier-transform infrared and Raman spectroscopic evidence for the incorporation of cinnamaldehydes into the lignin of transgenic tobacco (*Nicotiana tabacum* L.) plants with reduced expression of cinnamyl alcohol dehydrogenase. *Planta*, 201, 311–318.

Stewart, J.J., Akiyama, T., Chapple, C., *et al.* (2009) The effects on lignin structure of overexpression of ferulate 5-hydroxylase in hybrid poplar. *Plant Physiology*, 150, 621–635.

Sutela, S., Niemi, K., Edesi, J., *et al.* (2009) Phenolic compounds in ectomycorrhizal interaction of lignin modified silver birch. *BMC Plant Biology*, 9, 124.

Tamagnone, L., Merida, A., Parr, A., *et al.* (1998) The AmMYB308 and AmMYB330 transcription factors from antirrhinum regulate phenylpropanoid and lignin biosynthesis in transgenic tobacco. *Plant Cell*, 10, 135–154.

Terashima, N., Fukushima, K., He, L.-F. & Takabe, K. (1993) Comprehensive model of the lignified plant cell wall. In: *Forage Cell Wall Structure and Digestibility* (eds H.G. Jung, D.R. Buxton, R.D. Hatfield & J. Ralph), pp. 247–260. ASA-CSSA-SSA, Madison, WI.

Terashima, N., Nakashima, J. & Takabe, K. (1998) Proposed structure for protolignin in plant cell walls. In: *Lignin and Lignin Biosynthesis* (eds N.B. Lewis & S. Sarkanen), pp. 180–193. American Chemical Society Symposium Series 697. American Chemical Society, Washington, DC.

Teutsch, H.G., Hasenfratz, M.P., Lesot, A., *et al.* (1993) Isolation and sequence of a cDNA encoding the Jerusalem artichoke cinnamate 4-hydroxylase, a major plant cytochrome P450 involved in the general phenylpropanoid pathway. *Proceedings of the National Academy of Sciences USA*, 90, 4102–4106.

Thévenin, J., Pollet, B., Letarnec, B., *et al.* (2011) The simultaneous repression of CCR and CAD, two enzymes of the lignin biosynthetic pathway, results in sterility and dwarfism in *Arabidopsis thaliana*. *Molecular Plant*, 4, 70–82.

Tiimonen, H., Aronen, T., Laakso, T., *et al.* (2005) Does lignin modification affect feeding preference or growth performance of insect herbivores in transgenic silver birch (*Betula pendula* Roth)? *Planta*, 222, 699–708.

Tsai, C.J., Popko, J.L., Mielke, M.R., *et al.* (1998) Suppression of *O*-methyltransferase gene by homologous sense transgene in quaking aspen causes red-brown wood phenotypes. *Plant Physiology*, 117, 101–112.

Tu, Y., Rochfort, S., Liu, Z., *et al.* (2010) Functional analyses of caffeic acid *O*-methyltransferase and cinnamoyl-CoA-reductase genes from perennial ryegrass (*Lolium perenne*). *Plant Cell*, 22, 3357–3373.

Ulbrich, B. & Zenk, M.H. (1980) Partial purification and properties of *p*-hydroxycinnamoyl-CoA: shikimate *p*-hydroxycinnamoyl transferase from higher plants. *Phytochemistry*, 19, 1625–1629.

Umezawa, T. (2009) The cinnamate/monolignol pathway. *Phytochemical Reviews*, 9, 1–17.

Van der Rest, B., Danoun, S., Boudet, A.M. & Rochange, S.F. (2006) Down-regulation of cinnamoyl-CoA reductase in tomato (*Solanum lycopersicum* L.) induces dramatic changes in soluble phenolic pools. *Journal of Experimental Botany*, 57, 1399–1411.

Van Doorsselaere, J., Baucher, M., Chognot, E., *et al.* (1995) A novel lignin in poplar trees with a reduced caffeic acid/5–hydroxyferulic acid *O*-methyltransferase activity. *Plant Journal*, 8, 855–864.

Vermerris, W. & Boon, J.J. (2001) Tissue specific patterns of lignification are disturbed in the *brown midrib2* mutant of maize (*Zea mays* L.) *Journal of Agricultural and Food Chemistry*, 49, 721–728.

Voelker, A.L., Lachenbruch, B., Meinzer, F.C., *et al.* (2010) Antisense down-regulation of 4CL expression alters lignification, tree growth and saccharification potential of field grown poplar. *Plant Physiology*, 154, 874–886.

Wadenbäck, J., von Arnold, S., Egertsdotter, U., *et al.* (2008) Lignin biosynthesis in transgenic Norway spruce plants harboring an antisense construct for cinnamoyl CoA reductase (CCR). *Transgenic Research*, 17, 379–392.

Wagner, A., Ralph, J., Akiyama, T., *et al.* (2007) Exploring lignification in conifers by silencing hydroxycinnamoyl-CoA: shikimate hydroxycinnamoyltransferease in *Pinus radiata*. *Proceedings of the National Academy of Sciences USA*, 104, 11856–11861.

Wagner, A., Tobimatsu, Y., Phillips, L., *et al.* (2011) CCoAOMT suppression modifies lignin composition in *Pinus radiata*. *Plant Journal*, 67, 119–129.

Wang, H., Avci, U., Nakashima, J., *et al.* (2010) Mutation of WRKY transcription factors initiates pith secondary wall formation and increases stem biomass in dicotyledonous plants. *Proceedings of the National Academy of Sciences USA*, 107, 22338–22343.

Weng, J.K., Banks, J. & Chapple, C. (2008a) Parallels in lignin biosynthesis: a study in *Selaginella moellendorfii* reveals convergence across 400 million years of evolution. *Communications in Integrative Biology*, 1, 20–22.

Weng, J.K., Li, X., Stout, J. & Chapple, C. (2008b) Independent origins of syringyl lignin in vascular plants. *Proceedings of the National Academy of Sciences USA*, 105, 7887–7892.

Wróbel-Kwiatkowska, M., Starzycki, M., Zebrowski, J., *et al.* (2007) Lignin deficiency in transgenic flax resulted in plants with improved mechanical properties. *Journal of Biotechnology*, 128, 919–934.

Xu, Z., Zhang, D., Hu, J., *et al.* (2009) Comparative genome analysis of lignin biosynthesis gene families across the plant kingdom. *BMC Bioinformatics* 10 (Suppl. 11): S3.

Yamaguchi, M., Kubo, M., Fukuda, H. & Demura, T. (2008) Vascular-related NAC-DOMAIN7 is involved in the differentiation of all types of xylem vessels in Arabidopsis roots and shoots. *Plant Journal*, 55, 652–664.

Yang, F., Mitra, P., Zhang, L., *et al.* (2013) Engineering secondary cell wall deposition in plants. *Plant Biotechnology Journal*, 11, 325–335.

Ye, Z.H., Kneusel, R.E., Matern, U. & Varner, J.E. (1994) An alternative methylation pathway in lignin biosynthesis in Zinnia. *Plant Cell*, 6, 1427–1439.

Zandbergen, N. (2006) Brown mid-rib corn showcased. *AgriNews Interactive*, 30, http://www.agrinewsinteractive.com/archives/article-7899.htm (last accessed April 2, 2014).

Zeng, Y., Saar, B.G., Friedrich, M.G., *et al.* (2010) Imaging lignin-downregulated alfalfa using coherent anti-stokes Raman scattering microscopy. *BioEnergy Research*, 3, 272–277.

Zhang, K., Qian, Q., Huang, Z., *et al.* (2006) GOLD HULL AND INTERNODE2 encodes a primarily multifunctional cinnamyl-alcohol dehydrogenase in rice. *Plant Physiology*, 140, 972–983.

Zhao, Q. & Dixon, R.A. (2011) Transcriptional networks for lignin biosynthesis: more complex than we thought? *Trends in Plant Science*, 16, 227–233.

Zhao, Q., Gallego-Giraldo, L., Wang, H., *et al.* (2010a) A NAC transcription factor orchestrates multiple features of cell wall development in *Medicago truncatula*. *Plant Journal*, 63, 100–114.

Zhao, Q., Wang, H., Yin, Y., *et al.* (2010b) Syringyl lignin biosynthesis is directly regulated by a secondary cell wall master switch. *Proceedings of the National Academy of Sciences USA*, 107, 14496–14501.

Zhong, R. & Ye, Z.H. (2007) Regulation of cell wall biosynthesis. *Current Opinion in Plant Biology*, 10, 564–572.

Zhong, R. & Ye, Z.H. (2009) Secondary cell walls. In: *eLS*. John Wiley & Sons Ltd, Chichester. http://www.els.net, doi: 10.1002/9780470015902.a0021256 (last accessed April 2, 2014).
Zhong, R., Morrison, W.H., Negrel, J. & Ye, Z.-H. (1998) Dual methylation pathways in lignin biosynthesis. *Plant Cell*, 10, 2033–2046.
Zhong, R., Morrison, W.H. 3rd, Himmelsbach, D.S., *et al.* (2000) Essential role of caffeoyl coenzyme A *O*-methyltransferase in lignin biosynthesis in woody poplar plants. *Plant Physiology*, 124, 563–578.
Zhong, R., Demura, T. & Ye, Z.H. (2006) SND1, a NAC domain transcription factor, is a key regulator of secondary wall synthesis in fibers of Arabidopsis. *Plant Cell*, 18, 3158–3170.
Zhong, R., Richardson, E.A. & Ye, Z.H. (2007) The MYB46 transcription factor is a direct target of SND1 and regulates secondary wall biosynthesis in Arabidopsis. *Plant Cell*, 19, 2776–2792.
Zhong, R., Lee, C. & Ye, Z.H. (2010) Evolutionary conservation of the transcriptional network regulating secondary cell wall biosynthesis. *Trends in Plant Science*, 15, 625–632.
Zhou, J., Lee, C., Zhong, R. & Ye, Z.H. (2009) MYB58 and MYB63 are transcriptional activators of the lignin biosynthetic pathway during secondary cell wall formation in Arabidopsis. *Plant Cell*, 21, 248–266.
Zhou, R., Jackson, L., Shadle, G., *et al.* (2010) Distinct cinnamoyl CoA reductases involved in parallel routes to lignin in *Medicago truncatula*. *Proceedings of the National Academy of Sciences USA*, 107, 17803–17808.
Ziebell, A., Gracom, K., Katahira, R., *et al.* (2010) Increase in 4–coumaroyl alcohol units during lignification in alfalfa (*Medicago sativa*) alters extractability and molecular weight of lignin. *Journal of Biological Chemistry*, 285, 38961–38968.

# Chapter 2
# Perturbing Lignin Biosynthesis: Metabolic Changes in Response to Manipulation of the Phenylpropanoid Pathway

Nickolas A. Anderson and Clint Chapple
*Department of Biochemistry, Purdue University, West Lafayette, IN, USA*

**Abstract:** The phenylpropanoid pathway, which is largely conserved throughout land plants, is responsible for the biosynthesis of many compounds that are involved in plants′ structural integrity, water transport, UV protection, and defense against herbivores. The major end product of the phenylpropanoid pathway is lignin, a complex heteropolymer of the secondary cell wall. Decreasing or altering lignin structure provides enhanced cell-wall digestibility and can greatly increase the utilization of lignin itself or other cell-wall polysaccharides. Motivated by the agro-industrial importance of lignin, the genes participating in lignin biosynthesis have been identified and their manipulation has been undertaken in many plant species. Here we describe the metabolic changes in both the soluble and cell wall-bound phenylpropanoid pools in plants with perturbed lignin biosynthesis. Frequently, plants with disrupted lignin biosynthesis display sufficient metabolic plasticity that carbon flux, which is normally directed toward one or more metabolites in wild-type plants, is redirected to another biosynthetically related compound or group of compounds. This redistribution is best demonstrated in cases where plants with perturbed lignin biosynthesis accumulate atypical lignins that differ in monomer composition. In addition to metabolic changes, some perturbations in lignin biosynthesis affect development and can result in dwarfism. These observations may indicate a direct impact of altered lignin and cell-wall architecture on plant growth, or they may suggest that lignin biosynthesis is tied to other metabolic networks that affect plant growth and development.

**Keywords:** hydroxycinnamoyl esters, hydroxycinnamic acids, biofuel, metabolic plasticity, Arabidopsis

*Recent Advances in Polyphenol Research, Volume 4*, First Edition.
Edited by Annalisa Romani, Vincenzo Lattanzio, and Stéphane Quideau.

## 2.1 Introduction

Plants produce a wide array of specialized metabolites that are required for normal growth and development. A subset of these compounds is synthesized from phenylalanine via the phenylpropanoid pathway (Fig. 2.1). Phenylpropanoids aid in structural support, water transport, UV protection, pollen viability, and responses to biotic and abiotic stressors (Veit & Pauli, 1999; Dixon, 2001; Widarto *et al.*, 2006). Not only are phenylpropanoids important for plant vigor, but some are also commercially valuable as pharmaceuticals and dietary supplements, and for other applications in industry and agriculture (Kroon & Williamson, 1999). Processes including ruminant digestion, chemical pulping, and the saccharification of biomass for biofuel production are inhibited by lignin, a major product of the phenylpropanoid pathway (Schooneveld-Bergmans *et al.*, 1999; Grabber *et al.*, 2000). Directed manipulations of lignin composition and deposition, through the rerouting of the flow of carbon through the pathway, have been attempted with the ultimate goal of improving these processes (Chapple *et al.*, 2007). Plants that have been engineered in this way have revealed that phenylpropanoid metabolism is highly plastic, a topic that will be further expanded upon throughout this chapter.

Phenylpropanoid metabolites are found in most plant tissues, at most stages of development, and many are conserved evolutionarily across modern plant lineages (Carnachan & Harris, 2000; Popper & Fry, 2003, 2004; Bottcher *et al.*, 2008; Ligrone *et al.*, 2008). Research in model systems including Arabidopsis, Populus, tobacco, and others has revealed both conserved and unique aspects of phenylpropanoid metabolism within the plant kingdom.

A particularly fruitful approach to acquiring a deeper understanding of the genes and enzymes involved in phenylpropanoid biosynthesis has been the forward-genetic characterization of Arabidopsis mutants (Shirley *et al.*, 1995; Ruegger & Chapple, 2001). In Arabidopsis, sinapate esters, a class of hydroxycinnamoyl esters (HCEs), accumulate as end products of the phenylpropanoid pathway; they can be visualized *in vivo* under UV light by their bluish-green fluorescence (Chapple *et al.*, 1992) (Fig. 2.2). Decreases in sinapate esters as a result of pathway perturbation in mutant plants lead to a *reduced epidermal florescence* (*ref*) phenotype, and these mutants have contributed substantially to our current understanding of the phenylpropanoid pathway as it occurs in flowering plants (Fig. 2.1). This chapter elaborates on several of these *ref* mutants, as well as other relevant Arabidopsis phenylpropanoid mutants and transgenic plants, and how these findings compare with complementary work done in tobacco, Medicago, Populus, and several other species (Franke *et al.*, 2000; van der Rest *et al.*, 2006; Chen & Dixon, 2007).

### *2.1.1 Cell wall-bound phenylpropanoids*

Cell wall-bound phenylpropanoids can be categorized into hydroxycinnamyl alcohol-derived compounds, mainly lignins, and hydroxycinnamic acid (HCA)-derived compounds,

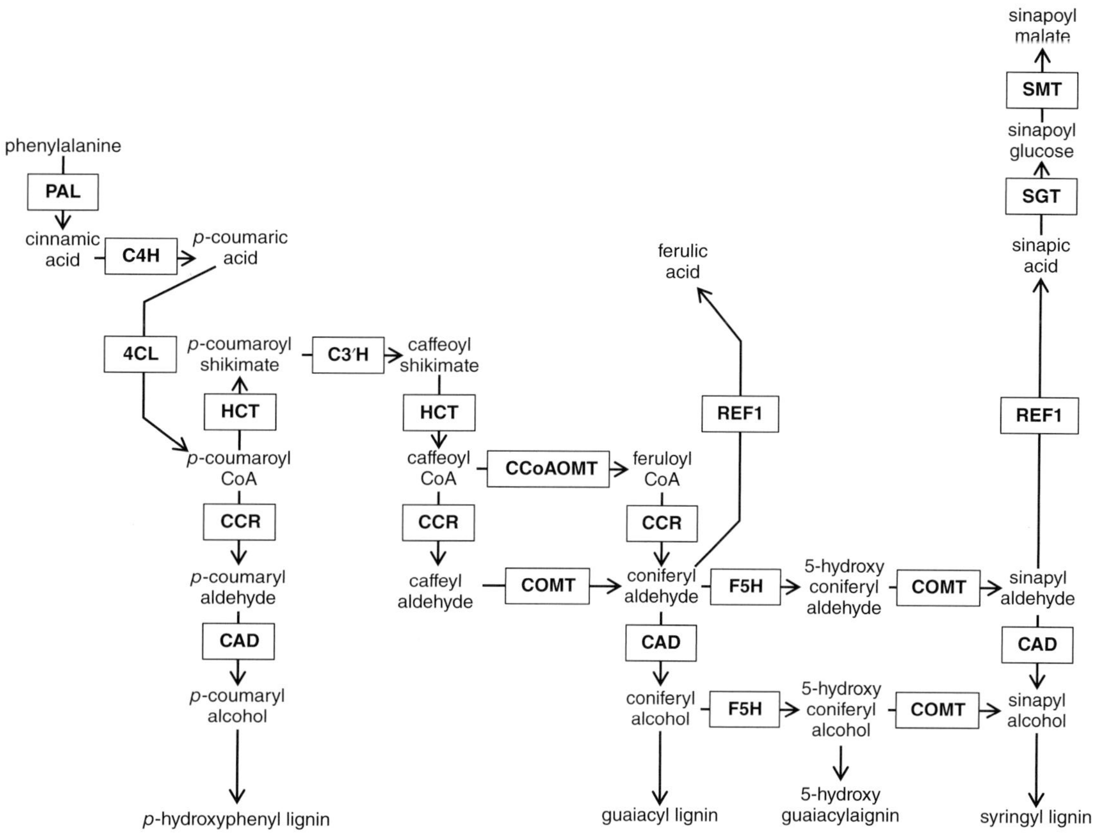


**Fig. 2.1** The current phenylpropanoid pathway as it occurs in flowering plants, with a specific emphasis on pathways of sinapate ester synthesis relevant to pathway elucidation in Arabidopsis. Enzymes of the pathway are phenylalanine ammonia lyase (PAL); cinnamate 4-hydroxylase (C4H); hydroxycinnamoyl CoA shikimate:quinate hydroxycinnamoyl transferase (HCT); *p*-coumaroylshikimate 3′-hydroxylase (C3′H); caffeoyl CoA *O*-methyl transferase (CCoAOMT); cinnamoyl CoA reductase (CCR); ferulic acid 5-hydroxylase (F5H); caffeic acid/5-hydroxyferulic acid *O*-methyltransferase (COMT); cinnamyl alcohol dehydrogenase (CAD); hydroxycinnamaldehyde dehydrogenase (REF1); sinapic acid: UDP-glucose dependent sinapoyltransferase (SGT); and sinapoylglucose malate: sinapoyltransferase (SMT).

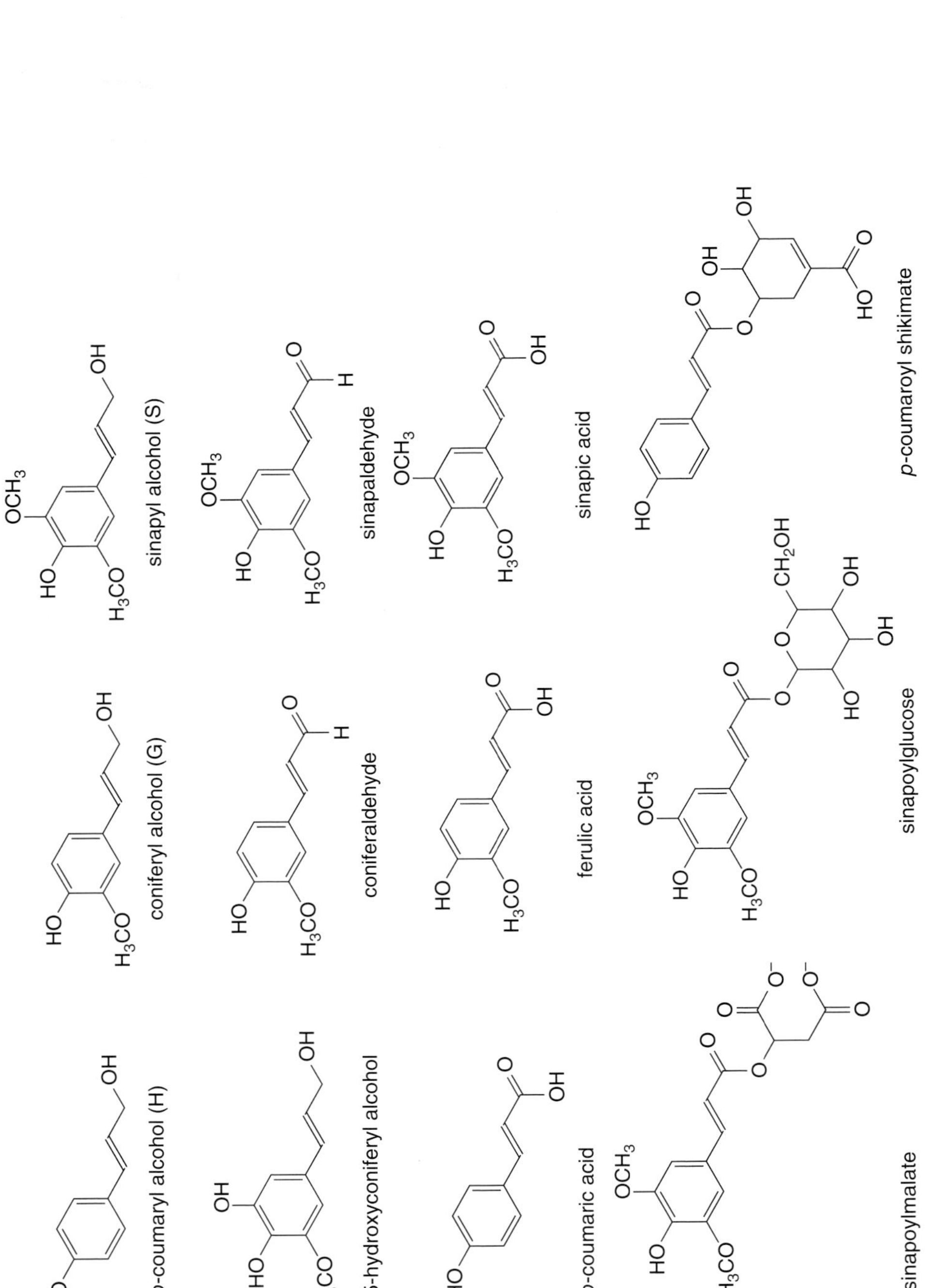


**Fig. 2.2** (a) Common hydroxycinnamyl alcohol monolignols. (b) Several uncommon lignin monomers. (c) Common hydroxycinnamic acids. (d) Several common soluble hydroxycinnamoyl esters. These compounds serve as lignin monomers – intermediates of the phenylpropanoid pathway – and are involved in responses to environmental stresses.

largely consisting of *p*-coumarate and ferulate conjugates (de Oliveira Buanafina, 2009; Bonawitz & Chapple, 2010). Lignin, a complex polymer of hydroxycinnamyl alcohols also referred to as monolignols, is deposited in the secondary cell wall and plays a major role in structural rigidity and in rendering tissues impermeable to water (Whetten *et al.*, 1998; Vanholme *et al.*, 2010a; Vogt, 2010). Lignin polymers are derived primarily from three hydroxycinnamyl alcohol moieties: *p*-coumaryl alcohol, coniferyl alcohol, and sinapyl alcohol, referred to as *p*-hydroxyphenyl (H), guaiacyl (G), and syringyl (S) subunits, respectively (Boerjan *et al.*, 2003) (Fig. 2.2). The relative abundance of these moieties in lignin differs from species to species and ranges from almost all G to different ratios of G and S (Campbell & Sederoff, 1996).

Both *p*-coumaric and ferulic acid are found in the primary and secondary cell walls of most plant species and can account for up to 3% of the dry weight of cell walls in graminaceous species (de Oliveira Buanafina, 2009). Ferulic acid plays an important role in structural support by crosslinking polysaccharides to lignin, serving as a lignin monomer, and potentially acting as a nucleation site for the formation of lignin polymers (Iiyama *et al.*, 1994; Hatfield *et al.*, 1999; Schooneveld-Bergmans *et al.*, 1999; Grabber *et al.*, 2000; Ralph *et al.*, 2008).

## 2.1.2 Soluble phenylpropanoids

Soluble phenylpropanoids include some of the most specialized and diverse metabolites in the plant kingdom (Tohge *et al.*, 2005; Wasson *et al.*, 2006). They can be grouped into a few main categories, including flavonoids, HCEs, lignans, and coumarins (Mauch-Mani & Slusarenko, 1996; Saindrenan *et al.*, 2002; Fraser *et al.*, 2007; Matsuno *et al.*, 2009).

Flavonoids are synthesized via a major branch of the phenylpropanoid pathway found in all plants examined to date. Plants employ *p*-coumaroyl CoA and malonyl CoA for the synthesis of over 6000 of these compounds, including chalcones, flavones, isoflavones, flavonols, flavandiols, aurones, anthocyanins, and condensed tannins (Winkel-Shirley, 2001). Flavonoids can be found in most plant species, where they contribute to the pigmentation observed in many tissue types and play a role in physiological processes including seed dormancy and viability, as well as in responses to different environmental stresses (Dixon & Paiva, 1995; Shirley *et al.*, 1995; Grotewold, 2006; Lepiniec *et al.*, 2006; Fini *et al.*, 2011).

In addition to flavonoids, phenylpropanoids, including HCEs such as chlorogenic acid, are present in many plant tissues and participate in physiological functions including responses to environmental stress (Kroon & Williamson, 1999; Hoffmann *et al.*, 2003; Cayuela *et al.*, 2006). Other HCEs, such as hydroxycinnamoyl CoA thioesters and *p*-coumaroyl and caffeoyl shikimate, serve as critical intermediates in phenylpropanoid biosynthesis. Of particular importance to this chapter are the HCEs known as hydroxycinnamoylmalate esters, which are largely conserved to members of Brassicaceae and include compounds that participate as UV-protectants in the epidermis (Grawe *et al.*, 1992; Ruegger & Chapple, 2001; Widarto *et al.*, 2006). Beyond the Brassicaceae, recent metabolomic analyses have revealed detectable levels of hydroxycinnamoylmalate esters in members of Fabaceae (Sullivan, 2009). HCE biosynthesis, starting with an HCA, occurs through one of two different modes. The first includes the glucosylation of the HCA by a glucosyltransferase to form a glucose ester, which can then be used by a serine carboxypeptidase-like acyltransferase to form the HCE (Lim *et al.*, 2001; Fraser *et al.*, 2007). The second requires an ATP-dependent CoA ligase to form a CoA thioester, which a BAHD acyltransferase (named for the first identified members of this class of enzyme, benzylalcohol *O*-acetyltransferase, anthocyanin *O*-hydroxycinnamoyltransferase, anthranilate *N*-hydroxycinnamoyl/benzoyltransferase, and deacetylvindoline 4-*O*-acetyltransferase) uses as an acyl donor to form the HCE (Ehlting *et al.*, 1999; D'Auria *et al.*, 2007). A wide variety of substrates, including malate, choline, and spermidine, are used by these enzymes as acyl acceptors in the transesterification of the glucose esters and thioesters (Milkowski *et al.*, 2004; Fraser *et al.*, 2007; Grienenberger *et al.*, 2009) (Fig. 2.2).

## 2.2 Changes in metabolism associated with phenylpropanoid-pathway disruptions

### 2.2.1 *Phenylalanine ammonia-lyase (PAL)*

PAL catalyzes the deamination of phenylalanine to generate cinnamic acid and ammonia, and is the first committed step of the phenylpropanoid pathway (Bate *et al.*, 1994). In Arabidopsis, four known PAL isoforms, named PAL1–4, have redundant activity towards phenylalanine, with PAL3 having the lowest catalytic efficiency among the four (Cochrane *et al.*, 2004). Furthermore, *PAL3* is expressed at very low levels, whereas *PAL1*, *2*, and *4* are highly expressed in stem tissue, leading to the suggestion that these three isoforms are responsible for the majority of the PAL activity observed in wild-type plants (Raes *et al.*, 2003).

Arabidopsis plants deficient in each individual PAL enzyme display a wild-type growth and metabolic phenotype that further support the hypothesis that the PAL isoforms have redundant activity (Rohde *et al.*, 2004; Huang *et al.*, 2010). *pal1 pal2* double mutants have reduced lignin, flavonoids, and soluble HCEs (Rohde *et al.*, 2004; Huang *et al.*, 2010). These metabolic phenotypes are more severe in the *pal1 pal2 pal3* triple mutants and the *pal1 pal2 pal3 pal4* quadruple mutants. In addition to their metabolic phenotypes, the triple and quadruple mutants are sterile and are more susceptible to pathogens and to damage resulting from exposure to UV light (Huang *et al.*, 2010). In addition, and like many other plants with lower levels of lignin than wild-type, the *pal* multiple mutants are dwarf and have collapsed vasculature, which has been attributed to the impact of tension generated by water transport in inadequately reinforced tracheary elements. Decreased PAL activity also leads to an increase in phenylalanine and several other amino acids (Rohde *et al.*, 2004). The increase in phenylalanine is most likely the result of carbon not being able to enter the phenylpropanoid pathway, and the changes in other amino acids may be the result of the indirect effects of phenylalanine over-accumulation. Similar effects have been observed previously in transgenic plants, in which artificial sinks were used to decrease tryptophan and/or tyrosine pools, which resulted in the indirect reduction of phenylalanine and altered the accumulation of other amino acids (Yao *et al.*, 1995; Guillet *et al.*, 2000).

Cosuppression of PAL isoforms in tobacco and *Salvia miltiorrhiza* results in plants with similar growth and metabolic phenotypes as the Arabidopsis PAL mutants (Elkind *et al.*, 1990; Song & Wang, 2011). These phenotypes include dwarfism, sterility due to decreased pollen production and reduced pollen viability, lower levels of lignin, and, in the case of *Salvia*, a total reduction in soluble phenylpropanoids (Elkind *et al.*, 1990; Bate *et al.*, 1994; Sewalt *et al.*, 1997a; Song & Wang, 2011). Unlike the PAL mutants in Arabidopsis, in which all of the levels of lignin monomers were reduced uniformly, PAL down-regulated tobacco transformants display a surprising increase in the S/G ratio of monolignols (Sewalt *et al.*, 1997a). This uneven reduction of phenylpropanoids in tobacco could be the result of different regulatory mechanisms, such as feed-forward down-regulation affecting genes involved specifically in G monolignol biosynthesis, metabolic channeling between the enzymes of the central phenylpropanoid pathway and those involved in monolignol biosynthesis, or a preferential impact of the down-regulation construct on cells that deposit G lignin. Regardless of the mechanism, down-regulation of PAL in tobacco results in increased enzymatic digestibility and supports the idea that manipulation of lignin biosynthesis will be an effective strategy for the development of more suitable biomass for industrial purposes (Sewalt *et al.*, 1997b).

### 2.2.2 *Cinnamate 4-hydroxylase (C4H)*

C4H – CYP73A5 in Arabidopsis – is a cytochrome P450-dependent monooxygenase (P450) that catalyzes the *p*-hydroxylation of *trans*-cinnamic acid to form *p*-coumaric acid (Mizutani *et al.*, 1997). It is widely expressed at all stages of development, but most abundantly in lignifying cell types (Bell-Lelong *et al.*, 1997). In Arabidopsis, several mutants of C4H, known as *ref3*, have been identified. The most severe mis-sense mutant, *ref3-2*, is both sterile and dwarf, and has a metabolic phenotype that includes reductions in lignin, flavonoids, seed-coat tannins, and soluble sinapate esters, consistent with its metabolic block very early in the phenylpropanoid pathway (Schilmiller *et al.*, 2009). Similarly, *C4H* down-regulation in tobacco, Medicago, and Populus results in plants with stunted growth and reduced total lignin (Sewalt *et al.*, 1997a; Reddy *et al.*, 2005; Bjurhager *et al.*, 2010). For unknown reasons, the down-regulation of C4H in Medicago results in a decrease in the S/G monomer ratio that is not observed in Arabidopsis, tobacco, or Populus (Reddy *et al.*, 2005).

In Arabidopsis, the *ref3-2* mutant accumulates about 20% of the total lignin observed in wild-type plants and exhibits extensive vascular collapse, which may account for its dwarf phenotype. On the other hand, or in addition, other phenotypes suggestive of altered auxin transport or signaling are also evident in *ref3* plants and may play a role in their altered stature (Schilmiller *et al.*, 2009). As in all *ref* mutants, sinapoylmalate levels are significantly reduced and cinnamoylmalate, a compound that is not detectable in wild-type plants, accumulates to significant levels, indicating that there is significant plasticity in the plants′ secondary metabolic catalytic machinery (Schilmiller *et al.*, 2009). Nevertheless, the total level of both sinapoylmalate and cinnamoylmalate is still less than the total wild-type HCE accumulation, an observation that parallels the decreased lignin deposition in these mutants.

### 2.2.3 4-coumarate: CoA ligase (4CL)

4CL catalyzes the ATP-dependent formation of a thioester with an HCA and CoA (Lee *et al.*, 1997). In Arabidopsis, there are four known 4CL isoforms, named 4CL1–4, with varying activity towards all known HCAs (Lee *et al.*, 1997; Hamberger & Hahlbrock, 2004). Both expression profiling and *in vitro* characterization suggest that of these four isoforms, 4CL1 and 2 are the primary enzymes that catalyze the formation of *p*-coumaroyl CoA and that this activity is required for wild-type lignin deposition (Lee *et al.*, 1995, 1997; Ehlting *et al.*, 1999). 4CL3 has a similar catalytic efficiency towards *p*-coumarate but is not expressed at high levels in lignifying tissues, and therefore may play a role in soluble metabolite biosynthesis (Ehlting *et al.*, 1999; Raes *et al.*, 2003). Unlike the other three isoforms, 4CL4 has a higher catalytic efficiency towards ferulate and sinapate, and as such its role is not fully understood (Hamberger & Hahlbrock, 2004). Arabidopsis *4CL* anti-sense lines, the most extreme of which contain only 8% residual 4CL activity, display decreased G lignin accumulation without disrupting S lignin content or growth (Lee *et al.*, 1997). The preferential decrease of G subunits is also observed in Populus 4CL knock-down lines (Voelker *et al.*, 2010). Unlike in Arabidopsis, knocking down 4CL expression in Populus resulted in plants with decreased above-ground biomass, irregular xylem, and an increased red color in the wood.

### 2.2.4 Hydroxycinnamoyl-coenzyme A: shikimate/quinate hydroxycinnamoyltransferase (HCT)/p-coumaroyl shikimate 3′-hydroxylase (C3′H)

HCT is a BAHD acyltransferase that catalyzes two nonconsecutive reactions in the phenylpropanoid pathway. The first is the transfer of *p*-coumaroyl CoA to form the corresponding shikimic or quinic acid ester. HCT can use both shikimate and quinate based on *in vitro* experiments using heterologously expressed HCT from tobacco and coffee; however, reactions with shikimate are favored in some plant species (Hoffmann *et al.*, 2003; Lallemand *et al.*, 2012). After this reaction, C3′H is responsible for the hydroxylation of *p*-coumaroyl shikimate to caffeoyl shikimate (Schoch *et al.*, 2001; Morant *et al.*, 2007). After this, HCT acts again to transfer caffeate from caffeoyl shikimate back on to CoA to form caffeoyl CoA (Hoffmann *et al.*, 2003). In contrast to this conventional model of the pathway, it has recently been reported that Populus contains an enzyme complex that can catalyze the 3-hydroxylation of *p*-coumaric acid to caffeic acid without using 4CL or HCT (Chen *et al.*, 2011a).

Arabidopsis and Medicago plants with reduced expression of *HCT* are dwarf and have collapsed vasculature, modified lignin composition, delayed flowering, and reduced lignin (Hoffmann *et al.*, 2004; Besseau *et al.*, 2007; Shadle *et al.*, 2007). These phenotypes are consistent with the previously described *pal* and *ref3* mutants, as well as plants defective in C3′H, referred to as *ref8* mutants in Arabidopsis. Both *HCT* knock-down plants and *C3′H* knock-down plants are distinct in that their lignin is derived primarily from H subunits (Franke *et al.*, 2002; Reddy *et al.*, 2005; Li *et al.*, 2010). A similar shift from G- to H-substituted compounds can be observed in the cell wall-bound HCA content in *ref8* mutants. Levels of cell wall-bound vanillin and ferulate released by base treatment of *ref8* cell walls are decreased, whereas levels of *p*-hydroxybenzaldehyde and *p*-coumaric acid are increased (Franke *et al.*, 2002).

In addition, *HCT*-RNA interference (RNAi) and *ref8* plants have increased soluble flavonoid accumulation compared to wild-type (Besseau *et al.*, 2007; Li *et al.*, 2010; Weng *et al.*, 2010a). The increase in flavonoid accumulation has been suggested to be the result of redirection of carbon from the lignin biosynthetic pathway into flavonoid biosynthesis; however, the elevated levels of flavonoids could also be derived from the pool of intermediates dedicated to soluble metabolite biosynthesis or simply have accumulated as part of a stress response associated with the dwarfism seen in these plants. Sinapoylmalate levels are decreased in the *ref8* mutant, and two compounds that are not detectable in wild-type plants, *p*-coumaroyl shikimate and *p*-coumaroylmalate, accumulate in *HCT*-RNAi plants (Franke *et al.*, 2002; Coleman *et al.*, 2008). The same shift in soluble metabolites is not observed in *ref8* seeds where both sinapoylcholine and sinapoylglucose are reduced to levels that are below detectable limits and no *p*-coumaroyl derivatives accumulate. These data would suggest that there are fewer fates for *p*-coumarate than for sinapate in seed metabolism.

### *2.2.5 Cinnamoyl CoA reductase (CCR)*

CCR catalyzes the NADPH-dependent reduction of the hydroxycinnamoyl CoA thioesters *p*-coumaroyl CoA, feruloyl CoA, and to a lesser extent caffeoyl CoA to their corresponding hydroxycinnamaldehydes, with a concomitant release of CoA (Goujon *et al.*, 2003a; Do *et al.*, 2007; Zhou *et al.*, 2010). CCR1, the most abundant CCR encoded by the members of the *CCR* gene family in Arabidopsis, has been characterized *in vitro* and displays the highest catalytic efficiency towards feruloyl CoA (Lauvergeat *et al.*, 2001; Raes *et al.*, 2003; Baltas *et al.*, 2005). Plants defective in CCR are dwarfed, but less so than plants defective in earlier steps of the pathway, although again they have collapsed vasculature (Pichon *et al.*, 1998; Jones *et al.*, 2001; van der Rest *et al.*, 2006; Dauwe *et al.*, 2007; Leple *et al.*, 2007; Mir Derikvand *et al.*, 2008).

Quantitative lignin analysis of *ccr1* in Arabidopsis revealed a 50% decrease in lignin deposition in mature inflorescence stems compared to wild-type plants (Jones *et al.*, 2001; Goujon *et al.*, 2003a; Mir Derikvand *et al.*, 2008). In contrast, lignin composition appears to be unaffected by the mutation, and *ccr1* plants retain a normal G/S ratio (Mir Derikvand *et al.*, 2008). The residual lignin in the *ccr1* mutant suggests that another reductase has redundant function with CCR1, and several *CCR*-like genes have been annotated based on sequence homology within the Arabidopsis genome (Goujon *et al.*, 2003a,b; Raes *et al.*, 2003). Of the candidate *CCR*-like genes, the enzyme encoded by *CCR2* has been characterized *in vitro*; it also shows reductase activity towards hydroxycinnamoyl CoA thioesters, albeit with a lower catalytic efficiency than CCR1 (Lauvergeat *et al.*, 2001). These data suggest that CCR2 may be one of the reductases that is functionally redundant with CCR1 and partially compensates for the lack of its activity in a *ccr1* mutant.

Cell wall-bound metabolite analysis of *ccr1* inflorescence stems revealed a fourfold increase of cell wall-bound ferulic acid and an almost 100-fold increase in soluble feruloylmalate compared to wild-type plants (Mir Derikvand *et al.*, 2008). Increased feruloylmalate is also observed in leaves where sinapoylmalate levels are decreased (Mir Derikvand *et al.*, 2008). In addition to accumulating higher levels of cell wall-bound ferulic acid, *ccr1* mutants also incorporate ferulate into the lignin polymer (Ralph *et al.*, 2008) (Fig. 2.2). The accumulation of ferulate derivatives is somewhat surprising since at least one of the direct biosynthetic precursors for ferulate is thought to be coniferaldehyde, which requires CCR1 for its biosynthesis (Nair *et al.*, 2004). These observations suggest that an intermediate upstream of CCR1, probably feruloyl CoA via thioesterase-mediated hydrolysis, is the precursor for ferulate derivatives in the *ccr1* mutant, and possibly also in wild-type plants (Meyer *et al.*, 1991; Mir Derikvand *et al.*, 2008). A similar accumulation of ferulate derivatives is observed in tobacco, Populus, maize, and Medicago plants that are defective in their respective CCR1 homologs (Pichon *et al.*, 1998; Piquemal *et al.*, 1998; Dauwe *et al.*, 2007; Leple *et al.*, 2007).

### 2.2.6 Ferulate 5-hydroxylase (F5H)

F5H is a P450 that can hydroxylate several G-substituted lignin biosynthetic intermediates at the 5 position. Hydroxylation at the 5 position was initially thought to occur at the level of the free acid (Meyer *et al.*, 1996), but subsequent *in vitro* analysis demonstrated that coniferaldehyde and coniferyl alcohol are the preferred substrates (Humphreys *et al.*, 1999; Osakabe *et al.*, 1999; Humphreys & Chapple, 2002).

Arabidopsis plants deficient in F5H are known as *ferulic acid hydroxylase* (*fah1*) mutants and display a wild-type growth phenotype (Chapple *et al.*, 1992; Ruegger & Chapple, 2001). Null *fah1* mutants lack all S-substituted compounds, including sinapate esters and S lignin (Chapple *et al.*, 1992; Meyer *et al.*, 1998). Although lignin composition is dramatically changed, total lignin deposition remains constant through a compensatory increase in G monolignols. A similar decrease in the S/G monomer ratio is observed in Medicago plants with decreased *F5H* expression (Reddy *et al.*, 2005). Conversely, increased F5H expression, using an alternate strong promoter, results in an almost complete redirection of flux to S lignin (Meyer *et al.*, 1998). Heterologous expression of the Arabidopsis F5H gene in tobacco and Populus results in a similar G to S shift in lignin composition and generates plants with superior characteristics in the pulping process (Franke *et al.*, 2000; Huntley *et al.*, 2003). Both *F5H*-deficient and *F5H*-over-expressing plants display a growth phenotype that is largely indistinguishable from wild-type, suggesting that extreme variations in G and S subunit composition are readily tolerated and are less important than lignin content to normal growth and development.

In addition to the changes in lignin composition, *fah1* mutants also display an increased accumulation of cell wall-bound ferulate (Franke *et al.*, 2002). Feruloylmalate accumulates in the *fah1* mutant, albeit at much lower levels than in *ccr1*, and the total HCE content of *fah1* plants is much less than in wild-type plants and the other mutants described in this chapter (Chapple *et al.*, 1992; Hemm *et al.*, 2003; Nair *et al.*, 2004). This is somewhat surprising since ferulate levels are increased, at least in the stem, and the enzymes that participate in the biosynthesis of sinapoylmalate from sinapic acid can make feruloylmalate from ferulic acid, at least *in vitro* (Lim *et al.*, 2001; Nair *et al.*, 2004; Sinlapadech *et al.*, 2007). These observations may indicate that G-substituted compounds in the *fah1* mutant are not available for soluble-ester synthesis like they are in the *ccr1* mutant, due to compartmentalization or feedback regulation.

### 2.2.7 *Caffeic acid/5-hydroxyferulic acid* O*-methyltransferase (COMT)/caffeoyl CoA 3-*O*-methyltransferase (CCoAOMT)*

COMT catalyzes the methylation of both 3-hydroxylated and 5-hydroxylated cinnamic acids, aldehydes, and alcohols and of quercetin *in vitro* (Inoue *et al.*, 1998, 2000; Goujon *et al.*, 2003b). Genetic analysis of *COMT* mutants and further *in vitro* characterization of several COMT homologs suggest that COMT mainly functions to methylate 5-hydroxyconiferaldehyde to sinapaldehyde, 5-hydroxyconiferyl alcohol to sinapyl alcohol, and quercetin to isorhamnetin (Humphreys *et al.*, 1999; Inoue *et al.*, 2000; Goujon *et al.*, 2003b; Do *et al.*, 2007). Like F5H, COMT has a low catalytic efficiency towards free HCAs, which is consistent with a model in which decoration of the 5-position of lignin precursors and sinapic acid occurs at the alcohol and aldehyde level (Goujon *et al.*, 2003b). The role of COMT in the methylation of 3-hydroxylated compounds *in vivo* has yet to be defined in wild-type plants. In Arabidopsis and Medicago, disrupting the expression of the known 3-*O*-methyltransferase CCoAOMT does not result in a substantial decrease in compounds downstream of CCoAOMT (Parvathi *et al.*, 2001; Do *et al.*, 2007). However, disrupting the expression of both *CCoAOMT* and *COMT* dramatically reduces G and S lignin and 3-*O*-methoxylated soluble metabolites, and the double mutants in Arabidopsis arrest early in seedling development (Parvathi *et al.*, 2001; Do *et al.*, 2007). These findings suggest that COMT contributes significant and important *O*-methyltransferase activity in plants deficient in CCoAOMT. Unlike Arabidopsis and Medicago, Populus and flax CCoAOMT down-regulated plants do display reduced lignin content and irregular xylem, suggesting that there may be less redundancy for 3-methylation in these species (Meyermans *et al.*, 2000; Day *et al.*, 2009).

Plants defective in COMT have a wild-type morphology. Both soluble and cell wall-bound sinapoylated compounds and isorhamnetin found in wild-type Arabidopsis are either decreased or absent in the *comt* mutant (Goujon *et al.*, 2003b). In addition, in Arabidopsis, tobacco, sugarcane, maize, and Populus COMT-deficient or down-regulated plants, S monolignols are either absent or reduced and 5-hydroxy G monolignols – monomers that are not observed in wild-type plants – are incorporated into lignin (Vailhe *et al.*,1996; Sewalt *et al.*, 1997b; He *et al.*, 2003; Marita *et al.*, 2003; Morreel *et al.*, 2004a; Lu *et al.*, 2010; Jung *et al.*, 2012) (Fig. 2.2). This shift in lignin composition probably explains why no change in total lignin deposition is observed in stem tissue in the Arabidopsis *comt* mutant. In switchgrass, down-regulation of COMT results in the atypical accumulation of isosinapate compounds (Tschaplinski *et al.*, 2012). The *para* methylation observed in these plants is dependent on the reduction of 5-methylation activity. In contrast to the lack of S units in the Arabidopsis *comt* mutant lignin, phenylpropanoid units that normally accumulate as sinapoylmalate are divided between 5-hydroxy feruloylmalate and sinapoylmalate in the *comt* mutant, with the result that the total HCE content is unchanged (Goujon *et al.*, 2003b; Weng *et al.*, 2011). The presence of sinapate esters indicates that a methyltransferase other than COMT can catalyze the methylation of 5-hydroxy G compounds; however, this redundancy is not observed for lignin biosynthesis. This observation strongly suggests that the redundant methyltransferase is expressed only in nonlignifying cells.

### 2.2.8 Cinnamyl alcohol dehydrogenases (CAD)

CADs catalyze the reduction of hydroxycinnamaldehydes to hydroxycinnamyl alcohols (Kim *et al.*, 2004). This reduction is required for the production of H, G, and S monolignols, as well as soluble hydroxycinnamyl alcohol-coupling products, including lignans (Boerjan *et al.*, 2003; Rohde *et al.*, 2004). There are at least six genes that encode enzymes with redundant activity for the reduction of hydroxycinnamaldehydes, but in Arabidopsis the majority of the activity is accounted for by two CAD enzymes, referred to as CAD C and CAD D (Costa *et al.*, 2003; Kim *et al.*, 2004; Sibout *et al.*, 2005). *cad c cad d* double mutants display a 40% reduction in total lignin content compared with wild-type plants and plants with either single mutation. These double mutants attain a wild-type height but have what is described as a "limp-floral-stem" phenotype, in which the inflorescences are flexible and recumbent (Sibout *et al.*, 2005). The limp-floral-stem phenotype is also observed in *Nicotiana attenuata CAD*-silenced plants and is thought to be dependent on the perturbation of lignin biosynthesis (Kaur *et al.*, 2012). Interestingly, the *CAD*-silenced plants do not display this recumbent phenotype when exposed to UV light and higher-temperature conditions, indicating that the metabolic block can be circumvented, or compensated for, under certain environmental conditions (Kaur *et al.*, 2012).

In addition to the decrease in total lignin content, the lignin composition of *cad c cad d* double mutants is altered. In these plants, 94% of G and S subunits are replaced with coniferaldehyde and sinapaldehyde units, and to a lesser extent ferulic and sinapic acid (Sibout *et al.*, 2005) (Fig. 2.2). The addition of these unconventional monolignols results in a reddish-orange color, and similar pigmentation is observed in Populus, Medicago, and tobacco lines that have decreased CAD expression (Halpin *et al.*, 1994; Baucher *et al.*, 1996, 1999; Vailhe *et al.*, 1998; Chabannes *et al.*, 2001; Sibout *et al.*, 2005). Cell wall-bound ferulate and *p*-coumarate levels of CAD-deficient plants have not been reported.

Soluble phenylpropanoids, including feruloyl and sinapoyl glucose and malate esters, accumulate to a significantly higher level in the *cad c cad d* double mutants than in wild-type plants or other plants with disruptions in lignin biosynthesis (Thevenin *et al.*, 2010). The accumulation of these soluble phenylpropanoids is not likely to be responsible for the limp-floral-stem phenotype, since other mutants with similar increases in these compounds do not recapitulate these growth phenotypes (Bonawitz *et al.*, 2012). The accumulation of higher levels of soluble metabolites may be the result of redirection of precursors that would normally accumulate as cell wall-bound HCAs or lignin.

## 2.3 Atypical lignins

As mentioned previously, lignins are primarily constructed from three hydroxycinnamyl alcohol monomers: H, G, and S (Boerjan *et al.*, 2003). Manipulation of the composition of these three monomers and the incorporation of atypical subunits demonstrates the uniqueness of the lignin polymer. Unlike other polymers found in the cell wall and throughout nature, lignin's polymerization is template-independent and its composition appears to be influenced primarily by the availability of subunits. This observation is supported by the increase in H content observed in plants defective in HCT, C3′H, and to a lesser extent 4CL, plants with a greatly increased G/S ratio resulting from a disruption in F5H and COMT, and plants with an increased S/G resulting from over-expression of F5H (Sewalt *et al.*, 1997a,b; Meyer *et al.*, 1998; Meyermans *et al.*, 2000; Franke *et al.*, 2002; Reddy *et al.*, 2005; Shadle *et al.*, 2007; Li *et al.*, 2010; Voelker *et al.*, 2010).

In addition to the manipulation of these three monomers, lignin polymer plasticity is also observed upon the incorporation of atypical lignin subunits not abundant in most plants. Additions of these atypical subunits include but are not limited to 5-OH guaiacyl monomers present in plants defective in COMT, hydroxycinnamaldehyde subunits observed in plants with modified CAD expression, and ferulic acid incorporation in plants deficient in CCR (Baucher *et al.*, 1996; Sibout *et al.*, 2005; Ralph *et al.*, 2008; Vanholme *et al.*, 2010b; Weng *et al.*, 2010b). Atypical lignin polymers are increasingly found to be present not just in mutant plants but also in nature. Examples include lignin consisting of almost pure caffeyl alcohol monomers in the seed coat of the vanilla orchid and several *Cactaceae* members, and the incorporation of tricin, a flavonoid, in coconut and several grass species (Chen *et al.*, 2011b; del Rio *et al.*, 2012; Rencoret *et al.*, 2013). In addition, acylated and dihydrohydroxylated forms of the tradition H, G, and S monolignols and hydroxybenzaldehyde derivatives also get incorporated into lignin (Lu & Ralph, 1999, 2008; Ralph *et al.*, 2001; Morreel *et al.*, 2004b; del Rio *et al.*, 2007). Numerous candidate compounds, identified based on the structures of these bona fide monolignols, have the potential to serve as lignin subunits and may produce lignins, which no longer pose challenges to the saccharification of biomass (Vanholme *et al.*, 2012).

## 2.4 Dwarfism

Lignin impedes cell-wall deconstruction and the saccharification of polysaccharides, key steps in biofuel production (Hamelinck *et al.*, 2005; Chen & Dixon, 2007; Ximenes *et al.*, 2010). Attempts to manipulate lignin content and to modify lignin composition have sought to address this challenge. Some efforts to reroute the phenylpropanoid pathway, as outlined in this chapter, result in plants that are dwarf, sterile, more susceptible to environmental stresses, have irregular growth or developmental patterns, or display a combination of these phenotypes (Dixon, 2001; Franke *et al.*, 2002; Mir Derikvand *et al.*, 2008). Dwarfism in these plants is generally associated with decreased lignin deposition, but the exact cause of the reduction in growth has not yet been conclusively determined (Bonawitz & Chapple, 2013). It may arise directly from alterations in lignin content as a result of the mechanical failure of an incompletely reinforced water-conducting system (Fraser & Chapple, 2011); alternatively, perturbations in phenylpropanoid metabolism may disrupt plant development. For example, in CCR-deficient plants, the low lignin content may be the cause of the collapsed vasculature, which in turn leads to the dwarf phenotype (Mir Derikvand *et al.*, 2008). On the other hand, the decrease in lignin deposition is also consistent with plants being developmentally delayed, which may itself alter lignin deposition (Patten *et al.*, 2005; Laskar *et al.*, 2006). Finally, dwarfism may instead arise from perturbations in the accumulation of some as yet unidentified phenylpropanoid-derived growth-regulating substance such as a dehydrodiconiferyl alcohol glucoside, a compound that has been shown to promote cell division in undifferentiated tobacco cells (Binns *et al.*, 1987). In this context, it is important to note that flavonoid hyper-accumulation has been suggested in one study to cause the dwarf phenotype observed in *HCT*-RNAi plants (Besseau *et al.*, 2007). This study employed RNAi lines for both HCT and chalcone synthase, the entry-point enzyme for flavonoid biosynthesis, but a more recent study using the chalcone synthase-deficient mutant *transparent testa 4* (*tt4*) revealed that HCT-deficiency-induced dwarfing occurs independently of flavonoid accumulation (Li *et al.*, 2010). It is still unknown whether these phenotypes arise from decreased lignin or from the perturbation of a phenylpropanoid other than a flavonoid.

Whatever the cause, yield penalties associated with lignin manipulation are highly undesirable in potential biofuel crops, and for this reason a thorough understanding of phenylpropanoid metabolism and its interaction(s) with plant development is essential in order to obviate any negative impact of lignin modification (Bonawitz & Chapple, 2013).

## 2.5 Conclusions

Plants synthesize hundreds of thousands of specialized compounds that are required for proper growth, development, and response to environmental conditions. Because many of these compounds are vital to plant health, it is easy to imagine that there has been strong selective pressure for strategies that maintain their relative amounts. Several of these strategies are observed in the phenylpropanoid pathway, including gene redundancy and the rerouting of metabolites within either the soluble or the cell wall-bound pool. Overall, the plasticity of phenylpropanoid accumulation presents multiple challenges to the rational manipulation of the pathway, and addressing these issues will be essential to the engineering of phenylpropanoid biosynthesis and other desirable pathways in plants.

## References

Baltas, M., Lapeyre, C., Bedos-Belval, F., *et al.* (2005) Kinetic and inhibition studies of cinnamoyl-CoA reductase 1 from *Arabidopsis thaliana*. *Plant Physiology and Biochemistry*, 43, 746–753.

Bate, N.J., Orr, J., Ni, W., *et al.* (1994) Quantitative relationship between phenylalanine ammonia-lyase levels and phenylpropanoid accumulation in transgenic tobacco identifies a rate-determining step in natural product synthesis. *Proceedings of the National Academy of Sciences USA*, 91, 7608–7612.

Baucher, M., Chabbert, B., Pilate, G., *et al.* (1996) Red xylem and higher lignin extractability by down-regulating a cinnamyl alcohol dehydrogenase in poplar. *Plant Physiology*, 112, 1479–1490.

Baucher, M., Bernard-Vailhe, M.A., Chabbert, B., *et al.* (1999) Down-regulation of cinnamyl alcohol dehydrogenase in transgenic alfalfa (*Medicago sativa* L.) and the effect on lignin composition and digestibility. *Plant Molecular Biology*, 39, 437–447.

Bell-Lelong, D.A., Cusumano, J.C., Meyer, K. & Chapple, C. (1997) Cinnamate-4-hydroxylase expression in Arabidopsis (regulation in response to development and the environment). *Plant Physiology*, 113, 729–738.

Besseau, S., Hoffmann, L., Geoffroy, P, *et al.* (2007) Flavonoid accumulation in Arabidopsis repressed in lignin synthesis affects auxin transport and plant growth. *Plant Cell*, 19, 148–162.

Binns, A.N., Chen, R.H., Wood, H.N. & Lynn, D.G. (1987) Cell division promoting activity of naturally occurring dehydrodiconiferyl glucosides: do cell wall components control cell division? *Proceedings of the National Academy of Sciences USA*, 84, 980–984.

Bjurhager, I, Olsson, A.M., Zhang, B., *et al.* (2010) Ultrastructure and mechanical properties of *Populus* wood with reduced lignin content caused by transgenic down-regulation of cinnamate 4-hydroxylase. *Biomacromolecules*, 11, 2359–2365.

Boerjan, W., Ralph, J. & Baucher, M. (2003) Lignin biosynthesis. *Annual Review of Plant Biology*, 54, 519–546.

Bonawitz, N.D. & Chapple, C. (2010) The genetics of lignin biosynthesis: connecting genotype to phenotype. *Annual Review of Genetics*, 44, 337–363.

Bonawitz, N.D. & Chapple, C. (2013) Can genetic engineering of lignin deposition be accomplished without an unacceptable yield penalty? *Current Opinion in Biotechnology*, 24, 336–343.

Bonawitz, N.D., Soltau, W.L., Blatchley, M.R., *et al.* (2012) REF4 and RFR1, subunits of the transcriptional coregulatory complex mediator, are required for phenylpropanoid homeostasis in Arabidopsis. *Journal of Biological Chemistry*, 287, 5434–5445.

Bottcher, C., Roepenack-Lahaye, E.V., Schmidt, J., *et al.* (2008) Metabolome analysis of biosynthetic mutants reveals diversity of metabolic changes and allows identification of a large number of new compounds in *Arabidopsis thaliana*. *Plant Physiology*, 147, 2107–2120.

Campbell, M.M. & Sederoff, R.R. (1996) Variation in lignin content and composition – mechanism of control and implications for the genetic improvement of plants. *Plant Physiology*, 110, 3–13.

Carnachan, S.M. & Harris, P.J. (2000) Ferulic acid is bound to the primary cell walls of all gymnosperm families. *Biochemical Systematics and Ecology*, 28, 865–879.

Cayuela, J.A., Rada, M., Rios, J.J, *et al.* (2006) Changes in phenolic composition induced by *Pseudomonas savastanoi pv. Savastanoi* infection in olive tree: presence of large amounts of verbascoside in nodules of tuberculosis disease. *Journal of Agricultural and Food Chemistry*, 54, 5363–5368.

Chabannes, M., Barakate, A., Lapierre, C., *et al.* (2001) Strong decrease in lignin content without significant alteration of plant development is induced by simultaneous down-regulation of cinnamoyl CoA reductase (CCR) and cinnamyl alcohol dehydrogenase (CAD) in tobacco plants. *Plant Journal*, 28, 257–270.

Chapple, C., Vogt, T., Ellis, B.E. & Somerville, C.R. (1992) An Arabidopsis mutant defective in the general phenylpropanoid pathway. *Plant Cell*, 4, 1413–1424.

Chapple, C., Ladisch, M. & Meilan, R. (2007) Loosening lignin's grip on biofuel production. *Nature Biotechnology*, 25, 746–748.

Chen, F. & Dixon, R.A. (2007) Lignin modification improves fermentable sugar yields for biofuel production. *Nature Biotechnology*, 25, 759–761.

Chen, H.C., Li, Q.Z., Shuford, C.M., *et al.* (2011a) Membrane protein complexes catalyze both 4- and 3-hydroxylation of cinnamic acid derivatives in monolignol biosynthesis. *Proceedings of the National Academy of Sciences USA*, 108, 21253–21258.

Chen, F., Tobimatsu, Y., Havkin-Frenkel, D, *et al.* (2011b) A polymer of caffeyl alcohol in plant seeds. *Proceedings of the National Academy of Sciences USA*, 109, 1772–1777.

Cochrane, F.C., Davin, L.B. & Lewis, N.G. (2004) The Arabidopsis phenylalanine ammonia lyase gene family: kinetic characterization of the four PAL isoforms. *Phytochemistry*, 65, 1557–1564.

Coleman, H.D., Park, J.Y., Nair, R, *et al.* (2008) RNAi-mediated suppression of p-coumaroyl-CoA 3'-hydroxylase in hybrid poplar impacts lignin deposition and soluble secondary metabolism. *Proceedings of the National Academy of Sciences USA*, 105, 4501–4506.

Costa, M.A., Collins, R.E., Anterola, A.M, *et al.* (2003) An *in silico* assessment of gene function and organization of the phenylpropanoid pathway metabolic networks in *Arabidopsis thaliana* and limitations thereof. *Phytochemistry*, 64, 1097–1112.

D'Auria, J.C., Pichersky, E., Schaub, A, *et al.* (2007) Characterization of a BAHD acyltransferase responsible for producing the green leaf volatile (Z)-3-hexen-1-yl acetate in *Arabidopsis thaliana. Plant Journal*, 49, 194–207.

Dauwe, R., Morreel, K., Goeminne, G., *et al.* (2007) Molecular phenotyping of lignin-modified tobacco reveals associated changes in cell-wall metabolism, primary metabolism, stress metabolism and photorespiration. *Plant Journal*, 52, 263–285.

Day, A., Neutelings, G., Nolin, F., *et al.* (2009) Caffeoyl coenzyme A *O*-methyltransferase down-regulation is associated with modifications in lignin and cell-wall architecture in flax secondary xylem. *Plant Physiology and Biochemistry*, 47, 9–19.

Dixon, R.A. (2001) Natural products and plant disease resistance. *Nature*, 411, 843–847.

Dixon, R.A. & Paiva, N.L. (1995) Stress-induced phenylpropanoid metabolism. *Plant Cell*, 7, 1085–1097.

Do, C.T., Pollet, B., Thevenin J., *et al.* (2007) Both caffeoyl coenzyme A 3-*O*-methyltransferase 1 and caffeic acid *O*-methyltransferase 1 are involved in redundant functions for lignin, flavonoids and sinapoyl malate biosynthesis in Arabidopsis. *Planta*, 226, 1117–1129.

Ehlting, J., Buttner, D., Wang, Q, *et al.* (1999) Three 4-coumarate: coenzyme A ligases in *Arabidopsis thaliana* represent two evolutionarily divergent classes in angiosperms. *Plant Journal*, 19, 9–20.

Elkind, Y., Edwards, R., Mavandad, M., *et al.* (1990) Abnormal-plant development and down-regulation of phenylpropanoid biosynthesis in transgenic tobacco containing a heterologous phenylalanine ammonia-lyase gene. *Proceedings of the National Academy of Sciences USA*, 87, 9057–9059.

Fini, A., Brunetti, C., Di Ferdinando, M, *et al.* (2011) Stress-induced flavonoid biosynthesis and the antioxidant machinery of plants. *Plant Signaling and Behavior*, 6, 709–711.

Franke, R., McMichael, C.M., Meyer, K, *et al.* (2000) Modified lignin in tobacco and poplar plants over-expressing the Arabidopsis gene encoding ferulate 5-hydroxylase. *Plant Journal*, 22, 223–234.

Franke, R., Hemm, M.R., Denault, J.W, *et al.* (2002) Changes in secondary metabolism and deposition of an unusual lignin in the *ref8* mutant of Arabidopsis. *Plant Journal*, 30, 47–59.

Fraser, C.M. & Chapple, C. (2011) The phenylpropanoid pathway in Arabidopsis. *The Arabidopsis Book/The American Society of Plant Biologists*, 9, e0152, doi: 10.1199:tab.0152.

Fraser, C.M., Thompson, M.G., Shirley, A.M., *et al.* (2007) Related Arabidopsis serine carboxypeptidase-like sinapoylglucose acyltransferases display distinct but overlapping substrate specificities. *Plant Physiology*, 144, 1986–1999.

Goujon, T., Ferret, V., Mila, I., *et al.* (2003a) Down-regulation of the AtCCR1 gene in *Arabidopsis thaliana*: effects on phenotype, lignins and cell wall degradability. *Planta*, 217, 218–228.

Goujon, T., Sibout, R., Pollet, B., *et al.* (2003b) A new *Arabidopsis thaliana* mutant deficient in the expression of *O*-methyltransferase impacts lignins and sinapoyl esters. *Plant Molecular Biology*, 51, 973–989.

Grabber, J.H., Ralph, J. & Hatfield, R.D. (2000) Cross-linking of maize walls by ferulate dimerization and incorporation into lignin. *Journal of Agricultural and Food Chemistry*, 48, 6106–6113.

Grawe, W., Bachhuber, P., Mock, H.P. & Strack, D. (1992) Purification and characterization of sinapolyglucose – malate sinapoyltransferase from *Raphanussativus L. Planta*, 187, 236–241.

Grienenberger, E., Besseau, S., Geoffroy, P., *et al.* (2009) A BAHD acyltransferase is expressed in the tapetum of Arabidopsis anthers and is involved in the synthesis of hydroxycinnamoyl spermidines. *Plant Journal*, 58, 246–259.

Grotewold, E. (2006) The genetics and biochemistry of floral pigments. *Annual Review of Plant Biology*, 57, 761–780.

Guillet, G., Poupart, J., Basurco, J. & De Luca, V. (2000) Expression of tryptophan decarboxylase and tyrosine decarboxylase genes in tobacco results in altered biochemical and physiological phenotypes. *Plant Physiology*, 122, 933–943.

Halpin, C., Knight, M.E., Foxon, G.A., *et al.* (1994) Manipulation of lignin quality by down-regulation of cinnamyl alcohol-dehydrogenase. *Plant Journal*, 6, 339–350.

Hamberger, B. & Hahlbrock, K. (2004) The 4-coumarate: CoA ligase gene family in *Arabidopsis thaliana* comprises one rare, sinapate-activating and three commonly occurring isoenzymes. *Proceedings of the National Academy of Sciences USA*, 101, 2209–2214.

Hamelinck, C.N., van Hooijdonk, G., Faaij, A.P.C. (2005) Ethanol from lignocellulosic biomass: techno-economic performance in short-, middle- and long-term. *Biomass & Bioenergy*, 28, 384–410.

Hatfield, R.D., Ralph, J. & Grabber, J.H. (1999) Cell wall cross-linking by ferulates and diferulates in grasses. *Journal of the Science of Food and Agriculture*, 79, 403–407.

He, X., Hall, M.B., Gallo-Meagher, M. & Smith, R.L. (2003) Improvement of forage quality by downregulation of maize *O*-methyltransferase. *Crop Science*, 43, 2240–2251.

Hemm, M.R., Ruegger, M.O. & Chapple, C. (2003) The Arabidopsis *ref2* mutant is defective in the gene encoding CYP83A1 and shows both phenylpropanoid and glucosinolate phenotypes. *Plant Cell*, 15, 179–194.

Hoffmann, L., Maury, S., Martz, F, *et al.* (2003) Purification, cloning, and properties of an acyltransferase controlling shikimate and quinate ester intermediates in phenylpropanoid metabolism. *Journal of Biological Chemistry*, 278, 95–103.

Hoffmann, L., Besseau, S., Geoffroy, P., *et al.* (2004) Silencing of hydroxycinnamoyl-coenzyme A shikimate/quinate hydroxycinnamoyltransferase affects phenylpropanoid biosynthesis. *Plant Cell*, 16, 1446–1465.

Huang, J., Gu, M., Lai, Z., *et al.* (2010) Functional analysis of the Arabidopsis PAL gene family in plant growth, development, and response to environmental stress. *Plant Physiology*, 153, 1526–1538.

Humphreys, J.M. & Chapple, C. (2002) Rewriting the lignin roadmap. *Current Opinion in Plant Biology*, 5, 224–229.

Humphreys, J.M., Hemm, M.R. & Chapple, C. (1999) New routes for lignin biosynthesis defined by biochemical characterization of recombinant ferulate 5-hydroxylase, a multifunctional cytochrome P450-dependent monooxygenase. *Proceedings of the National Academy of Sciences USA*, 96, 10045–10050.

Huntley, S.K., Ellis, D., Gilbert, M, *et al.* (2003) Significant increases in pulping efficiency in C4H-F5H-transformed poplars: improved chemical savings and reduced environmental toxins. *Journal of Agricultural and Food Chemistry*, 51, 6178–6183.

Iiyama, K., Lam, T.B.T. & Stone, B.A. (1994) Covalent cross-links in the cell-wall. *Plant Physiology*, 104, 315–320.

Inoue, K., Sewalt, V.J., Murray, G.B, *et al.* (1998) Developmental expression and substrate specificities of alfalfa caffeic acid 3–*O*-methyltransferase and caffeoyl coenzyme A 3-*O*-methyltransferase in relation to lignification. *Plant Physiology*, 117, 761–770.

Inoue, K., Parvathi, K. & Dixon, R.A. (2000) Substrate preferences of caffeic acid/5-hydroxyferulic acid 3/5-*O*-methyltransferases in developing stems of alfalfa (*Medicago sativa L.*). *Archives of Biochemistry and Biophysics*, 375, 175–182.

Jones, L., Ennos, A.R. & Turner, S.R. (2001) Cloning and characterization of irregular xylem4 (*irx4*): a severely lignin-deficient mutant of Arabidopsis. *Plant Journal*, 26, 205–216.

Jung, J.H., Fouad, W.M., Vermerris, W, *et al.* (2012) RNAi suppression of lignin biosynthesis in sugarcane reduces recalcitrance for biofuel production from lignocellulosic biomass. *Plant Biotechnology Journal*, 10, 1067–1076.

Kaur, H., Shaker, K., Heinzel, N, *et al.* (2012) Environmental stresses of field growth allow cinnamyl alcohol dehydrogenase-deficient *Nicotiana attenuata* plants to compensate for their structural deficiencies. *Plant Physiology*, 159, 1545–1570.

Kim, S.J., Kim, M.R., Bedgar, D.L., *et al.* (2004) Functional reclassification of the putative cinnamyl alcohol dehydrogenase multigene family in Arabidopsis. *Proceedings of the National Academy of Sciences USA*, 101, 1455–1460.

Kroon, P.A. & Williamson, G. (1999) Hydroxycinnamates in plants and food: current and future perspectives. *Journal of the Science of Food and Agriculture*, 79, 355–361.

Lallemand, L.A., Zubieta, C., Lee, S.G., *et al.* (2012) A structural basis for the biosynthesis of the major chlorogenic acids found in coffee. *Plant Physiology* 160, 249–260.

Laskar, D.D., Jourdes, M., Patten, A.M, *et al.* (2006) The Arabidopsis cinnamoyl CoA reductase *irx4* mutant has a delayed but coherent (normal) program of lignification. *Plant Journal*, 48, 674–686.

Lauvergeat, V., Lacomme, C., Lacombe, E, *et al.* (2001) Two cinnamoyl-CoA reductase (CCR) genes from *Arabidopsis thaliana* are differentially expressed during development and in response to infection with pathogenic bacteria. *Phytochemistry*, 57, 1187–1195.

Lee, D., Ellard, M., Wanner, L.A, *et al.* (1995) The *Arabidopsis thaliana* 4-coumarate: CoA ligase (4CL) gene: stress and developmentally regulated expression and nucleotide sequence of its cDNA. *Plant Molecular Biology*, 28, 871–884.

Lee, D., Meyer, K., Chapple, C. & Douglas, C.J. (1997) Antisense suppression of 4-coumarate: coenzyme A ligase activity in Arabidopsis leads to altered lignin subunit composition. *Plant Cell*, 9, 1985–1998.

Lepiniec, L., Debeaujon, I., Routaboul, J.M., *et al.* (2006) Genetics and biochemistry of seed flavonoids. *Annual Review of Plant Biology*, 57, 405–430.

Leple, J.C., Dauwe, R., Morreel, K., *et al.* (2007) Downregulation of cinnamoyl-coenzyme A reductase in poplar: multiple-level phenotyping reveals effects on cell wall polymer metabolism and structure. *Plant Cell*, 19, 3669–3691.

Li, X., Bonawitz, N.D., Weng, J.K. & Chapple, C. (2010) The growth reduction associated with repressed lignin biosynthesis in *Arabidopsis thaliana* is independent of flavonoids. *Plant Cell*, 22, 1620–1632.

Ligrone, R., Carafa, A., Duckett, J.G, *et al.* (2008) Immunocytochemical detection of lignin-related epitopes in cell walls in bryophytes and the charalean alga *Nitella. Plant Systematics and Evolution*, 270, 257–272.

Lim, E.K., Li, Y., Parr, A, *et al.* (2001) Identification of glucosyltransferase genes involved in sinapate metabolism and lignin synthesis in Arabidopsis. *Journal of Biological Chemistry*, 276, 4344–4349.

Lu, F. & Ralph, J. (1999) Detection and determination of *p*-coumaroylated units in lignins. *Journal of Agricultural and Food Chemistry*, 47, 1988–1992.

Lu, F. & Ralph, J (2008) Novel tetrahydrofuran structures derived from beta-beta-coupling reactions involving sinapyl acetate in Kenaf lignins. *Organic & Biomolecular Chemistry*, 6, 3681–3694.

Lu, F., Marita, J.M., Lapierre, C., *et al.* (2010) Sequencing around 5–hydroxyconiferyl alcohol-derived units in caffeic acid *O*-methyltransferase-deficient poplar lignins. *Plant Physiology*, 153, 569–579.

Marita, J.M., Vermerris, W., Ralph, J. & Hatfield, R.D. (2003) Variations in the cell wall composition of maize brown midrib mutants. *Journal of Agricultural and Food Chemistry* 51, 1313–1321.

Matsuno, M., Compagnon, V., Schoch, G.A., *et al.* (2009) Evolution of a novel phenolic pathway for pollen development. *Science*, 325, 1688–1692.

Mauch-Mani, B. & Slusarenko, A.J. (1996) Production of salicylic acid precursors is a major function of phenylalanine ammonia-lyase in the resistance of Arabidopsis to *Peronospora parasitica. Plant Cell*, 8, 203–212.

Meyer, K., Kohler, A. & Kauss, H. (1991) Biosynthesis of ferulic acid-esters of plant-cell wall polysaccharides in endomembranes from parsley cells. *FEBS Letters*, 290, 209–212.

Meyer, K., Cusumano, J.C., Somerville, C. & Chapple, C.C. (1996) Ferulate-5-hydroxylase from *Arabidopsis thaliana* defines a new family of cytochrome P450-dependent monooxygenases. *Proceedings of the National Academy of Sciences USA*, 93, 6869–6874.

Meyer, K., Shirley, A.M., Cusumano, J.C, *et al.* (1998) Lignin monomer composition is determined by the expression of a cytochrome P450-dependent monooxygenase in Arabidopsis. *Proceedings of the National Academy of Sciences USA*, 95, 6619–6623.

Meyermans, H., Morreel, K., Lapierre, C., *et al.* (2000) Modifications in lignin and accumulation of phenolic glucosides in poplar xylem upon down-regulation of caffeoyl-coenzyme A

*O*-methyltransferase, an enzyme involved in lignin biosynthesis. *Journal of Biological Chemistry*, 275, 36899–36909.

Milkowski C., Baumert A., Schmidt D, *et al.* (2004) Molecular regulation of sinapate ester metabolism in *Brassica napus*: expression of genes, properties of the encoded proteins and correlation of enzyme activities with metabolite accumulation. *Plant Journal*, 38, 80–92.

Mir Derikvand, M., Sierra, J.B., Ruel, K., *et al.* (2008) Redirection of the phenylpropanoid pathway to feruloyl malate in Arabidopsis mutants deficient for cinnamoyl-CoA reductase 1. *Planta*, 227, 943–956.

Mizutani, M., Ohta, D. & Sato, R. (1997) Isolation of a cDNA and a genomic clone encoding cinnamate 4–hydroxylase from Arabidopsis and its expression manner in planta. *Plant Physiology*, 113, 755–763.

Morant, M., Schoch, G.A., Ullmann, P., *et al.* (2007) Catalytic activity, duplication and evolution of the CYP98 cytochrome P450 family in wheat. *Plant Molecular Biology* 63, 1–19.

Morreel, K., Ralph, J., Lu, F., *et al.* (2004a) Phenolic profiling of caffeic acid *O*-methyltransferase-deficient poplar reveals novel benzodioxane oligolignols. *Plant Physiology*, 136, 4023–4036.

Morreel, K., Ralph, J., Kim, H., *et al.* (2004b) Profiling of oligolignols reveals monolignol coupling conditions in lignifying poplar xylem. *Plant Physiology*, 136, 3537–3549.

Nair, R.B., Bastress, K.L., Ruegger, M.O, *et al.* (2004) The *Arabidopsis thaliana REDUCED EPIDERMAL FLUORESCENCE1* gene encodes an aldehyde dehydrogenase involved in ferulic acid and sinapic acid biosynthesis. *Plant Cell*, 16, 544–554.

de Oliveira Buanafina, M.M. (2009) Feruloylation in grasses: current and future perspectives. *Molecular Plant*, 2, 861–872.

Osakabe, K., Tsao, C.C., Li, L., *et al.* (1999) Coniferyl aldehyde 5-hydroxylation and methylation direct syringyl lignin biosynthesis in angiosperms. *Proceedings of the National Academy of Sciences USA*, 96, 8955–8960.

Parvathi, K., Chen, F., Guo, D, *et al.* (2001) Substrate preferences of *O*-methyltransferases in alfalfa suggest new pathways for 3-*O*-methylation of monolignols. *Plant Journal*, 25, 193–202.

Patten, A.M., Cardenas, C.L., Cochrane, F.C., *et al.* (2005) Reassessment of effects on lignification and vascular development in the *irx4* Arabidopsis mutant. *Phytochemistry*, 66, 2092–2107.

Pichon, M., Courbou, I., Beckert, M, *et al.* (1998) Cloning and characterization of two maize cDNAs encoding cinnamoyl-CoA reductase (CCR) and differential expression of the corresponding genes. *Plant Molecular Biology*, 38, 671–676.

Piquemal, J., Lapierre, C., Myton, K., *et al.* (1998) Down-regulation of cinnamoyl-CoA reductase induces significant changes of lignin profiles in transgenic tobacco plants. *Plant Journal*, 13, 71–83.

Popper, Z.A. & Fry, S.C. (2003) Primary cell wall composition of bryophytes and charophytes. *Annals of Botany*, 91, 1–12.

Popper, Z.A. & Fry, S.C. (2004) Primary cell wall composition of pteridophytes and spermatophytes. *New Phytologist*, 164, 165–174.

Raes, J., Rohde, A., Christensen, J.H, *et al.* (2003) Genome-wide characterization of the lignification toolbox in Arabidopsis. *Plant Physiology*, 133, 1051–1071.

Ralph, J., Lapierre, C., Marita, J.M., *et al.* (2001) Elucidation of new structures in lignins of CAD- and COMT-deficient plants by NMR. *Phytochemistry*, 57, 993–1003.

Ralph, J., Kim, H., Lu, F., *et al.* (2008) Identification of the structure and origin of a thioacidolysis marker compound for ferulic acid incorporation into angiosperm lignins (and an indicator for cinnamoyl CoA reductase deficiency). *Plant Journal*, 53, 368–379.

Reddy, M.S.S., Chen, F., Shadle, G, *et al.* (2005) Targeted down-regulation of cytochrome P450 enzymes for forage quality improvement in alfalfa (*Medicago sativa* L.). *Proceedings of the National Academy of Sciences USA*, 102, 16573–16578.

Rencoret, J., Ralph, J., Marques, G, *et al.* (2013) Structural characterization of lignin isolated from coconut (*Cocos nucifera*) coir fibers. *Journal of Agricultural and Food Chemistry*, 61, 2434–2445.

del Rio, J.C., Marques, G., Rencoret, J, *et al.* (2007) Occurrence of naturally acetylated lignin units. *Journal of Agricultural and Food Chemistry*, 55, 5461–5468.

del Rio, J.C., Prinsen, P., Rencoret, J., *et al.* (2012) Structural characterization of the lignin in the cortex and pith of elephant grass (*Pennisetum purpureum*) stems. *Journal of Agricultural and Food Chemistry*, 60, 3619–3634.

Rohde, A., Morreel, K., Ralph, J., *et al.* (2004) Molecular phenotyping of the *pal1* and *pal2* mutants of *Arabidopsis thaliana* reveals far-reaching consequences on phenylpropanoid, amino acid, and carbohydrate metabolism. *Plant Cell*, 16, 2749–2771.

van der Rest, B., Danoun, S., Boudet, A.M. & Rochange, S.F. (2006) Down-regulation of cinnamoyl-CoA reductase in tomato (*Solanum lycopersicum L.*) induces dramatic changes in soluble phenolic pools. *Journal of Experimental Botany*, 57, 1399–1411.

Ruegger, M. & Chapple, C. (2001) Mutations that reduce sinapoylmalate accumulation in *Arabidopsis thaliana* define loci with diverse roles in phenylpropanoid metabolism. *Genetics*, 159, 1741–1749.

Saindrenan, P., Chong, J., Baltz, R, *et al.* (2002) Downregulation of a pathogen-responsive tobacco UDP-Glc : phenylpropanoid glucosyltransferase reduces scopoletin glucoside accumulation, enhances oxidative stress, and weakens virus resistance. *Plant Cell*, 14, 1093–1107.

Schilmiller, A.L., Stout, J., Weng, J.K, *et al.* (2009) Mutations in the cinnamate 4–hydroxylase gene impact metabolism, growth and development in Arabidopsis. *Plant Journal*, 60, 771–782.

Schooneveld-Bergmans, M.E.F., Dignum, M.J.W., Grabber, J.H, *et al.* (1999) Studies on the oxidative cross-linking of feruloylated arabinoxylans from wheat flour and wheat bran. *Carbohydrate Polymers*, 38, 309–317.

Schoch, G., Goepfert, S., Morant, M., *et al.* (2001) CYP98A3 from *Arabidopsis thaliana* is a 3′-hydroxylase of phenolic esters, a missing link in the phenylpropanoid pathway. *Journal of Biological Chemistry*, 276, 36566–36574.

Sewalt, V.J.H., Ni, W.T., Blount, J.W., *et al.* (1997a) Reduced lignin content and altered lignin composition in transgenic tobacco down-regulated in expression of L-phenylalanine ammonia-lyase or cinnamate 4-hydroxylase. *Plant Physiology*, 115, 41–50.

Sewalt, V.J.H., Ni, W.T., Jung, H.G. & Dixon, R.A. (1997b) Lignin impact on fiber degradation: increased enzymatic digestibility of genetically engineered tobacco (*Nicotiana tabacum*) stems reduced in lignin content. *Journal of Agricultural and Food Chemistry*, 45, 1977–1983.

Shadle, G., Chen, F., Reddy, M.S.S, *et al.* (2007) Down-regulation of hydroxycinnamoyl CoA: shikimate hydroxycinnamoyl transferase in transgenic alfalfa affects lignification, development and forage quality. *Phytochemistry*, 68, 2023–2023.

Shirley, B.W., Kubasek, W.L., Storz, G., *et al.* (1995) Analysis of Arabidopsis mutants deficient in flavonoid biosynthesis. *Plant Journal*, 8, 659–671.

Sibout, R., Eudes, A., Mouille, G., *et al.* (2005) *CINNAMYL ALCOHOL DEHYDROGENASE-C* and *-D* are the primary genes involved in lignin biosynthesis in the floral stem of Arabidopsis. *Plant Cell*, 17, 2059–2076.

Sinlapadech, T., Stout, J., Ruegger, M.O, *et al.* (2007) The hyper-fluorescent trichome phenotype of the *brt1* mutant of Arabidopsis is the result of a defect in a sinapic acid: UDPG glucosyltransferase. *Plant Journal*, 49, 655–668.

Song, J. & Wang, Z.Z. (2011) RNAi-mediated suppression of the phenylalanine ammonia-lyase gene in *Salvia miltiorrhiza* causes abnormal phenotypes and a reduction in rosmarinic acid biosynthesis. *Journal of Plant Research*, 124, 183–192.

Sullivan, M.L. (2009) A novel red clover hydroxycinnamoyl transferase has enzymatic activities consistent with a role in phaselic acid (2–*O*-[caffeoyl]-L-malate) biosynthesis. *Plant Physiology*, 150, 1866–1879.

Thevenin, J., Pollet, B., Letarnec, B., *et al.* (2010) The simultaneous repression of CCR and CAD, two enzymes of the lignin biosynthetic pathway, results in sterility and dwarfism in *Arabidopsis thaliana. Molecular Plant*, 4, 70–82.

Tohge, T., Nishiyama, Y., Hirai, M.Y., *et al.* (2005) Functional genomics by integrated analysis of metabolome and transcriptome of Arabidopsis plants over-expressing an MYB transcription factor. *Plant Journal*, 42, 218–235.

Tschaplinski, T.J., Standaert, R.F., Engle, N.L., *et al.* (2012) Down-regulation of the caffeic acid *O*-methyltransferase gene in switchgrass reveals a novel monolignol analog. *Biotechnology for Biofuels*, 5.1, 1–15.

Vailhe, M.A.B., Migne, C., Cornu, A., *et al.* (1996) Effect of modification of the *O*-methyltransferase activity on cell wall composition, ultrastructure and degradability of transgenic tobacco. *Journal of the Science of Food and Agriculture*, 72, 385–391.

Vailhe, M.A.B., Besle, J.M., Maillot, M.P, *et al.* (1998) Effect of down-regulation of cinnamyl alcohol dehydrogenase on cell wall composition and on degradability of tobacco stems. *Journal of the Science of Food and Agriculture*, 76, 505–514.

Vanholme, R., Demedts, B., Morreel, K, *et al.* (2010a) Lignin biosynthesis and structure. *Plant Physiology*, 153, 895–905.

Vanholme, R., Ralph, J., Akiyama, T., *et al.* (2010b) Engineering traditional monolignols out of lignin by concomitant up-regulation of F5H1 and down-regulation of COMT in Arabidopsis. *Plant Journal*, 64, 885–897.

Vanholme, R., Morreel, K., Darrah, C., *et al.* (2012) Metabolic engineering of novel lignin in biomass crops. *New Phytologist*, 196, 978–1000.

Veit, M. & Pauli, G.F. (1999) Major flavonoids from *Arabidopsis thaliana* leaves. *Journal of Natural Products*, 62, 1301–1303.

Voelker, S.L., Lachenbruch, B., Meinzer, F.C., *et al.* (2010) Antisense down-regulation of *4CL* expression alters lignification, tree growth, and saccharification potential of field-grown poplar. *Plant Physiology*, 154, 874–886.

Vogt, T. (2010) Phenylpropanoid biosynthesis. *Molecular Plant*, 3, 2–20.

Wasson, A.P., Pellerone, F.I. & Mathesius, U. (2006) Silencing the flavonoid pathway in *Medicago truncatula* inhibits root nodule formation and prevents auxin transport regulation by rhizobia. *Plant Cell*, 18, 1617–1629.

Weng, J.K., Akiyama, T., Bonawitz, N.D, *et al.* (2010a) Convergent evolution of syringyl lignin biosynthesis via distinct pathways in the lycophyte Selaginella and flowering plants. *Plant Cell*, 22, 1033–1045.

Weng, J.K., Mo, H.P. & Chapple, C. (2010b) Over-expression of F5H in COMT-deficient Arabidopsis leads to enrichment of an unusual lignin and disruption of pollen wall formation. *Plant Journal*, 64, 898–911.

Weng, J.K., Akiyama, T., Ralph, J. & Chapple, C. (2011) Independent recruitment of an *O*-methyltransferase for syringyl lignin biosynthesis in *Selaginella moellendorffii*. *Plant Cell*, 23, 2708–2724.

Whetten, R.W., MacKay, J.J. & Sederoff, R.R. (1998) Recent advances in understanding lignin biosynthesis. *Annual Review of Plant Physiology and Plant Molecular Biology*, 49, 585–609.

Widarto, H.T., Van Der Meijden, E., Lefeber, A.W., *et al.* (2006) Metabolomic differentiation of *Brassica rapa* following herbivory by different insect instars using two-dimensional nuclear magnetic resonance spectroscopy. *Journal of Chemical Ecology*, 32, 2417–2428.

Winkel-Shirley, B. (2001) Flavonoid biosynthesis. A colorful model for genetics, biochemistry, cell biology, and biotechnology. *Plant Physiology*, 126, 485–493.

Ximenes, E., Kim, Y., Mosier, N, *et al.* (2010) Inhibition of cellulases by phenols. *Enzyme and Microbial Technology*, 46, 170–176.

Yao, K., De Luca, V. & Brisson, N. (1995) Creation of a metabolic sink for tryptophan alters the phenylpropanoid pathway and the susceptibility of potato to *Phytophthora infestans*. *Plant Cell*, 7, 1787–1799.

Zhou, R., Jackson, L., Shadle, G., *et al.* (2010) Distinct cinnamoyl CoA reductases involved in parallel routes to lignin in *Medicago truncatula*. *Proceeding of the National Academy of Sciences USA*, 107, 17803–17808.

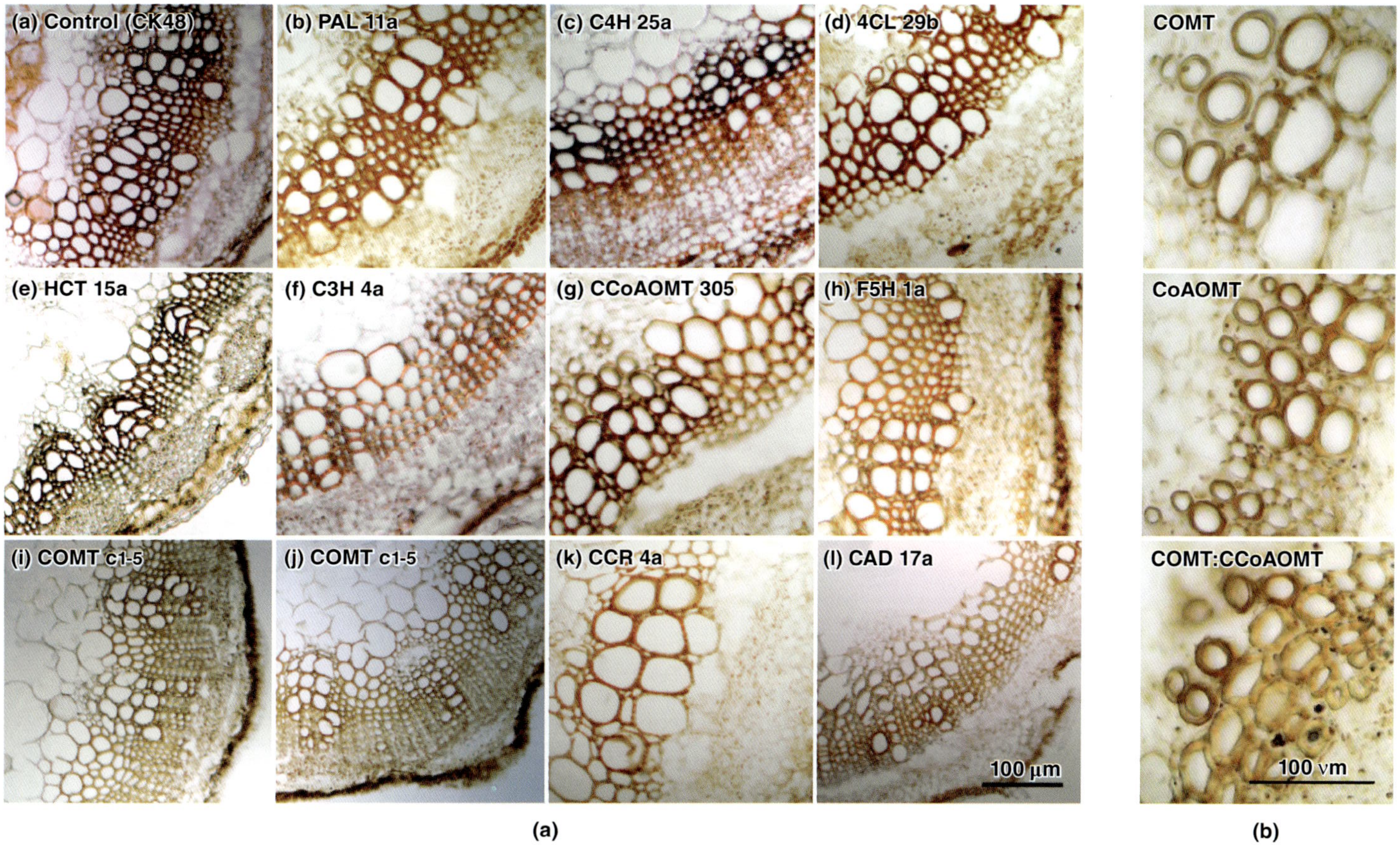


**Plate 1.3** Cellular impacts of the targeting of monolignol biosynthesis in alfalfa through RNAi-mediated silencing of specific enzymatic steps. (a) Micrographs showing cross-sections through the fifth internodes of transgenic alfalfa plants down-regulated independently at each of the enzymatic steps shown. (b) As in (a), but showing a comparison of single down-regulation of COMT or CCoAOMT with the result of down-regulating both genes simultaneously. (Nakashima *et al.*, 2008. Reproduced with permission of John Wiley & Sons.)

**Plate 1.4** Growth phenotypes of alfalfa and *M. truncatula* plants as a result of down-regulation (silencing or insertional mutagenesis) of monolignol biosynthetic enzymes or regulatory TFs. In each case, a corresponding wild-type control grown in parallel is included for comparison. (a,b) RNAi-mediated down-regulation of HCT and C3′H, respectively, in alfalfa. (c–e) Transposon insertion mutants in CCR1, CCoAOMT, and CAD, respectively, in *M. truncatula*. (f,g) Lignin autofluorescence in cross-sections of the sixth internodes of wild-type *M. truncatula* and a line harboring a transposon insertion in the NST1 NAC TF gene. (h) Growth phenotypes of the plants in (f,g) showing lodging in the *nst1* mutant. (i,j) Lignin autofluorescence in cross-sections of the sixth internodes of wild-type *M. truncatula* and a line harboring a transposon insertion in the WRKY-12 TF gene (Wang *et al.*, 2010. Reproduced with permission of National Academy of Sciences of the United States of America). (k) Growth phenotypes of the plants in (i,j).

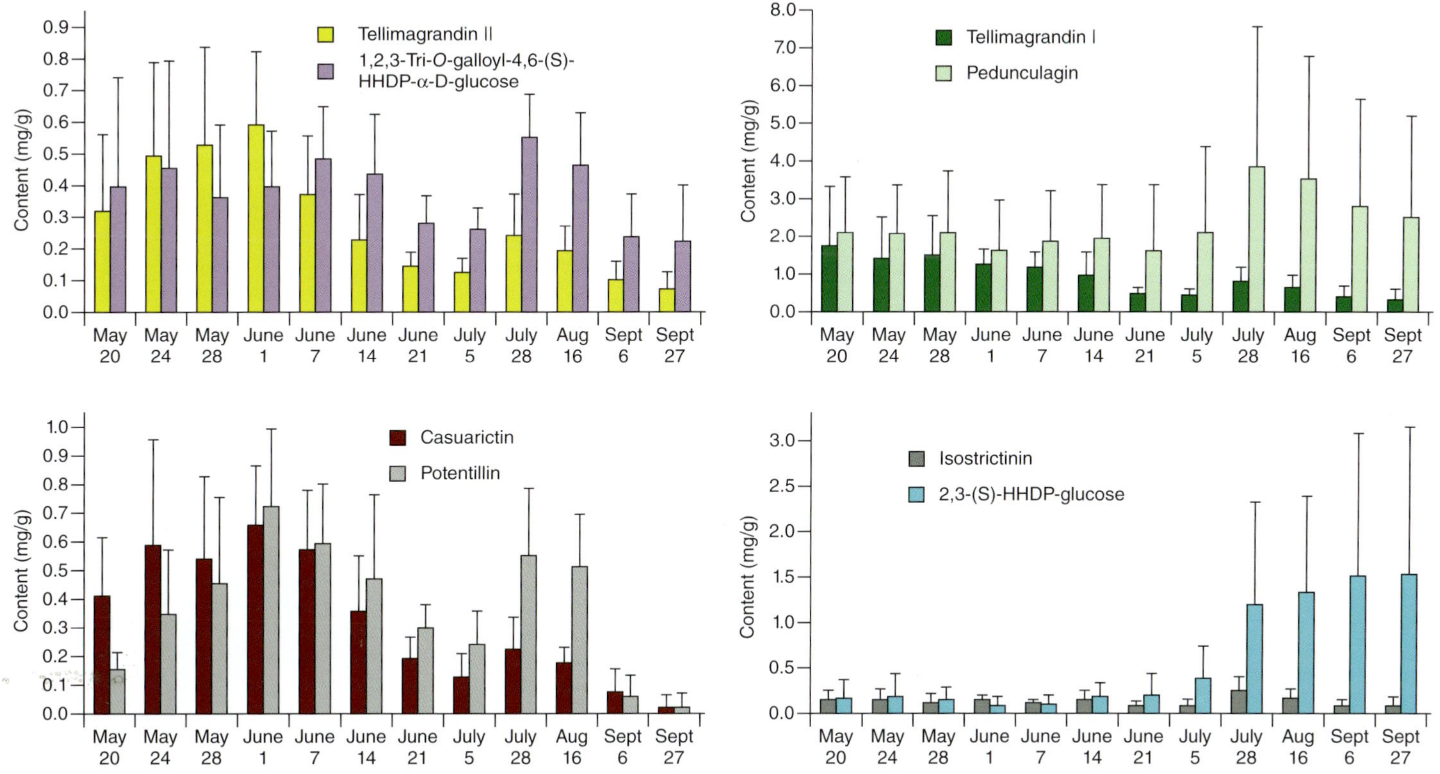


**Plate 4.10** Seasonal variation of individual ETs in leaves of *Betula pubescens*. (Adapted from Salminen *et al.*, 2001. Reproduced with permission of Elsevier.)

**Plate 4.11** HPLC-UV traces of polyphenols found in the foliage of *Betula pubescens* (top left), *Quercus robur* (top right), *Epilobium angustifolium* (bottom left), and *Acer platanoides* (bottom right).

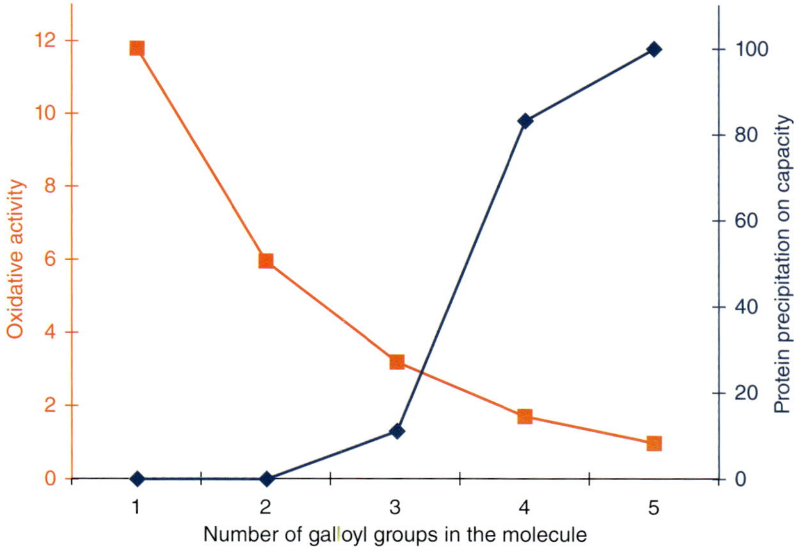


**Plate 4.13** The addition of galloyl groups into an HT increases its PPC but decreases its pro-oxidant activity. As these two activities are negatively correlated, it is difficult to find a single tannin that is efficient at both activity mechanisms. (Adapted from Salminen *et al.*, 2011. Reproduced with permission of John Wiley & Sons.)

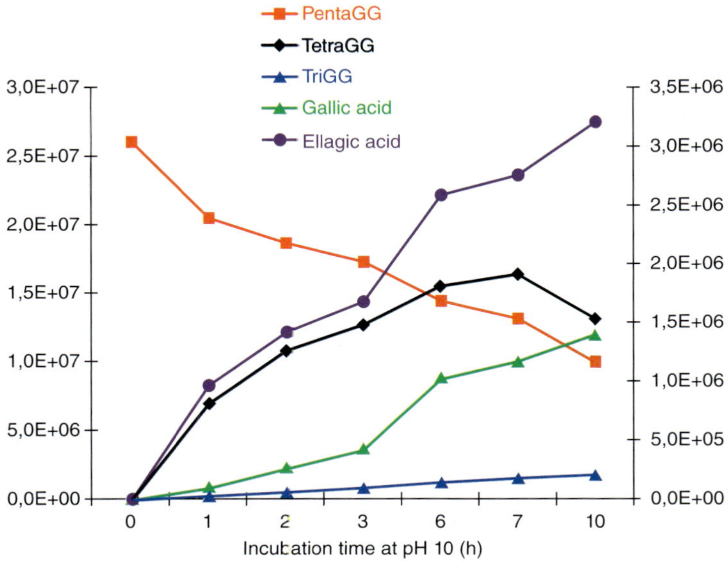


**Plate 4.16** Incubation or pentagalloylglucose at pH 10 decreases its concentration and causes the simultaneous accumulation of the hydrolysis products tetragalloyl glucose, trigalloyl glucose, gallic acid, and ellagic acid.

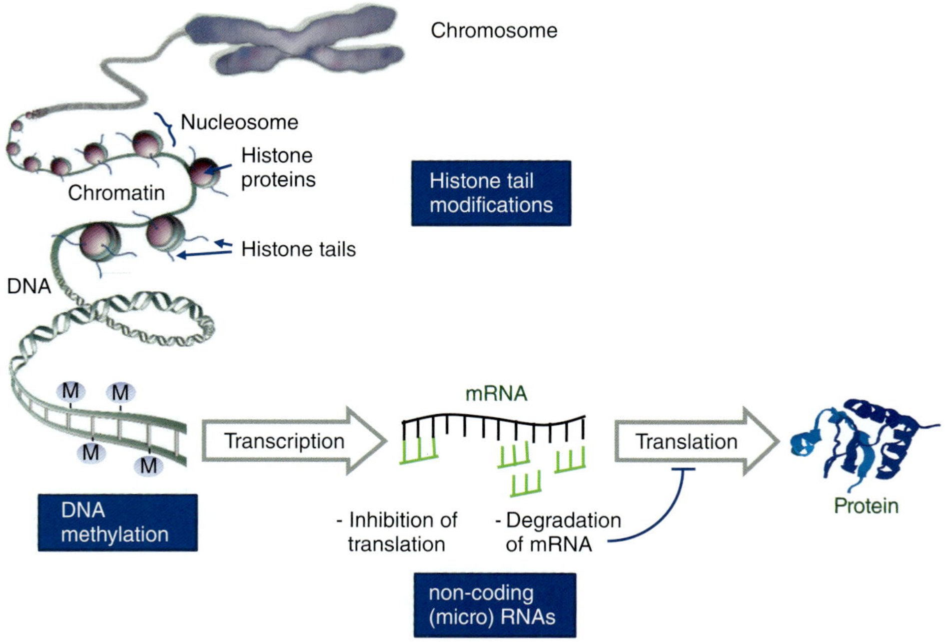


**Plate 6.1** Overview of chromatin organization and epigenetic mechanisms, including DNA methylation, histone modifications, and noncoding miRNAs, targeting DNA, N-terminal histone tails, and mRNA. (Adapted from Qiu, 2006. Reproduced with permission of Nature Publishing Group.) See text for details.

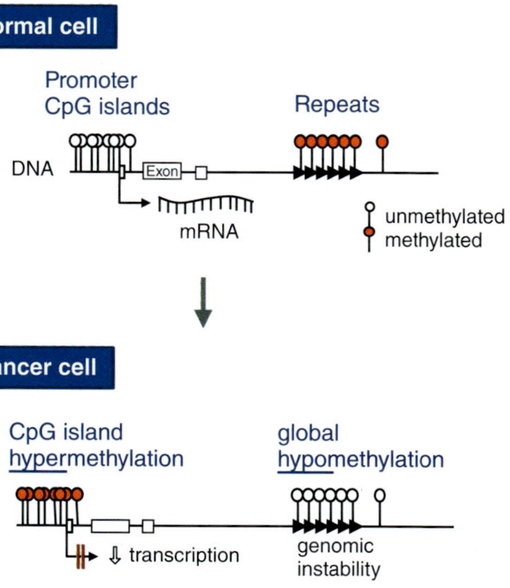


**Plate 6.2** Overview of DNA methylation changes during carcinogenesis, including gain of methylation (red circles) at CpG islands (CGIs), often located in gene promoter regions (CGI hypermethylation) and associated with epigenetic silencing of gene transcription and global loss of DNA methylation (DNA hypomethylation, open circles) at intergenic regions and repetitive sequences, contributing to genomic instability.

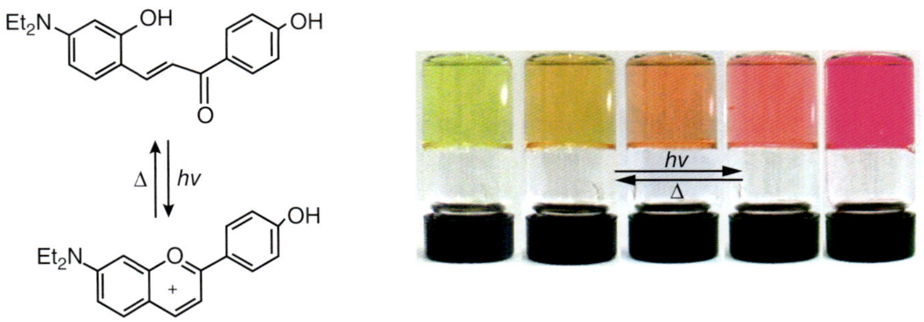


**Plate 11.12** Highly transparent photochromic gels in the visible region can be prepared by the incorporation of flavylium compounds in Pluronic F-127 matrices. The chromophore 7-(N,N-diethylamino)-4-hydroxyflavylium switches from its yellow chalcone form to the red flavylium cation upon irradiation and reverts back to the chalcone form in the dark. (Pina and Hatton, 2008. Reproduced with permission of American Chemical Society.)

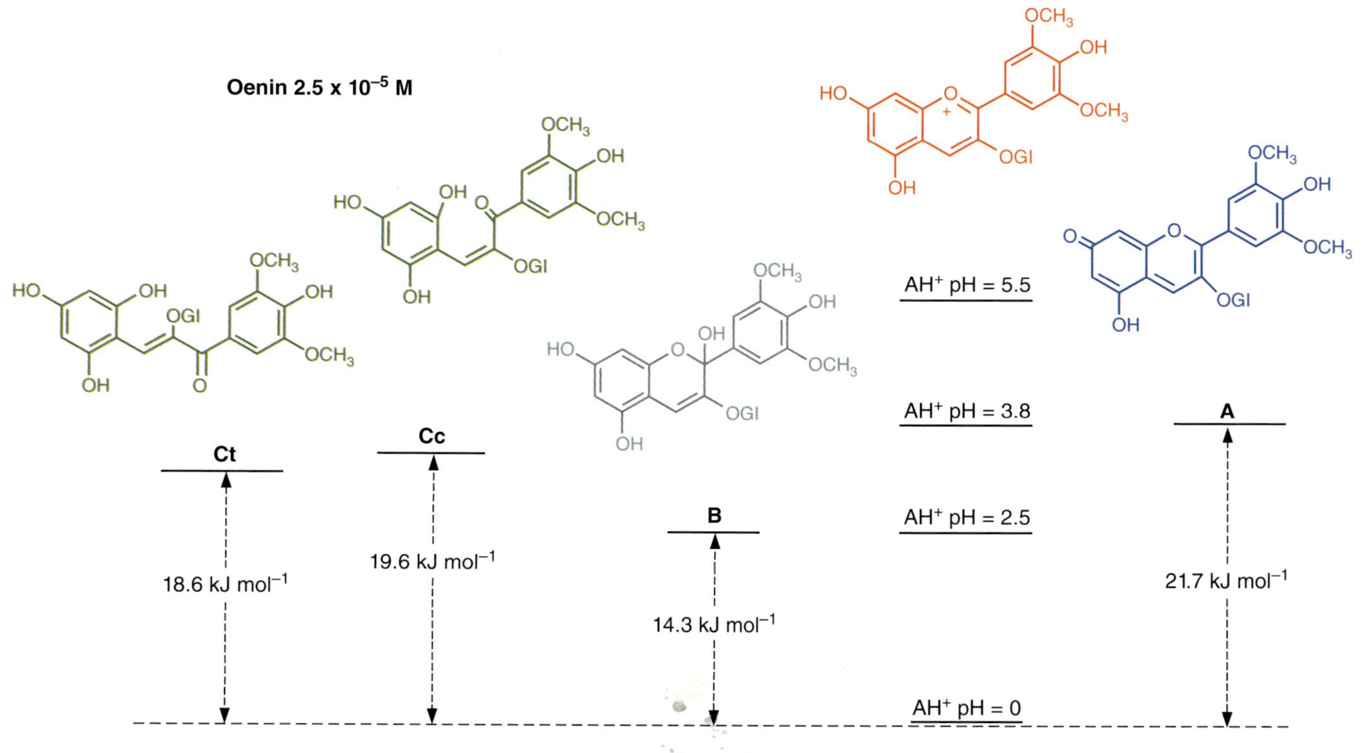


**Plate 11.1** Energy-level diagram of oenin. At this concentration the self-aggregation can be neglected. (Leydet *et al.*, 2012. Reproduced with permission of Elsevier.)

Chapter 3
# Function, Structure, and Evolution of Flavonoid Glycosyltransferases in Plants

Keiko Yonekura-Sakakibara[1] and Kazuki Saito[1,2]
[1]*RIKEN Center for Sustainable Resource Science, Yokohama, Japan*
[2]*Graduate School of Pharmaceutical Sciences, Chiba University, Chiba, Japan*

**Abstract:** Glycosylation plays an important role in the stable accumulation of flavonoids and serves as a preparative step for the acylation that is involved in flower color variation. Generally, glycosylation of plant secondary metabolites is catalyzed by UDP-dependent glycosyltransferases (UGTs). In this chapter, we describe recent progress in the study of plant-flavonoid UGTs, including a *C*-glycosyltransferase involved in the biosynthesis of *C*-glycosylflavones, a citrus glycosyltransferase directing biosynthesis of tasteless flavanones, and novel anthocyanin glycosyltransferases of the family 1 glycoside hydrolase (GH1)-type. We also discuss structural and evolutionary aspects of the flavonoid glycosyltransferases, especially glycoside glycosyltransferases (GGTs) (which have glycosylate sugar moieties attached to a flavonoid aglycone).

**Keywords:** glycosyltransferase, glycoside hydrolase, flavonoid, anthocyanin, plant secondary metabolites, evolution

## 3.1 Introduction

Flavonoids are one of the major classes of secondary metabolites, with ~9000 identified compounds (Harborne & Williams, 2000; Anderson & Markham, 2006; Buer *et al.*, 2010) (Fig. 3.1). The core flavonoid skeleton is relatively limited in complexity, so the huge diversity of flavonoid structures generally results from modification mechanisms such as hydroxylation, glycosylation, acylation, and methylation. Glycosylation is important for enhancing molecular stability and water solubility. Alteration of flavonoid chemical properties affects their transport within the cell, their biological activities, and, most notably for the consumer, fruit flavor (Bowles *et al.*, 2006; Horowitz & Gentili, 1969). Glycosylation

*Recent Advances in Polyphenol Research, Volume 4*, First Edition.
Edited by Annalisa Romani, Vincenzo Lattanzio, and Stéphane Quideau.

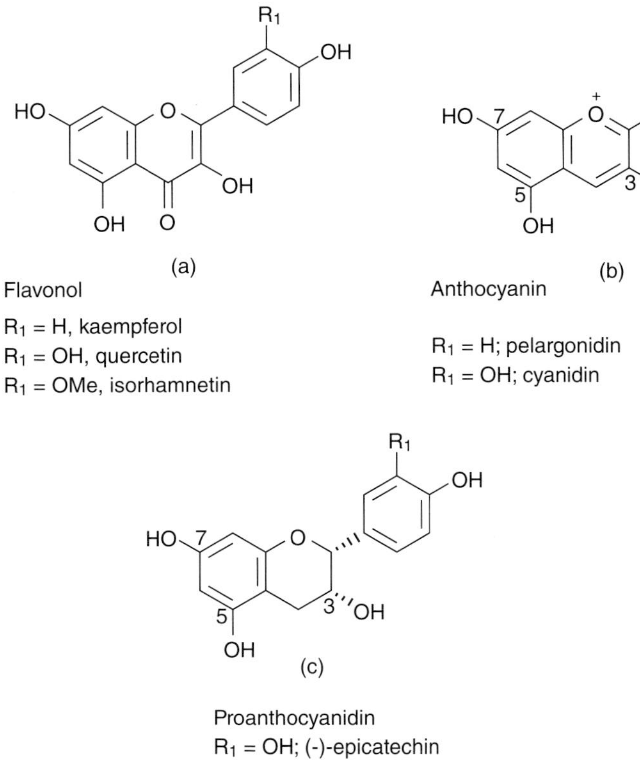


**Fig. 3.1** Flavonol and anthocyanin structures: (a) flavonol, (b) anthocyanin, (c) proanthocyanidin.

is also involved in detoxification and deactivation of xenobiotics and in plant–pathogen interactions.

A glycosyltransferase is generally an enzyme which catalyzes the transfer of sugar moieties from activated donor molecules to acceptor molecules such as sugars, lipids, nucleic acids, proteins, antibiotics, and other small molecules (Bowles *et al.*, 2006). As of September 2012, the known glycosyltransferases were divided into 94 families (GT1–94) based on amino acid sequence similarities (http://www.cazy.org/GlycosylTransferases.html). Family 1 glycosyltransferases (GT1), also referred to as UDP-dependent glycosyltransferases (UGTs), are encoded by a large multigene family and make up the largest glycosyltransferase family in plants (Yonekura-Sakakibara & Hanada, 2011). UGTs transfer a glycosyl moiety from UDP-sugar to a wide range of acceptor molecules, including flavonoids and other plant secondary metabolites (Gachon *et al.*, 2005; Bowles *et al.*, 2006; Yonekura-Sakakibara *et al.*, 2009). Recently it was reported that glycoside hydrolase-type glycosyltransferases can transfer the sugar moiety from an acyl-glucose to anthocyanins (Matsuba *et al.*, 2010; Miyahara *et al.*, 2012, 2013).

In this chapter, we describe recent progress in the study of plant-flavonoid UGTs, such as a *C*-glycosyltransferase involved in the biosynthesis of *C*-glycosylflavones, flavonoid glycoside glycosyltransferases (GGTs) including a citrus glycosyltransferase involved in fruit flavor, and novel anthocyanin glycosyltransferases classified as family 1 glycoside hydrolase (GH1)-types. Structural and evolutionary aspects of the flavonoid glycosyltransferases, especially GGTs (which have glycosylate sugar moieties attached to a flavonoid aglycone), are also discussed.

## 3.2 UDP-dependent glycosyltransferases

### *3.2.1 Functional identification of flavonoid UGTs*

A number of flavonoid UGTs, including flavonoid 3-*O*-glycosyltransferases, 5-*O*-glycosyltransferases, 7-*O*-glycosyltransferases, 3′-*O*-glycosyltransferases, and GGTs, have been identified (Gachon *et al.*, 2005; Bowles *et al.*, 2006). A phylogenetic tree of flavonoid UGT genes indicates that they diverged from an ancestral gene into isogenes that encode enzymes with specificity for the sugar-attachment position (Fig. 3.2). In this section we describe recently identified genes encoding flavonoid 3-*O*-glycosyltransferases, a flavonoid 3′-*O*-glycosyltransferase, a flavonoid 7-*O*-glycosyltransferase, and flavonoid GGTs (Table 3.1).

#### *3.2.1.1 Flavonoid 3-O-glycosyltransferases*

***Flavonol 3-*O*-glycosyltransferase from grape (Vitis vinifera)***

The flavonoid 3-*O*-glycosyltransferases flavonol 3-*O*-glucopyranosyltransferase and flavonol 3-*O*-glucosyltransferase/galactosyltransferase, have been identified in *Vitis vinifera* L. (common grapevine) (Ono *et al.*, 2010). *In silico* screening of the grapevine genome with the DNA sequence of the gene (*VvGT1*) encoding anthocyanidin 3-*O*-glucosyltransferase (Ford *et al.*, 1998) indicated that the genome contains seven additional UGT-related genes: *VvGT2*, *VvGT2*-like, *VvGT3*, *VvGT4*, *VvGT4*-like, *VvGT5*, and *VvGT6*. Biochemical analyses using recombinant proteins indicated that *VvGT5* and *VvGT6* encode flavonol 3-*O*-glucopyranosyltransferase and flavonol 3-*O*-glucosyltransferase/galactosyltransferase, respectively. The tandem arrangement of *VvGT5* and *VvGT6* and their high sequence similarity suggest that one of these UGT genes was duplicated from the other by a gene-duplication event. Substitution of amino acid residues based on molecular modeling indicates that Arg140 is crucial to determining the substrate specificity of VvGT5 for UDP-glucuronic acid. Gln373 and Pro19 are important to the bifunctional glucosyltransferase/galactosyltransferase activity of VvGT6.

***Anthocyanin 3-*O*-glactosyltransferase from kiwifruits***

See "Cyanidin 3-*O*-galactoside: 2″-*O*-xylosyltransferase from kiwifruit" in Section 3.2.1.3.

#### *3.2.1.2 Flavonoid 7-O-glycosyltransferases*

***Flavonoid 7-*O*-glucuronosyltransferase from Lamiales***

In order to discover how plants acquire sugar-donor specificity during evolution, flavonoid 7-*O*-glucuronosyltransferases (F7GATs) from *Lamiales* were biochemically characterized

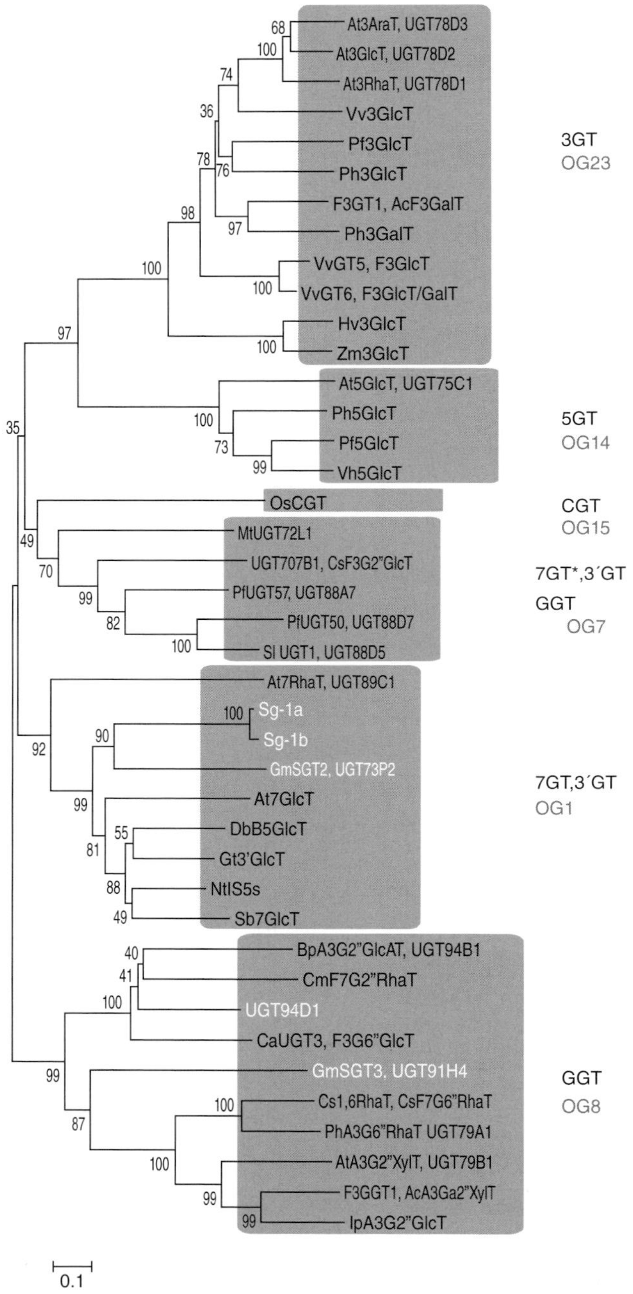

At3AraT, UGT78D3
At3GlcT, UGT78D2
At3RhaT, UGT78D1
Vv3GlcT
Pf3GlcT
Ph3GlcT
F3GT1, AcF3GalT
Ph3GalT
VvGT5, F3GlcT
VvGT6, F3GlcT/GalT
Hv3GlcT
Zm3GlcT
At5GlcT, UGT75C1
Ph5GlcT
Pf5GlcT
Vh5GlcT
OsCGT
MtUGT72L1
UGT707B1, CsF3G2"GlcT
PfUGT57, UGT88A7
PfUGT50, UGT88D7
Sl UGT1, UGT88D5
At7RhaT, UGT89C1
Sg-1a
Sg-1b
GmSGT2, UGT73P2
At7GlcT
DbB5GlcT
Gt3'GlcT
NtIS5s
Sb7GlcT
BpA3G2"GlcAT, UGT94B1
CmF7G2"RhaT
UGT94D1
CaUGT3, F3G6"GlcT
GmSGT3, UGT91H4
Cs1,6RhaT, CsF7G6"RhaT
PhA3G6"RhaT UGT79A1
AtA3G2"XylT, UGT79B1
F3GGT1, AcA3Ga2"XylT
IpA3G2"GlcT
3GT
OG23
5GT
OG14
CGT
OG15
7GT*,3´GT
GGT
OG7
7GT,3´GT
OG1
GGT
OG8
0.1

◀ **Fig. 3.2** Phylogenetic tree of the glycosyltransferases for plant secondary metabolites. The evolutionary history was inferred using the neighbor-joining method (Saitou & Nei, 1987). The optimal tree with sum of branch length = 14.16 792 972 is shown. The percentage of replicate trees in which the associated taxa clustered together in the bootstrap test (1000 replicates) is indicated next to the branches (Felsenstein, 1985). The tree is drawn to scale, with branch lengths in the same units as those of the evolutionary distances used to infer the phylogenetic tree. The evolutionary distances were computed using the Poisson correction method (Zuckerkandl & Pauling, 1965) and are in units of the number of amino acid substitutions per site. The analysis involved 41 amino acid sequences. All positions containing gaps and missing data were eliminated. There were a total of 262 positions in the final dataset. Evolutionary analyses were conducted in MEGA5 (Tamura *et al.*, 2011). Glycosyltransferases for flavonoids are shown in black and lignan/terpenoids are in white. An asterisk indicates Lamiales-specific 7-*O*-glycosyltrnasferases. OG, orthologous group; 3GT, 3-*O*-glycosyltransferase; 5GT, 5-*O*-glycosyltransferase; 7GT, 7-*O*-glycosyltransferase; 3′GT, 3′-*O*-glycosyltransferase; CGT, *C*-glycosyltransferase; GGT, glycoside glycosyltransferase (with glycosylate sugar moieties attached to a flavonoid aglycone). The GenBank accession numbers for the sequences are shown in parentheses: Ac3GalT F3GT1 (GU079683); AcA3Ga2″XylT AcF3GGT1 (FG404013); AtA3G2″XylT UGT79B1 (NM_124785); At7GlcT (NM_129234); At3AraT (NM_121709); At3GlcT (NM_121711); At3RhaT (NM_102790); At5GlcT (NM_117485); At7RhaT (NM_100480); BpA3G2″GlcAT (AB190262); CaF3G6″GlcT CaUGT3 (AB443870); CsF3G2″GlcT UGT707B1 (HE793682); CsF7G6″RhaT Cs1,6RhaT (DQ119035); CmF7G2″RhaT (AY048882); DbB5GlcT (Y18871); Gt3′GlcT (AB076697); Hv3GT (X15694); IpA3G2″GlcT/UGT79G16 (AB192315); MtUGT72L1 (EU434684); NtIS5s (AF346431); OsCGT (FM179712); PhA3G6″RhaT (Z25802); Ph3GalT (AF316552); Ph3GlcT (AB027454); Pf3GlcT (AB002818); Pf5GlcT (AB013596); Ph5GlcT (AB027455); Sb7GlcT (AB031274); Vv3GlcT (AF000371); VvGT5 3GlcT (AB499074); VvGT6 3GlcT/GalT (AB499075); Vh5GlcT (AB013598); Zm3GlcT (AF000371). 3GlcT, 3-*O*-glucosyltransferase; 3′GlcT, 3′-*O*-glucosyltransferase; 3GalT, 3-*O*-galactosyltransferase; 3RhaT, 3-*O*-rhamnosyltransferase; 5GlcT, 5-*O*-glucosyltransferase; 7GlcT, 7-*O*-glucosyltransferase; 7RhaT, 7-*O*-rhamnosyltransferase; A3G2″GlcAT, anthocyanidin 3-*O*-glucoside 2″-*O*-glucuronosyltransferase; A3G2″GlcT, anthocyanidin 3-*O*-glucoside 2″-*O*-glucosyltransferase; A3G6″RhaT, anthocyanidin 3-*O*-glucoside 6″-*O*-rhamnosyltransferase; A3G2″XylT, anthocyanidin 3-*O*-glucoside 2″-*O*-xylosyltransferase; A3Ga2″XylT, anthocyanidin 3-*O*-galactoside 2″-*O*-xylosyltransferase; F3G6″GlcT, flavonoid 3-*O*-glucoside 6″-*O*-glucosyltransferase; F7G2″RhaT, flavanone 7-*O*-glucoside 2″-*O*-rhamnosyltransferase; F7G6″RhaT, flavanone 7-*O*-glucoside 6″-*O*-rhamnosyltransferase; NtIS5a, salicylate-induced glucosyltransferase. Abbreviations for species: Ac, *Actinidia chinensis*; At, *Arabidopsis thaliana*; Bp, *Bellis perennis*; Ca, *Catharanthus roseus*; Cm, *Citrus maxima*; Cs, *Citrus sinensis*; Db, *Dorotheanthus bellidiformis*; Gt, *Gentiana triflora*; Hv, *Hordeum vulgare*; Ip, *Ipomoea purpurea*; Mt, *Medicago truncatula*; Nt, *Nicotiana tabacum*; Os, *Oryza sativa*; Pf, *Perilla frutescens*; Ph, *Petunia hybrida*; Sb, *Scutellaria baicalensis*; Vh, *Verbena hybrida*; Vv, *Vitis vinifera*; Zm, *Zea mays*.

(Noguchi *et al.*, 2009). All F7GATs from *Lamiales* belong to the UGT88 subfamily. Arginine residues in F7GATs and serine residues in the UGT88D subfamily (e.g. Arg339 and Ser111 in SbUBGAT, UGT88D1) were found by aligning the amino acid sequences of *Lamiales* F7GATs and other flavonoid glucosyltransferases in the UGT88 subfamily with the homologous enzymes of other plant species. The Arg and Ser residues correspond to Trp353 and Thr141 in VvGT1, a flavonoid 3-*O*-glucosyltransferase from *Vitis vinifera*. These residues interact with the O4 and O6 hydroxyl groups, respectively, of the glucose moiety of UDP-glucose. The sugar donor specificities of F7GATs were changed simply by the substitution of Arg to Trp and of Ser to Thr. The Arg residue conserved in *Lamiales* F7GATs plays a dominant role in determining sugar donor specificity. These results are consistent with the observation that sugar donor specificity differentiates after the establishment of regiospecificity (Ono *et al.*, 2006; Yonekura-Sakakibara *et al.*, 2007, 2008).

**Table 3.1** Flavonoid UGT genes reported in the literature since 2008.

| Type | Plant species | Common name | Gene | Function | Reference |
|---|---|---|---|---|---|
| UGT | *Actinidia chinensis* | Kiwifruit | F3GT1 | anthocyanin 3-*O*-galactosyltransferase | Montefiori *et al.* (2011) |
| | | | F3GGT1 | anthocyanin 3-*O*-galactoside: 2′′-*O*-xylosyltransferase | |
| | *Arabidopsis thaliana* | Arabidopsis | UGT79B1 | anthocyanin 3-*O*-glucoside: 2′′-*O*-xylosyltransferase | Yonekura-Sakakibara *et al.* (2012) |
| | *Catharanthus roseus* | | CaUGT3 | flavonoid glucoside: 6′′-*O*-glucosyltransferase activity | Masada *et al.* (2009) |
| | *Citrus sinensis* | Orange | Cs1,6RhaT | flavanone 7-*O*-glucoside: 6′′-*O*-rhamnosyltransferase | Frydman *et al.* (2012) |
| | *Crocus sativus* | Saffron | UGT707B1 | flavonol 3-*O*-glucoside: 2′′-*O*-glucosyltransferase | Trapero *et al.* (2012) |
| | *Medicago truncatula* | | UGT72L1 | proanthocyanidin 3′-*O*-glucosyltransferase | Pang *et al.* (2008) |
| | *Oryza sativa* | Rice | OsCGT | *C*-glucosyltransfease | Brazier-Hicks *et al.* (2009) |
| | *Perilla frurtescens* | | PfUGT50 | flavonoid 7-*O*-glucuronosyltransferase | Noguchi *et al.* (2009) |
| | | | PfUGT57 | flavonoid 7-*O*-glucosyltransferase | |
| | *Scutellaria laeteviolacea* | Skullcap | Sl UGT1 | flavonoid 7-*O*-glucuronosyltransferase | Noguchi *et al.* (2009) |
| | *Vitis vinifera* | Grapevine | VvGT5 | flavonol 3-*O*-glucopyranosyltransferase | Ono *et al.* (2010) |
| | | | VvGT6 | flavonol 3-*O*-glucosyltransferase/galactosyltransferase | |
| GH1 | *Dianthus caryophyllus* | Carnation | DcAA5GT | anthocyanin 5-*O*-glucosyltransferase | Matsuba *et al.* (2010) |
| | *Delphinium grandiflorum* | Delphinium | DgAA7GT | anthocyanin 7-*O*-glucosyltransferase | Matsuba *et al.* (2010) |
| | *Agapanthus africanus* | | AaAA7GT | anthocyanin 7-*O*-glucosyltransferase | Miyahara *et al.* (2012) |
| | *Arabidopsis thaliana* | | AtBGLU10 | anthocyanin 3-*O*-6′′-*O*-coumaroylglucoside: glucosyltransferase | Miyahara *et al.* (2013) |

### 3.2.1.3 Flavonoid glycosyltransferases that glycosylate the sugar moiety attached to a flavonoid aglycone

***Flavanone 7-O-glucoside: 6′′-O-rhamnosyltransferase from sweet orange (Citrus sinensis)***

Bitterness is an important agronomic trait for citrus breeding. Flavanone 7-*O*-neohesperidosides are perceived as a bitter flavor but flavanone 7-*O*-rutinosides are tasteless. A gene encoding flavanone 7-*O*-glucoside: 6′′-*O*-rhamnosyltransferase (1,6-rhamnosyltransferase, Cs1,6RhaT) that converts flavanone 7-*O*-glucoside to flavanone 7-*O*-rutinoside was identified in *Citrus sinensis* (Frydman *et al.*, 2012). Expressed sequence tag (EST) databases of non-bitter citrus species (*C. sinensis*, *C. clementina*, and *C. reticulata*) were screened based on expression profiles in bitter and non-bitter citrus species for sequence similarity with other flavonoid GGTs. 1,6-rhamnosyltransferase activity was confirmed using transgenic tobacco BY2 cells over-expressing full-length candidate genes. Cs1,6RhaT can utilize a broad range of flavonoids, including flavanone 7-*O*-gucosides, flavone 7-*O*- glucosides, flavonol 3-*O*-glucosides and 7-*O*-glucosides, and anthocyanidin 3-*O*-glucosides, as sugar acceptors. By contrast, a previously identified flavanone 7-*O*-glucoside: 2′′-*O*-rhamnosyltransferase (1,2-rhamnosyltransferase, Cm1,2RhaT) from *C. maxima* (pummelo) was specific for 7-*O*-glucosides of flavanones and flavones (Frydman *et al.*, 2004). Phylogenetic analysis of flavonoid UGTs suggested that 1,6RhaT and 1,2RhaT diverged at an early stage of *Citrus* evolution (Frydman *et al.*, 2012).

***Cyanidin 3-O-galactoside: 2′′-O-xylosyltransferase from kiwifruit (Actinidia chinensis)***

The anthocyanin biosynthetic pathway genes CHS, F3H, F3′H, DFR, ANS, and the UGTs were isolated from *Actinidia chinensis* (kiwifruit) by a homology search of kiwifruit ESTs (Montefiori *et al.*, 2011). Of three UGT candidates, F3GT1 was identified as an anthocyanin 3-*O*-galactosyltransferase by an enzymatic assay using recombinant F3GT1 protein and a transient RNA interference (RNAi) assay in fruit of *A. eriantha* (Montefiori *et al.*, 2011). F3GT1 also has a small amount of anthocyanin 3-*O*-glucosyltransferase activity (<5% of anthocyanin 3-*O*-galactosyltransferase activity). Recombinant F3GGT1 protein converted cyanidin 3-*O*-galactoside to cyanidin 3-*O*-xylosyl(1→2)galactoside. F3GGT1 can also recognize cyanidin 3-*O*-glucoside, although no cyanidin 3-*O*-xylosyl(1→2)glucoside was detected in *A. chinensis*. Transient assay in fruit of *A. eriantha* using RNAi showed that F3GGT1 is involved in 2′′-*O*-xylosylation of cyanidin 3-*O*-galactoside. Thus, F3GGT1 is an anthocyanin 3-*O*-galactoside: 2′′-*O*-xylosyltransferase (Montefiori *et al.*, 2011).

***Cyanidin 3-O-glucoside: 2′′-O-xylosyltransferase from Arabidopsis thaliana***

The largest number of flavonoid glycosyltransferases were identified from *Arabidopsis thaliana* by taking advantage of the current wealth of omics-based resources. Of the genes induced by the MYB transcription factor PAP1, those encoding flavonoid 3-*O*-glucosyltransferase and anthocyanin 5-*O*-glucosyltransferase were identified by enzymatic assays with the recombinant proteins and by the analysis of the anthocyanin accumulation pattern in T-DNA knock-out mutants (Tohge *et al.*, 2005). Flavonol 3-*O*-rhamnosyltransferase and flavonol 7-*O*-glucosyltransferase candidates were selected based on sequence similarity with known flavonoid glycosyltransferases, and enzymatic functions were identified by *in vitro* enzymatic assays and by the analysis of the flavonol profiles of gene knock-out mutants (Jones *et al.*, 2003). Transcriptome coexpression analysis using flavonoid biosynthetic genes as queries efficiently identified a flavonol 7-*O*-rhamnosyltransferase and a flavonol 3-*O*-arabinosyltransferase (Yonekura-Sakakibara *et al.*, 2007, 2008). Thus, six genes encoding flavonoid glycosyltransferases have been identified.

In order to identify additional candidate flavonoid glycosyltransferase genes, independent component analysis (ICA) was used to analyze microarray data (Yonekura-Sakakibara *et al.*, 2012). ICA, a form of unsupervised algorithm, is based on the assumption that a given level of gene expression is determined by a linear combination of some set of independent components (ICs). ICA is used to extract and characterize informative features from microarray data (Kong *et al.*, 2008). A total of 1877 genes, including flavonoid-related genes and all genes in AraCyc, were analyzed by ICA on >1000 microarray data from ATTED-II (Obayashi *et al.*, 2009). A hierarchical clustering of data obtained by ICA established two major gene clusters, consisting of anthocyanin-pathway and flavonoid-pathway genes. UGT79B1 and UGT84A2 clustered with the anthocyanin biosynthetic genes (Yonekura-Sakakibara *et al.*, 2012). Enzymatic assays with recombinant proteins and anthocyanin analysis of T-DNA knock-out mutants indicated that UGT79B1 is an anthocyanin 3-*O*-glucoside: 2′′-*O*-xylosyltransferase. UGT84A2 is already known to be a sinapic acid glucosyltransferase that converts sinapic acid to 1-*O*-sinapoylglucose (Lim *et al.*, 2001; Sinlapadech *et al.*, 2007), but there is no direct evidence for a relationship between UGT84A2 and anthocyanin biosynthesis. Anthocyanin analysis of *ugt84a*s knock-out mutants grown under anthocyanin-induced conditions indicated that UGT84A2 is part of the 1-*O*-sinapoylglucose supply chain for sinapoylation of anthocyanin.

ICA provides a different perspective on microarray data depending on the number of ICs. For example, flavonoid 3-*O*-glucosyltransferase, which converts anthocyanidins and flavonols to their cognate 3-*O*-glucosides, falls into the anthocyanin cluster based on 10 ICs but into the flavonol cluster based on 8 ICs (Yonekura-Sakakibara *et al.*, 2012). Bifunctional genes like these are likely to be identified by ICA.

#### *Flavonoid glucoside: 6′′-glucosyltransferase from Madagascar periwinkle (Catharanthus roseus)*

Methyl jasmonate-treated suspension cultures of *Catharanthus roseus* (Madagascar periwinkle) convert exogenously added flavonoid aglycones to flavonoid gentiobiosides (Masada *et al.*, 2009). A gene encoding flavonoid glucoside: 6′′-glucosyltransferase, *CaUGT3*, was identified from *C. roseus* (Masada *et al.*, 2009). Phylogenetic analysis showed that *CaUGT3* belongs to the flavonoid GGT family. Recombinant CaUGT3 protein can convert quercetin 3-*O*-glucoside to quercetin 3-*O*- gentiobiosides, and prolonged incubation yields the further glucosylated products quercetin 3-*O*-gentiotriosides and quercetin 3-*O*-gentiotetrosides. Thus, CaUGT3 has sequential flavonoid glucoside: 6′′-*O*-glucosyltransferase activity.

***Flavonol 3-O-glucoside: 2′′-O-glucosyltransferase from saffron crocus (Crocus sativus)***

UGT707B1 was isolated from *Crocus sativus* (saffron crocus) using degenerate primers based on UGT conserved regions (Trapero *et al.*, 2012). The accumulation of UGT707B1 transcripts in tepals of various saffron species was consistent with the presence of flavonol 3-*O*-glucosyl(1→2)glucoside, although no recombinant UGT707B1 protein was recovered as soluble protein. The content of kaempferol 3-*O*-glucosyl(1→2)glucoside and quercetin 3-*O*-glucosyl(1→2)glucoside is more than double in transgenic *A. thaliana* lines over-expressing UGT707B1. These results suggest that UGT707B1 is a flavonol 3-*O*-glucoside: 2′′-*O*-glucosyltransferase, although a phylogenetic tree of UGTs indicates that UGT707B1 belongs to a different cluster than the flavonol glycosyltransferases UGT79 and UGT94, which glycosylate glycosyl moieties attached to flavonoid aglycones.

### *3.2.1.4 Flavonoid 3′-O-glycosyltransferase*

***Flavan-3-ol 3′-O-glucosyltransferase from Medicago truncatula***

Transparent testa 2 (TT2) is an R2R3 MYB transcription factor that is required for proanthocyanidin biosynthesis in *Arabidopsis* (Nesi *et al.*, 2001). Among the genes that are induced more than twofold by ectopic expression of *TT2* in hairy roots of *Medicago truncatula* (Barrel medic, *Fabaceae*), UGT72L1 genes that are preferentially expressed in seed coat have high glucosyltransferase activity towards the flavan-3-ol (−)-epicatechin (Pang *et al.*, 2008). The resultant product has been identified by NMR as (−)-epicatechin 3′-*O*-glucoside. UGT72L1 also has weak activity on (+)-epicatechin and cyanidin (<15%). Kinetic analysis of UGT72L1 showed lower Km values for epicatechin than with other flavonoids. These data suggest that UGT72L1 is an epicatechin 3′-*O*-glucosyltransferase. Interestingly, epicatechin glucoside is detected only in immature seeds.

### *3.2.1.5 Flavonoid C-glycosyltransferase*

***C-glycosyltransferase from rice (Oryza sativa)***

The flavonoid *C*-glycosides, a group that includes the *C*-glycosylflavones, *C*-glycosylflavanones, *C*-glycosylflavanols, *C*-glycosylchalcones, and *C*-glycosyldihydrochalcones, are found in plants as taxonomically diverse as the Bryophyta (mosses) and Pteridophyta (ferns), as well as the monocots and dicots (Jay *et al.*, 2006). As with other flavonoids, *C*-glycosylflavonoids have antioxidant, antibiotic, and antibacterial activity and function as insect attractants, UV protectants, and signaling molecules (Jay *et al.*, 2006). *C*-glycosylflavones also contribute bluing to flower color by acting as copigments of anthocyanins in *Iris ensata* (Yabuya *et al.*, 1997).

It has been reported that 2-hydroxyflavanone-6(or 8)-*C*-glucosyltransferase catalyzes glucose transfer to 2-hydroxynaringenin, but not to naringenin or apigenin (Kerscher & Franz, 1987). UDP-glucose: 2-hydroxyflavanone-6(or 8)-*C*-glucosyltransferase was purified from *Fagopyrum esculentum* (buckwheat) (Kerscher & Franz, 1988), but no genes that could encode a *C*-glycosyltransferase were identified until recently. *C6*-Glucosyltransferase was purified from *Oryza sativa* (rice) and *Triticum aestivum* (wheat) using

2,5,7-trihydroxyflavanone as a substrate (Brazier-Hicks *et al.*, 2009). The corresponding family 1 glycosyltransferase gene, *OsCGT*, was isolated based on the amino acid sequences of purified proteins, and its function was identified by enzymatic characterization of recombinant OsCGT protein (Brazier-Hicks *et al.*, 2009). OsCGT recognizes 2,5,7-trihydroxyflavanone, 2-hydroxynaringenin, 2′,4′,6′-trihydroxydihydrochalcone, phoretin, and 2,4,6-trihydroxybenzophenone as substrates. Incubation of 2,5,7-trihydroxyflavanone and UDP-glucose with OsCGT *in vitro* yielded a single product with a mass corresponding to 2,5,7-trihydroxyflavanone *C*-glucoside. The product was spontaneously converted into two compounds with a total mass corresponding to dehydrated 2-hydroxyflavanone *C*-glucosides, chrysin 6-*C*-glucoside, and chrysin 8-*C*-glucoside. Dehydration to 6-*C*-glucoside flavones was preferred *in planta*, suggesting that dehydratase selectively converts 2-hydroxyflavanone *C*-glucoside to the corresponding flavone 6-*C*-glucosides (Fig. 3.3). Assays using crude protein extracts of rice and wheat demonstrated the specificity of the dehydratase.

### *3.2.2 3D structures of flavonoid UGTs*

Amino acid sequence identity among UGTs is relatively low (~20–30%), but their 3D structures are conserved. All UGTs contain a GT-B fold consisting of two $\beta/\alpha/\beta$ Rossmann-like domains. These domains are involved in the binding of the sugar donor and sugar acceptor.

So far, crystal structures of five plant UGTs – UGT71G1, UGT72B1, UGT78A5 (VvGT1), UGT78G1, and UGT85H2 – have been resolved (Table 3.2). Of these, two are involved in flavonoid glycosylation *in planta*. VvGT1 (UGT78A5) was identified as an anthocyanin 3-*O*-glucosyltransferase from grapevine (Ford *et al.*, 1998). UGT78G1 from *M. truncatula* appears to act as an anthocyanin glucosyltransferase *in planta* (Peel *et al.*, 2009), although recombinant UGT78G1 protein gave a higher $K_{cat}/K_m$ value for flavones and isoflavones than for anthocyanidins (Modolo *et al.*, 2007). Interestingly, UGT78G1 also has deglycosylation activity.

The crystal structures of VvGT1 (UGT78A5) and UGT78G1 suggest the presence of several key amino acid residues that interact with UDP-glucose and flavonoids (Offen *et al.*, 2006; Modolo *et al.*, 2009b). In VvGT1, His20 deprotonates the O3 hydroxyl group of the flavonoid and Asp119 supports the role of His20. Mutagenesis of H20A and/or D119A abolishes detectable glycosyltransferase activity (Offen *et al.*, 2006). Asp374 and Gln375 interact with the O3/O4 and O2/O3 hydroxyl groups of UDP-2-deoxy-2-fluoro glucose, respectively. Mutation of D374 causes loss of the activity, while Q375N and Q375H mutations result in a drastic decrease in catalytic activity. His26 and Asp124, corresponding to His20 and Asp119 in VvGT1, are conserved in UGT78G1 (Modolo *et al.*, 2009b). Substitution of these residues to Ala also leads to the complete loss of catalytic activity. According to an analysis using E192A and D376A mutants, Glu192 and Asp376 play an important role in deglycosylation activity. His26 is required for both glycosylation and deglycosylation reactions. There is a V200E mutation in UGT85H2 that confers deglycosylation activity for kaempferol 3-*O*-glucoside and isoflavonoid biochanin A 7-*O*-glucoside (Modolo *et al.*, 2009a). The detailed mechanism of plant secondary-product

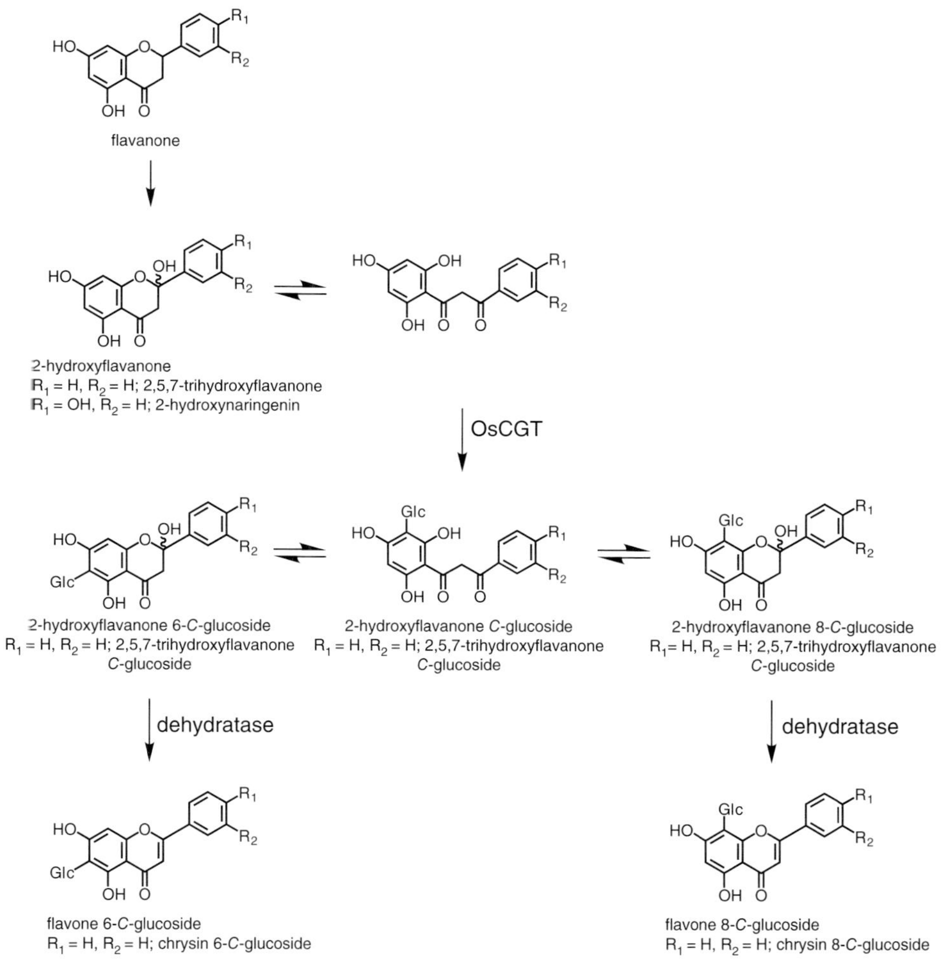


**Fig. 3.3** Proposed *C*-glycosylflavone biosynthetic pathway in rice. Flavanone is converted to 2-hydroxyflavanone, which exists in equilibrium with its open-chain form, by flavanone 2-hydroxylase. 2-hydroxyflavanone in the open-chain form is *C*-glucosylated by OsCGT. 2-hydroxyflavanone 6-*C*-glucoside and 2-hydroxyflavanone 8-*C*-glucoside are dehydrated to the flavone 6-*C*-glucoside and flavone 8-*C*-glucoside, respectively. (Adapted from Brazier-Hicks *et al.*, 2009.)

glycosyltransferases based on crystal structures and modeling is well summarized by Osmani *et al.* (2009) and Wang (2009).

Structural models of flavonoid glycosyltransferases (VvGT5 and UGT88D7) and triterpenoid glycosyltransferases (Sg-1s) based on the crystal structure of VvGT1 have been used to determine the key residues for substrate specificity (Noguchi *et al.*, 2009; Ono *et al.*, 2010; Sayama *et al.*, 2012). At present, crystal-structure information about the flavonoid 3-*O*-glycosyltransferases is limited. Comparative structural and genetic analyses for flavonoid glycosyltransferases that recognize other flavonoid positions may provide useful tools for understanding the reaction mechanisms of these enzymes, including substrate specificity. Taken together, this information may eventually be used to engineer

**Table 3.2** Known crystal structures of UDP-dependent glycosyltransferases.

| Plant UGT | Function | PDB# | In complex with | Reference |
|---|---|---|---|---|
| MtUGT71G1 | Saponin biosynthesis | 2ACV (molecules A and B) | UDP | Shao *et al.* (2005) |
| | | 2ACW (molecules A and B) | UDP-Glc | |
| AtUGT72B1 | *O*-GT/*N*-GT | 2VCE | UDP-Glc | Brazier-Hicks *et al.* (2007) |
| | | 2VCH | UDP-Tris | |
| | | 2VG8 | UDP-2-fluoro-Glc, TCP | |
| VvGT1(UGT78A5) | Anthocyanin 3GlcT | 2C1X | UDP | Offen *et al.* (2006) |
| | | 2C1Z | UDP-2-fluoro-Glc, Kae | |
| | | 2C9Z | UDP | |
| MtUGT78G1 | Flavonoid 3GlcT | 3HBJ | UDP | Modolo *et al.* (2009b) |
| | | 3HBF | UDP, myricetin | |
| MtUGT85H2 | Flavonoid 3GlcT activity | 2PQ6 | None | Li *et al.* (2007) |

UDP-Glc, UDP-glucose; TCP, 2,4,5-trichlorophenol; Kae, Kaempferol.

glycosyltransferases that alter flower color, affect disease resistance, or change other plant agronomic characteristics.

### 3.2.3 *Functional evolution in UGTs*

UGTs encoded by a multigene family utilize a wide variety of small molecules as sugar acceptors. Phylogenetic analysis of UGTs from *Arabidopsis*, rice, and flax (*Linum usitatissimum*) revealed the overall organization and evolutionary relationships of UGTs in each plant lineage (Li *et al.*, 2001; Cao *et al.*, 2008; Barvkar *et al.*, 2012). In order to understand how UGTs have expanded in number and function over the course of plant evolution, genome-wide comparative analyses of UGT sequences have been examined (Nishimura *et al.*, 2010; Yonekura-Sakakibara & Hanada, 2011; Caputi *et al.*, 2012). The conserved UGT related domain UDPGT (PROSITE accession number PS00375) and a corresponding sequence called PSPG (plant secondary product glycosyltransferase) motif were used to isolate putative UGTs from the genomic sequences of *Chlamydomonas reinhardtii*, *Physcomitrella patens*, *Selaginella moellendorffii*, and a number of angiosperms. *C. reinhardtii* contains one UGT. *P. patens* has at least 12, *S. moellendorffii* at least 74, and most of the angiosperms contain more than 100, suggesting that the UGT family began to expand early in the land-plant lineage. The ratio of UGT genes to total genes in vascular plants (0.18–0.72%) is higher than those in *C. reinhardtii* (0.01/0.006%), *P. patens* (0.06/0.03%), and humans (0.07%). The number of UGT genes per genome size in vascular plants is also significantly higher than in *P. patens* (Caputi *et al.*, 2012), suggesting that plant UGT genes likely expanded rapidly after divergence from *P. patens* and continue to expand independently at various rates among vascular plant lineages.

It is generally thought that gene-duplication events increase the number of genes over evolutionary time, but that most of them are quickly lost (Lynch & Conery, 2000; Lockton & Gaut, 2005; Hanada *et al.*, 2008). A phylogenetic tree for putative UGTs from six plants (*P. patens*, *S. moellendorffii*, *O. sativa*, *Populus trichocarpa*, *A. thaliana*, and *A. lyrata*) was constructed and the gain and loss events in each plant lineage were identified by reconciling the phylogenetic and species trees (Yonekura-Sakakibara & Hanada, 2011) (Fig. 3.4). Most branches, except for the *P. patens*, *A. thaliana*, and *A. lyrata* lineages, had high gain/loss ratios, indicating that the number of common ancestral UGT genes increased as the vascular plants diverged, resulting in a remarkable lineage-specific expansion in vascular plants.

UGT genes from these 6 plants were classified into 24 orthologous groups using phylogenetic analysis. An orthologous group is defined as a group containing genes that had diverged from a common ancestor (Yonekura-Sakakibara & Hanada, 2011). The distribution of UGT genes into orthologous groups varies in terms of the number of UGTs, UGT family composition, and plant species. Orthologous groups include UGTs from all six plant species, the vascular plants only, angiosperms only, single-plant species, or specified plant species. Some groups contain multiple UGT families and others contain a single UGT family. The distribution of UGTs with known functions among orthologous groups suggests that the diversity of UGT functions was acquired after gene expansion in vascular-plant lineages. UGT families from 11 plant species (*P. patens*, *S. moellendorffii*, *Sorghum bicolor*, *O. sativa*, *Mimulus guttatus*, *Cucumis sativus*, *Glycine max*, *P. trichocarpa*, *V. vinifera*, *Malus domestica*, and *A. thaliana*) also clustered into 18 phylogenetic groups, A–P, plus two groups containing UGTs unique to *P. patens* and *S. moellendorffii* (Caputi *et al.*, 2012).

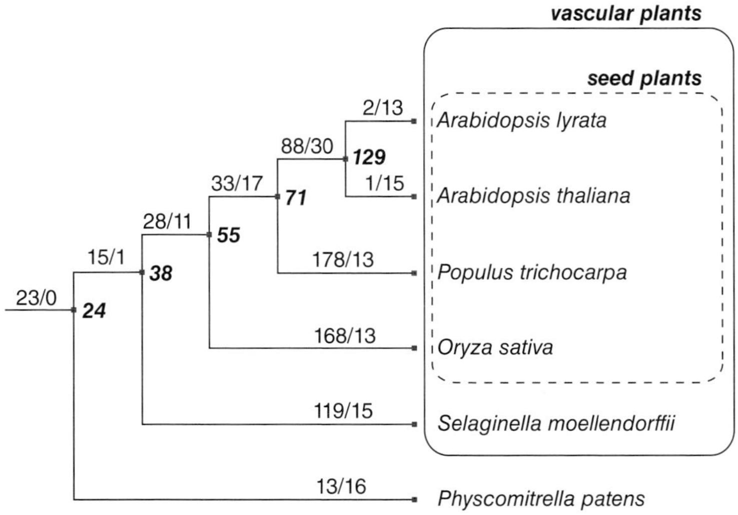


**Fig. 3.4** Expansion of UGT genes in six land plants (*P. patens*, *S. moellendorffii*, *O. sativa*, *P. trichocarpa*, *A. thaliana*, and *A. lyrata*). The numbers of common ancestral genes are shown in *italics*. The proportions of gain (numerator) versus loss (denominator) events are shown at the nodes of each branch. (Adapted from Yonekura-Sakakibara and Hanada, 2011. Reproduced with permission of John Wiley & Sons.)

A majority of UGTs used in these analyses were annotated based on sequence similarity and their functions have not been identified. Even in *A. thaliana*, the functions of only ~25% of UGT genes have been established by experimental evidence (Yonekura-Sakakibara, 2009). The progress of UGT functional identification may provide useful information about the evolutionary relationships between structure and function.

### *3.2.3.1 Functional evolution in flavonoid UGTs*

Flavonoid UGTs are characteristically distributed across 24 orthologous groups determined by genome-wide phylogenetic analyses of plant UGTs (Fig. 3.2) (Yonekura-Sakakibara & Hanada, 2011).

Flavonoid 3-*O*-glycosyltransferases belong to orthologous group OG23, containing a single UGT family, UGT78. Other orthologous groups are also made up solely by one UGT family. *UGT80B1*, the sole UGT gene of the UGT80B family in Arabidopsis, encodes a sterol glucosyltransferase that catalyzes the biosynthesis of steryl glycosides, a group of membrane-bound lipids. Mutation of *UGT80B1* alters embryonic development and reduces suberization and cutin formation in the seed coat (DeBolt *et al.*, 2009). Steryl glycosides are widely distributed in plants including moss and among bacteria, fungi, and animals (DeBolt *et al.*, 2009) but UGT80Bs are found only in seed plants. *UGT80A*s also encode sterol glycosyltransferases and are found in all land plants tested, suggesting that UGT80As play a common role as sterol glycosyltransferases. Glycosylation at the *C*-3 position of flavonoids is the first modification step and is essential for stable accumulation of flavonoids. No orthologous UGTs from *P. patens* are found in OG23, suggesting that flavonoid accumulation was crucial for evolution of the vascular plants. Thus, orthologous groups containing only a single UGT family tend to play an important role in plants, suggesting that such UGT families may have not expanded because of evolutionary constraints.

In contrast, flavonoid 5-*O*-glycosyltransferases belong to OG14, a group that contains multiple UGT families, including UGT74, UGT75, and UGT84 (Yonekura-Sakakibara & Hanada, 2011). Glycosyltransferases for auxin, anthranilate, phenylpropanoids, and glucosinolates are also classified into OG14. Flavonoid 7-*O*-glycosyltransferases belong to OG1, which contains multiple UGT families, including UGT73, UGT89, and UGT90. Flavonoid 7-*O*-glycosyltransferases specific to the Lamiales belong to the UGT88 family in OG7. The flavonoid 3′-*O*-glycosyltransferases proanthocyanindin 3′-*O*-glucosyltransferase (MtUGT72L1) and anthocyanin 3′-*O*-glucosyltransferase (Gt3′GlcT) also belong within OG7 and OG1, respectively. Given that genes in an orthologous group diverged from a common ancestor and the number of UGT genes was significantly expanded in a plant lineage-specific manner in vascular plant lineages, UGTs in the orthologous groups containing multiple UGT families may discriminate compounds as substrates during evolution in each plant lineage. The distribution of 3′-*O*-glucosyltransferases and 7-*O*-glycosyltransferases into two orthologous groups suggests that the functions of 3′-*O*-glucosyltransferases and 7-*O*-glycosyltransferases had not been established before the divergence of OG7 and OG1 and may have increased in a plant lineage-specific manner after the divergence.

Flavonoid GGTs other than UGT707B1 belong to the UGT79 or UGT94 families in OG8 (Yonekura-Sakakibara & Hanada, 2011). UGT707B1 presumably belongs to OG7, based on the phylogenetic tree (Trapero *et al.*, 2012). Recently, several GGTs for plant natural products – triterpenoids and lignans – have been identified. *UGT94D1* in OG8 encodes lignan GGT (Noguchi *et al.*, 2008). GmSGT2 (UGT73P2) and GmSGT3 (UGT91H4) catalyze the second and third glycosylations of the glucuronosyl moiety at position *C*-3 of soyasapogenol B, a triterpene saponin (Shibuya *et al.*, 2010). Sg-1a (UGT73F4) and Sg-1b (UGT73F2) glycosylate an arabinosyl moiety at the *C*-22 position of soyasapogenol A (Sayama *et al.*, 2012). These triterpenoid GGTs fall within OG1 (UGT73, UGT89, UGT90) or OG8 (UGT79, UGT91, UGT92, UGT94). The convergence of plant natural-product GGTs into the UGT79, UGT91, and UGT94 families of OG8 suggests that the sugar moiety glycosylation of natural product glycosides may have been established before the divergence of UGTs in OG8 and that GGTs obtained the ability to distinguish among the natural product glycosides as substrates during divergence within OG8. In the case of GGTs in OG1 and OG7, there are two possibilities. First, UGTs in these orthologous groups have binding pockets large enough to utilize the glycosides as sugar acceptors, and thus acquire the GGT function during evolution. Flavonoid 7-*O*-glycosytransferases which glycosylate flavonoid 3-*O*-glycosides, the same substrates as GGTs, also belong to the OG1 group, supporting this hypothesis. The second possibility is that GGTs in OG1 and OG7 are restricted to UGT73 families from *G. max* and UGT707B1 from *Crocus sativus*, respectively. These genes may have arisen by convergent evolution. Identification of additional GGTs for plant natural products may be needed to shed light on the mechanism, or mechanisms, of GGT function acquisition.

## 3.3 Glycoside hydrolase-type glycosyltransferases

### *3.3.1 Functional identification of flavonoid GH1-type glycosyltransferases*

It had until recently been assumed that glycosylation was catalyzed solely by UGTs. Novel anthocyanin glycosyltransferases that utilize acyl-glucose as a sugar donor have now been identified from *Dianthus caryophyllus* (carnation), *Delphinium grandiflorum* (delphinium), *Agapanthus africanus*, and *A. thaliana* (Matsuba *et al.*, 2010; Miyahara *et al.*, 2012, 2013) (Table 3.1). These novel glycosyltransferases have high sequence similarity to members of GH1, which typically act as $\beta$-glucosidases at the amino acid level. These GH1-type glycosyltransferases are thought to be localized in vacuoles and contribute to the formation of stored anthocyanins with serine carboxypeptidase-like (SCPL) type acyltransferases.

#### *3.3.1.1 Anthocyanin 5/7-O-glycosyltransferases*

The major anthocyanins in the pink and red petals of carnations are 3,5-di-*O*-(glucopyranosyl) pelargonidin 6″-*O*-4,6‴-*O*-1-cyclic malate (Pg3,5cMdG) and pelargonidin 3-*O*-malylglucoside, respectively (Nakayama *et al.*, 2000). A cDNA clone that encodes anthocyanin 3-*O*-glucosyltransferase has been isolated from carnation (Ogata *et al.*, 2004), and 1-*O*-malylglucose-dependent anthocyanin 3-*O*-glucoside: 6″-*O*-malyltransferase activity has been reported in this flower (Abe *et al.*, 2008). However, no anthocyanin 5-*O*-glucosyltransferase (AA5GT) activity has been detected with UDP-glucose as the sugar donor (Matsuba *et al.*, 2010). Anthocyanin 3-*O*-glucoside: 6″-*O*-malyltransferase much prefers anthocyanidin 3-*O*-glucoside over anthocyanidin 3,5-di-*O*-glucoside as a sugar acceptor. Anthocyanin 3-*O*-glucoside: 6″-*O*-malyltransferase is a member of the SCPL family of plant acyltransferases, suggesting that anthocyanin 3-*O*-glucoside: 6″-*O*-malyltransferase is localized in vacuoles. These results indicate that pelargonidin 3-*O*-glucoside is modified to Pg3,5cMdG via pelargonidin 3-*O*-malylglucoside and pelargonidin 3-*O*-malylglucoside-5-*O*-glucoside, rather than pelargonidin 3,5-di-*O*-glucoside, and that AA5GT is a novel type of vacuolar glycosyltransferase. In fact, AA5GT activity was detected using the low-molecular-weight acyl-glucose compound 1-*O*-*β*-D-vanillyl glucose (VG), which was isolated from carnation petals (Matsuba *et al.*, 2010) (Fig. 3.5). A full-length cDNA (*DcAA5GT*) was isolated based on the amino acid sequence of purified AA5GT enzyme. The amino acid sequence deduced from *DcAA5GT* had high sequence similarity with GH1. Crude extracts of *E. coli* expressing DcAA5GT that lacked a putative signal peptide with VG converted cyanidin 3-*O*-glucoside to cyanidin 3,5-di-*O*-glucoside. The expression pattern of *DcAA5GT* transcripts during carnation floral development is consistent with the accumulation pattern of Pg3,5cMdG. Fluorescence attributed to the accumulation of anthocyanin 5-*O*-glucoside was induced by transient expression of *DcAA5GT* under the control of the cauliflower mosaic virus 35S promoter in carnation petals. These results indicate that *DcAA5GT* encodes an acyl-glucose-dependent anthocyanin 5-*O*-glucosyltransferase. cDNAs encoding the acyl-glucose-dependent anthocyanin 7-*O*-glucosyltransferases *DgAA7GT* and *AaAA7GT* were also identified from delphinium and *A. africanus*, respectively, in a similar way (Matsuba *et al.*, 2010; Miyahara *et al.*, 2012).

### *3.3.1.2 Anthocyanin 3-*O*-6″-*O*-coumaroylglucoside: glucosyltransferase*

In *A. thaliana*, glucosylation of the *p*-coumaroyl moiety and sinapoylation of anthocyanin A11, cyanidin 3-*O*-[2″-*O*-(2‴-*O*-(sinapoy)-xylosyl)-6″-*O*-(4-*O*-(glucosyl)-*p*-coumaroyl) -glucoside] -5-*O*-(6⁗-*O*-malonyl)-glucoside, are considered to be the last two steps of anthocyanin modification (Fraser *et al.*, 2007; Yonekura-Sakakibara *et al.*, 2012; Saito *et al.*, 2013). Sinapoylation is catalyzed by a vacuolar SCPL-type acyltransferase, suggesting that GH1-type glycosyltransferases may glucosylate the *p*-coumaroyl moiety of anthocyanin A11. Based on sequence similarity, 11 *β*-glucosidase genes (*AtBGLU1–11*) were recognized as candidate orthologs in *A. thaliana* (Miyahara *et al.*, 2011). Anthocyanin analyses of an *atbglu10* knock-out mutant and the mutant that is complemented by *AtBGLU10* cDNA showed that ATBGLU10 encodes anthocyanin 3-*O*-6″-*O*-coumaroylglucoside: glucosyltransferase, and Arabidopsis may thus have functionally redundant genes *in vivo* (Miyahara *et al.*, 2013). These results suggest that glycosylation by a GH1-type glycosyltransferase is a common reaction mechanism among monocots and dicots and may be conserved among other plant divisions.

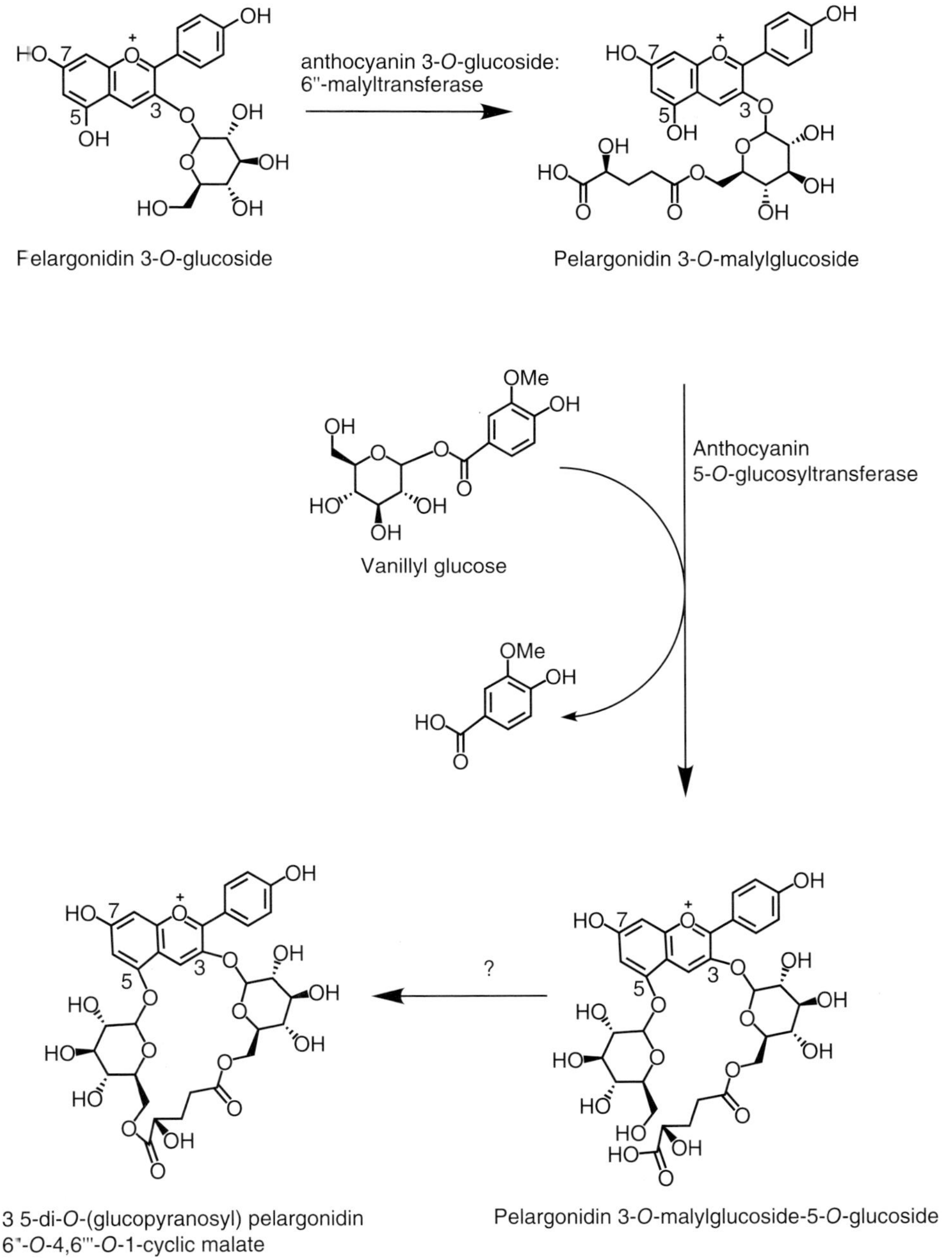


**Fig. 3.5** Proposed anthocyanin modification pathway in carnations. Pelargonidin 3-*O*-glucoside is converted to pelargonidin 3-*O*-malylglucoside by a vacuolar anthocyanin malyltransferase. GH1-type anthocyanin 5-*O*-glucosyltransferases catalyze the transfer of glucose to pelargonidin 3-*O*-malylglucoside. The cyclization mechanism of pelargonidin 3-*O*-malylglucoside-5-*O*-glucoside remains unknown.

### 3.3.2 The reaction mechanism of GH1-type glycosyltransferases

In general, GH1 enzymes function as $\beta$-glucosidases that hydrolyze glycosidic bonds (Marana, 2006). A number of flavonoid glycosides, such as isoflavonoid 7-*O*-$\beta$-apiosyl-glucoside and 8-*O*-$\beta$-glucoside, have been reported as substrates of plant $\beta$-glucosidases (Svasti *et al.*, 1999; Chuankhayan *et al.*, 2005). Within the GH1 conserved motifs of DcAA5GT and DgAA7GT, two glutamic acid residues required for hydrolysis of a $\beta$-glycosidic bond are also conserved. In fact, DcAA5GT and DgAA7GT have glycosidase activities, but their efficiency is lower than that of those with glycosyltransferase activity (Matsuba *et al.*, 2010). So far, the reaction mechanism of GH1-type glycosyltransferases remains unknown.

## 3.4 Conclusions

Flavonoids are among the most well studied plant secondary metabolites in terms of chemistry, functional identification of biosynthetic genes, and gene regulation. However, our understanding of their metabolism remains incomplete. A quantitative trait locus (QTL) analysis using two recombinant inbred line (RIL) sets detected 22 flavonoid QTLs in *Arabidopsis* (Routaboul *et al.*, 2012). Of the 16 QTLs confirmed using heterogeneous inbred families, three corresponded to known flavonoid genes, but the others did not colocalize with any known flavonoid-related genes. In maize, *Pericarp Color1* (*P1*) encoding an R2R3-type MYB transcription factor is a positive regulator involved in the biosynthesis of a wide range of phenolic compounds, including anthocyanins, flavones, and 3-deoxyflavonoids. A combination of RNA-Seq and chromatin immunoprecipitation methods coupled with high-throughput sequencing using maize with a contrasting *P1* genotype may indicate which open reading frames (ORFs) encode missing genes in the metabolic pathways, as well as the additional role of P1 as a negative regulator (Morohashi *et al.*, 2012). Thus, the flavonoids remain ripe for exploration. Completion of the flavonoid metabolic map, including modifications, gene regulation, crosstalk with other metabolic pathways, and signaling pathways from environmental stimuli, will require a comprehensive approach based on all the 'omics' approaches coupled with bioinformatics techniques.

## References

Abe, Y., Tera, M., Sasaki, N., *et al.* (2008) Detection of 1-*O*-malylglucose: pelargonidin 3-*O*-glucose-6″-*O*-malyltransferase activity in carnation (*Dianthus caryophyllus*). *Biochemical and Biophysical Research Communications*, 373, 473–477.

Anderson, Ø.M. & Markham, K.R. (2006) *Flavonoids: Chemistry, Biochemistry, and Applications*. CRC Press, Boca Raton, FL.

Barvkar, V.T., Pardeshi, V.C., Kale, S.M, *et al.* (2012) Phylogenomic analysis of UDP glycosyltransferase 1 multigene family in *Linum usitatissimum* identified genes with varied expression patterns. *BMC Genomics*, 13, 175.

Bowles, D., Lim, E.K., Poppenberger, B. & Vaistij, F.E. (2006) Glycosyltransferases of lipophilic small molecules. *Annual Review of Plant Biology*, 57, 567–597.

Brazier-Hicks, M., Offen, W.A., Gershater, M.C., *et al.* (2007) Characterization and engineering of the bifunctional *N*- and *O*-glucosyltransferase involved in xenobiotic metabolism in plants. *Proceedings of the National Academy of Sciences USA*, 104, 20238–20243.

Brazier-Hicks, M., Evans, K.M., Gershater, M.C., *et al.* (2009) The C-glycosylation of flavonoids in cereals. *Journal of Biological Chemistry*, 284, 17926–17934.

Buer, C.S., Imin, N. & Djordjevic, M.A. (2010) Flavonoids: new roles for old molecules. *Journal of Integrative Plant Biology*, 52, 98–111.

Cao, P.J., Bartley, L.E., Jung, K.H. & Ronald, P.C. (2008) Construction of a rice glycosyltransferase phylogenomic database and identification of rice-diverged glycosyltransferases. *Molecular Plant*, 1, 858–877.

Caputi, L., Malnoy, M., Goremykin, V., *et al.* (2012) A genome-wide phylogenetic reconstruction of family 1 UDP-glycosyltransferases revealed the expansion of the family during the adaptation of plants to life on land. *The Plant Journal*, 69, 1030–1042.

Chuankhayan, P., Hua, Y., Svasti, J., *et al.* (2005) Purification of an isoflavonoid 7-O-beta-apiosyl-glucoside beta-glycosidase and its substrates from Dalbergia nigrescens Kurz. *Phytochemistry*, 66, 1880–1889.

DeBolt, S., Scheible, W.R., Schrick, K., *et al.* (2009) Mutations in UDP-glucose:sterol glucosyltransferase in Arabidopsis cause transparent testa phenotype and suberization defect in seeds. *Plant Physiology*, 151, 78–87.

Felsenstein, J. (1985) Confidence limits on phylogenies: an approach using the bootstrap. *Evolution*, 39, 783–791.

Ford, C.M., Boss, P.K. & Hoj, P.B. (1998) Cloning and characterization of *Vitis vinifera* UDP-glucose:flavonoid 3-O-glucosyltransferase, a homologue of the enzyme encoded by the maize *Bronze*-1 locus that may primarily serve to glucosylate anthocyanidins in vivo. *The Journal of Biological Chemistry*, 273, 9224–9233.

Fraser, C.M., Thompson, M.G., Shirley, A.M., *et al.* (2007) Related Arabidopsis serine carboxypeptidase-like sinapoylglucose acyltransferases display distinct but overlapping substrate specificities. *Plant Physiology*, 144, 1986–1999.

Frydman, A., Weisshaus, O., Bar-Peled, M., *et al.* (2004) Citrus fruit bitter flavors: isolation and functional characterization of the gene Cm1,2RhaT encoding a 1,2 rhamnosyltransferase, a key enzyme in the biosynthesis of the bitter flavonoids of citrus. *The Plant Journal*, 40, 88–100.

Frydman, A., Liberman, R., Huhman, D.V., *et al.* (2012) The molecular and enzymatic basis of bitter/non-bitter flavor of citrus fruit: evolution of branch-forming rhamnosyltransferases under domestication. *The Plant Journal*, doi: 10.1111/tpj.12030.

Gachon, C.M., Langlois-Meurinne, M. & Saindrenan, P. (2005) Plant secondary metabolism glycosyltransferases: the emerging functional analysis. *Trends in Plant Science*, 10, 542–549.

Hanada, K., Zou, C., Lehti-Shiu, M.D., *et al.* (2008) Importance of lineage-specific expansion of plant tandem duplicates in the adaptive response to environmental stimuli. *Plant Physiology*, 148, 993–1003.

Harborne, J.B. & Williams, C.A. (2000) Advances in flavonoid research since 1992. *Phytochemistry*, 55, 481–504.

Horowitz, R.M. & Gentili, B. (1969) Taste and structure in phenolic glycosides. *Journal of Agricultural and Food Chemistry*, 17, 696–700.

Jay, M., Viricel, M.-R. & Gonnet, J.-F. (2006) *C*-glycosylflavonoids. In: *Flavonoids: Chemistry, Biochemistry, and Applications* (eds Ø.M. Anderson & K.R. Markham), pp. 857–916. CRC Press, Boca Raton, FL.

Jones, P., Messner, B., Nakajima, J., Schaffner, A.R. & Saito, K. (2003) UGT73C6 and UGT78D1, glycosyltransferases involved in flavonol glycoside biosynthesis in *Arabidopsis thaliana*. *The Journal of Biological Chemistry*, 278, 43910–43918.

Kerscher, F. & Franz, G. (1987) Biosynthesis of Vitexin and Isovitexin – enzymatic-synthesis of the C-glucosylflavones vitexin and isovitexin with an enzyme preparation from *Fagopyrum-esculentum* M seedlings. *Zeitschrift Fur Naturforschung C*, 42, 519–524.

Kerscher, F. & Franz, G. (1988) Isolation and some properties of an UDP-glucose – 2-hydroxy-flavanone-6(or 8)-C-glucosyltransferase from *Fagopyrum-esculentum* M cotyledons. *Journal of Plant Physiology*, 132, 110–115.

Kong, W., Vanderburg, C.R., Gunshin, H., *et al.* (2008) A review of independent component analysis application to microarray gene expression data. *Biotechniques*, 45, 501–520.

Li, Y., Baldauf, S., Lim, E.K. & Bowles, D.J. (2001) Phylogenetic analysis of the UDP-glycosyltransferase multigene family of *Arabidopsis thaliana*. *Journal of Biological Chemistry*, 276, 4338–4343.

Li, L., Modolo, L.V., Escamilla-Trevino, L.L., *et al.* (2007) Crystal structure of *Medicago truncatula* UGT85H2 – insights into the structural basis of a multifunctional (iso)flavonoid glycosyltransferase. *Journal of Molecular Biology*, 370, 951–963.

Lim, E.K., Li, Y., Parr, A., *et al.* (2001) Identification of glucosyltransferase genes involved in sinapate metabolism and lignin synthesis in Arabidopsis. *Journal of Biological Chemistry*, 276, 4344–4349.

Lockton, S. & Gaut, B.S. (2005) Plant conserved non-coding sequences and paralogue evolution. *Trends in Genetics*, 21, 60–65.

Lynch, M. & Conery, J.S. (2000) The evolutionary fate and consequences of duplicate genes. *Science*, 290, 1151–1155.

Marana, S.R. (2006) Molecular basis of substrate specificity in family 1 glycoside hydrolases. *IUBMB Life*, 58, 63–73.

Masada, S., Terasaka, K., Oguchi, Y., *et al.* (2009) Functional and structural characterization of a flavonoid glucoside 1,6-glucosyltransferase from *Catharanthus roseus*. *Plant and Cell Physiology*, 50, 1401–1415.

Matsuba, Y., Sasaki, N., Tera, M., *et al.* (2010) A novel glucosylation reaction on anthocyanins catalyzed by acyl-glucose-dependent glucosyltransferase in the petals of carnation and delphinium. *Plant Cell*, 22, 3374–3389.

Miyahara, T., Matsuba, Y., Ozeki, Y. & Sasaki, N. (2011) Identification of genes in *Arabidopsis thaliana* with homology to a novel acyl-glucose dependent glucosyltransferase of carnations. *Plant Biotechnology*, 28, 311–315.

Miyahara, T., Takahashi, M., Ozeki, Y. & Sasaki, N. (2012) Isolation of an acyl-glucose-dependent anthocyanin 7-O-glucosyltransferase from the monocot *Agapanthus africanus*. *Journal of Plant Physiology*, 169, 1321–1326.

Miyahara, T., Sakiyama, R., Ozeki, Y. & Sasaki, N. (2013) Acyl-glucose-dependent glucosyltransferase catalyzes the final step of anthocyanin formation in Arabidopsis. *Journal of Plant Physiology*, 170, 619–624.

Modolo, L.V., Blount, J.W., Achnine, L., *et al.* (2007) A functional genomics approach to (iso)flavonoid glycosylation in the model legume *Medicago truncatula*. *Plant Molecular Biology*, 64, 499–518.

Modolo, L.V., Escamilla-Trevino, L.L., Dixon, R.A. & Wang, X. (2009a) Single amino acid mutations of Medicago glycosyltransferase UGT85H2 enhance activity and impart reversibility. *FEBS Letters*, 583, 2131–2135.

Modolo, L.V., Li, L., Pan, H., *et al.* (2009b) Crystal structures of glycosyltransferase UGT78G1 reveal the molecular basis for glycosylation and deglycosylation of (iso)flavonoids. *Journal of Molecular Biology*, 392, 1292–1302.

Montefiori, M., Espley, R.V., Stevenson, D., *et al.* (2011) Identification and characterisation of F3GT1 and F3GGT1, two glycosyltransferases responsible for anthocyanin biosynthesis in red-fleshed kiwifruit (*Actinidia chinensis*). *The Plant Journal*, 65, 106–118.

Morohashi, K., Casas, M.I., Ferreyra, L.F., *et al.* (2012) A genome-wide regulatory framework identifies maize pericarp color1 controlled genes. *Plant Cell*, 24, 2745–2764.

Nakayama, M., Koshioka, M., Yoshida, H., *et al.* (2000) Cyclic malyl anthocyanins in *Dianthus caryophyllus*. *Phytochemistry*, 55, 937–939.

Nesi, N., Jond, C., Debeaujon, I., *et al.* (2001) The Arabidopsis TT2 gene encodes an R2R3 MYB domain protein that acts as a key determinant for proanthocyanidin accumulation in developing seed. *Plant Cell*, 13, 2099–2114.

Nishimura, Y., Tokimatsu, T., Kotera, M., *et al.* (2010) Genome-wide analysis of plant ugt family based on sequence and substrate information. *Genome Inform*, 24, 127–138.

Noguchi, A., Fukui, Y., Iuchi-Okada, A., *et al.* (2008) Sequential glucosylation of a furofuran lignan, (+)-sesaminol, by Sesamum indicum UGT71A9 and UGT94D1 glucosyltransferases. *The Plant Journal*, 54, 415–427.

Noguchi, A., Horikawa, M., Fukui, Y., *et al.* (2009) Local differentiation of sugar donor specificity of flavonoid glycosyltransferase in Lamiales. *Plant Cell*, 21, 1556–1572.

Obayashi, T., Hayashi, S., Saeki, M., *et al.* (2009) ATTED-II provides coexpressed gene networks for Arabidopsis. *Nucleic Acids Research*, 37, D987–991.

Offen, W., Martinez-Fleites, C., Yang, M., *et al.* (2006) Structure of a flavonoid glucosyltransferase reveals the basis for plant natural product modification. *The EMBO Journal*, 25, 1396–1405.

Ogata, J., Itoh, Y., Ishida, M., *et al.* (2004) Cloning and heterologous expression of cDNAs encoding flavonoid glucosyltransferases from *Dianthus caryophyllus*. *Plant Biotechnology*, 21, 367–375.

Ono, E., Fukuchi-Mizutani, M., Nakamura, N., *et al.* (2006) Yellow flowers generated by expression of the aurone biosynthetic pathway. *Proceedings of the National Academy of Sciences USA*, 103, 11075–11080.

Ono, E., Homma, Y., Horikawa, M., *et al.* (2010) Functional differentiation of the glycosyltransferases that contribute to the chemical diversity of bioactive flavonol glycosides in grapevines (*Vitis vinifera*). *Plant Cell*, 22, 2856–2871.

Osmani, S.A., Bak, S. & Moller, B.L. (2009). Substrate specificity of plant UDP-dependent glycosyltransferases predicted from crystal structures and homology modeling. *Phytochemistry*, 70, 325–347.

Pang, Y., Peel, G.J., Sharma, S.B., *et al.* (2008) A transcript profiling approach reveals an epicatechin-specific glucosyltransferase expressed in the seed coat of *Medicago truncatula*. *Proceedings of the National Academy of Sciences USA,* 105, 14210–14215.

Peel, G.J., Pang, Y., Modolo, L.V. & Dixon, R.A. (2009) The LAP1 MYB transcription factor orchestrates anthocyanidin biosynthesis and glycosylation in Medicago. *The Plant Journal*, 59, 136–149.

Routaboul, J.M., Dubos, C., Beck, G., *et al.* (2012) Metabolite profiling and quantitative genetics of natural variation for flavonoids in Arabidopsis. *Journal of Experimental Botany*, 63, 3749–3764.

Saito, K., Yonekura-Sakakibara, K., Nakabayashi, R., *et al.* (2013) *The flavonoid biosynthetic pathway in Arabidopsis: structural and genetic diversity. Plant Physiology and Biochemistry.*

Saitou, N. & Nei, M. (1987) The neighbor-joining method: a new method for reconstructing phylogenetic trees. *Molecular Biology and Evolution*, 4, 406–425.

Sayama, T., Ono, E., Takagi, K., *et al.* (2012) The Sg-1 glycosyltransferase locus regulates structural diversity of triterpenoid saponins of soybean. *Plant Cell*, 24, 2123–2138.

Shao, H., He, X., Achnine, L., *et al.* (2005) Crystal structures of a multifunctional triterpene/flavonoid glycosyltransferase from *Medicago truncatula*. *Plant Cell*, 17, 3141–3154.

Shibuya, M., Nishimura, K., Yasuyama, N. & Ebizuka, Y. (2010) Identification and characterization of glycosyltransferases involved in the biosynthesis of soyasaponin I in Glycine max. *FEBS Letters*, 584, 2258–2264.

Sinlapadech, T., Stout, J., Ruegger, M.O., *et al.* (2007) The hyper-fluorescent trichome phenotype of the brt1 mutant of Arabidopsis is the result of a defect in a sinapic acid: UDPG glucosyltransferase. *The Plant Journal*, 49, 655–668.

Svasti, J., Srisomsap, C., Techasakul, S. & Surarit, R. (1999) Dalcochinin-8′-*O*-$\beta$-d-glucoside and its $\beta$-glucosidase enzyme from *Dalbergia cochinchinensis*. *Phytochemistry*, 50, 739–743.

Tamura, K., Peterson, D., Peterson, N., *et al.* (2011) MEGA5: molecular evolutionary genetics analysis using maximum likelihood, evolutionary distance, and maximum parsimony methods. *Molecular Biology and Evolution*, 28, 2731–2739.

Tohge, T., Nishiyama, Y., Hirai, M.Y., *et al.* (2005) Functional genomics by integrated analysis of metabolome and transcriptome of Arabidopsis plants over-expressing an MYB transcription factor. *The Plant Journal*, 42, 218–235.

Trapero, A., Ahrazem, O., Rubio-Moraga, A., *et al.* (2012) Characterization of a glucosyltransferase enzyme involved in the formation of kaempferol and quercetin sophorosides in *Crocus sativus*. *Plant Physiology*, 159, 1335–1354.

Wang, X. (2009) Structure, mechanism and engineering of plant natural product glycosyltransferases. *FEBS Letters*, 583, 3303–3309.

Yabuya, T., Nakamura, M., Iwashina, T., *et al.* (1997) Anthocyanin-flavone copigmentation in bluish purple flowers of Japanese garden iris (*Iris ensata* Thunb.). *Euphytica*, 98, 163–167.

Yonekura-Sakakibara, K. (2009) Functional genomics of family 1 glycosyltransferases in Arabidopsis. *Plant Biotechnology*, 26, 267–274.

Yonekura-Sakakibara, K. & Hanada, K. (2011) An evolutionary view of functional diversity in family 1 glycosyltransferases. *The Plant Journal*, 66, 182–193.

Yonekura-Sakakibara, K., Tohge, T., Niida, R. & Saito, K. (2007) Identification of a flavonol 7-O-rhamnosyltransferase gene determining flavonoid pattern in *Arabidopsis* by transcriptome coexpression analysis and reverse genetics. *The Journal of Biological Chemistry*, 282, 14932–14941.

Yonekura-Sakakibara, K., Tohge, T., Matsuda, F., *et al.* (2008) Comprehensive flavonol profiling and transcriptome coexpression analysis leading to decoding gene-metabolite correlations in *Arabidopsis*. *Plant Cell*, 20, 2160–2176.

Yonekura-Sakakibara, K., Nakayama, T., Yamazaki, M. & Saito, K. (2009) Modification and Stabilization of Anthocyanins. In: *Anthocaynins: Biosynthesis, Functions, and Applications* (eds K. Gould, K. Davis & C. Winefield), pp. 169–190. Springer, New York.

Yonekura-Sakakibara, K., Fukushima, A., Nakabayashi, R., *et al.* (2012) Two glycosyltransferases involved in anthocyanin modification delineated by transcriptome independent component analysis in Arabidopsis thaliana. *The Plant Journal*, 69, 154–167.

Zuckerkandl, E. & Pauling, L. (1965) Evolutionary divergence and convergence in proteins. In: *Evolving Genes and Proteins* (eds V. Bryson & H.J. Vogel), pp. 97–166. Academic Press, New York.

Chapter 4

# The Chemistry and Chemical Ecology of Ellagitannins in Plant–Insect Interactions: From Underestimated Molecules to Bioactive Plant Constituents

Juha-Pekka Salminen
*Laboratory of Organic Chemistry and Chemical Biology, Department of Chemistry, University of Turku, Turku, Finland*

**Abstract:** This chapter reviews our current knowledge of the chemistry and chemical ecology of ellagitannins (ETs), placing particular emphasis on the role these under-appreciated polyphenols play in plant–herbivore interactions. Instead of the traditional tannin-linked activity that comes from their complex formation with proteins, it discusses tannins' less well-known pro-oxidant activity. ETs are likely the most active tannin pro-oxidants. Plant cells synthesize ETs via the shikimate pathway, utilizing the enzymatic conversion of dehydroshikimic acid into gallic acid and further to galloyl glucoses. During the biosynthetic steps from monogalloyl glucose towards pentagalloyl glucose, the pro-oxidant activity of the polyphenols decreases, but it begins to increase again as soon as pentagalloyl glucose is transformed to glucopyranose-based ETs and further to very active C-glycosidic ETs. Thus, the tannin-based pro-oxidant capacity of a plant cell is highly dependent on the total and relative concentrations of different types of products or intermediates of the hydrolyzable tannin pathway. Within plant species that accumulate different types of galloyl glucoses and ETs differing in pro-oxidant activities, this balance typically varies significantly. The situation is further complicated by the variation in ETs observed seasonally and between individual plants. In addition, ETs may also derive their antiherbivore activities via their hydrolysis products, which is largely an unexplored area of research. Historically, ETs have not been taken specifically into account as plant-defense compounds, mainly due to difficulties in their compound-specific analysis from the plant cell. Recent developments in tannin chemistry have diminished these problems and the joint effort of ecologists and

*Recent Advances in Polyphenol Research, Volume 4*, First Edition.
Edited by Annalisa Romani, Vincenzo Lattanzio, and Stéphane Quideau.

chemists can be used to verify the true ecological and evolutionary significance of ETs in plant–herbivore interactions.

**Keywords:** ellagitannins, mass spectrometry, biological activity, oxidation, biosynthesis, plant–herbivore interactions, phytochemistry, plant defense

## 4.1 Introduction

The tannins are a widespread family of phenolic secondary metabolites present in the leaves, bark, wood, and fruits of many higher plants (Haslam, 1989). On the basis of their chemical structures, the tannins of terrestrial plants can be divided into two major groups: proanthocyanidins (also known as condensed tannins, CTs) and hydrolyzable tannins (HTs). One of the earliest human uses of plant materials rich in tannins, such as oak galls, was in the conversion of animal hides into leather; archeological records show such processes in operation in Mediterranean regions around BCE 1500 (Haslam, 1992). Leather-making, also known as "tanning," is based on the unique and characteristic ability of tannins to bind and precipitate proteins and other macromolecules (Niehaus & Gross, 1997). The protein precipitation capacity (PPC) of tannins also contributes to the taste of foods and beverages by providing astringency, a sensation of dryness that is especially pronounced in some red wines, tea, and unripe fruit (Hagerman *et al.*, 1997). Tannins greatly affect the quality, color, and composition of whiskeys, brandies, and other alcoholic beverages conditioned in oak barrels (Hervé du Penhoat *et al.*, 1991; Viriot *et al.*, 1993; Vivas *et al.*, 1995; Mosedale *et al.*, 1996; Cadahía *et al.*, 2001). Through the human diet, tannins and other antioxidant polyphenols can cause many health-promoting actions; they have activity against cardiovascular and neurodegenerative diseases, cancer, and diabetes, for example (Quideau *et al.*, 2011).

Aside from their anthropocentric functions, tannins are also believed to participate in the defense of plants against herbivores by rendering their tissues unpalatable and non-nutritious, thereby making them unacceptable as food sources (Baxter *et al.*, 1997). Following the classic experiments of Paul Feeny (1968, 1970) on the assumed deleterious effects of oak-leaf tannins on the feeding of the larvae of the winter moth (*Operophtera brumata*), the view that tannins are uniquely a quantitative defense was formulated (Luck *et al.*, 1994). However, since then the ecological significance and functions of tannins have come into question, due to the publication of many contradictory results. On the far end of this spectrum, some studies have even shown that dietary tannins have positive effects on herbivores consuming them (Bernays *et al.*, 1980; Bernays & Woodhead, 1982; Appel & Schultz, 1994). These positive effects are perhaps not unexpected, considering that HTs have been shown to have antioxidant, antiviral, antibacterial, and antimicrobial activities (Burapadaja & Bunchoo, 1995; Feldman *et al.*, 1999; Kolodziej *et al.*, 1999; Niemetz & Gross, 1999; Wang *et al.*, 1999), potentially contributing to the well-being of some herbivores (see also Schultz *et al.*, 1992).

Even though individual tannins differ greatly in their biological activities (Ozawa *et al.*, 1987; Kilkowski & Gross, 1999; Barbehenn *et al.*, 2006a; Moilanen & Salminen, 2008), most previous ecological studies have concentrated on the quantification of so-called total tannins only. These methods are typically based on the precipitation of proteins by tannins present in a plant sample; that is, they measure a functional property of the plant sample, not specifically the tannin content. With these functional assays, two samples might show very different tannin contents even though their true tannin content is the same: the assay result depends on the types of tannin (active or less active protein precipitants) present. This problem is especially pronounced with samples containing mainly monomeric *C*-glycosidic ellagitannins (ETs), since it is known that the loss of conformational flexibility decreases the affinity of tannins towards proteins and that *C*-glycosidic ETs present one of the most rigid tannin subgroups (Baxter *et al.*, 1997; Kilkowski & Gross, 1999; Hofmann *et al.*, 2006; Deaville *et al.*, 2007). For this reason, samples that mainly contain gallotannins (GTs) or proanthocyanidins are expected to give a much higher yield in the protein-precipitation assays than *C*-glycosidic ETs. In addition, there have not been any such widely used methods for ETs as, for example, the HCl-butanol assay for total proanthocyanidins. The HCl-butanol assay is very specific to proanthocyanidins and unfortunately the protein-precipitation assays lack this specificity towards any given tannin group. Therefore, if ETs have been quantified with the functional assays, this has most likely happened without knowledge of the presence of ETs in the samples analyzed. It is easy to agree with Edwin Haslam (1988), who pointed out that "The chemical complexity and heterogeneity of plant tannins means that they do not lend themselves to ready quantitative assessment and this has produced a confused picture of their real significance – both evolutionary and ecological."

This chapter gives an up-to-date assessment of the potential ecological roles for ETs in plant–insect interactions. For the most part – if not almost entirely – the case will rest upon the evidence acquired by utilizing chemical methods that are specific to ETs. This is a prerequisite for the formation of any true understanding of their functions. As a downside, the evidence is based on work done by a very limited number of laboratories during the last decade. Hopefully, the potential ecological roles of ETs presented in this chapter will be verified by other groups as well. This possibility is becoming more and more probable each year, since many ecologically oriented research groups are currently equipping their laboratories with modern instruments that enable compound-specific ET measurements.

## 4.2 Definitions and chemical structures of hydrolyzable tannins

According to a biochemical definition, tannins (including both CTs and HTs) are compounds that have a tendency to interact with aqueous solutions of proteins and other biological macromolecules to form insoluble precipitates (Bate-Smith, 1962; Gross, 1999a,b). The simplest chemical definition of HTs, which are esters of polyols such as glucose and phenolic acids derived from gallic acid, classifies them on the basis of their hydrolysis products: those yielding gallic acid are gallotannins (GTs) and those yielding ellagic acid are ellagitannins (ETs) (Fig. 4.1).

Unfortunately, both of these definitions are far too simple to describe this large group of several hundreds of natural products; some HTs are not able to precipitate proteins

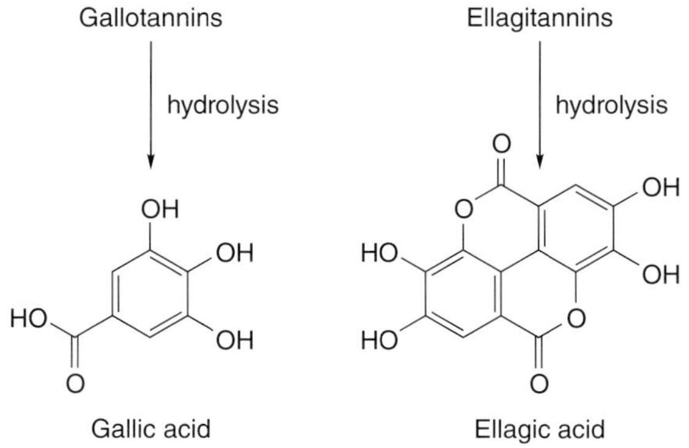


**Fig. 4.1** Simple classification of hydrolyzable tannins on the basis of their hydrolysis products as either gallotannins (GTs) or ellagitannins (ETs).

and some individual HTs may yield both gallic and ellagic acid or else more complicated degradation products upon hydrolysis. A more in-depth examination of their chemical features provides an inclusive definition of these diverse groups of compounds. All HTs are characterized by a central polyol moiety that is most often $\beta$-D-glucose, although HTs carrying hamamelose, shikimic acid, quinic acid, and cyclitol cores have also been found. The hydroxyl groups of the glucose core are typically esterified with gallic acid, the simplest galloylglucose (GG) thus being 1-*O*-galloyl-$\beta$-D-glucose (Fig. 4.2a). When the glucose core carries one to five galloyl groups, the compounds are defined as simple GGs. As the degree of galloylation exceeds five, as observed in 1,2,3,4,6-penta-*O*-galloyl-$\beta$-D-glucose (PGG, shown in Fig. 4.2b), the compounds are defined as GTs. For example, 3,4-bis-*O*-digalloyl-1,2,6-tri-*O*-galloyl-$\beta$-D-glucose (Fig. 4.2c), with seven galloyl groups, is characterized by the attachment of additional galloyl residues to the PGG core via so-called *meta*-depside bonds, thus forming digalloyl groups as shown in Fig. 4.2c. However, the commonly used term "*meta*-depside" may be a slight oversimplification, as NMR spectroscopy studies have indicated the existence of both *meta*- and *para*-depsides in some GTs (Gross, 1992b).

ETs differ from GGs and GTs in having hexahydroxydiphenoyl (HHDP) units in their structures, as in tellimagrandin II (Fig. 4.3a). ETs may also have acyclic glucose cores, for example in vescalagin (Fig. 4.3b), that are further characterized by the presence of a nonahydroxytriphenoyl (NHTP) group in addition to the HHDP group. The glucose core may also be inverted ($^1C_4$ instead of $^4C_1$), as in geraniin (Fig. 4.3c) and chebulagic acid (Fig. 4.3d). Both of these ETs carry modified HHDP groups in their structures, geraniin a dehydrohexahydroxydiphenoyl (DHHDP) and chebulagic acid an additionally transformed DHHDP group (Okuda *et al.*, 2000). While GG and GT dimers are scarce, monomeric ETs may couple intermolecularly, yielding ET dimers and subsequently oligomers with molecular weights of up to 8624 Da (Clifford & Scalbert, 2000; Karonen *et al.*, 2010; Salminen *et al.*, 2011). Due to the oligomerization, modifications in or even hydrolysis of the HHDP units, and opening of the glucose core, many ETs may only distantly resemble their biosynthetic precursor PGG. This inevitably results in large differences between the biological activities

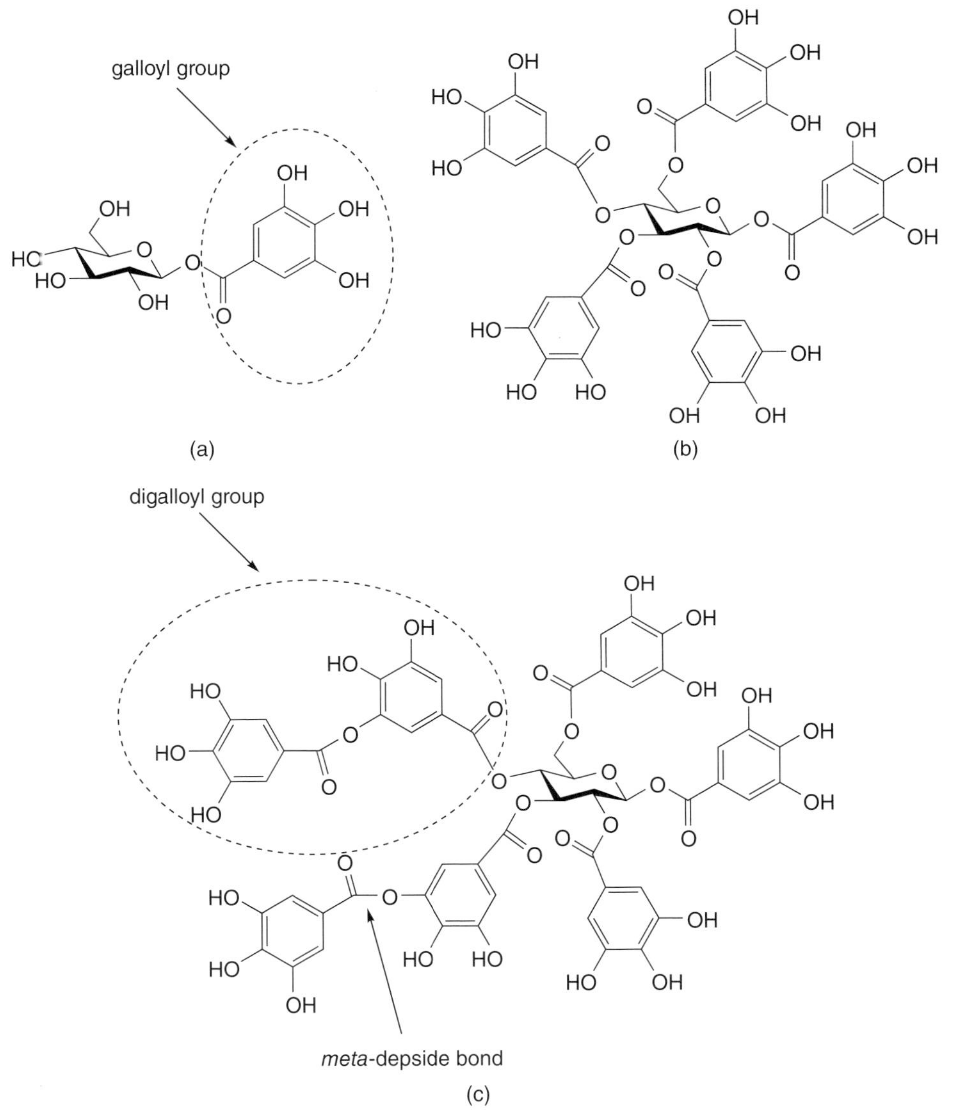


**Fig. 4.2** Chemical structures of the simple GGs (a) 1-*O*-galloyl-*β*-D-glucose and (b) 1,2,3,4,6-penta-*O*-galloyl-*β*-D-glucose, and of (c) gallotannin 3,4-bis-*O*-digalloyl-1,2,6-tri-*O*-galloyl-*β*-D-glucose.

of PGG and individual ETs (Kilkowski & Gross, 1999; Barbehenn *et al.*, 2006a; Moilanen & Salminen, 2008). Despite this, PGG has been widely used as a model HT for *in vitro* work.

## 4.3 Biosynthetic pathways of hydrolyzable tannins in plants

As has been shown, HTs are structurally diverse compounds. This fact, together with the unequal distribution of individual HTs in plants (Haddock *et al.*, 1982a,b), suggests that their biosynthesis follows a complex multistep pathway involving species-specific enzymes that determine the direction of the reaction steps (Hatano *et al.*, 1992; Gross, 1999a,b). A closer look at the biosynthetic pathway of HTs divides it into three steps: (1) the origin of gallic acid as the starting substrate of HT biosynthesis, (2) the formation of GGs, and (3) the formation of either (3a) GTs or (3b) monomeric ETs, and further their dimerization and oligomerization.

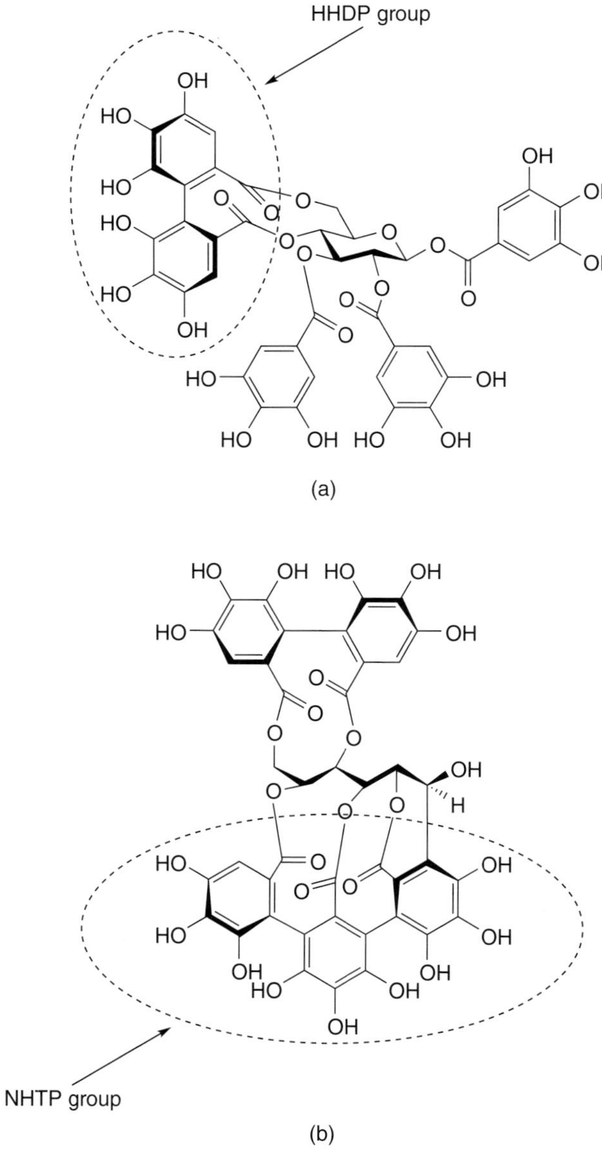


**Fig. 4.3** Chemical structures of the ETs (a) tellimagrandin II, (b) vescalagin, (c) geraniin, and (d) chebulagic acid.

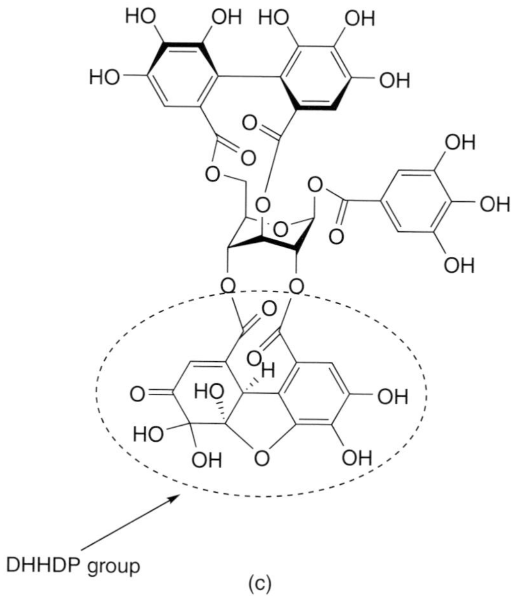


(c)

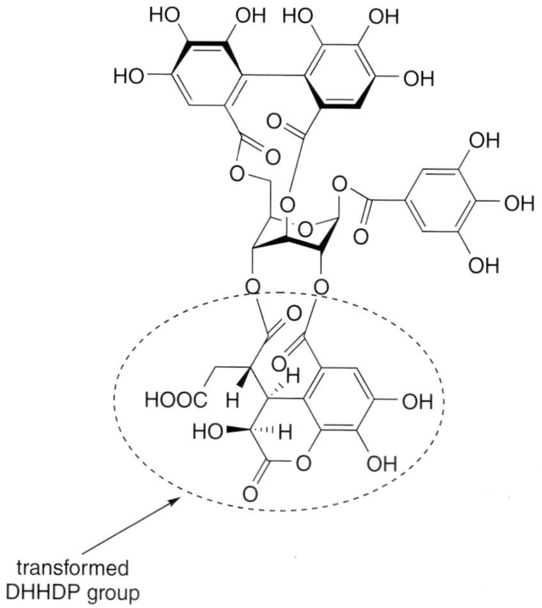


(d)

**Fig. 4.3** *Continued*

### 4.3.1 Tannin biosynthetic pathways have many branching points that affect the flux of biosynthetic energy towards different tannins

Although most of the details of the HT biosynthetic pathway are still unknown to us, the biosynthetic branches prior to HT formation are well known. Fig. 4.4 summarizes the biosynthetic scheme, showing the branch points in the biosynthesis of different types of polyphenol, including tannins (Salminen & Karonen, 2011). This shows that HTs and CTs utilize some of the same early building blocks: (i) phosphoenolpyruvate from glycolysis and (ii) 3-dehydroshikimic acid from the shikimate pathway. If high levels of phosphoenolpyruvate are directed towards the acetate/malonate pathway, HT synthesis may suffer due to a partial lack of the 3-dehydroshikimic needed for the production of gallic acid. On the other hand, a single flavonoid or CT monomer will need one *p*-coumaroyl-CoA together with three malonyl-CoA for its biosynthesis. Therefore, about 75% of the biosynthetic CHO required for flavonoid biosynthesis comes from the acetate/malonate pathway and 25% from the shikimate pathway. This is in contrast with HTs, which derive almost 100% of their phenolic CHO from the shikimate pathway, with the carbohydrate skeleton derived from central carbohydrate metabolism. A further look at the HT pathway reveals that it branches after pentagalloyl glucose to produce either GTs or ETs. Already this knowledge is useful in providing insight into the differences in tannin content among species and their tissues, and the ecological and evolutionary consequences of these. For example, a

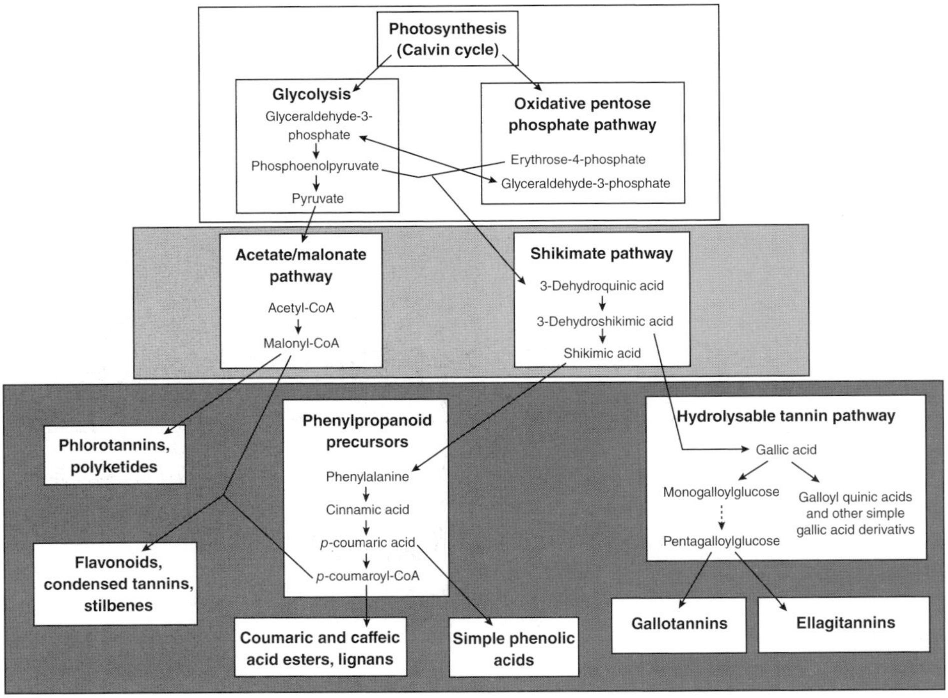


**Fig. 4.4** Biosynthetic origins of different types of polyphenol classes in plants. (Adapted from Salminen and Karonen, 2011. Reproduced with permission of John Wiley & Sons.)

typical plant cell does not produce both GTs and ETs simultaneously, and the very efficient production of ETs must have direct negative effects for CT and flavonoid biosynthesis. This is likely why ETs and CTs have not been documented to accumulate together in high concentrations in the same tissue: they typically show opposite seasonal patterns in their contents, such that ETs peak in young tissues and CTs are most abundant in mature tissues (Ossipov *et al.*, 2001; Salminen *et al.*, 2001, 2004).

### 4.3.2 Biosynthesis of gallic acid, galloylglucoses, and gallotannins

In spite of numerous investigations over the decades, the biosynthesis of gallic acid, the starting material for the synthesis of HTs, was for many years one of the major enigmas of plant phenolic metabolism. According to the current consensus, gallic acid is formed from an intermediate compound of the upstream reactions of the shikimate pathway, most probably from 3-dehydroshikimic acid, and the reaction is catalyzed by a specific dehydrogenase (Gross, 1992a,b; Werner *et al.*, 1999). However, this enzyme was not detected in plants or microorganisms until 2003, when Ossipov *et al.* (2003) found a highly active enzyme, dehydroshikimate dehydrogenase (DSDG), that catalyzed the direct conversion of 3-dehydroshikimic acid into gallic acid. There have not been any studies linking the activity of this enzyme to observed tannin levels, but it is possible that part of the high between-tree, between-species, and seasonal variation in the foliar content of HTs (shown for example for birches in Salminen *et al.*, 2001, 2002) could be attributed to the activity of DSDG.

The biogenesis of GGs (the main pathway is shown in Fig. 4.5) starts with 1-*O*-galloyl-*β*-D-glucose (1-GG) as the first metabolite. This compound – more familiarly known as *β*-glucogallin – is formed via the esterification of gallic acid and activated UDP-glucose. In addition to being the first metabolite in the GG pathway, *β*-glucogallin also serves as

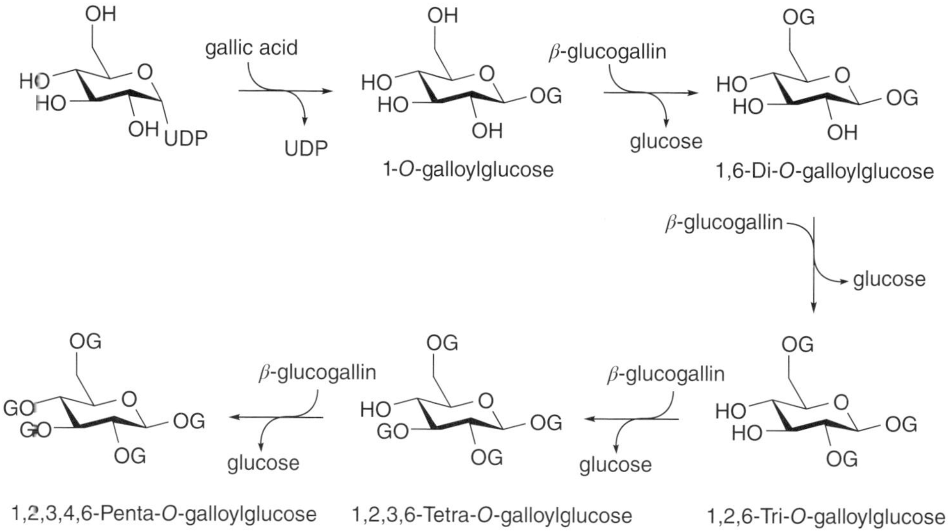


**Fig. 4.5** Main pathway in the formation of 1,2,3,4,6-penta-*O*-galloylglucose. G denotes a galloyl unit.

the main acyl donor in the formation of other higher galloylated GGs (Niemetz & Gross, 1999); gallic acid as such is used only in the biosynthesis of *β*-glucogallin.

For instance, the biosynthesis of 1,6-di-*O*-galloyl-*β*-D-glucose (1,6-di-GG) needs two molecules of *β*-glucogallin, one of which serves as a galloyl donor and the other as an acceptor (Gross, 1999b). Interestingly, 1,6-di-GG has also been shown to have the potential to donate its 1-galloyl moiety, especially for the formation of the next GG metabolite, 1,2,6-tri-*O*-galloyl-*β*-D-glucose (1,2,6-tri-GG) (Denzel & Gross, 1991). Since both *β*-glucogallin and 1,6-di-GG require different acyltransferases in order to function as acyl donors in the galloylation processes (Denzel & Gross, 1991; Gross & Denzel, 1991), the choice of acyl donor is based partly on the enzymes present in a plant. However, it has been suggested that the significance of *β*-glucogallin-independent reactions is negligible in comparison to the reactions involved in the main pathway (Gross, 1999a,b). Furthermore, Hagenah & Gross (1993) found that the enzyme that catalyzed the *β*-glucogallin-dependent galloylation of 1,2,6-tri-GG to 1,2,3,6-tetra-*O*-galloyl-*β*-D-glucose (1,2,3,6-tetra-GG) also catalyzed the galloylation of 1,3,6-tri-*O*-galloylglucose to the same tetragalloylglucose, but with 12-fold efficiency. Because of these differences in reaction rates, it is possible that some intermediates of the main pathway are formed via an alternative route to that shown in Fig. 4.5, but these alternative intermediates cannot be found as major metabolites in plants due to their high turnover rates.

The formation of GTs from PGG has been evidenced by several enzymological studies (Hofmann & Gross, 1990; Niemetz & Gross, 1998, 1999, 2001). In all cases, the enzymes depend on the presence of 1-GG as the acyl donor. However, the high position specificity found in the consecutive galloylations of simple GGs (1-galloyl > 6-galloyl > 2-galloyl > 3-galloyl ≥ 4-galloyl) is not as characteristic for the formation of GTs. Niemetz & Gross (1999) found that the galloyltransferase most reactive in acylating PGG in the 4 position acylated the 2 position instead if the 4 position was already substituted with a digalloyl group. On the other hand, the galloyltransferase that preferentially galloylated PGG in the 3 position further catalyzed galloylation of 3-*O*-digalloyl-1,2,4,6-tetra-*O*-galloylglucose to afford the corresponding heptagalloylglucose, characterized by a 3-*O*-trigalloylgroup (Niemetz & Gross, 1999). The expected sequential synthesis of hexa-, hepta-, and octa-galloylglucoses has also been proved by time-course experiments *in vitro* using enzyme preparations from leaves of *Rhus typhina* L. in the presence of PGG and *β*-glucogallin (Hofmann & Gross, 1990).

### *4.3.3 Biosynthesis of ellagitannins*

In contrast to the rather limited distribution of GTs in nature, ETs are typical constituents of many plant families (Grundhöfer *et al.*, 2001); over 500 structurally characterized ETs have been identified (Khanbabaee *et al.*, 1997). This structural diversity is a consequence of the almost inexhaustible number of combinations and permutations involving intra-(C-C) and inter-(C-O) molecular oxidative couplings available to the assumed ET precursor PGG (Feldman & Smith, 1996). However, the biosynthetic origin of the first ET in the ET pathway, tellimagrandin II (shown in Fig. 4.3a), was proved experimentally only very recently. Using radiolabeled PGG as a substrate, Niemetz *et al.* (2001) showed that the formation of a 3,4,5,3′,4′,5′-HHDP residue of tellimagrandin II originates from the dehydrogenation of two spatially adjacent galloyl groups of PGG. All previous attempts to solve this biosynthetic enigma had resulted only in the formation of free ellagic acid from GGs (Gross, 1992a,b, 1999a,b). Although the subsequent biosynthetic steps from tellimagrandin II onwards have not been proved, some proposed biosynthetic steps are largely accepted. For example, it is generally assumed that casuarictin is a product of the further dehydrogenation of two spatially adjacent galloyl groups of tellimagrandin II (Fig. 4.6). Similarly, the anomeric center of ETs can be degalloylated to produce tellimagrandin I, dehydrogenation of which produces pedunculagin (Haddock *et al.*, 1982a,b) (Fig. 4.6).

Monomeric ETs can be divided into three groups according to their subsequent biogenetic transformations. The first biosynthetic group consists of compounds with simple HHDP groups such as tellimagrandin II and pedunculagin. The second, containing dehydroellagitannins such as geraniin (a key metabolite in the biosynthesis of ETs with a $^1C_4$ glucopyranose core; see Fig. 4.3c), is characterized by the presence of a DHHDP group (Okuda *et al.*, 1992, 2000) with a unique hydrated cyclohexenetrione ring that usually exists in solutions as an equilibrium mixture of five- and six-membered hemiketal forms (Tanaka *et al.*, 1992). The third group consists of transformed dehydroellagitannins with a transformed DHHDP group, such as chebulagic acid (Okuda *et al.*, 2000) (see Fig. 4.3). Moreover, all of the ETs in these three groups can be further modified. For example, ETs of the first group can have acyclic glucose cores as in vescalagin (Fig. 4.3b), which is considered to be formed biosynthetically from pedunculagin (Helm *et al.*, 1997), or they may be complexed with catechin residues, as in camelliatannin A (Okuda *et al.*, 2000) (Fig. 4.7).

Oxidative couplings between two galloyl units, or between galloyl and HHDP units, can lead to more complex structures in ETs, particularly in oligomeric ones. Since the complexation of different monomeric ETs can lead to innumerable dimeric and subsequently oligomeric ETs with molecular weights up to 8624 Da (Clifford & Scalbert, 2000;

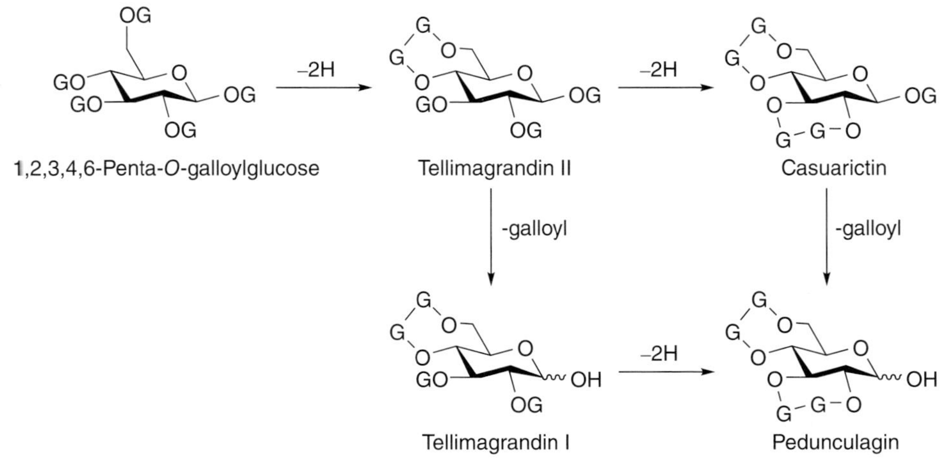


**Fig. 4.6** First biosynthetic steps in the formation of ETs from 1,2,3,4,6-penta-*O*-galloyl-*β*-D-glucose onwards. G denotes a galloyl unit, G-G an HHDP unit.

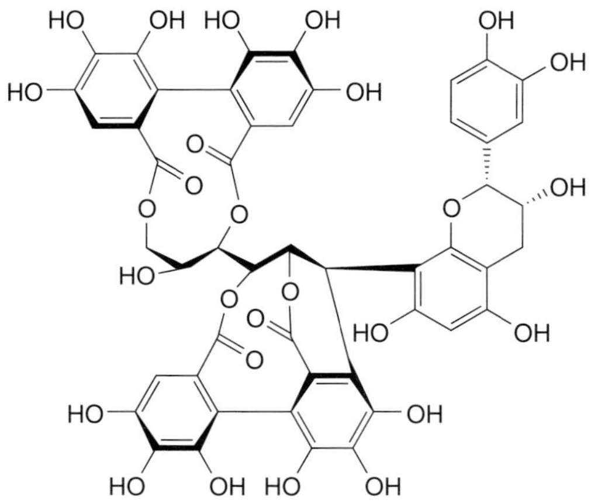


**Fig. 4.7** Chemical structure of camelliatannin A.

Salminen *et al.*, 2011), Fig. 4.8 shows just a few examples of the linking units between the monomers. As in the formation of GGs and GTs, the intermolecular oxidative coupling that produces oligomeric ETs in plants is not dependent on a simple self-association of constituent monomeric tannins, but is controlled regiospecifically by enzymatic systems that vary depending on the plant species (Hatano *et al.*, 1992).

## 4.4 Distributions of different types of tannin in plants

Many articles have been published describing the distribution of ETs in the plant kingdom. In addition, our laboratory has screened more than 1000 species for their tannin composition. It appears that proanthocyanidins are the most widely distributed group of plant tannins, but they are directly followed by ETs. Simple gallic-acid derivatives (e.g. mono- to pentagalloyl glucoses and mono- to trigalloyl quinic acids) lag behind ETs in abundance, followed by true GTs, which are relatively scarce. However, these accurate species-specific findings naturally depend on multiple details, such as the age of the plant tissue used in the analysis (Salminen *et al.*, 2001), the plant individual, chemotype, or genotype chosen for the analysis (Johnson *et al.*, 2009), and the methodology used to analyze the tannin composition (Karonen *et al.*, 2010).

Fig. 4.9 shows how the overall tannin composition of the foliage of English oak (*Quercus robur*) changed during a growing season. HTs (consisting 95% of ETs) dominated in the spring, when proanthocyanidins were absent. Proanthocyanidins did accumulate in the autumn, but ETs remained the main polyphenols. The overall pattern was dramatically different when individual ETs of white birch (*Betula pubescens*) were examined. As can be seen in Fig. 4.10, some of the ETs accumulated in young foliage, while others showed fluctuating patterns or accumulated in mature foliage. It is thus clear that the tannin composition and biological activity of a plant tissue may dramatically change even in the space of a month.

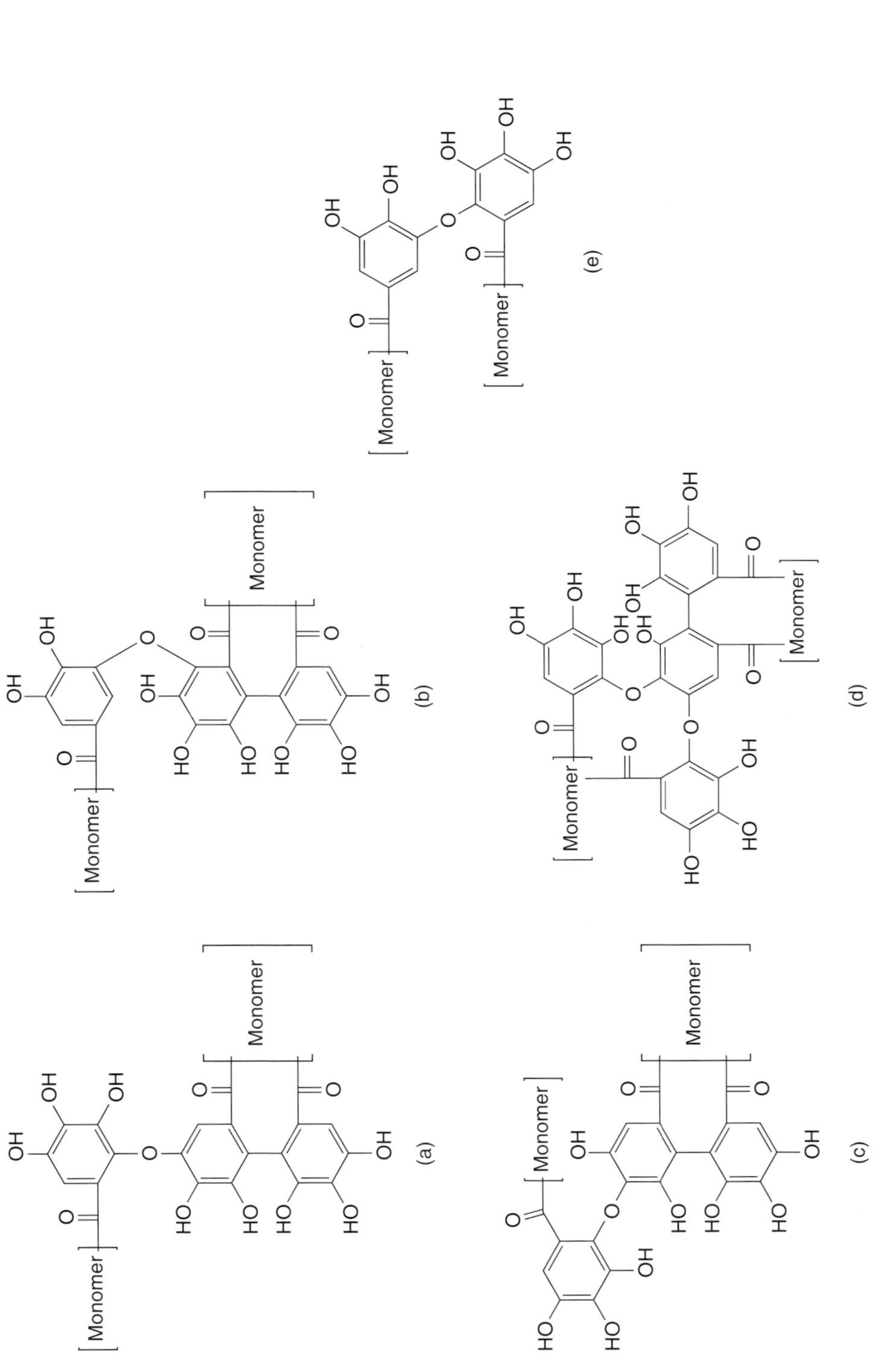


**Fig. 4.8** Examples of the linking units between monomers within dimeric ellagitannins: (a) valoneoyl group, (b) sanguisorboyl group, (c) tergalloyl group, (d) doubly C-O coupled linking unit, (e) dehydrodigalloyl group (modified from Nonaka, 1989).

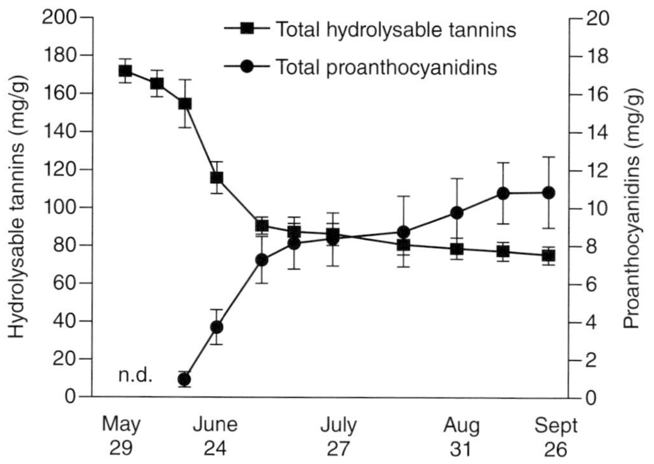


**Fig. 4.9** Seasonal variation of HTs and proanthocyanidins in leaves of *Quercus robur* (Adapted from Salminen *et al.*, 2004. Reproduced with permission of Springer Science+Business Media).

At the same time, Fig. 4.10 shows that ET concentration varies a lot between individual birch trees, as evidenced by the relatively large standard-deviation bars (Salminen *et al.*, 2001, 2002; Ossipov *et al.*, 2003). It is fully possible that two white birches growing side by side might have over 50-fold differences in their foliar concentrations of monomeric ETs. In fact, it appears that species like white birch which synthesize monomeric HHDP esters but no oligomer have a higher within-species variation in their ET content than do species that accumulate oligomeric ETs (Salminen *et al.*, 2001, 2002, 2004; Ossipov *et al.*, 2003; Salminen, unpublished data from *Geum*, *Fragaria*, *Rubus*, *Epilobium*, *Potentilla*, and *Alchemilla* species). Another apparent difference in these types of species (ET monomers versus ET oligomers) is that individual galloyl glucoses tend to accumulate in the ET monomer-specific species but not in the ET oligomer-specific species. This could simply be an indication that monomer-specific species have less efficient turnover rates of galloyl glucoses into ETs and indirect evidence for higher enzyme activities in the galloyl glucose and ET pathway of the oligomer-specific species (Salminen *et al.*, 2001, 2002, 2004; Salminen, unpublished observations). However, little is known about the enzymes controlling the ET pathways, and much of this variation could depend solely on the activity of the DSDG enzyme that directly controls the production of gallic acid, the necessary starting material for all the HTs (Ossipov *et al.*, 2003).

The discussed variation in the tannin content, or the differential accumulation of individual tannins in different plant species, is directly reflected in the chromatographic profiles of plant polyphenolic extracts. Fig. 4.11 shows how *Betula pubescens* produces relatively complicated HPLC-UV traces for its polyphenols, which mainly consist of galloyl glucoses, glucopyranose-based monomeric ETs, and flavonoid glycosides; the chromatographic hump typical of polymeric proanthocyanidins is visible under the individual polyphenol peaks as well. *Quercus robur* (Fig. 4.11) tends to accumulate a number of C-glycosidic monomeric ETs together with one rarely found dimer, cocciferin

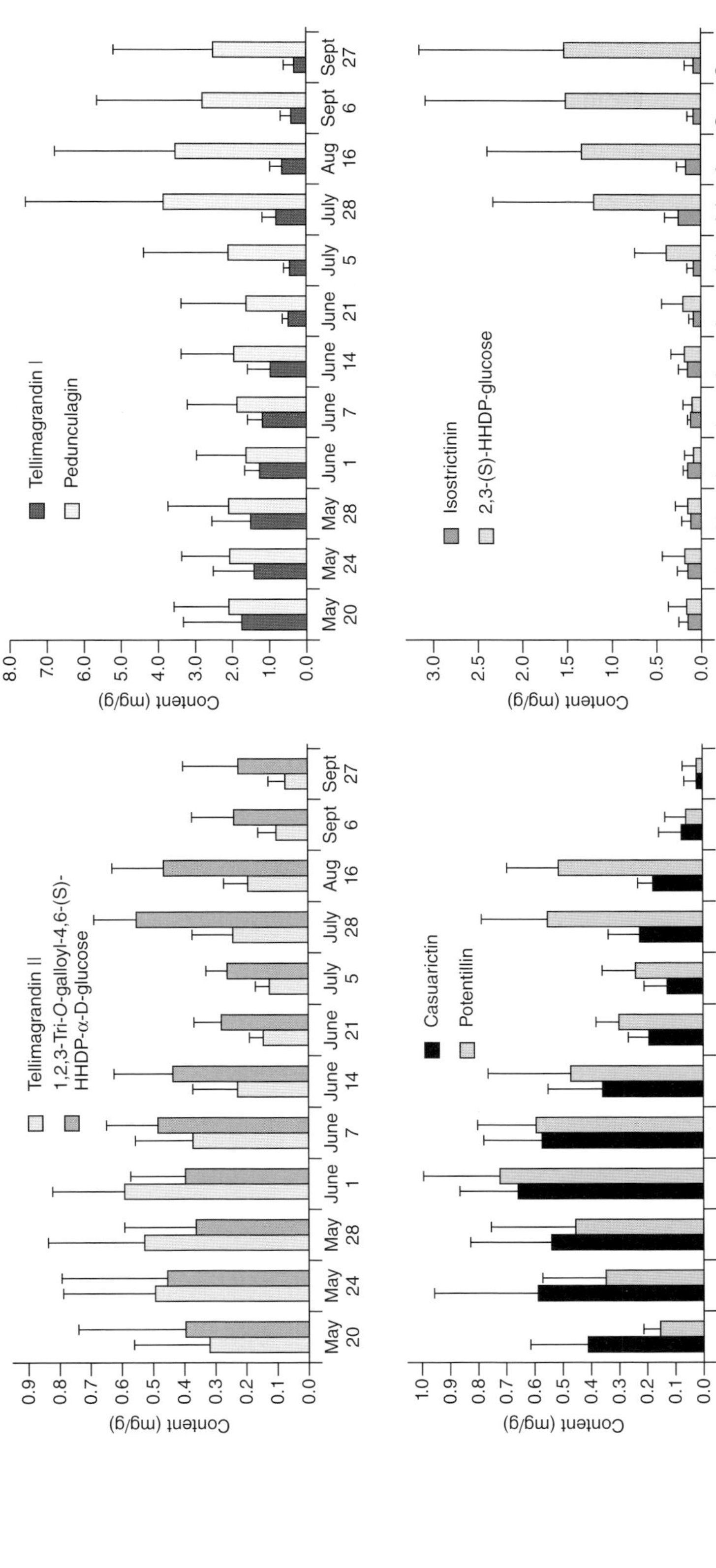


**Fig. 4.10** Seasonal variation of individual ETs in leaves of *Betula pubescens*. (Adapted from Salminen *et al.*, 2001. Reproduced with permission of Elsevier.) For color details, please see Plate 4.10.

$D_2$. Willow herb (*Epilobium angustifolium*) instead accumulates the dimeric ET oenothein B, while monomeric ETs or galloyl glucoses are found at trace levels, although they must be produced in the biosynthetic pathway that leads to oenothein B. Norway maple (*Acer platanoides*) utilizes a separate branch of the HT pathway by accumulating hexa-, hepta-, and octagalloyl glucoses (i.e. GTs), together with flavonoid glycosides. These examples highlight the heterogeneity of the tannin types present in different species and show that species that accumulate oligomeric ETs tend to show simpler chromatographic profiles for their ETs than species that produce monomeric ETs only (Fig. 4.11). Given the earlier-mentioned trend for monomer-specific species to have higher between-plant variation in ET content, we propose that the ecological roles of ETs might be more easily studied or verified with oligomer-specific species than with monomer-specific ones. At the very least, their HPLC analysis should be more robust and reliable, due to the simpler chromatography.

By utilizing more sensitive analysis methods (QTOF-MS), Karonen *et al.* (2010) reported for the first time hexa- and heptameric ET oligomers in the common evening primrose (*Oenothera biennis*); later, Salminen *et al.* (2011) updated these findings by reporting polymers (undecamers) from the same species. Prior to this, the level of detection for ET oligomers had been in the pentamer range. This emphasizes the need for even more accurate phytochemical methods if the whole tannin composition of a plant is to be revealed. Interestingly, Johnson *et al.* (2009) had earlier noted that *O. biennis* populations contain two extreme types of genotype: those that accumulate the dimeric oenothein B (Fig. 4.12a) and those that accumulate the trimeric oenothein A (Fig. 4.12b). Other genotypes have diverse proportions of dimer to trimer, suggesting a microevolutionary gradient in the populations/genotypes of *O. biennis*. Recent evidence suggests that the evolution of this gradient towards more trimer-specific genotypes may be due to high insect herbivore pressure in certain *O. biennis* populations (Agrawal *et al.*, 2012). When the effects of the presence or absence of insect herbivores on the population genotype composition were tested across 5 years, it was shown that insects caused an 89% change in the dimer/trimer ratio of the genotypes, the more complex trimers accumulating in the presence of herbivores. Most interestingly, the effects of insects were revealed only at the molecular level, as the total concentration of ETs did not change (Agrawal *et al.*, 2012). This means that without compound-specific tannin quantifications, the results of this study would have erroneously shown that insect herbivory had no effect on the tannin production of *O. biennis*.

## 4.5 Tannins in plant–herbivore interactions

### *4.5.1 General aspects of tannins and plant–herbivore interactions*

In Section 4.4, we noted that plant cells are able to synthesize a remarkably high number of different tannin structures. On looking at their structural diversity, one might assume that their effects against herbivores are compound-specific rather than specific to tannins *per se*. Moreover, species profiles are highly heterogeneous. No single plant species contains all

**Fig. 4.11** HPLC-UV traces of polyphenols found in the foliage of *Betula pubescens* (top left), *Quercus robur* (top right), *Epilobium angustifolium* (bottom left), and *Acer platanoides* (bottom right). For color details, please see Plate 4.11.

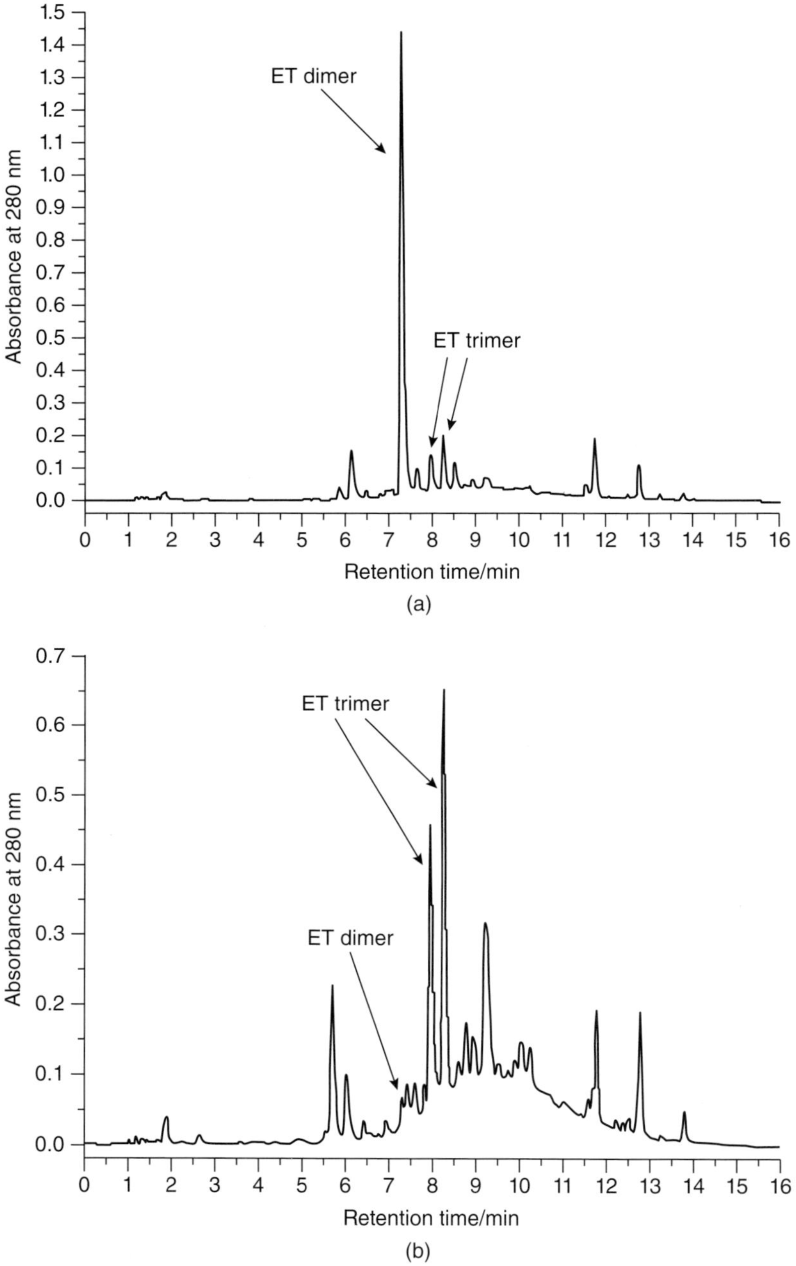


**Fig. 4.12** Content of dimer-specific (top) and trimer-specific (bottom) genotypes in *Oenothera biennis*. (Adapted from Johnson *et al*., 2009. Reproduced with permission of John Wiley & Sons.)

of the tannin molecules, and every species has its own unique tannin composition. It seems likely that these unique compositions of tannin molecules with variable activities contribute to specific plant–herbivore interactions. Unfortunately, the results of one plant–herbivore system cannot be directly extrapolated to explain the results of another if the plant species to be compared have different tannin compositions. The situation is complicated by the fact that herbivores also exhibit a remarkable degree of variation in their ability to tolerate tannins in their diet. This variation is best demonstrated by comparisons between herbivores that differ in their diet breadth: specialist herbivores that consume only a restricted set of species and generalist herbivores that consume a wide range of species. Specialist herbivores can evolve greater tolerance to certain types of tannin found in their host plant than their generalist counterparts, which may encounter those types of tannin in their diet infrequently. This scenario was demonstrated by Roslin & Salminen (2008), who showed that two specialist herbivores of *Quercus robur* were not affected by the addition of either purified vescalagin (a monomeric C-glycosidic ET) or a mixture of polymeric proanthocyanidins into their diet, while generalists species clearly suffered from the addition of vescalagin, although not from the addition of proanthocyanidins. It is possible that the wider distribution of proanthocyanidins in plants in comparison to vescalagin-type C-glycosidic ETs is an explanatory factor: specialist herbivores of *Quercus robur* typically consume both types of tannin (Salminen *et al.*, 2004), while generalists often face proanthocyanidins in their multispecies diet but vescalagin only in oak. Thus, the constant exposure to C-glycosidic ETs by the specialist species has selected for tolerance to C-glycosidic ETs, while such mechanisms are absent for the generalists (Barbehenn *et al.*, 2001).

Despite the complexity in tannin structures and distribution it has been very common to try to find simplified answers for their ecological roles. To date, the most commonly ascribed explanation of tannins' ecological function relates to their ability to complex with and precipitate proteins out of a solution. By precipitating proteins out of a solution, it is believed tannins decrease the nutritive value of consumed plant tissue, conferring a defensive role for plants. Due to this putative mode of action, protein precipitation-based assays have been commonly utilized to quantify total tannins in the field of chemical ecology. Using these methods, compounds with high PPC are easily detected, while compounds with low or negligible PPC, such as many simple galloyl glucoses (up to trigalloyl glucoses) and small or C-glycosidic ETs (HHDP-glucose- or vescalagin-type ETs), have moderate or limited detection (Haslam, 1989; Kilkowski & Gross, 1999). This is why protein precipitation-based assays have been advocated in the literature rather than spectrophotometric methods that measure tannin building blocks after hydrolysis, because these protein-precipitation methods were thought to be more closely related to the biological effects of tannins (Martin & Martin, 1982; Reed, 1994). A central tennet behind this explanation is that for a long time it was thought that tannins precipitate proteins in all herbivores, including insects. However, this has not been evidenced in practice with insect herbivores, most probably because protein precipitation by tannins may be inhibited in the gut by surfactants and high pH, both of which are believed to be common in the guts of insects, especially lepidopteran insects. It is therefore possible that the PPC of a plant sample *per se* does not reveal its biological activity against those herbivores with high pH and/or surfactants in the gut.

This is not to say that PPC measurements do not have any relevance in plant–herbivore research. When used cautiously, they may be very useful in estimating the tannin content of a plant species if one has *a priori* knowledge of the types of tannin that are estimated. For example, it is possible that a species could contain only C-glycosidic ETs that are weakly detected by PPC but are active against insects via mechanisms other than protein precipitation (Moilanen & Salminen, 2008). In this case, PPC-based activity comparisons between individuals within this specific species will give relevant information on the concentrations of this one type of tannin, although it will be only weakly detected. But if PPC is used in activity comparisons between separate species then the relevance may be lost: a lower PPC value of one species may in fact come from 10 times as many active tannins (with another activity mechanism) than are found in another species with a higher PPC value. The possible activity behind these contrasting PPC trends is discussed in the next section.

### 4.5.2 The tannin oxidation hypothesis and its verification in plant–herbivore interactions

It was hypothesized since at least the early 1990s that tannins could be auto-oxidized in the alkaline gut of insect herbivores (Appel, 1993). If this were true then tannin oxidation products could (i) reduce the protein digestion of insects by covalent complex formation between the amino acid residues of proteins and the quinone-form tannins and/or (ii) harm biomolecules in the gut by accumulation of reactive oxygen species (ROS). From a chemical point of view, the tannin oxidation hypothesis (auto-oxidation of tannins at high pH rather than protein precipitation) was highly plausible, but it was difficult to confirm experimentally. This was largely due to the lack of commercially available pure tannins and a correct methodology for verifying the tannin oxidation. Both of these shortcomings were put aside by Barbehenn *et al.* (2006a): the oxidations at high pH of five individual galloyl glucoses, four individual ETs, four different types of proanthocyanidin monomer, oligomer, and polymer mixture, and a mixture of GTs were studied by both EPR spectroscopy and the spectrophotometric browning assay. The first method is able to detect the initial tannin oxidation products (i.e. semiquinone radicals), while the second measures the accumulation of colorful quinones, the further oxidation products of semiquinone radicals. Generally speaking, the results obtained by both of these methods suggest that the pro-oxidant activity of a tannin (the ease of tannin oxidation) is inversely related to its PPC (Haslam, 1996; Kilkowski & Gross, 1999; Barbehenn *et al.*, 2006a). For instance, a tannin's protein affinity (PPC) is known to increase as the degree of galloylation increases and to decrease as HHDP groups are formed from two galloyls. Within the tested tannin series, the pro-oxidant activity functioned in the opposite direction: the ease of oxidation was enhanced by removing galloyl groups from the molecule (Fig. 4.13) and by forming HHDP groups from the galloyls (Moilanen & Salminen, 2008). Thus, it seemed that these two potential modes of activity were negatively correlated: tannins with high PPC activity had low pro-oxidant activity, while tannins with low PPC had high pro-oxidant activity.

All these results were based only on *in vitro* experiments, so additional evidence on tannin oxidation was needed *in vivo*. This was obtained by Barbehenn *et al.* (2008), who showed that the foliar tannin composition of tree species had a direct effect on the level

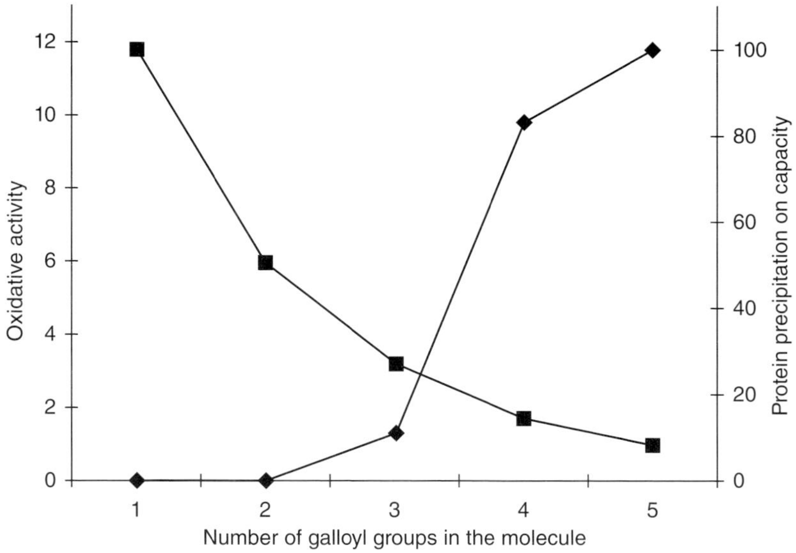


**Fig. 4.13** The addition of galloyl groups into an HT increases its PPC but decreases its pro-oxidant activity. As these two activities are negatively correlated, it is difficult to find a single tannin that is efficient at both activity mechanisms. (Adapted from Salminen *et al.*, 2011. Reproduced with permission of John Wiley & Sons.) For color details, please see Plate 4.13.

of polyphenol oxidation in the gut of *Orgyia leucostigma* larvae (Fig. 4.14). The three tree species that contained only low-molecular-weight phenolics or proanthocyanidins produced low levels of semiquinone radicals in the *O. leucostigma* gut, suggesting that proanthocyanidins of these tree species were not easily oxidized. With other species, a trend was detected showing that as the proportional level of proanthocyanidins decreased in the foliage and that of ETs increased (*Q. rubra* → *C. glabra* → *A. saccharum* → *Q. alba*) a clear increase was also apparent in the phenolic oxidation level in the gut. This clearly suggested that ETs are more prone to oxidation than proanthocyanidins at high pH *in vivo* as well as *in vitro*.

The high semiquinone radical concentration caused by an ET-free *A. rubrum* diet clearly showed that we should not focus on ETs as the sole source of active pro-oxidants; instead, we should be looking at individual molecules, regardless of the polyphenol group to which they belong. This compound-specific approach is not yet a common method of polyphenol analysis in the field of chemical ecology, but it must gain in popularity if we want to progress in our understanding of polyphenol activities in ecological systems.

Another important factor to consider is the observation that plants contain antioxidants in addition to pro-oxidants, and antioxidants may turn into pro-oxidants given the right conditions and concentrations. Other studies have shown that foliar polyphenols are able to oxidize to semiquinone radicals only after the foliar antioxidant reserves (e.g. ascorbic acid) are depleted below a certain threshold (Barbehenn *et al.*, 2006b), indicating that ascorbic acid is more easily oxidized than polyphenols. In addition to the obvious antioxidants, such as ascorbic acid, one must take into account that some foliar polyphenols may also

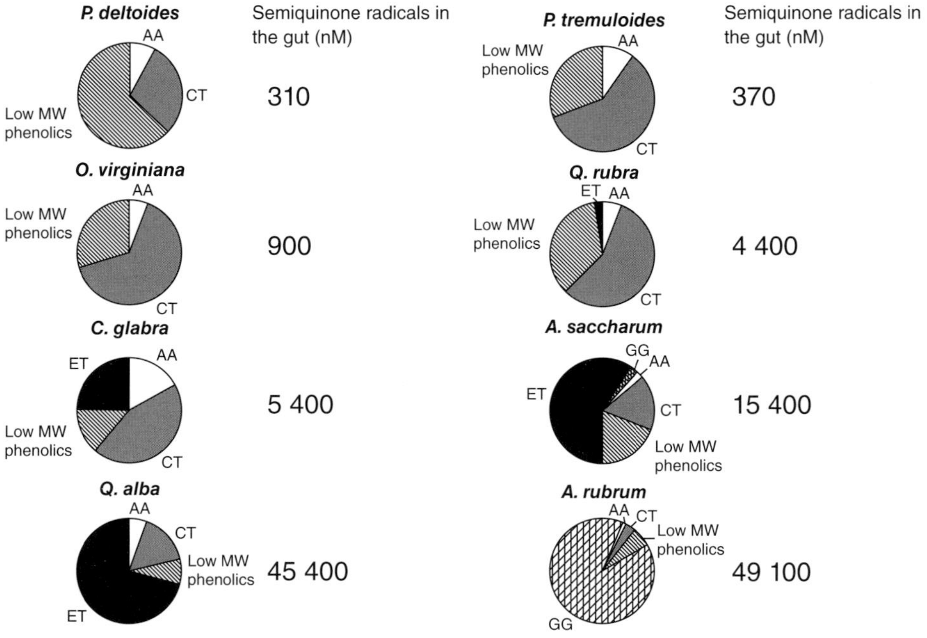


**Fig. 4.14** Effect of the relative tannin composition of plant foliage on the levels of semiquinone radicals detected in the gut of insect herbivores feeding on that foliage. (Adapted from Barbehenn *et al.*, 2008. Reproduced with permission of Springer Science+Business Media.)

function as biological antioxidants (Hagerman *et al.*, 1998). For instance, previous studies have shown that certain types of proanthocyanidin can prevent the oxidation of ETs at high pH by as much as 90% (Barbehenn *et al.*, 2006b). This might help explain why the highest semiquinone radical levels were observed in the gut of larvae fed with *A. rubrum*, since this species produced the lowest contents of both ascorbic acid and proanthocyanidins of all the species studied (Fig. 4.14). These observations emphasize the need for accurate analysis to identify antioxidant compounds before the reasons behind high or low *in vivo* activities can be understood. The analysis of total tannins alone could not have revealed any of the patterns shown in Fig. 4.14.

### *4.5.3 The ease of oxidation of individual ellagitannins can be predicted by their chemical structures and chromatographic properties*

Evidence that ETs might be the most potent tannin-type pro-oxidants both *in vitro* and *in vivo* has been presented in the previous sections. Here, "the most potent tannin pro-oxidant" means the tannin with the highest capacity to produce semiquinone radicals upon its oxidation. As with the protein-precipitation hypothesis, it is yet to be shown whether this potent tannin oxidation results in a decrease in insect performance. First, one would need to know which plant species contain these potent pro-oxidants and whether these species are more harmful to insects than species with less potent oxidants. But first one would need to know more of the potent pro-oxidants themselves.

Earlier, Barbehenn *et al.* (2006a) were able to show that individual ETs do differ in their pro-oxidant activities, but it was not fully clear what the exact structural causes for these differences were. This gap in knowledge was narrowed by Moilanen & Salminen (2008), who utilized 27 individually purified ETs to clarify structure–activity relationships among ETs. The most unambiguous findings were that C-glycosidic ETs are far more active than their glucopyranose-based counterparts and that the activity is roughly doubled if a C-C bond is added between two galloyls to form an HHDP group. Using this feature, along with other structural features correlated with activity, Moilanen & Salminen (2008) were able to construct a formula for the calculation of the activity (Fig. 4.15). This shows a clear structure–activity relationship between ETs and their pro-oxidant activity, indicating that, with the help of the formula, the pro-oxidant activities of many ETs can be quite accurately estimated.

One of the shortcomings of this approach is that the ET structure must be exactly known. In most cases, this requires the isolation of the compound, followed by structure elucidation with NMR techniques. However, there are other, more simple tools available for a rougher estimation of an ET's pro-oxidant activity (Salminen *et al.*, 2011; Moilanen *et al.* 2013). First, it appears that the most water-soluble ETs are more active as pro-oxidants than the less water-soluble ones. Interestingly, Haslam and coworkers had earlier proposed the opposite to be true for the PPC activity of tannins; that is, that the less hydrophilic the tannin, the better its ability to complex with proteins (Beart *et al.*, 1985; Haslam, 1996). Thus, a simple fractionation of an aqueous plant extract with ethyl acetate will find the most efficient

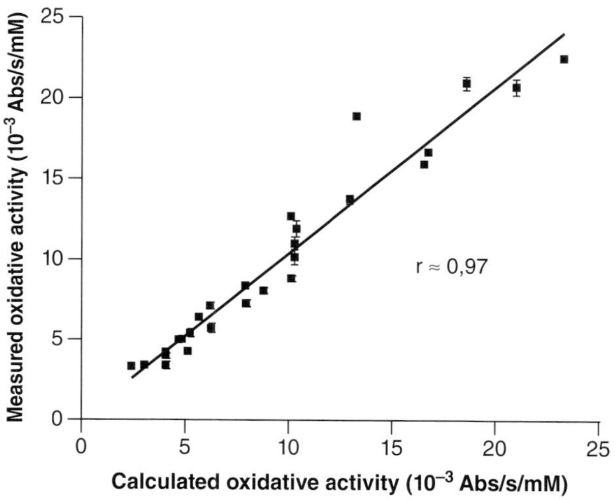


$$\text{Oxidative activity} = \frac{15.5 \times A + 6.6 \times B + 4.9 \times C + 4.5 \times D + 3.6 \times E + 1.9 \times F + 1.1 \times G}{\text{Molecular weight of the whole ellagitannin (g/mol)}}$$

**Fig. 4.15** Estimation of the oxidative activity of an ellagitannin on the basis of its structural components. A, number of valoneoyl groups (free -COOH); B, number of NHTP groups; C, number of valoneoyl (bound -COOH) or sanguisorboyl groups; D, number of xylose/lyxose groups; E, number of acyclic castalagin-type glucoses; F, number of HHDP groups; G, number of acyclic vescalagin-type glucoses. (Adapted from Moilanen and Salminen, 2008. Reproduced with permission of Springer Science+Business Media.)

ET pro-oxidants primarily in the water phase and the poorest ET pro-oxidants in the ethyl acetate phase (for PPC-active ETs the situation is reversed). Second, the retention time of efficient ET pro-oxidants in reversed-phase LC analysis is shorter than that of the less efficient ones (for PPC-active ETs the retention time is longer than for the less effective ones). Simultaneously, an ET's UV spectrum can provide information about the number of HHDP or NHTP groups, where the number of these groups is positively correlated with activity (Salminen *et al.*, 2011; Moilanen *et al.*, 2013). By coupling these two pieces of information (Rt and UV spectrum), the relative activities of co-existing ETs in any plant species can be estimated simply by RP-HPLC-DAD analysis. Finally, the most pro-oxidant ETs show the lowest affinity towards the Sephadex LH-20 gel, while the least active ones are bound to the gel more effectively (for PPC-active ETs the situation is reversed). Thus, a traditional Sephadex LH-20 fractionation of a polyphenol extract followed by RP-HPLC-DAD analysis will provide an even better tool for the estimation of the relative activities of ETs found in plants (Salminen & Karonen, 2011; Moilanen *et al.*, 2013).

These structure–activity patterns and tools are able to explain some of Barbehenn *et al.*'s (2008) results, which showed that insects feeding on specific leaf diets differed in the ease of formation of semiquinone radicals in their guts. White oak (*Q. alba*) contained only half of the ET levels found in sugar maple (*A. saccharum*), but white oak still produced three times the concentration of semiquinone radicals in the gut. This could be explained by the presence of mainly C-glycosidic ETs (efficient pro-oxidants) in *Q. alba* and their absence in the sugar maple, which contained glucopyranose-based ETs (less efficient pro-oxidants). The functioning of the previously mentioned simple tools was confirmed by the facts that (i) the oak ETs had shorter retention times than the maple ETs, (ii) they eluted earlier during the Sephadex LH-20 chromatography, and (iii) they had more HHDP/NHTP-rich UV spectra. In fact, the functioning of these rough activity-estimation methods has already been verified with a number of plant species (Salminen, unpublished data).

The white oak/sugar maple comparison also illustrates another common problem in studies of plant–herbivore interactions. In most cases where tannins are considered antiherbivore agents it is thought that a plant species with higher tannin content should be better defended against herbivores than a plant species with lower tannin content. We showed that the 50% lower ET concentration of oak produced 300% levels of semiquinone radicals compared to those produced by ET-richer maple. Moilanen & Salminen (2008) have already shown that individual ETs may have sixfold differences in their pro-oxidant activities. This means that it is very difficult to repeatedly prove that tannins are active (or nonactive) antiherbivore agents if we stick with the analysis of total tannins: even an analysis of total ETs would not have helped in understanding the oak/maple comparison. Thus, very specific qualitative analysis of plant tannin types will be crucial if future studies in this area are to increase our true knowledge of different types of tannin defense in plants: concentration is important, but tannin identity is even more so. Most importantly, these types of specific quantitative analysis of potent pro-oxidants in many ecological systems will help to solve an outstanding puzzle: are the potent pro-oxidants able to decrease insect performance *in vivo*, and if so what concentrations are needed? Clearly the tannin oxidation hypothesis needs an accumulation of repeated evidence by multiple research groups before its function in plant–herbivore systems can be fully accepted.

### *4.5.4 Other factors that may affect ellagitannin activities against insect herbivores*

We have shown how ET structure, absolute concentration, and relative concentration to assumed biological antioxidants (e.g. vitamin C and even proanthocyanidins) together determine the level of semiquinone radical formation that takes place in the guts of herbivorous insects. To measure all of these parameters in plant foliage is possible – although tedious – but unfortunately is still not quite enough to determine whether a decrease in insect performance is caused by foliar ETs or not. In addition to auto-oxidation at high pH, other post-ingestive reactions may either enhance or decrease ET activity. Two of the most relevant reactions in this respect are the oxidation of ETs due to plant enzymes (e.g. polyphenol oxidases and peroxidases) and the hydrolysis of ETs due to alkaline gut conditions. The hydrolysis aspect has been experimentally proven (Salminen & Lempa, 2002; Barbehenn *et al.*, 2009), but so far there is no published evidence of ET oxidation by plant oxidative enzymes. However, this could be one of the first reactions that takes place after cell disruption by leaf-chewing insects.

Hydrolysis of ETs in the gut could be important for insect performance, since it both decreases the concentration of the foliar ETs in the gut and produces new metabolites that may have been absent in the foliar diet. The hydrolysis metabolites identified to date have included gallic acid, ellagic acid, and the ET residue left over after partial hydrolysis of the parent ET. The removal of galloyl groups from ETs by hydrolysis is known to increase the pro-oxidant activity of the ETs (Moilanen & Salminen, 2008), and the simultaneously produced gallic acid has been found to have significant pro-oxidant activity as well (Barbehenn *et al.*, 2006b, 2008). On the other hand, the removal of HHDP groups by hydrolysis should have the opposite effect, since the resulting ETs will not have higher pro-oxidant activity than the parent compound (Moilanen & Salminen, 2008) and the simultaneously produced ellagic acid is quite stable at pH 10 (i.e. it is not easily oxidized). To further complicate the scenario, gallic and ellagic acids may be small enough to cross the peritrophic envelope of the insect gut (Barbehenn & Martin, 1992, 1994; Barbehenn *et al.*, 1996, 2001) and thus affect insect performance in ways that are still unknown to us (but see Bernays & Woodhead, 1982; Bernays *et al.*, 1983). Given these facts, it would not be surprising if many of the effects of ETs on insects were derived from ET metabolites rather than from the original ET present in the leaf. Nevertheless, this area of research needs considerably more effort before any clear conclusions can be made.

The situation is further complicated by the unpredicted magnitude of the joint effect of hydrolysis and oxidation on the concentration of the original tannin. For instance, when pentagalloyl glucose was incubated at pH 10 for 10 hours its concentration dropped more than 60% despite its low pro-oxidant activity *in vitro* (Salminen, unpublished data) (Fig. 4.16). Much of the decline of the pentagalloyl glucose concentration could be explained by the accumulation of tri- and tetragalloyl glucoses together with gallic acid and – surprisingly – ellagic acid. This was a similar finding to the results of historic attempts to produce ETs from pentagalloyl glucose *in vitro* without enzymatic help: no ETs were formed, only ellagic acid. Whether ellagic acid is formed from pentagalloyl glucose or related compounds also found *in vivo* in insects is an unexplored and intriguing

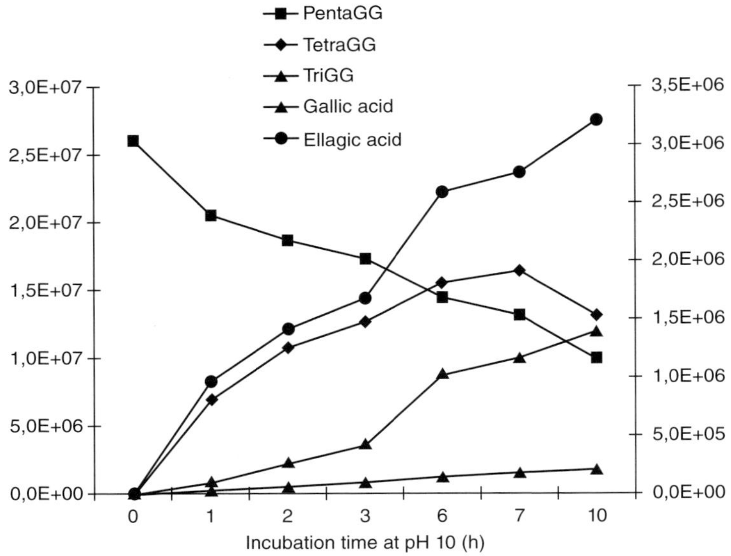


**Fig. 4.16** Incubation or pentagalloylglucose at pH 10 decreases its concentration and causes the simultaneous accumulation of the hydrolysis products tetragalloyl glucose, trigalloyl glucose, gallic acid, and ellagic acid. For color details, please see Plate 4.16.

question. This example shows that tannin metabolism in insects may not necessarily follow the most obvious routes predicted directly from the original structure. If we do not know the details of tannin metabolism and how this varies between individual tannins we may lack one of the major components in our understanding of tannin effects on insect herbivores. In addition to knowing the initial reactions of tannins in the gut (e.g. oxidation and hydrolysis), we need to know the fates and effects of the metabolites formed; this is currently one of the major routes to understanding the effects of polyphenols on human health (Del Rio *et al.*, 2010) and should be a priority in chemical ecology as well.

## 4.6 Conclusions

This chapter showed that plant cells synthesize ETs via the shikimate pathway, utilizing the enzymatic conversion of dehydroshikimic acid into gallic acid and further to galloyl glucoses, and provided evidence in support of the hypothesis that ETs are the most active tannin pro-oxidants. The pro-oxidant activity of the polyphenols decreases during the steps from monogalloyl glucose towards pentagalloyl glucose, only to begin increasing once pentagalloyl glucose is transformed to glucopyranose-based ETs and further to very active C-glycosidic ETs. Thus the tannin-based pro-oxidant capacity of a plant cell will greatly depend on its balance of different types of product or of intermediates of the HT or ET pathway. This balance of more and less active pro-oxidants must further take into account the coexisting foliar antioxidants, which may inhibit tannin oxidation until the antioxidant reserves are consumed. Without this type of detailed knowledge it will be difficult to estimate whether the detailed tannin composition of one plant species should have better antiherbivore effect than that of another. Most importantly, in the absence of detailed and repeated studies we find ourselves in the same situation with the tannin oxidation hypothesis as we were in for a long time with the protein-precipitation hypothesis: we still do not know for sure if it is really tannin oxidation that causes antiherbivore effects. For instance, ETs may derive their antiherbivore activities from their hydrolysis products; this is a largely unexplored area of research. All these aspects show the need for accurate metabolite-specific chemical analysis to be accompanied by more ecologically oriented tannin research if we truly want to unravel the ecological roles and modes of action of ETs in plant–herbivore interactions.

## Acknowledgments

Our current knowledge of tannin chemistry and chemical ecology would be much reduced without the group effort of our Natural Chemistry Research Group and the help of numerous chemically and ecologically oriented collaborators. Especial thanks to Kyösti Lempa, Erkki Haukioja, Ray Barbehenn, Tomas Roslin, Anurag Agrawal, and Marc Johnson for sharing an enthusiastic interest in ET research. Jeff Ahern kindly checked the language, and two reviewers gave constructive comments on the earlier version of the manuscript. Funding by the Academy of Finland is acknowledged (grants no. 119659 and 258992).

## References

Agrawal, A.A., Hastings, A., Johnson, M.T.J., *et al.* (2012) Insect herbivores drive real-time ecological and evolutionary change in plant populations. *Science*, 338, 113–116.

Appel, H.M. (1993) Phenolics in ecological interactions: the importance of oxidation. *Journal of Chemical Ecology*, 19, 1521–1552.

Appel, H.M. & Schultz, J.C. (1994) Oak tannins reduce effectiveness of thuricide (*Basillus thuringiensis*) in the gypsy moth (Lepidoptera: Lymantriidae). *Plant Resistance*, 87, 1736–1742.

Barbehenn, R.V. & Martin, M.M. (1992) The protective role of the peritrophic membrane in the tannin-tolerant larvae of *Orgyia leucostigma* (Lepidoptera). *Journal of Insect Physiology*, 38, 973–980.

Barbehenn, R.V. & Martin, M.M. (1994) Tannin sensitivity in larvae of *Malacosoma disstria* (lepidoptera): roles of the peritrophic envelope and midgut oxidation. *Journal of Chemical Ecology*, 20, 1985–2001.

Barbehenn, R.V., Martin, M.M. & Hagerman, A.E. (1996) Reassessment of the roles of the peritrophic envelope and hydrolysis in protecting polyphagous grasshoppers from ingested hydrolyzable tannins. *Journal of Chemical Ecology*, 22, 1901–1919.

Barbehenn, R.V., Bumgarner, S.L., Roosen, E.F. & Martin, M.M. (2001) Antioxidant defenses in caterpillars: role of the ascorbate-recycling system in the midgut lumen. *Journal of Insect Physiology*, 47, 349–357.

Barbehenn, R.V., Jones, C.P., Hagerman, A.E., *et al.* (2006a) Ellagitannins have greater oxidative activities than condensed tannins and galloylglucoses at high pH: potential impact on caterpillars. *Journal of Chemical Ecology*, 32, 2253–2267.

Barbehenn, R.V., Jones, C.P., Karonen, M. & Salminen, J.-P. (2006b) Tannin composition affects the oxidative activities of tree leaves. *Journal of Chemical Ecology*, 32, 2235–2251.

Barbehenn, R., Weir, Q. & Salminen, J.-P. (2008) Oxidation of ingested phenolics in the tree feeding caterpillar *Orgyia leucostigma* depends on foliar chemical composition. *Journal of Chemical Ecology*, 34, 748–756.

Barbehenn, R., Jaros, A., Lee, G., *et al.* (2009) Hydrolyzable tannins as "quantitative defenses": limited impact against *Lymantria dispar* caterpillars on hybrid poplar. *Journal of Insect Physiology*, 55, 297–304.

Bate-Smith, E.C. (1962) Phenolic constituents of plants and their taxonomic significance. *Journal of the Linnean Society of London*, 58, 95–174.

Baxter, N.J., Lilley, T.H., Haslam, E. & Williamson, M.P. (1997) Multiple interactions between polyphenols and a salivary proline-rich protein repeat result in complexation and precipitation. *Biochemistry*, 36, 5566–5577.

Beart, J.E., Lilley, T.H. & Haslam, E. (1985) Plant polyphenols – secondary metabolism and chemical defence: some observations. *Phytochemistry*, 24, 33–38.

Bernays, E.A. & Woodhead, S. (1982) Incorporation of dietary phenols into the cuticle in the tree locust *Anacridium melanorhodon*. *Journal of Insect Physiology*, 28, 601–606.

Bernays, E.A., Chamberlain, D. & McCarthy, P. (1980) The differential effects of ingested tannic acid on different species of Acridoidea. *Entomologia Experimentalis et Applicata*, 28, 158–166.

Bernays, E.A., Chamberlain, D.J. & Woodhead, S. (1983) Phenols as nutrients for a phytophagous insect *Anacridium melanorhodon*. *Journal of Insect Physiology*, 29, 535–539.

Burapadaja, S. & Buncho, A. (1995) Antimicrobial activity of tannins from *Terminalia citrina*. *Planta Medica*, 61, 365–366.

Cadahía, E., Varea, S., Muñoz, L., *et al.* (2001) Evolution of ellagitannins in Spanish, French, and American oak woods during natural seasoning and toasting. *Journal of Agricultural and Food Chemistry*, 49, 3677–3684.

Clifford, M.N. & Scalbert A. (2000) Ellagitannins – nature, occurrence and dietary burden. *Journal of the Science of Food and Agriculture*, 80, 1118–1125.

Deaville, E.R., Green, R.J., Mueller-Harvey, I., *et al.* (2007) Hydrolyzable tannin structures influence relative globular and random coil protein binding strengths. *Journal of Agricultural and Food Chemistry*, 55, 4554–4561.

Del Rio, D., Costa, L.G., Lean, M.E.J. & Crozier, A. (2010) Polyphenols and health: what compounds are involved? *Nutrition, Metabolism & Cardiovascular Diseases*, 20, 1–6.

Denzel, K. & Gross, G.G. (1991) Biosynthesis of gallotannins: enzymatic "disproportionation" of 1,6-digalloylglucose to 1,2,6-trigalloylglucose and 6-galloylglucose by an acyltransferase from leaves of *Rhus typhina* L. *Planta*, 184, 285–289.

Feeny, P. (1968) Effect of oak leaf tannins on larval growth of the winter moth *Operophtera brumata*. *Journal of Insect Physiology*, 14, 805–817.

Feeny, P. (1970) Seasonal changes in oak leaf tannins and nutrients as a cause of spring feeding by winter moth caterpillars. *Ecology*, 51, 565–581.

Feldman, K.S. & Smith, R.S. (1996) Ellagitannin chemistry. First total synthesis of the 2,3- and 4,6-coupled ellagitannin pedunculagin. *Journal of Organic Chemistry*, 61, 2606–2612.

Feldman, K.S., Sambandam, A., Lemon, S.T., *et al.* (1999) Binding affinities of gallotannin analogs with bovine serum albumin: ramifications for polyphenol-protein molecular recognition. *Phytochemistry*, 51, 867–872.

Gross, G.G. (1992a) Enzymes in the biosynthesis of hydrolyzable tannins. In: *Plant Polyphenols* (eds R.W. Hemingway & P.E. Laks), pp. 43–60. Plenum Press, New York.

Gross, G.G. (1992b) Enzymatic synthesis of gallotannins and related compounds. In: *Recent Advances in Phytochemistry, Vol. 26. Phenolic Metabolism in Plants* (eds H. Stafford & R. Ibrahim), pp. 297–324. Plenum Press, New York and London.

Gross, G.G. (1999a) Biosynthesis, biodegradation, and cellular localization of hydrolyzable tannins. In: *Recent Advances in Phytochemistry, Vol. 34. Phytochemicals in Human Health Protection,*

*Nutrition, and Plant Defense* (eds N.G. Lewis, J.T. Romeo & G.H.N. Towers), pp. 185–213. Plenum Press, New York.

Gross, G.G. (1999b) Biosynthesis of hydrolyzable tannins. In: *Comprehensive Natural Products Chemistry, Vol. 3. Carbohydrates and Their Derivatives Including Tannins, Cellulose and Related Lignins* (ed. B.M. Pinto), pp. 799–826. Elsevier, Amsterdam.

Gross, G.G. & Denzel, K. (1991) Biosynthesis of gallotannins. $\beta$-glucogallin-dependent galloylation of 1,6,-digalloylglucose to 1,2,6-trigalloylglucose. *Zeitschrift für Naturforschung*, 46c, 389–394.

Grundhöfer, R., Niemetz, R., Schilling, G. & Gross, G.G. (2001) Biosynthesis and subcellular distribution of hydrolyzable tannins. *Phytochemistry*, 57, 915–927.

Haddock, E.A., Gupta, R.K., Al-Shafi, S.M.K. & Haslam, E. (1982a) The metabolism of gallic acid and hexahydroxydiphenic acid in plants. Part 1. Introduction: naturally occurring galloyl esters. *Journal of the Chemical Society, Perkin Transactions* 1, 2515–2524.

Haddock, E.A., Gupta, R.K., Al-Shafi, S.M.K., *et al.* (1982b) The metabolism of gallic acid and hexahydroxydiphenic acid in plants: biogenetic and molecular taxonomic considerations. *Phytochemistry*, 21, 1049–1062.

Hagenah, S. & Gross, G.G. (1993) Biosynthesis of 1,2,3,6-tetra-*O*-galloyl-$\beta$-D-glucose. *Phytochemistry*, 32, 637–641.

Hagerman, A.E., Zhao, Y. & Johnson, S. (1997) Methods for determination of condensed and hydrolyzable tannins. *ACS Symposium Series*, 662, 209–222.

Hagerman, A.E., Riedl, K.M., Jones, G.A., *et al.* (1998) High molecular weight plant polyphenolics (tannins) as biological antioxidants. *Journal of Agricultural and Food Chemistry*, 46, 1887–1892.

Haslam, E. (1988) Plant polyphenols (syn. vegetable tannins) and chemical defence – a reappraisal. *Journal of Chemical Ecology*, 14, 1789–1805.

Haslam, E. (1989) *Plant Polyphenols: Vegetable Tannins Revisited*. Cambridge University Press, Cambridge.

Haslam, E. (1992) Gallic acid and its metabolites. In: *Plant Polyphenols* (eds R.W. Hemingway & P.E. Laks), pp. 169–194. Plenum Press, New York.

Haslam, E. (1996) Natural polyphenols (vegetable tannins) as drugs: possible modes of action. *Journal of Natural Products*, 59, 205–215.

Hatano, T., Okonogi, A. & Okuda, T. (1992) Oligomeric hydrolyzable tannins from *Liquidambar formosana* and spectral analysis of the orientation of valoneoyl groups in their molecules. In: *Plant Polyphenols* (eds R.W. Hemingway & P.E. Laks), pp. 195–207. Plenum Press, New York.

Helm, R.F., Ranatunga, T.D. & Chandra, M. (1997) Lignin-hydrolyzable tannin interactions in wood. *Journal of Agricultural and Food Chemistry*, 45, 3100–3106.

Hervé du Penhoat, C.L.M., Michon, V.M.F., Peng, S., *et al.* (1991) Structural elucidation of new dimeric ellagitannins from *Quercus robur* L. *Journal of the Chemical Society, Perkin Transactions* 1, 1653–1660.

Hofmann, A.S. & Gross, G.G. (1990) Biosynthesis of gallotannins: Formation of polygalloylglucoses by enzymatic acylation of 1,2,3,4,6-penta-*O*-galloylglucose. *Archives of Biochemistry and Biophysics*, 283, 530–532.

Hofmann, T., Glabasnia, A., Schwarz, B., *et al.* (2006) Protein binding and astringent taste of a polymeric procyanidin, 1,2,3,4,6-penta-*O*-galloyl-$\beta$-d-glucopyranose, castalagin, and grandinin. *Journal of Agricultural and Food Chemistry*, 54, 9503–9509.

Johnson, M., Agrawal, A., Maron, J. & Salminen, J.-P. (2009) Heritability, covariation, and natural selection on 24 traits of common evening primrose (*Oenothera biennis*) from a field experiment. *Journal of Evolutionary Biology*, 22, 1295–1307.

Karonen, M., Parker, J., Agrawal, A. & Salminen, J.-P. (2010) First evidence of hexameric and heptameric ellagitannins in plants detected by liquid chromatography/electrospray ionization mass spectrometry. *Rapid Communications in Mass Spectrometry*, 24, 3151–3156.

Khanbabaee, K., Schulz, C. & Lötzerich, K. (1997) Synthesis of enantiomerically pure strictinin using a stereoselective esterification reaction. *Tetrahedron Letters*, 38, 1367–1368.

Kilkowski, W.J. & Gross, G.G. (1999) Color reaction of hydrolyzable tannins with Bradford reagent, Coomassie brilliant blue. *Phytochemistry*, 51, 363–366.

Kolodziej, H., Kayser, O., Latté, K.P. & Ferreira, D. (1999) Evaluation of the antimicrobial potency of tannins and related compounds using the microdilution broth method. *Planta Medica*, 65, 444–446.

Luck, G., Liao, H., Murray, N.J., *et al.* (1994) Polyphenols, astringency and proline-rich proteins. *Phytochemistry*, 37, 357–371.

Martin, J.S. & Martin, M.M. (1982) Tannin assays in ecological studies: Lack of correlation between phenolics, proanthocyanidins and protein-precipitating constituents in mature foliage of six oak species. *Oecologia*, 54, 205–211.

Moilanen, J. & Salminen, J.-P. (2008) Ecologically neglected tannins and their biologically relevant activity: chemical structures of plant ellagitannins reveal their in vitro oxidative activity at high pH. *Chemoecology*, 18, 73–83.

Moilanen, J., Sinkkonen, J. & Salminen, J.-P. (2013) Characterization of bioactive plant ellagitannins by chromatographic, spectroscopic and mass spectrometric methods. *Chemoecology*, doi: 10.1007/s00049-013-0132-3.

Mosedale, J.R., Charrier, B. & Janin, G. (1996) Genetic control of wood colour, density and heartwood ellagitannin concentration in European oak (*Quercus petraea* and *Q. robur*). *Forestry*, 69, 111–124.

Niehaus, J.U. & Gross, G.G. (1997). A gallotannin degrading esterase from leaves of pedunculate oak. *Phytochemistry*, 45, 1555–1560.

Niemetz, R. & Gross, G.G. (1998) Gallotannin biosynthesis: purification of *β*-glucogallin: 1,2,3,4,6-pentagalloyl-*β*-D-glucose galloyltransterase from sumac leaves. *Phytochemistry*, 49, 327–332.

Niemetz, R. & Gross, G.G. (1999) Gallotannin biosynthesis: a new *β*-glucogallin-dependent galloyltransferase from sumac leaves acylating gallotannins at positions 2 and 4. *Journal of Plant Physiology*, 155, 441–446.

Niemetz, R. & Gross, G.G. (2001) Gallotannin biosynthesis: *β*-glucogallin: hexagalloyl 3-*O*-galloyltransferase from *Rhus typhina* leaves. *Phytochemistry*, 58, 657–661.

Niemetz, R., Schilling, G. & Gross, G.G. (2001) Ellagitannin biosynthesis: oxidation of pentagalloylglucose to tellimagrandin II by an enzyme from *Tellima grandiflora* leaves. *Chemical Communications*, 2001, 35–36.

Nonaka, G. (1989) Isolation and structure elucidation of tannins. *Pure & Applied Chemistry*, 61, 357–360.

Okuda, T., Yoshida, T. & Hatano, T. (1992) Polyphenols from Asian plants. Structural diversity and antitumor and antiviral activities. In: *Phenolic Compounds in Food and Their Effects on Health I: Analysis, Occurrence, and Chemistry* (eds C.-T. Ho, C.Y. Lee & M.-T. Huang), pp. 160–183. American Chemical Society, Washington, DC.

Okuda, T., Yoshida, T. & Hatano, T. (2000) Correlation of oxidative transformations of hydrolyzable tannins and plant evolution. *Phytochemistry*, 55, 513–529.

Ossipov, V., Haukioja, E., Ossipova, S., *et al.* (2001) Phenolic and phenolic-related factors as determinants of suitability of mountain birch leaves to an herbivorous insect. *Biochemical Systematics and Ecology*, 29, 223–240.

Ossipov, V., Salminen, J.-P., Ossipova, S., *et al.* (2003) Gallic acid and hydrolysable tannins are formed in birch leaves from an intermediate compound of the shikimate pathway. *Biochemical Systematics and Ecology*, 31, 3–16.

Ozawa, T., Lilley, T.H. & Haslam, E. (1987) Polyphenol interactions: astringency and the loss of astringency in ripening fruit. *Phytochemistry*, 26, 2937–2942.

Quideau, S., Deffieux, D., Douat-Casassus, C. & Pouységu, L. (2011) Plant polyphenols: chemical properties, biological activities, and synthesis. *Angewandte Chemie International Edition*, 50, 586–621.

Reed, J.D. (1994) Nutritional toxicology of tannins and related polyphenols in forage legumes. *Journal of Animal Science*, 73, 1516–1528.

Roslin, T. & Salminen, J.-P. (2008) Specialization pays off: contrasting effects of two types of tannins on oak specialist and generalist moth species. *Oikos*, 117, 1560–1568.

Salminen, J.-P. & Karonen, M. (2011) Chemical ecology of tannins and other phenolics: we need a change in approach. *Functional Ecology*, 25, 325–338.

Salminen, J.-P. & Lempa, K. (2002) Effects of hydrolysable tannins on an herbivorous insect: fate of individual tannins in insect digestive tract. *Chemoecology*, 12, 203–211.

Salminen, J.-P., Ossipov, V., Haukioja, E. & Pihlaja, K. (2001) Seasonal variation in the content of hydrolysable tannins in leaves of *Betula pubescens*. *Phytochemistry*, 57, 15–22.

Salminen, J.-P., Ossipov, V. & Pihlaja, K. (2002) Distribution of hydrolysable tannins in the foliage of Finnish birch species. *Zeitschrift für Naturforshung*, 57c, 248–256.

Salminen, J.-P., Roslin, T., Karonen, M., *et al.* (2004) Seasonal variation in the content of hydrolysable tannins, flavonoid glycosides and proanthocyanidins in oak leaves. *Journal of Chemical Ecology*, 30, 1693–1711.

Salminen, J.-P., Karonen, M. & Sinkkonen, J. (2011) Chemical ecology of tannins: recent developments in tannin chemistry reveal new structures and structure-activity patterns. *Chemistry – A European Journal*, 17, 2806–2816.

Schultz, J.C, Hunter, M.D. & Appel, H.M. (1992) Antimicrobial activity of polyphenols mediates plant-herbivore interactions. In: *Plant Polyphenols* (eds R.W. Hemmingway & P.E. Laks), pp. 521–637. Plenum Press, New York.

Tanaka, T., Fujisaki, H., Nonaka, G. & Nishioka, I. (1992) Reaction of dehydroellagitannins with L-cysteine methyl ester. *Heterocycles*, 33, 375–383.

Viriot, C., Scalbert, A., Lapierre, C. & Moutounet, M. (1993) Ellagitannins and lignins in aging of spirits in oak barrels. *Journal of Agricultural and Food Chemistry*, 41, 1872–1879.

Vivas, N., Laguerre, M., Glories, Y., *et al.* (1995) Structure simulation of two ellagitannins from *Quercus robur* L. *Phytochemistry*, 39, 1193–1199.

Wang, C.-C., Chen, L.-G. & Yang, L.-L. (1999) Antitumor activity of four macrocyclic ellagitannins from *Cuphea hyssopifolia*. *Cancer Letters*, 140, 195–200.

Werner, I., Bacher, A. & Eisenreich, W. (1999) Analysis of gallic acid biosynthesis via quantitative prediction of isotope labeling patterns. In: *Plant Polyphenols. 2. Chemistry, Biology, Pharmacology, Ecology* (eds G.G. Gross, R. Hemingway & T. Yoshida), pp. 43–61. Kluwer Academic/Plenum, New York.

Chapter 5

# Diverse Ecological Roles of Plant Tannins: Plant Defense and Beyond

C. Peter Constabel, Kazuko Yoshida, and Vincent Walker
*Centre for Forest Biology, Department of Biology, University of Victoria, Victoria, BC, Canada*

**Abstract:** Tannins are polyphenolic compounds found throughout the plant kingdom and occurring commonly in woody plants and forest trees. They are functionally defined by their ability to precipitate proteins, but can also be oxidatively active. Since tannins are found at high concentrations in forest plants, they can have far-reaching ecological effects. This chapter provides an overview of their diverse functions in defense against leaf-eating herbivores and pathogens, in modulating dispersal in fruits and seeds, and as regulators of nutrient cycling and abiotic stress tolerance.

**Keywords:** polyphenols, proanthocyanidin, plant defense, antimicrobial, plant stress, pro-oxidant, antioxidant

## 5.1 Introduction

The study of plant tannins has a long history among ecologists interested in plant defense, as well as among food scientists interested in the effects of tannins on human health and nutrition. Tannins are chemically diverse, but are functionally defined by their ability to bind protein. Although much research has been driven by early work postulating that the protein-binding ability of tannins is important in defense against insect pests, currently the best evidence for the antiherbivore role of tannins comes from work with mammals. The role of tannins in defense against insects and mammals has been reviewed recently (Barbehenn & Constabel, 2011), and in this chapter we will only highlight the key findings in this area. Our aim is to give a broad overview that illustrates the diverse ecological functions that have been demonstrated in different biological systems. Many of these functions relate to the tannins' broad antimicrobial nature. By inhibiting bacteria in complex environments,

*Recent Advances in Polyphenol Research, Volume 4*, First Edition.
Edited by Annalisa Romani, Vincenzo Lattanzio, and Stéphane Quideau.

including water and soils, tannins can have pronounced effects on nutrient cycling and can affect entire ecosystems. At the scale of the individual plant, it is becoming clear that tannin synthesis is precisely regulated and is often restricted to specific cells or tissues. Therefore, knowledge about a tannin's localization will provide clues as to its function in the plant. Similarly, the induction of tannin synthesis in response to environmental stresses may also indicate potential functions, and tannin-specific genes and molecular probes are very useful tools in such studies. Finally, the well-known effects of dietary tannins on human health have spawned many studies in the realm of food science, and this literature contains a wealth of information relevant to the ecology of tannins. In this chapter we consider these diverse perspectives on tannins in the context of their biological roles in plants and in ecosystems.

## 5.2 Overview of tannin structure and function in defense

### *5.2.1 Structural diversity and distribution*

Tannins are defined as water-soluble polyphenolic compounds that precipitate proteins and thus comprise several distinct types of phenolic compound. Two main groups are recognized: the hydrolyzable tannins (HTs) and the condensed tannins (CTs; also known as proanthocyanidins, PAs). Galloyl glucoses (GGs) and ellagitannins, which include significant structural diversity, are all considered HTs. GGs are esterified with multiple galloyl units (Fig. 5.1), while the ellagitannins are characterized by the presence of at least one hexahydroxyldiphenoyl (HDDP) group, formed by the oxidative coupling of adjacent galloyl groups. Examples of ellagitannins include peduncalagin and vescalagin (Okuda *et al.*, 2009; Yoshida *et al.*, 2009; Salminen *et al.*, 2011). The CTs are oligomers and polymers of flavan-3-ols (Fig. 5.2); while structurally more uniform than the HTs, CTs from different species may differ depending on hydroxylation pattern, stereochemistry, type of interflavan linkage, and degree of polymerization (Haslam, 1975; Porter, 1988). The best-characterized CTs are the procyanidins and prodelphinidins, which are linked via C4–C8 linkages of the flavan-3-ol monomers but differ in the degree of hydroxylation of their flavonoid B-ring (Fig. 5.2). Other CT variants may have different linkage types or substituents (Ferreira *et al.*, 2006). The mean degree of polymerization (mDP) of CTs can vary widely, ranging from dimers to polymers of up to 30 or more subunits depending on the plant species (Barbehenn & Constabel, 2011). Even within one genus, there can be dramatic variation in chain length. In *Populus fremonti*, the mDP is two, compared to a mean of 22 in the closely related *P. angustifolia* (Scioneaux *et al.*, 2011). The diversity of tannin structure can complicate studies of biological functions, since structural variants may differ in activity. Furthermore, structurally distinct tannins respond differently in commonly used assays such as the butanol HCl test (Porter *et al.*, 1986), and thus a careful choice of standard is essential for accurate quantitative data (Schofield *et al.*, 2001).

Tannins are among the most widely occurring secondary metabolites of plants, although they are not evenly distributed. They have been reported within the ferns, fern allies, gymnosperms, and many dicotyledonous and monocotyledonous flowering plants (Mole, 1993; Porter, 1988). While the CTs can be found in all these groups, HTs are restricted to the

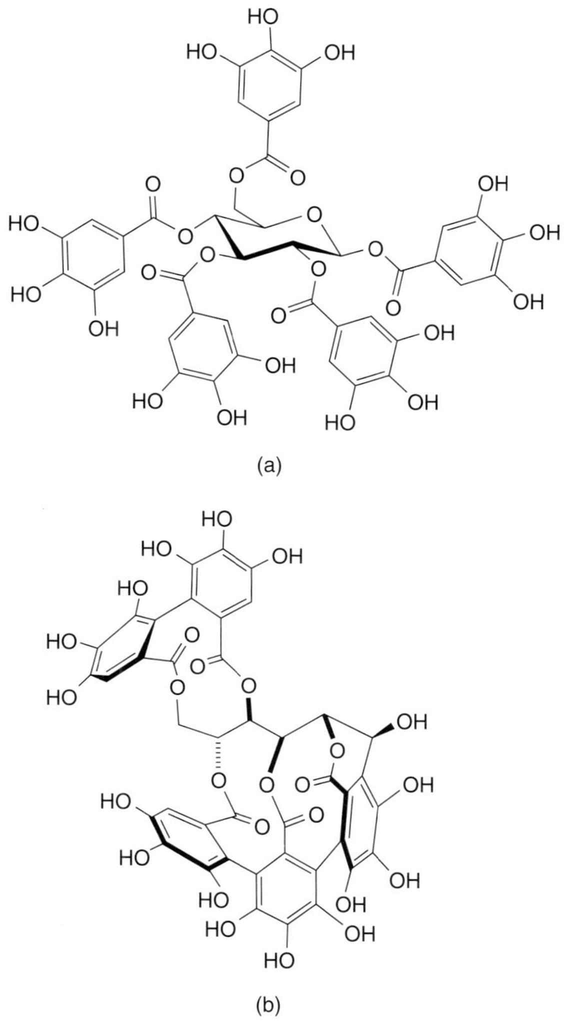


**Fig. 5.1** Common hydrolyzable tannins. (a) $\beta$-pentagalloylglucose. (b) Vescalagin, an example of an ellagitannin derived from $\beta$-pentagalloylglucose.

dicotyledonous plants (Iwashina, 2000; Okuda *et al.*, 2009). Tannins are often found in the vegetative tissues of woody plants, including roots, bark, stems, and leaves (Porter, 1988). Many herbaceaous plants also accumulate tannins. Some plant families appear not to make tannins at all, whereas others contain many tannin-rich representatives (Mole, 1993). Data regarding the absence of tannins should be treated with caution, since the analytical method used must be kept in mind (see Section 5.3.1).

In leaves of forest trees, tannin concentrations can differ remarkably, varying with developmental stage, tissue, and environmental conditions (Barbehenn & Constabel, 2011). Furthermore, there is considerable genotype-dependent variation in tannin content within a species; in trembling aspen, for example, CT concentrations among genotypes can range

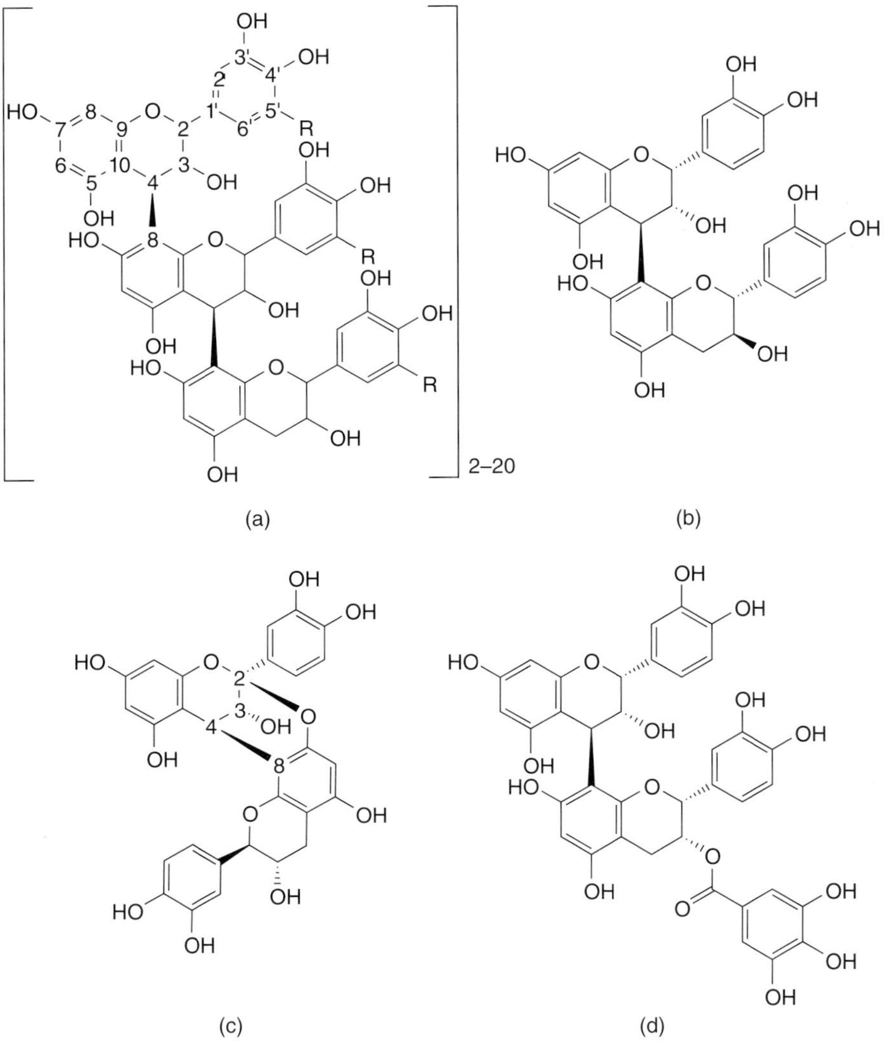


**Fig. 5.2** Common condensed-tannin structures. (a) Typical CT showing C4–C8 linkages and positions. The degree of polymerization is typically 2–20, but many short tannins are found as well. Prodelphinidins (R=OH) are more hydroxylated than procyanidins (R=H). (b) Procyanidin dimer B1. (c) Procyanidin dimer A2, showing additional linkage to C2. (d) Galloylated B2 dimer, typical for grape-seed CTs.

from 2 to 25% d.w. (Hwang & Lindroth, 1997; Osier & Lindroth, 2006). In aspen and poplar, an increase in CT concentration with leaf maturity is often observed, although the significance of this trend is not clear (Donaldson *et al.*, 2006; Rehill *et al.*, 2006). By contrast, HTs typically decline in abundance during the growing season, for example with aging of sessile oak (Peng *et al.*, 1991) and white birch (Laitinen *et al.*, 2002). In *Liquidambar formosana*, galloylglucoses are synthesized in spring, oxidated to ellagitannins in summer, and degraded in fall (Hatano *et al.*, 1986). It is important to note that a given species may contain both CTs and HTs. For example, *Acer saccharum* leaves have been reported to contain 3%

CT and 3.5% gallotannin; *Liquidambar styractiflua* had 3.3 and 1.5%, respectively (Inoue & Hagerman, 1998).

While the leaves of many herbaceous plants do not synthesize tannins, in some legume species foliar tannin levels are important. For example, the forage legumes *Lotus corniculatus* and *Onobrychis viciifolia* can accumulate CTs at levels of up to 9%. These are of interest for agriculture, since moderate levels of tannin can be beneficial for cattle and other ruminants, reducing the risk of bloat caused by high-protein diets and reducing internal parasite loads (see Section 5.3.1). By contrast, other legumes such as alfalfa (*Medicago sativa*) contain no detectable tannins; this important forage crop is thus a major target for genetic engineering of enhanced tannins (Dixon *et al.*, 2005).

## *5.2.2* **In vitro** ***biochemical activities***

A broad array of *in vitro* biochemical activities has been described for the tannins. Their ability to bind proteins and other macromolecules is well known, and forms the basis of many practical applications, for example in tanning of leather. This property is also the basis for some of their biological functions. Tannins bind proteins principally via hydrogen bonds at their many phenolic hydroxyls, although hydrophobic interactions with aromatic rings also play a role. At pH values above 8–9, the phenolic groups are oxidized and protein precipitation no longer occurs. It is thus unlikely that precipitation of dietary protein is an effective defense against leaf-eating lepidopterans, which have extremely alkaline guts (Martin *et al.* 1987). Despite this, it is often incorrectly assumed that it is the protein-precipitating capacity of tannins that is important for defense against insects. By contrast, at the more acidic pH values present within vertebrate herbivore digestive systems, tannin-mediated protein binding appears to be an effective mechanism of defense (Barbehenn & Constabel, 2011). The efficiency of protein precipitation by tannins has been extensively investigated and depends on tannin molecular weight and concentration, as well as the tannin and protein structure; furthermore, conformational flexibility appears to be an important contributor to efficacy of binding (for a review, see Quideau *et al.*, 2011). The binding of tannins to extracellular enzymes or secreted proteins may give rise to some of their antimicrobial effects (see Section 5.2.4).

The idea that the pro-oxidant properties of tannins and other phenolics could be the basis of their activity against insects was first proposed in a landmark paper by H. Appel (1993). Under both alkaline and oxidizing conditions, auto-oxidation of the phenolic leads to formation of semiquinone radicals, which can be converted to quinones (Barbehenn & Constabel, 2011). Enzymes such as peroxidases and polyphenol oxidases can also participate in these processes. The propensity of tannins to oxidize depends on the redox conditions of the solution, its pH, and the concentrations of oxidants and antioxidants. Thus, ascorbic acid can be an important counter-adaptation of herbivores, affecting the defense potential of tannins. The type and structure of the tannin is clearly important, and Barbehenn *et al.* (2006) suggest that ellagitannins have greater pro-oxidant activity than the CTs. Tannins with *ortho*-tri-hydroxylation, such as prodelphinidins, oxidize more readily than the di-hydroxylated procyanidins. Salminen *et al.* (2011) systematically compared a series of ellagitannins and gallotannins, and correlated their potential for pro-oxidant activity with the structural features of substitution patterns. Interestingly, pro-oxidant activity was inversely correlated with protein-binding activity.

Because of the potential health benefits of dietary tannins, most *in vitro* investigations have focused on their antioxidant properties. The ability of tannins, like other polyphenols, to act as antioxidants is rooted in their ability to directly quench free radicals. This can occur via donation of hydrogen atoms from the phenolic to a free radical or through a single electron transfer (Quideau *et al.*, 2011). The key structural features that appear to be important are the number and proximity of additional hydroxyls and aromatic rings; for example, the number of galloyl subunits and the degree or hydroxylation of the B-ring in PAs are important parameters determining *in vitro* antioxidant activity (Hagerman *et al.*, 1998). Furthermore, the tannins have greater antioxidant activity than structurally similar simple phenolics, indicating that higher degree of polymerization and molecular weight are important features for this property. Several different techniques and assays have been used to demonstrate the *in vitro* antioxidant activity of a variety of tannins (Hagerman *et al.*, 1998; Okuda *et al.*, 2009), but whether this is important for their biological and ecological functions still needs to be resolved. Interestingly, tannic acid and catechin can act as free-radical scavengers when added to hemolymph and midgut fluids collected from *Manduca sexta* caterpillars (Johnson, 2005).

An important biochemical property of tannins is their ability to chelate metal ions, based on their many *ortho*-dihydroxyl groups. Tannins can bind metals (Fe, Pb, Cu, and Zn) *in vitro* (Chin *et al.*, 2009; McDonald *et al.*, 1996). In particular, their ability to bind Fe ions may be important for their antimicrobial properties (Scalbert, 1991; see Section 5.2.4). Complexation of heavy metals may play a role in other biological processes, as outlined in Section 5.4. Thus, tannins interact with organisms and molecules via multiple potential mechanisms, and ultimately their ecological effects will reflect this diversity. We note that caution must be exercised when extrapolating *in vitro* activities to *in vivo* functions or ecological roles.

### 5.2.3 *Old and new views on tannins in defense*

The publication of widely cited papers by Paul Feeny first showed a negative correlation of leaf tannin content with insect herbivore performance (Feeny, 1968; 1970). Since tannins were known to bind proteins, these publications led to the widespread idea that tannins act as general defenses against insect pests by binding dietary protein or digestive enzymes. However, while tannins can have negative effects on insects, these appear not to be related to their protein-binding potential but rely on other mechanisms. Many insect pests are lepidopterans, which have a very basic gut pH, and tannins do not bind proteins under these conditions (reviewed in Barbehenn & Constabel, 2011). Experiments with *Malacosoma disstria* indicated that tannic acid decreases the efficiency of conversion of digested food, reduces growth, and increases mortality, but does not reduce digestibility (Karowe, 1989). This suggests that tannic acid acts as a toxin and feeding deterrent. Likewise, adding CTs or HTs to the diet does not reduce protein utilization by *Lymantria dispar* (Barbehenn *et al.*, 2009a,2009b). Bernays *et al.* (1981) also found no evidence of tannins affecting the digestibility of proteins in grasshoppers, even at high concentrations.

As a result of such studies, which make tannin-protein binding in insect guts very unlikely, the most recent work on tannin–insect herbivore interactions stresses the importance of the' oxidative potential of tannins (Barbehenn & Constabel, 2011; Salminen *et al.*, 2011). Electron spin resonance shows higher levels of semiquinone radicals in the midguts of insects fed on high-tannin leaves versus control leaves (Barbehenn *et al.*, 2009a). Tannin oxidation occurs in caterpillar midguts despite the extremely low oxygen levels, possibly with $Fe^{3+}$ acting as an oxidant. The dismutation of the semiquinones to quinones can lead to accumulation of the more stable oxidant hydrogen peroxide, which can enter and produce oxidative damage within epithelial cells (Barbehenn & Constabel, 2011). In these experiments, ellagitannins were oxidatively more active in caterpillar guts than CTs, presumably due to the presence of HHDP groups (Salminen *et al.*, 2011).

Many non-adapted insects do appear to be deterred from consuming plants high in tannins, although it is difficult to extrapolate from artificial diet studies, which may exaggerate the tannins' effects (Barbehenn & Constabel, 2011). On the other hand, many studies with CTs in foliage have shown no effect on insects. For example, Ayres *et al.* (1997) compared a suite of tannins from different species against six different lepidopteran pests and found that most were not affected by tannins at natural concentrations, including insect species that normally do not encounter tannins. In hybrid aspen, high levels of CTs generated by transgenic over-expression of a CT regulatory gene (Mellway *et al.*, 2009) did not deter the polyphagous *Lymantria disspar* or the oligophagous *Malacosoma disstria* (Boeckler *et al.*, 2014); rather, these lepidopteran pests appear to be more sensitive to the salicinoid phenolic glycosides that are found in the Salicaceae. Therefore, it appears likely that there are metabolic adaptations, such as antioxidants in the midgut, which prevent oxidative damage and help buffer pest insects from a diet high in tannins (Barbehenn & Constabel, 2011). A key question that remains to be answered is whether pests that do not feed on high-tannin plants are prevented from doing so by the presence of tannins. This can perhaps be addressed using gene knock-outs that produce tannin-deficient mutant plants.

In vertebrates, tannins act via a different mechanism: they can bind proteins in the acidic environment of the gut, which can have significant impact on health and reproduction (Aerts *et al.*, 1999; Wallis *et al.*, 2012). Binding of dietary proteins by leaf tannins in the gut may make the proteins unavailable to the animal, and thus reduce available nitrogen. DeGabriel *et al.* (2009) demonstrated a negative correlation between tannin levels and the reproductive success of eucalyptus-feeding Australian ringtail possum, using an assay that measures the tannin protein-binding capacity *in vitro* (DeGabriel *et al.*, 2008). Likewise, nutritionally available nitrogen correlates well with the reproductive fitness of larger herbivores such as moose (*Alces alces*) (McArt *et al.*, 2009). Here, numbers of offspring were significantly affected by the nitrogen availability of summer diet (*Salix* and *Betula*) in Alaska. Browsing mammals such as moose, which feed on woody plants in ecosystems that are typically nitrogen-limited, are most likely to be impacted by tannins. The effect of tannins on vertebrates has been extensively studied in acorns, which contain diverse HTs and CTs. Oak tannins (found in acorns at up to 10% d.w.) act as deterrents to feeding by the Japanese wood mouse, and also reduce its survival (Shimada *et al.*, 2006). Most acorn-feeding specialists cannot survive on acorn-only diets and must supplement their food intake (Vander Wall, 2001).

Many browsing mammals are known to produce tannin-binding proteins in the saliva. These proline-rich proteins alleviate the negative effects of tannins. Grazing mammals, which typically feed on grasses and forbs with low tannin levels, generally do not produce such tannin-binding proteins. Omnivores and frugivores are exposed to tannins in the diet, and like browsers, produce proline-rich salivary proteins, supporting the idea that tannins can negatively affect mammals (Shimada, 2006). The effects of tannins on vertebrates can also be seen with domestic animals; goats feed on woody shrubs and are tolerant of tannins, but their performance is improved by the addition of polyethylene glycol (PEG) to the diet. The effects of PEG (which binds tannins) are analogous to those of the proline-rich salivary proteins. In pigs, the experimental addition of fava bean CT to the diet reduced the digestibility of crude protein by 7–9% (Jansman *et al.*, 1995).

It must be noted that under some conditions tannins can also positively impact mammals, specifically cows and other domesticated ruminants. Forage crops with high protein content, such as alfalfa, promote animal growth and milk production, but their high nitrogen content stimulates excessive microbial growth in the rumen. Such rapid microbial activity leads to foaming, and potentially to cattle bloat and death. Moderate concentrations of tannin (2–4% d.w.) in forages such as sainfoin can suppress this excessive rumen bacterial activity, via the antimicrobial activity of tannins (see Section 5.2.4). In addition to preventing bloat, tannins also prevents rapid proteolysis in the rumen, shifting protein digestion into the intestine (Aerts *et al.*, 1999; McSweeney *et al.*, 2001). The resulting enhancement of nitrogen utilization from protein and reduction of loss of nitrogen as urea is also desirable from an economic perspective.

### 5.2.4 The antimicrobial nature of tannins

Tannins often exhibit antimicrobial activity, which leads to multiple potential biological and ecological effects. In pure culture, inhibition of growth by tannins has been documented for a wide array of microbes and filamentous fungi, with minimum inhibitory concentrations ranging from 0.5–20.0 g/l for bacteria and reaching as low as 0.012 g/l for fungi (Scalbert, 1991). Antimicrobial effects have been documented for many types of microbes, including human pathogens, free-living soil microbes, plant pathogens, food-spoiling saprophytes, and rumen symbionts (Mila *et al.*, 1996; Puupponen-Pimiä *et al.*, 2005; Engels *et al.*, 2009). The sensitivity of microbes to tannins can vary substantially, depending on both the microbial strain and the specific tannin. For example, the gastrointestinal bacterium *Staphylococcus aureus*, which can cause food poisoning, was inhibited by berry ellagitannins but not by CTs or other phenolics tested (Puupponen-Pimiä *et al.*, 2005). By contrast, the potentially pathogenic *Listeria monocytogenes* was not affected by any of the berry phenolics or extracts tested, with the exception of cranberry. Anderson *et al.* (2012) showed that the inhibitory effects of several CTs and HTs on cultures of *Camphylobacter jejuni* were distinct. The effects of CTs could be mitigated by the supplementing of casamino acids into the medium, suggesting that CTs act on this bacterium via protein-binding, whereas HTs act via a different mechanism.

Many species of rumen bacteria have also been tested for sensitivity to tannins. Such studies confirm that tannins inhibit microbial growth, but that their effect varies depending on the bacterial species (Nelson *et al.*, 1997; Sivakumaran *et al.*, 2004). Again, the specific tannin is important in determining its antimicrobial activity. For example, purified *Desmodium ovalifolium* and *Mirtus communis* CTs are more inhibitory than the commonly used quebracho tannin. These *in vitro* studies with pure bacterial cultures are the most direct demonstrations of the antibiotic potential of tannins. It should be emphasized that the impact of tannins may be manifested differently in mixed populations, as are found *in vivo*, or in ecologically relevant situations. Min *et al.* (2002) showed that while the growth of several species of proteolytic rumen bacteria was reduced by *Lotus corniculatus* CTs, overall proteolytic activity was unchanged.

The mechanisms by which tannins exert antimicrobial effects are not well understood. They could involve the binding or inactivation of secreted enzymes or substrates, and many studies show reduced breakdown of cellulose, pectins, or hemicelluloses in the presence of tannins. In some cases, proportionally greater inhibition is observed using higher-molecular-weight tannins, suggesting that protein-binding is involved. *In vitro*, a wide variety of enzyme activities (cellulases, pectinases, peroxidases, and others) are negatively affected by tannins (Scalbert, 1991). The proteolytic activity of rumen bacteria is reduced in the presence of tannins (Min *et al.*, 2003). In addition, tannins are able to bind to membrane or cell-wall components of Gram-positive bacteria, potentially disrupting transport or cell integrity (Smith *et al.*, 2005). In a comparative study, Nelson *et al.* (1997) found that all species of rumen bacteria tested were bound by tannin. This binding occurred rapidly (<10 minutes), although different bacteria were bound at different rates. In other experiments, smaller phenolics that do not precipitate protein were as effective as the larger polymers, suggesting that antimicrobial mechanisms besides protein-binding are involved (Scalbert, 1991). In addition to their inhibitory effects, tannins may also be utilized by some microbial species. Certain species within *Bacillus*, *Staphylococcus*, and *Klebsiellia*, as well as some fungal species (*Aspergillum niger*), were shown to be able to grow with tannin as the sole source of carbon. This ability is conferred by the production of tannase, which is active on gallotannins (Mingshu *et al.* 2006; Ascacio-Valdés *et al.* 2014).

One feature of tannins that is important for their interaction with microbes is their metal-binding activity. Since tannins, like most polyphenols, contain many *ortho*-dihydroxyl groups on aromatic rings, they can bind to or chelate metal ions, particularly iron (Scalbert, 1991). Iron is typically a limiting nutrient for microbes, and chelation is a common antimicrobial strategy. The involvement of iron can be seen in the induction of genes that encode proteins for the binding and transportation of metals in *Escherichia coli* treated with inhibitory levels of tannin (Lin *et al.*, 2011). Furthermore, Lim *et al.* (2013) found that *Pseudomonas protegens* responds to tannic acid with transcription of ion homeostasis-related genes, including siderophores. The Fe chelation activity of tannins was also shown by experiments in which the inhibitory effects of tannins on bacterial growth in culture were reversed by Fe supplementation. Gallotannin from mango fruit inhibited the growth of *Staphyllococcus aureus*, *E. coli*, or *Bacillus subtilis*, but did not inhibit growth of other species (Engels *et al.*, 2009). The toxicity of the tannins in these cultures, however, was counteracted by the addition of $Fe^{3+}$ and $Fe^{2+}$ to the medium, suggesting that tannins act either by chelating iron or via a redox-cycle mechanism (Hagerman *et al.*, 1998). Fe chelation by tannins has also been shown to affect the plant pathogen *Dickeya dadantii* (formerly *Erwinia chrysanthemi*); significantly, strains deficient in iron transport were most sensitive to the tannin (Mila *et al.*, 1996).

The *in vitro* antimicrobial activity of many tannins suggests they could play a role in defense against plant pathogens. To date, few studies have directly tested this idea. However, Holeski *et al.* (2009) found a strong correlation between naturally varying tannin levels in *P. tremuloides* clones with resistance to a *Venturia* shoot blight in a common garden experiment. A potential role of CTs in pathogen defense is consistent with the observed induction of the CT pathway in hybrid poplar after infection with *Melampsora medusae* (Miranda *et al.*, 2007). In plum (*Prunus domesticus*) fruit, the pathogen *Taphrina pruni* stimulates CT and chlorogenic acid accumulation (Fuchs & Spiteller, 1998). In acorns from *Quercus serrata*, Takahashi *et al.* (2010) found a correlation between resistance to the fruticolous fungus *Ciboria batschiana* and high levels of tannins in the cotyledons. Similarly, plant endophytic fungi appear to be affected by CTs; in twig bark, endophyte infection rates were negatively correlated to tannin concentration (Bailey *et al.*, 2005), further suggesting that PAs have the potential to influence plant-associated fungi.

## 5.3 Tissue localization and ecological function

The occurrence of tannins varies widely among tissues, organs, and cell types within plants. Extensive lists of occurrences in tissues and plant species have been compiled (e.g. Bate-Smith & Metcalfe, 1957; Haddock *et al.*, 1982; Porter, 1988). However, our knowledge about tannin distribution in nature and of tissue-specific occurrence must be evaluated in the context of the analytical tools available for detection and quantitation. CTs can be assayed spectrophotometrically with relatively simple techniques, such as the butanol HCl assay developed by Porter *et al.* (1986). This test is considered to give reliable CT estimates as long as the appropriate standard is used (Schofield *et al.*, 2001). No similar methods are available for the HTs and ellagitannins, which require more sophisticated technical approaches for accurate detection and quantitation, such as LC-MS (Mueller-Harvey, 2001; Salminen *et al.*, 2011). Although these methods are rapidly evolving and becoming more and more sensitive, they often require more material, and to date less information is available about the distribution of HTs. LC-MS is also useful to quantify the oligomeric CTs, but is not as accurate for larger CT polymers (Santos-Buelga & Scalbert, 2000; Prior & Gu, 2005). The development of thiolysis- and phloroglucinolysis-based assays that depolymerize CTs to derivatized monomers quantified by HPLC (Guyot *et al.*, 2001; Tarascou *et al.*, 2010) is an excellent approach, as they provide both quantitative and structural information about CTs. The general phenolic assays (e.g. Folin–Ciocalteu) are of more limited utility because they are very nonspecific. The analytical methods available are discussed in detail in excellent reviews by Schofield *et al.* (2001) and Tarascou *et al.* (2010).

In addition, the availability of CT-specific stains such as 4-dimethylaminocinnamaldehyde (DMACA) (Feucht *et al.*, 1986) and vanillin-HCl facilitates a histochemical approach for CTs. Therefore, much more is known about localization CTs than of HTs. It should be noted that DMACA reacts only with CTs (polymeric flavan-3-ols), while vanillin also detects the monomeric flavan-3-ols. Because of the lack of information about HT localization, the following section deals specifically with CTs.

### 5.3.1 *Distribution of tannins in vegetative tissues*

Tannins have been investigated intensively in many trees, where they are typically found in vegetative tissues including roots, bark, and leaves. Bark is often a source of tannins in conifers, and in *Acacia* species the presence of CTs in heartwood is also well documented (Haslam, 1975; Porter, 1988). In herbs, tannins are often reported in leaves and sometimes in roots, as well as in reproductive structures (see next paragraph). Tannin accumulation is sometimes specific to certain cell types. For example, in young *O. viciifolia* leaves, vanillin staining detected flavan-3-ols in subepidermal cells of both abaxial and adaxial surfaces; this is consistent with the accumulation of tannins in these cells, observed by electron microscopy (Lees *et al.*, 1993). During leaf development, the distribution of the CTs shifts from the abaxial to the adaxial side of the leaf, and also into specialized cells called idioblasts in the epidermis (Lees *et al.*, 1993). Such specialized CT-containing cells have also been described in the phloem parenchyma of young poplar (Kao *et al.*, 2002), where both vanillin-HCl and DMACA were used to show that CTs typically accumulate in the epidermal cells of young stems and in the hypodermal cells of older stems. In young leaves, the lower epidermis cells stain for CTs, as do leaf mesophyll cells. The tendency of epidermal layers to accumulate CTs is suggestive of a protective function against pathogens, as discussed before. The epidermal localization would also be consistent with a function in protection against UV stress and associated antioxidant activity (Close & McArthur, 2002). This function has never been demonstrated *in vivo*, however.

The roots of some woody plants can contain substantial amounts of tannin, and root turn-over contributes to the accumulation of tannins in soil (see Section 5.4). In *P. tremuloides* root tips, DMACA staining localized CTs to the cortex, lateral root cap, and epidermal tissues (Kao *et al.*, 2002). In jack pine (*Pinus banksiana*) and eucalyptus (*Eucalyptus pilularis*), tannins were localized to a discrete region between the root tip and the suberised zone (McKenzie & Peterson, 1995). The tannins appear to be deposited in the walls of cortical cells as these undergo cell death. The precise functions of the tannins in this root zone are not known but deserve more attention (Peterson *et al.*, 1999). In the camphor tree (*Cinnamomum camphora*), CTs were observed by DMACA staining in the outer epidermal cells of the root tip, where they contribute to tolerance against Al toxicity (see Section 5.5).

Stress-induced accumulation of tannins can be an indicator that there are additional roles for tannins in vegetative tissue such as leaves. This has been particularly well documented in *P. tremuloides*, where leaf CT levels increase after wounding and insect herbivory (Osier & Lindroth, 2001; Peters & Constabel, 2002). Molecular studies indicate that these increases are the result of rapid transcriptional activation of the entire CT biosynthetic pathway in response to stress, and that the PtMYB134 transcription factor plays a key role in this induction (Mellway *et al.*, 2009). Trancriptome analysis also revealed that in hybrid poplar, the CT pathway is induced by UV-B exposure, as well as by infection with poplar leaf rust (*Melampsora medusae*) (Miranda *et al.*, 2007; Mellway *et al.*, 2009). High visible-light exposure and nutrient deprivation also lead to enhanced tannin accumulation (Osier & Lindroth, 2001). All these responses could indicate a functional role in general stress protection, perhaps against reactive oxygen species, which are often generated during plant stress responses. To date, there are only a few reports of tannin induction in other species. Enhanced levels of tannins due to herbivory were observed in the marine vascular plant *Thalassia testudium* (Arnold *et al.*, 2008), and, in birch, CT accumulation after UV-B exposure was reported (Lavola, 1998). In persimmon (*Diospyros kaki*), along with the CT biosynthetic genes, the DkMYB2 CT regulator responds to wounding, suggesting that the persimmon CTs may be inducible (Akagi *et al.*, 2010). In the CT-containing legume forages, CTs are known to be affected by growing conditions, including temperature and precipitation (Anuraga *et al.*, 1993; Berard *et al.*, 2011).

### 5.3.2 Tannins in seeds and fruit

The reproductive propagules of plants, particularly seeds and fruits, can be a rich source of tannins. Because fruit tannins are important in the human diet, providing flavor and promoting health, the food science literature contains many detailed studies on the qualitative and quantitative aspects of tannins in fruit and seeds (Prior & Gu, 2005; Rasmussen *et al.*, 2005). The structure and quantity of tannins in fruits such as grapes, apples, and blueberries has been extensively investigated (Prior *et al.*, 2001; Kennedy *et al.*, 2002; Hellström *et al.*, 2009).

The seed coats of *Phaseolus vulgaris* and other beans, particularly colored varieties, are often rich in tannins (Ma & Bliss, 1978; Jin *et al.*, 2012), as are many nuts (Prior & Gu, 2005). Within the monocots, the seeds of the cereals sorghum and barley contain CTs (Strumeyer & Malin, 1975; Brandon *et al.*, 1980; Aastrup *et al.*, 1984), but maize, rice, and wheat do not (Prior & Gu, 2005). In high-tannin sorghum varieties, the immature grains contain flavan-3-ols, which are polymerized during seed maturation (Butler, 1982). Similarly, in Arabidopsis seed coats PAs are oxidized and polymerized into brown pigments, which are deposited on to the cell walls (Debeaujon *et al.*, 2001; Pourcel *et al.*, 2005). CTs in the seed coat may help protect the embryo and endosperm from biotic stresses by serving as physical and/or chemical barriers, and from early germination by blocking the movement of signaling molecules that counteract seed dormancy (Debeaujon *et al.*, 2001; Lepiniec *et al.*, 2006). In cowpea seed, high tannin content correlates with resistance to bruchid beetles during storage (Lattanzio *et al.*, 2005).

The occurrence of tannins in fruit is common, and many fruits are reported to contain them (Prior & Gu, 2005). Significant CT content has been reported in berries, such as rowanberry, blackcurrant, saskatoon berry, and sea buckthorn, as well as in many *Vaccinium* species (Hellström *et al.*, 2009; Landete, 2011). In a survey of 99 common food plants, about half of the species tested contained tannins, typically berries and fruit; vegetables usually did not contain detectable tannin (Hellström *et al.*, 2009; Gu *et al.*, 2004). Significant amounts of ellagitannins were detected in 5 species of berry (cloudberry, red raspberry, rose hip, strawberry, and sea buckthorn), among 33 food items screened by Koponen *et al.* (2007); these compounds were the most abundant phenolics in cloudberry and red raspberry. Grape (*Vitis vinifera* L.) berries have been intensively investigated for tannin content, as the tannins contribute to the taste and stability of wines. Grape skins and seeds have the highest levels of CTs and flavan-3-ols; in seeds, procyanidins are predominant, while prodelphinidins are also abundant in skins (Souquet *et al.*, 1996, 2000). It is important to note that tannin levels can vary tremendously among fruit varieties and cultivars (Tarascou *et al.*, 2010). In some fruit, such as persimmon, both high- and low-CT varieties are known. Fruits of the low-tannin varieties stop producing CTs at an early stage of development, while fruits of high tannins continue accumulating them until fully ripened, giving them a correspondingly astringent taste (Akagi *et al.*, 2009).

Despite the wealth of information on the structures and concentrations of tannins in economically important fruit, how these change with fruit maturation has been studied in only a few species. It is generally assumed that the astringent taste typical of unripe fruit is caused by higher concentrations or changes in the solubility or composition (i.e. lower mDP) of tannins at later developmental stages. In grape, some reports describe a decrease in DP of CTs over ripening, but others have found both increases and decreases (Bindon & Kennedy, 2011). In blueberry fruit, a recent study detailed molecular and histochemical changes in CTs and flavonoids during berry development (Zifkin *et al.*, 2012). Immature berries had the highest CT content (25 mg/g d.w.), which reduced (to <5 mg/g d.w.) as the fruit expanded. The early phase of synthesis and accumulation is reflected in patterns of gene expression, and the CT-specific flavonoid genes are expressed only in senescent flowers and very young green berries. When total CT content per fruit was determined, however, CT amounts per berry were found to increase through all stages of development. This suggests that sufficient pools of biosynthetic precursors and/or enzymes are present even after transcription is turned off (Zifkin *et al.*, 2012). Unlike the CT genes, anthocyanin-specific genes are expressed only in late stages of ripening, just prior to color development. These strict differences in gene regulation suggest that CT levels are ultimately controlled at the level of gene expression. Early CT synthesis and higher CT content are also seen in grape seeds and skins, as well as in apple fruit (Bogs *et al.*, 2005; Takos *et al.*, 2006).

In parallel with strict temporal control of tannin gene expression and tannin accumulation, tissue- and cell-specific regulation of the pathway has also been studied. In the early stage of blueberry fruit development, CTs are detected in ovary tissues, the developing seed coat, and the fruit flesh (mesocarp). As the fruit matures, the CTs are concentrated in placentae and the epidermal layers (Zifkin *et al.*, 2012). In grape, the skin contains over half of the extractable CT content of the berry, with seed and flesh accounting for 30 and 15%, respectively (Bindon *et al.*, 2010). Likewise, in apple the seed coat and hypodermis beneath the skin show clear tannin staining, but not the flesh (Lees *et al.*, 1995). In mangosteen fruit, the CTs are localized to the pericarp (Zhou *et al.*, 2011). Thus, at least in mature fruit, the localization of CTs is commonly in skin and periderm, which may be related to their defensive functions in these tissues.

### *5.3.3 Ecology of fruit tannins*

Compared to the large number of reports on tannin content in fruits and foods in the context of human health, the ecological functions of fruit tannins have been much less investigated.

Edible fruit evolved for fruit dispersal by frugivores, typically vertebrate animals, which obtain nutrients while dispersing the seeds. Frugivores generally prefer ripe to unripe fruit (Wrangham & Waterman, 1983), but even among ripe fruit they exhibit clear preferences. These decisions are presumably based on trade-offs in nutrition and levels of deterrent secondary metabolites. This provides the ecological context for studies of tannins in fruit.

In a comparison of 33 plant species with fleshy fruit and a complex frugivore community, there was a correlation between higher nutritive content (carbohydrate, lipid, and nitrogen) and rapid fruit consumption by frugivores (Schaefer *et al.*, 2003). On the other hand, secondary metabolites (phenols and tannins) were negatively correlated with removal rates, indicating that these chemicals have a deterrent function for fruit dispersers. Wrangham & Waterman (1983) examined fruit from 37 plant species and showed that unripe fleshy fruits are commonly protected by high CTs and are less preferred by frugivous animals. It is thus likely that the higher tannin levels observed in unripe fruit contribute to the deterrence against premature dispersal. A threshold level of deterrence of 0.2% tannin in food has been suggested for a variety of animals (Oates *et al.*, 1977). As described previously, in browsing vertebrates tannins can interfere with digestion and lead to reduced protein availability. Likewise, in frugivores, tannins are expected to be a factor in food choice and palatability (Schaefer *et al.*, 2003).

On the other hand, even ripe fruits often contain tannins and other secondary metabolites. Extreme examples are the highly astringent persimmon, as well as chokeberry (Hellström *et al.*, 2009). If the biological purpose of fruit is to facilitate consumption and seed dispersal by vertebrates, tannins in fruit would appear to be counter-adaptive. This apparent contradiction is resolved when considering the potential role of tannins as defenses against fruit-rotting fungi. These compete with fruit-dispersing vertebrates and may therefore reduce fruit consumption and seed dispersal (Cipollini & Stiles, 1993). The reduction of overall tannin levels during ripening and the apparent concentration of tannins in the epidermal layers of fruit (Lees *et al.*, 1995; Zifkin *et al.*, 2012) are consistent with an antifungal defensive function. However, the overall tannin content for the fruit remains low enough not to deter frugivores, since the skin makes up only a small fraction of fruit mass (Janzen, 1983; Cipollini & Stiles, 1992). Therefore, the defensive potential and secondary-product content of ripe fruit may reflect the trade-off in defending against fungi while permitting consumption by vertebrate dispersers.

This trade-off may result in divergent evolutionary strategies for fruit dispersal. Fruits can be nutrient-rich and poorly defended, leading to rapid removal by vertebrate dispersers. Alternatively, they can be nutrient-poor and tannin-rich, and thus less attractive to herbivores but more resistant to pathogenic fungi and fruit rot. It is believed that this strategy is advantageous for plants that depend on rare or unreliable dispersers (Schaefer *et al.*, 2003; Cazetta *et al.*, 2008). The importance of antifungal defense in fruit physiology is emphasized by the extremely high levels of the antifungal pathogenesis-related (PR) proteins that accumulate in fruit. The expression of PR-1 and PR-5 genes, as well as of those encoding chitinases and glucanases, is high throughout grape berry development (Grimplet *et al.*, 2007).

## 5.4 Tannins in plant–soil–environment interactions

The prevalence of tannins in trees and the global importance of forests to biogeochemical processes lead to tannins having potential impacts in forest dynamics (Whitham *et al.*, 2006). In the soil, tannins have strong effects on litter decomposition, nutrient cycling, carbon storage, and heavy-metal binding. By connecting leaf chemistry to soil processes, they provide potential feedbacks in plant nutrition and competitive ability. Tannins may also leach into surface water and affect aqueous ecosystems. Here we discuss the role of tannins in soils, with emphasis on their functions in nutrient cycling and their impact on microbial ecology.

### *5.4.1 Tannin distribution and stability in soil*

Tannins enter the soil from leaf litter or senescing roots, bark, and woody tissue. Since tannin concentration and composition in leaves can vary widely, inputs from leaf litter to the soil are highly variable (Amatangelo & Vitousek, 2009). In temperate trees, tannin contents of 2–4% d.w. for the senescent leaves are typically reported (Smolander *et al.*, 2005; Kanerva *et al.*, 2008), but for some trees, such as *P. tremuloides*, the CT content can be as high as 25% (Lindroth *et al.*, 2002). In plants growing on nutrient-poor soils, typically ericaceous shrubs and conifers, high tannin levels in the range of 10–30% can be observed (Kraus *et al.*, 2004a). Litter usually has a lower tannin content than the corresponding fresh leaves, but generally litter and below-ground tannin levels reflect the input material (Hättenschwiler *et al.*, 2003; Kurokawa *et al.*, 2010). Roots also contribute substantially to soil tannin inputs, although fewer data are available; in *Metrosideros polymorpha*, for example, fine roots were found to have a tannin content comparable to that of leaves (Hättenschwiler *et al.*, 2003).

The tannin concentration in different soil horizons is variable. The proportion of high-molecular-weight phenolics, including tannins, typically decreases as one moves from litter to the humus fraction, as seen in soils under silver birch, Norway spruce, and Scots pine (Adamczyk *et al.*, 2008; Kanerva *et al.*, 2008). In the humus layer, tannin contents of 2–4% d.w. have been reported (Smolander *et al.*, 2005). Lorenz *et al.* (2000) estimated a loss of 75–89% and 40–70%, respectively, of extractable and residual tannins within the first year of litter deposition (Lorenz *et al.*, 2000; Stark *et al.*, 2007). However, most workers agree that in such decomposition experiments it is not easy to determine if the tannins are chemically modified, are lost to leaching, or are sorbed on to soil mineral particles where they cannot be detected with standard methods (Maie *et al.*, 2003). Different types of tannin are retained with different efficiencies by different soils (Halvorson *et al.*, 2009). CTs and tannic acid can directly adsorb to quartz particles (Kaal *et al.*, 2005). In mangrove swamps, where leaching is likely to be important, during litter decomposition extractable tannins decreased rapidly by 90% in the first 84 days, but stabilized thereafter (Zhou *et al.*, 2012). Interestingly, the accumulation of bound CTs increased during this time period. Overall, while tannin inputs from litter into soils are significant and may substantially influence interactions between plants and microbes and microbial processes, the details of tannin fate and cycling in soil systems are poorly understood.

### 5.4.2 Impact of tannins on soil nitrogen cycling and microbial activity

The impact of soil tannins on nitrogen cycling in soils has been discussed in reviews but only demonstrated in a few specialized systems (Kraus *et al.*, 2003; Schweitzer *et al.*, 2008). The most extensively documented impacts of tannins are on litter decomposition and nitrogen mineralization. Schweitzer *et al.* (2004) found that in *Populus* hybrids with a wide range of CT concentrations, the best predictor of decomposition rate was the ratio of CT to nitrogen in the soil, rather than lignin or total nitrogen or phosphorous content. In *P. tremuloides* leaf litter, the rate of decomposition was negatively correlated with CT concentration (Madritch *et al.*, 2006); interestingly, microbial communities beneath specific *P. tremuloides* clones were adapted to the local CT levels (Madritch & Lindroth, 2011) (see Section 5.4.3). Coq *et al.* (2010) found that in a Hawaiian rainforest soil, rapid decomposition rates correlated with low CT content and with a rich microfauna and high earthworm activity. Reduced rates of decomposition in forest ecosystems could impact the availability of nitrogen, a limiting nutrient in these systems (Kraus *et al.*, 2003).

Following litter decomposition, additional transformations convert dissolved organic nitrogen (DON), which includes proteins and amino acids, into inorganic forms such as ammonia ($NH_4^+$) and nitrate ($NO_3^-$). These reactions are the result of microbial activity and are referred to as "nitrogen mineralization." Organic nitrogen is first converted to $NH_4^+$, then under aerobic conditions nitrifying bacteria transform it into $NO_3^-$ via nitrite ($NO_2^-$). Both $NH_4^+$ and $NO_3^-$ are major nitrogen sources for plants and can be taken up by them directly, but nitrate is more susceptible to loss by leaching. Denitrifying bacteria also remove nitrate from the soil. The rate of nitrogen mineralization has been correlated to tannin levels in many experimental systems and addition experiments (Schimel *et al.*, 1998; Kraus *et al.*, 2004b; Kanerva *et al.*, 2006; Nierop *et al.*, 2006a,b). Fierer *et al.* (2001) found that balsam poplar CTs reduced mineralization rates in both poplar and alder soils, and Bradley *et al.* (2000) showed reduced rates of nutrient cycling after addition of CTs from *Kalmia angustifolia* and *Abies balsamea* to a black spruce soil. Kraus *et al.* (2004b) tested tannins from several Ericaceous and coniferous plants and found that all significantly decreased nitrogen mineralization rates in a forest soil.

While tannins reduce the rate of nitrogen mineralization overall, the underlying mechanisms differ depending on soil and tannin type. The study by Bradley *et al.* (2000) suggested that the main effect of tannins in the soil is to sequester organic nitrogen and make it unavailable for mineralization, for example by forming tannin–protein complexes that resist mineralization. In that system, soil respiration was not affected, suggesting no direct effect on microbial activity. By contrast, the work of Kraus *et al.* (2004b) showed that added tannins are a carbon source and stimulate microbial respiration, leading to reduced mineralization via microbial nitrogen immobilization. Fierer *et al.* (2001) observed both mechanisms with poplar tannins: the higher-molecular-weight fractions form stable tannin–protein complexes, while the lower-molecular-weight tannins and phenolics tend to stimulate bacterial respiration. In addition, direct antimicrobial properties of tannins may lead to the inhibition of soil microbes or inhibit microbial exoenzymes by protein binding. Kanerva *et al.* (2006) suggested that direct toxic effects of high-molecular-weight tannins from Norway spruce needles caused an increase in nitrogen mineralization.

Nitrification, the microbial conversion of $NH_4^+$ to $NO_3^-$, may also be affected by tannins, although the evidence is more mixed. Some authors have shown that nitrifying bacteria can be inhibited in either culture or soil by secondary metabolites, including tannins (Thibault *et al.*, 1982; Baldwin *et al.*, 1983). However, in other cases nitrifiers had no specific effects on activity (Schimel *et al.*, 1996; Bradley *et al.*, 2000; Fierer *et al.*, 2001). It has been suggested that since tannins may be used as a growth substrate by microbes, their apparent inhibitory action on nitrification is due to the increased $NO_3^-$ consumption (Bradley *et al.*, 2000).

In summary, many studies have shown various effects of tannins on rates of both soil decomposition and nitrogen mineralization. It is important to recall the diversity in tannin structure when considering these effects. As just mentioned, the effects of high- and low-molecular-weight tannins on microbial respiration and nitrogen mineralization are different (Fierer *et al.*, 2001; Kanerva *et al.*, 2006). Tannic acid, used as a proxy for HTs, was rapidly metabolized by soil microbes, but CTs were not (Nierop *et al.*, 2006b). The study by Kraus *et al.* (2004b) found that unlike several other tannins tested, huckleberry (*Vaccinium ovatum*) tannin did not stimulate microbial respiration, and that it thus inhibits nitrogen mineralization via a different mechanism. Huckleberry contains CTs but not HTs and is characterized by very high prodelphinidin content (98%). Nierop *et al.* (2006a) specifically investigated the effects of B-ring hydroxylation by comparing CTs from different sources on rates of mineralization. CTs with additional B-ring hydroxylation (prodelphinidins) inhibited nitrogen mineralization more strongly than procyanidin-type CTs. This is consistent with the observation that prodelphinidins generally have higher biological activities against microbes and herbivores (Scalbert, 1991; Ayres *et al.*, 1997). In addition, the stereochemistry of CTs is also important, and *cis*-oriented subunits of CT have a positive influence on net nitrogen mineralization rates (Norris *et al.*, 2011).

### 5.4.3 Interaction with community and ecosystem processes

As described in the previous section, tannins can impact soil microbial processes directly via toxic effects and by binding soil proteins and exoenzymes, or alternatively they may be used as a carbon source. In addition, tannins can have indirect effects on microbes, for example on bacterial quorum sensing. Many microbial populations regulate themselves via this density-dependent process, which is based on bacterial secretion and sensing of small molecules such as N-acyl-homoserine lactones. The plant growth-promoting rhizobacteria (PGPR) make use of quorum sensing to regulate pectinase, siderophore, and phytohormone production (Boyer *et al.*, 2008; Eberl & Riedel, 2011). Likewise, *Rhizobium* and *Agrobacterium* species can use quorum sensing to optimize their host interactions. Although the effects of tannins on this process in soil bacteria have not been extensively investigated, Taganna *et al.* (2011) showed that a tannin-rich leaf extract impacts the quorum-sensing system of the plant pathogen *Chomobacterium violaceum*. The process of swarming, by which soil bacteria move in the direction of nutrients using a flagellum, can be inhibited by tannins. Ellagic acid, tannic acids, and epigallocatechin gallate all reduced swarming in several bacterial species (Huber *et al.*, 2003). Bacteria typically use lipopolysaccharides for motility and swarming; O'May & Tufenkji (2011) showed that tannins from different plants (cranberry, pomegranate, tea) inhibited swarming of the plant pathogenic bacterium *Pseudomonas aeruginosa* by binding such lipopolysaccharides.

The direct and indirect effects of tannins can lead to shifts or alterations of microbial populations in soil. Madritch *et al.* (2006) found that microbial communities in soils associated with *P. tremuloides* with varying litter tannin contents were significantly different from each other, based on common marker enzymes. Furthermore, following litter-transplant experiments, such communities shifted in composition to more closely resemble those in the soil from which the litter originated. Schweitzer *et al.* (2011) also found significant differences in microbial populations associated with different poplar genotypes of varying CT content, indicating that tannins exert selective pressure on bacterial populations, which then adapt to the tannin content. Recently, selection of bacterial communities by specific phenolics added to soil microcosms was directly demonstrated using denaturing gradient gel electrophoresis (Schmidt *et al.*, 2013). Advanced genomics techniques will enable more detailed studies of microbial populations to be carried out in the future.

Since a major source of below-ground tannins is leaves that fall as litter, tannins may provide a direct link between an observable leaf phenotype and soil processes. Tannins in leaves can be quantified, and in some trees, such as *Populus*, foliar tannin concentration is a strongly heritable trait. This has allowed the application of population-genetics approaches to the study of the biological effects of tannins. *Populus* is a keystone tree species, rich in CTs but with much heritable variation for this trait in natural populations. Whitham *et al.* (2006) were among the first to use such population techniques to test for correlations of foliar CT levels in *Populus* with community structure and "ecosystem phenotypes" in both natural stands and common garden experiments. CT levels correlate with above-ground community phenotypes, such as arthropod and endophyte communities supported by the poplar trees, as well as below-ground phenotypes, specifically net nitrogen mineralization (Whitham *et al.*, 2006). Potential mechanisms for the effect of tannins on nitrogen cycles via the soil microbiota have been outlined in this chapter, but given the central importance of nitrogen, this work puts the functions of tannins in a broad context with ecosystem-level consequences. Key genes regulating CT levels in *Populus* have been identified (Mellway *et al.*, 2009), and it is interesting to speculate that variation in the expression of key regulatory genes in poplar could have far-reaching ecological effects.

A potential consequence of high soil tannins and reduced rates of decomposition and nitrogen mineralization is the shift in dissolved organic nitrogen away from mineral forms ($NO_3^-$, $NH_4^+$) to primarily organic pathways (Northup *et al.*, 1995). In some older acidic soils, more than 90% of total dissolved nitrogen can be in organic form (Kraus *et al.*, 2003), perhaps due to protein–polyphenol complexes recalcitrant to mineralization or bacterial utilization. Interestingly, acidic and nutrient-poor soils are typically dominated by Ericaceous plants, which form associations with ericoid mycorrhizae. These and some ectomycorrhizal fungi produce extracellular oxidative enzymes capable of degrading protein–tannin complexes, making organic forms of nitrogen available to the fungus and plant (Bending & Read, 1996). Although speculative, the idea that tannins can help the plant compete for soil nitrogen is appealing. Utilization of organic nitrogen either directly or via mycorrhizal partners would allow mycorrhizal plants to access nitrogen that is not accessible to soil bacteria. Furthermore, boreal forest trees are known to take up nitrogen in organic form, particularly as amino acids (Näsholm *et al.*, 1998). Inselsbacher & Näsholm (2012) determined that in boreal forests 80% of the soil nitrogen supply was in the form of amino acids, with only 10% occurring as $NH_4^+$ or $NO_3^-$. In a mixed deciduous forest study, Gallet-Budynek *et al.* (2009) found that uptake of glycine was greater on sites with recalcitrant litter than on labile-litter sites. Although there is no direct evidence linking differences in litter tannins with nitrogen uptake preferences, it is tempting to speculate that tannins are a means by which the plants can influence soil nutrient cycles.

### *5.4.4 Tannins and other plant stress adaptations*

Tannins generally contain *ortho*-dihydroxyl substitution patterns and can chelate metals including $Fe^{3+}$, $Pb^{2+}$, $Al^{3+}$, and $Zn^{2+}$. Indeed, CTs bind $Al^{3+}$ in proportion to polymer length and B-ring hydroxylation (Yoneda & Nakatsubo, 1998). Iron chelation may underlie some of the antimicrobial effects of tannins described earlier, and thus contribute to plant defense (Scalbert, 1991). However, this property of the tannins may also help to protect plants against heavy-metal toxicity. Certain plant species called hyperaccumulators can tolerate toxic levels of metals, including $Al^{3+}$ and $Pb^{2+}$, and are thus often found at mine sites or on contaminated soils. While the secretion of organic acid cations is a typical adaptation to heavy metals, in some plants metal tolerance is correlated with tannin content. The gametophyte of the fern *Athyrium yokoscense* tolerates $Pb^{2+}$ at much higher concentrations (10 mg/g) than other ferns, which it accumulates in the rhizoids where the CT concentration is highest (Kamachi *et al.*, 2005). Likewise, in the aluminum-tolerant forage legume *Lotus pedunculatus*, $Al^{3+}$ accumulates near the root tips in association with vacuoles that also stain for CTs (Stoutjesdijk *et al.*, 2001). A detailed study of the camphor tree (*Cinnamomum camphora*) also linked tannins to $Al^{3+}$ tolerance. Here, CT-accumulating cells were identified at the expanding root tip. These cells proliferate and are rapidly sloughed off, especially during exposure to $Al^{3+}$. It is suggested that they shield the expanding cells of the root tip against $Al^{3+}$ (Osawa *et al.*, 2011). The addition of tea flavan-3-ols to solutions containing aluminum prevented $Al^{3+}$ accumulation in the root tips of soybean, which are normally $Al^{3+}$-sensitive (Osawa *et al.*, 2011). Other woody plants contain high levels of tannins in a discrete zone just behind the growing tip, but their role in metal-ion sequestration or mineral nutrition has not been studied (Peterson *et al.*, 1999).

Additional evidence for a protective effect of tannins against heavy-metal stress comes from studies in wheat roots (Kinraide & Hagerman, 2010). While wheat does not synthesize significant amounts of tannins, the roots can be used as an assay system for the interaction of tannic acid with heavy metals. The toxicities of $Al^{3+}$, $Cu^{2+}$, and $SeO_4^{2-}$ could all be alleviated by co-incubation with tannic acid, and reciprocally $Al^{3+}$ and $SeO_4^{2-}$ reduced the toxicity of the tannic acid. The stochiometric association of $Al^{3+}$ and tannic acid suggests that there is direct binding; this would be consistent with other *in vitro* studies, which indicate that immobilized plant tannins bind heavy metals (Chin *et al.*, 2009; Sakaguchi & Nakajima, 1994). However, it is also possible that the tannins are acting as antioxidants, and thus counteract oxidative damage caused by the presence of heavy metals (Hagerman *et al.*, 1998; Kinraide & Hagerman, 2010).

One further potential ecological role for tannins is in allelopathic interactions. "Allelopathy" refers to the detrimental effects of one plant species on the germination, development, and growth of another. The biological activities of tannins and their occurrence in roots make them potential candidates for allelopathic agents. Rawat *et al.* (1998) investigated the effect of *Prunus armeniaca*, a tree known to contain CTs, on wheat, and found that the growth and development of wheat were affected at a distance of 6.5 m. The most active inhibitory compounds were four PAs (Maphuannin A, Epphedrannin A, Mahuannin B, and 13′-hydroxy mahuannin). A more recent study on the Amazonian tree *Alchroneopsis floribunda* showed that the major phytochemical compound in leaves and stem was isocorilagin, an ellagitannin with a strong allelopathic effect on weed seeds (*M. pudica*) (Batista *et al.*, 2011). How widespread such allelopathic tannin effects are remains to be determined.

## 5.5 Conclusions

Tannins have been implicated in diverse ecological interactions and can have a bewildering range of effects on other organisms. These effects result from the ability of tannins to bind proteins, to act as antioxidants or pro-oxidants, and to chelate iron and other metals. The best-supported ecological functions of tannins are their roles as feeding deterrents for vertebrate herbivores and as modulators of decomposition and nutrient cycling in soil. Through their impacts on nitrogen mineralization and nutrient cycling, and their presence in large quantities in keystone species such as poplar, tannins influence the nitrogen cycle and energy flow in higher-scale processes and ecosystems. Tannins have been identified and quantified in many common fruits and may be beneficial for human health, yet the ecology of fruit tannins has scarcely been investigated. More detailed work on where and when tannins are synthesized will provide further clues as to their functions, as will studies on tannin-related gene expression.

## Acknowledgments

Work in the Constabel laboratory is generously supported by NSERC (Natural Science and Engineering Research Council) of Canada and the University of Victoria's Centre for Forest Biology.

## References

Aastrup, S., Outtrup, H. & Erdal, K. (1984) Location of the proanthocyanidins in the barley grain. *Carlsberg Research Communications*, 49, 105–109.

Adamczyk, B., Kitunen, V. & Smolander, A. (2008) Protein precipitation by tannins in soil organic horizon and vegetation in relation to tree species. *Biology and Fertility of Soils*, 45, 55–64.

Aerts, R.J., Barry, T.N. & Mcnabb, W.C. (1999) Polyphenols and agriculture: beneficial effects of proanthocyanidins in forages. *Agriculture Ecosystems & Environment*, 75, 1–12.

Akagi, T., Ikegami, A., Suzuki, Y., *et al.* (2009) Expression balances of structural genes in shikimate and flavonoid biosynthesis cause a difference in proanthocyanidin accumulation in persimmon (*Diospyros kaki*) fruit. *Planta*, 230, 899–915.

Akagi, T., Ikegami, A. & Yonemori, K. (2010) DkMyb2 wound-induced transcription factor of persimmon (*Diospyros kaki*), contributes to proanthocyanidin regulation. *Planta*, 232, 1045–1059.

Amatangelo, K.L. & Vitousek, P.M. (2009) Contrasting predictors of fern versus angiosperm decomposition in a common garden. *Biotropica*, 41, 154–161.

Anuraga, M., Duarsa, P., Hill, M. & Lovett, J. (1993) Soil moisture and temperature affect condensed tannin concentrations and growth in *Lotus corniculatus* and *Lotus pedunculatus*. *Australian Journal of Agricultural Research*, 44, 1667–1681.

Anderson, R.C., Vodovnik, M., Min, B.R., *et al.* (2012) Bactericidal effect of hydrolysable and condensed tannin extracts on *Campylobacter jejuni in vitro*. *Folia Microbiologica*, 57, 253–258.

Appel, H.M. (1993) Phenolics in ecological interactions – the importance of oxidation. *Journal of Chemical Ecology*, 19, 1521–1552.

Arnold, T.A., Tanner, C.E., Rothen, M. & Bullington, J. (2008) Wound-induced accumulations of condensed tannins in turtlegrass, *Thalassia testudinum*. *Aquatic Botany*, 89, 27–33.

Ascacio-Valdés, J.A., Buenrostro, J.J., De la Cruz, R., *et al.* (2014) Fungal biodegradation of pomegranate ellagitannins. *Journal of Basic Microbiology*, 54, 28–34.

Ayres, M.P., Clausen, T.P., Maclean, S.F., *et al.* (1997) Diversity of structure and antiherbivore activity in condensed tannins. *Ecology*, 78, 1696–1712.

Bailey, J.K., Deckert, R., Schweitzer, J.A., *et al.* (2005) Host plant genetics affect hidden ecological players: links among *Populus*, condensed tannins, and fungal endophyte infection. *Canadian Journal of Botany*, 83, 356–361.

Baldwin, I.T., Olson, R.K. & Reiners, W.A. (1983) Protein binding phenolics and the inhibition of nitrification in subalpine balsam fir soils. *Soil Biology and Biochemistry*, 15, 419–423.

Barbehenn, R.V. & Constabel, C.P. (2011) Tannins in plant–herbivore interactions. *Phytochemistry*, 72, 1551–1565.

Barbehenn, R.V., Jones, C.P., Hagerman, A.E., *et al.* (2006) Ellagitannins have greater oxidative activities than condensed tannins and galloyl glucoses at high pH: potential impact on caterpillars. *Journal of Chemical Ecology*, 32, 2253–2267.

Barbehenn, R.V., Jaros, A., Lee, G., *et al.* (2009a) Hydrolyzable tannins as "quantitative defenses": limited impact against *Lymantria dispar* caterpillars on hybrid poplar. *Journal of Insect Physiology*, 55, 297–304.

Barbehenn, R.V., Jaros, A., Lee, G., *et al.* (2009b) Tree resistance to *Lymantria dispar* caterpillars: importance and limitations of foliar tannin composition. *Oecologia*, 159, 777–788.

Bate-Smith, E.C. & Metcalfe, C.R. (1957) Leuco-anthocyanins. 3. The nature and systematic distribution of tannins in dicotyledonous plants. *Journal of the Linnean Society of London, Botany*, 55, 669–705.

Batista, E.F., Costa, D.M., Guilhon, G. M., *et al.* (2011) Chemical constituents and allelopathic and antioxidant activities of *Alchorneopsis floribunda* Müll. Arg. (Euphorbiaceae). *Natural Product Research*, 25, 1–8.

Bending, G.D. & Read, D.J. (1996) Effects of the soluble polyphenol tannic acid on the activities of ericoid and ectomycorrhizal fungi. *Soil Biology and Biochemistry*, 28, 1595–1602.

Berard, N.C., Wang, Y., Wittenberg, K.M., *et al.* (2011) Condensed tannin concentrations found in vegetative and mature forage legumes grown in western Canada. *Canadian Journal of Plant Science*, 91, 669–675.

Bernays, E.A., Chamberlain, D.J. & Leather, E.M. (1981) Tolerance of acridids to ingested condensed tannin. *Journal of Chemical Ecology*, 7, 247–256.

Bindon, K.A. & Kennedy, J.A. (2011) Ripening-induced changes in grape skin proanthocyanidins modify their interaction with cell walls. *Journal of Agricultural and Food Chemistry* 59, 2696–2707.

Bindon, K.A., Smith, P.A., Holt, H. & Kennedy, J.A. (2010) Interaction between grape-derived proanthocyanidins and cell wall material. 2. Implications for vinification. *Journal of Agricultural and Food Chemistry* 58, 10736–10746.

Boeckler, G.A., Towns, M., Unsicker S.B., *et al.* (2014) Transgenic upregulation of the condensed tannin pathway in poplar leads to dramatic shift in leaf palatability for two tree-feeding Lepidoptera. *Journal of Chemical Ecology*, 40, 150–158.

Bogs, J., Downey, M.O., Harvey, J.S., *et al.* (2005) Proanthocyanidin synthesis and expression of genes encoding leucoanthocyanidin reductase and anthocyanidin reductase in developing grape berries and grapevine leaves. *Plant Physiology*, 139, 652–663.

Boyer, M., Bally, R., Perrotto, S., *et al.* (2008) A quorum-quenching approach to identify quorum-sensing-regulated functions in *Azospirillum lipoferum*. *Research in Microbiology*, 159, 699–708.

Bradley, R.L., Titus, B.D. & Preston, C.P. (2000) Changes to mineral N cycling and microbial communities in black spruce humus after additions of $(NH_4)_2SO_4$ and condensed tannins extracted from *Kalmia angustifolia* and balsam fir. *Soil Biology and Biochemistry*, 32, 1227–1240.

Brandon, M.J., Foo, L.Y., Porter, L.J. & Meredith, P. (1980) Proanthocyanidins of barley and sorghum; composition as a function of maturity of barley ears. *Phytochemistry*, 21, 2953–2957.

Butler, L.G. (1982) Relative degree of polymerization of sorghum tannin during seed development and maturation. *Journal of Agricultural and Food Chemistry*, 30, 1090–1094.

Cazetta, E., Schaefer, H. & Galetti, M. (2008) Does attraction to frugivores or defense against pathogens shape fruit pulp composition? *Oecologia*, 155, 277–286.

Chin, L., Leung, D.W.M. & Taylor, H.H. (2009) Lead chelation to immobilised *Symphytum officinale* L. (comfrey) root tannins. *Chemosphere*, 76, 711–715.

Cipollini, M.L. & Stiles, E.W. (1992) Relative risks of fungal rot for temperate ericadeous fruits – effects of seasonal variation on selection for chemical defense. *Canadian Journal of Botany*, 70, 1868–1877.

Cipollini, M.L. & Stiles, E.W. (1993) Fruit rot, antifungal defense, and palatability of fleshy fruits for frugivorous birds. *Ecology*, 74, 751–762.

Close, D.C. & McArthur, C. (2002) Rethinking the role of many plant phenolics – protection from photodamage not herbivores? *Oikos*, 99, 166–172.

Coq, S., Souquet, J.-M., Meudec, E., *et al.* (2010) Interspecific variation in leaf litter tannins drives decomposition in a tropical rain forest of French Guiana. *Ecology*, 91, 2080–2091.

Debeaujon, I., Peeters, A.J.M., Léon-Kloosterziel, K.M. & Koornneef, M. (2001) The TRANSPARENT TESTA12 gene of Arabidopsis encodes a multidrug secondary transporter-like protein required for flavonoid sequestration in vacuoles of the seed coat endothelium. *Plant Cell*, 13, 853–872.

DeGabriel, J.L., Wallis, I.R., Moore, B.D. & Foley, W.J. (2008) A simple, integrative assay to quantify nutritional quality of browses for herbivores. *Oecologia*, 156, 107–116.

DeGabriel, J.L., Moore, B.D., Foley, W.J. & Johnson, C.N. (2009) The effects of plant defensive chemistry on nutrient availability predict reproductive success in a mammal. *Ecology*, 90, 711–719.

Dixon, R.A., Xie, D.-Y. & Sharma, S.B. (2005) Proanthocyanidins – a final frontier in flavonoid research? *New Phytologist*, 165, 9–28.

Donaldson, J.R., Stevens, M.T., Barnhill, H.R. & Lindroth, R.L. (2006) Age-related shifts in leaf chemistry of clonal aspen (*Populus tremuloides*). *Journal of Chemical Ecology*, 32, 1415–1429.

Eberl, L. & Riedel, K. (2011) Mining quorum sensing regulated proteins – role of bacterial cell-to-cell communication in global gene regulation as assessed by proteomics. *Proteomics*, 11, 3070–3085.

Engels, C., Knoedler, M., Zhao, Y.-Y., *et al.* (2009) Antimicrobial activity of gallotannins isolated from mango (*Mangifera indica* L.) kernels. *Journal of Agricultural and Food Chemistry*, 57, 7712–7718.

Feeny, P.P. (1968) Effect of oak leaf tannins on larval growth of winter moth *Operophthera brumata*. *Journal of Insect Physiology*, 14, 805–817.

Feeny, P. (1970) Seasonal changes in oak leaf tannins and nutrients as a cause of spring feeding by winter moth caterpillars. *Ecology*, 51, 565–581.

Feucht, W., Schmid, P.P.S. & Christ E. (1986) Distribution of flavanols in meristematic and mature tissues of *Prunus avium* shoots. *Journal of Plant Physiology*, 125, 1–8.

Ferreira, D., Slade D. & Marais, J.P.J (2006) Flavans and pranthocyanididns. In: The Flavonoids – Chemistry, Biochemistry and Applications (eds O. M. Andersen & K. R. Markham), pp. 553–616. CRC Taylor & Francis, Boca Raton, FL.

Fierer, N., Schimel, J.P., Cates, R.G. & Zou, J. (2001) Influence of balsam poplar tannin fractions on carbon and nitrogen dynamics in Alaskan taiga floodplain soils. *Soil Biology and Biochemistry*, 33, 1827–1839.

Fuchs, C.T. & Spiteller, G. (1998) Accumulation of caffeoyl-D-quinic acids and catechins in plums affected by the fungus *Taphrina pruni*. *Zeitschrift für Naturforschung*, 53c, 799–805.

Gallet-Budynek, A., Brzostek, E., Rodgers, V.L., *et al.* (2009) Intact amino acid uptake by northern hardwood and conifer trees. *Oecologia*, 160, 129–138.

Grimplet, J., Deluc, L.G., Tillett, R.L., *et al.* (2007) Tissue-specific mRNA expression profiling in grape berry tissues. *BMC Genomics*, 8, 187.

Gu, L., Kelm, M.A., Hammerstone, J.F., *et al.* (2004) Concentrations of proanthocyanidins in common foods and estimations of normal consumption. *The Journal of Nutrition*, 134, 613–617.

Guyot, S., Marnet, N., Sanoner, P., *et al.* (2001) Direct thiolysis on crude apple materials for high-performance liquid chromatography characterization and quantification of polyphenols in cider apple tissues and juices. *Methods in Enzymology*, 335, 57–70.

Haddock E.A., Gupta, R.K., Alshafi, S.M. K., *et al.* (1982) The metabolism of gallic acid and hexahydroxydiphenic acid in plants. 4. Biogenic and molecular taxonomic considerations. *Phytochemistry* 21, 1049–1062.

Hagerman, A.E., Riedl, K.M., Jones, G.A., *et al.* (1998) High molecular weight plant polyphenolics (tannins) as biological antioxidants. *Journal of Agricultural and Food Chemistry*, 46, 1887–1892.

Halvorson, J.J., Gonzalez, J.M., Hagerman, A.E. & Smith, J.L. (2009) Sorption of tannin and related phenolic compounds and effects on soluble-N in soil. *Soil Biology and Biochemistry*, 41, 2002–2010.

Haslam, E. (1975) Natural proanthocyanidins. In: *The Flavonoids* (eds J. B. Harborne, T. J. Mabry & H. Mabry), pp. 505–559. Chapman and Hall, London.

Hatano, T., Kira, R., Yoshizaki, M. & Okuda, T. (1986) Seasonal changes in the tannins of *Liquidambar formosana* reflecting their biogenesis. *Phytochemistry*, 25, 2787–2789.

Hättenschwiler, S., Hagerman, A.E. & Vitousek, P. M. (2003) Polyphenols in litter from tropical montane forests across a wide range in soil fertility. *Biogeochemistry*, 64, 129–148.

Hellström, J.K., Törrönen, A.R. & Mattila, P.H. (2009) Proanthocyanidins in common food products of plant origin. *Journal of Agricultural and Food Chemistry*, 57, 7899–7906.

Holeski, L.M., Vogelzang, A., Stanosz, G. & Lindroth, R.L. (2009) Incidence of Venturia shoot blight in aspen (*Populus tremuloides* Michx.) varies with tree chemistry and genotype. *Biochemical Systematics and Ecology*, 37, 139–145.

Huber, B., Eberl, L., Feucht, W. & Polster, J. (2003) Influence of polyphenols on bacterial biofilm formation and quorum-sensing. *Zeitschrift für Naturforschung*, 58c, 879–884.

Hwang, S.Y. & Lindroth, R.L. (1997) Clonal variation in foliar chemistry of aspen: Effects on gypsy moths and forest tent caterpillars. *Oecologia*, 111, 99–108.

Inoue, K.H. & Hagerman A.E. (1998) Determination of gallotannin with rhodanine. *Analytical Biochemistry*, 169, 363–369.

Inselsbacher, E. & Näsholm, T. (2012) The below-ground perspective of forest plants: soil provides mainly organic nitrogen for plants and mycorrhizal fungi. *New Phytologist*, 195, 329–334.

Iwashina, T. (2000) The structure and distribution of the flavonoids in plants. *Journal of Plant Research*, 113, 287–299.

Jansman, A.J.M., Verstegen, M.W. A., Huisman, J. & Vandenberg, J.W.O. (1995) Effects of hulls of faba beans (*Vicia faba* L.) with a low or high content of condensed tannins on the apparent ileal and fecal digestibility of nutrients and the excretion of endogenous protein in ileal digesta and feces of pigs. *Journal of Animal Science*, 73, 118–127.

Janzen, D.H. (1983) Seed and pollen dispersal by animals: convergence in the ecology of contamination and sloppy harvest. *Biological Journal of the Linnean Society*, 20, 103–113.

Jin, A., Ozga, J.A., Lopes-Lutz, D., Schieber, A. & Reinecke, D.M. (2012) Characterization of proanthocyanidins in pea (*Pisum sativum* L.), lentil (*Lens culinaris* L.), and faba bean (*Vicia faba* L.) seeds. *Food Research International*, 46, 528–535.

Johnson, K.S. (2005) Plant phenolics behave as radical scavengers in the context of insect (*Manduca sexta*) hemolymph and midgut fluid. *Journal of Agricultural and Food Chemistry*, 53, 10120–10126.

Kaal, J., Nierop, K.G. J. & Verstraten, J.M. (2005) Retention of tannic acid and condensed tannin by Fe-oxide-coated quartz sand. *Journal of Colloid and Interface Science*, 287, 72–79.

Kamachi, H., Komori, I., Tamura, H., *et al.* (2005) Lead tolerance and accumulation in the gametophytes of the fern *Athyrium yokoscense*. *Journal of Plant Research*, 118, 137–145.

Kanerva, S., Kitunen, V., Kiikkilä, O., *et al.* (2006) Response of soil C and N transformations to tannin fractions originating from Scots pine and Norway spruce needles. *Soil Biology and Biochemistry*, 38, 1364–1374.

Kanerva, S., Kitunen, V., Loponen, J. & Smolander, A. (2008) Phenolic compounds and terpenes in soil organic horizon layers under silver birch, Norway spruce and Scots pine. *Biology and Fertility of Soils*, 44, 547–556.

Kao, Y.Y., Harding, S.A. & Tsai, C.J. (2002) Differential expression of two distinct phenylalanine ammonia-lyase genes in condensed tannin-accumulating and lignifying cells of quaking aspen. *Plant Physiology*, 130, 796–807.

Karowe, D.N. (1989) Differential effect of tannic acid on two tree-feeding Lepidoptera – implications for theories of plant anti-herbivore chemistry. *Oecologia*, 80, 507–512.

Kennedy, J.A., Matthews, M.A. & Waterhouse, A.L. (2002) Effect of maturity and vine water status on grape skin and sine flavonoids. *American Journal of Enology and Viticulture*, 53, 268–274.

Kinraide, T.B. & Hagermann, A.E. (2010) Interactive intoxicating and ameliorating effects of tannic acid, aluminum ($Al^{3+}$), copper ($Cu^{2+}$), and selenate ($SeO_4^{2-}$) in wheat roots: a descriptive and mathematical assessment. *Physiologia Plantarum*, 139, 68–79.

Koponen, J.M., Happonen, A.M., Mattila, P.H. & Törrönen, A.R. (2007) Contents of anthocyanins and ellagitannins in selected foods consumed in Finland. *Journal of Agricultural and Food Chemistry*, 55, 1612–1619.

Kraus, T.E.C., Dahlgren, R.A. & Zasoski, R.J. (2003) Tannins in nutrient dynamics of forest ecosystems – a review. *Plant and Soil*, 256, 41–66.

Kraus, T.E.C., Zasoski, R.J. & Dahlgren, R.A. (2004a) Fertility and pH effects on polyphenol and condensed tannin concentrations in foliage and roots. *Plant and Soil*, 262, 95–109.

Kraus, T.E.C., Zasoski, R.J., Dahlgren, R.A., *et al.* (2004b) Carbon and nitrogen dynamics in a forest soil amended with purified tannins from different plant species. *Soil Biology and Biochemistry*, 36, 309–321.

Kurokawa, H., Peltzer, D.A. & Wardle, D.A. (2010) Plant traits, leaf palatability and litter decomposability for co-occurring woody species differing in invasion status and nitrogen fixation ability. *Functional Ecology*, 24, 513–523.

Laitinen, M.-L., Julkunen-Tiitto, R. & Rousi, M. (2002) Foliar phenolic composition of European white birch during bud unfolding and leaf development. *Physiologia Plantarum*, 114, 450–460.

Landete, J. M. (2011) Ellagitannins, ellagic acid and their derived metabolites: A review about source, metabolism, functions and health. *Food Research International*, 44, 1150–1160.

Lattanzio, V., Terzano, R., Cicco, N., *et al.* (2005) Seed coat tannins and bruchid resistance in stored cowpea seeds. *Journal of the Science of Food and Agriculture*, 85, 839–846.

Lavola, A. (1998) Accumulation of flavonoids and related compounds in birch induced by UV-B irradiance. *Tree Physiology*, 18, 53–58.

Lees, G.L., Suttill, N.H. & Gruber, M.Y. (1993) Condensed tannins in sainfoin.1. A histological and cytological survey of plant tissues. *Canadian Journal of Botany*, 71, 1147–1152.

Lees, G.L., Suttill, N.H., Wall, K.M. & Beveridge, T.H. (1995) Localization of condensed tannins in apple fruit peel, pulp, and seeds. *Canadian Journal of Botany*, 73, 1897–1904.

Lepiniec, L., Debeaujon, I., Routaboul, J.-M., *et al.* (2006) Genetics and biochemistry of seed flavonoids. *Annual Review of Plant Biology*, 57, 405–430.

Lim, C.K., Penesyan, A., Hassan, K.A., *et al.* (2013) Effect of tannic acid on the transcriptome of the soil bacterium *Pseudomonas protegens* Pf-5. *Applied and Environmental Microbiology*, 79, 3141–3145.

Lin, B., Johnson, B.J., Rubin, R.A., *et al.* (2011) Iron chelation by cranberry juice and its impact on *Escherichia coli* growth. *Biofactors*, 37, 121–130.

Lindroth, R.L., Osier, T.L., Barnhill, H.R.H. & Wood, S.A. (2002) Effects of genotype and nutrient availability on phytochemistry of trembling aspen (*Populus tremuloides* Michx.) during leaf senescence. *Biochemical Systematics and Ecology*, 30, 297–307.

Lorenz, K., Preston, C.M., Raspe, S., *et al.* (2000) Litter decomposition and humus characteristics in Canadian and German spruce ecosystems: information from tannin analysis and $^{13}C$ CPMAS NMR. *Soil Biology and Biochemistry*, 32, 779–792.

Ma, Y. & Bliss, F.A. (1978) Tannin content and inheritance in common bean. *Crop Science*, 18, 201–204.

Madritch, M.D. & Lindroth, R.L. (2011) Soil microbial communities adapt to genetic variation in leaf litter inputs. *Oikos*, 120, 1696–1704.

Madritch, M., Donaldson, J.R. & Lindroth, R.L. (2006) Genetic identity of *Populus tremuloides* litter influences decomposition and nutrient release in a mixed forest stand. *Ecosystems*, 9, 528–537.

Maie, N., Behrens, A., Knicker, H. & Kögel-Knabner, I. (2003) Changes in the structure and protein binding ability of condensed tannins during decomposition of fresh needles and leaves. *Soil Biology and Biochemistry*, 35, 577–589.

Martin, J.S., Martin, M.M. & Bernays, E.A., (1987) Failure of tannic acid to inhibit digestion or reduce digestibility of plant protein in gut fluids of insect herbivores: implications for theories of plant defense. *Journal of Chemical Ecology*, 13, 605–621.

McArt, S.H., Spalinger, D.E., Collins, W.B., *et al.* (2009) Summer dietary nitrogen availability as a potential bottom-up constraint on moose in south-central Alaska. *Ecology*, 90, 1400–1411.

McDonald, M., Mila, I. & Scalbert, A. (1996) Precipitation of metal ions by plant polyphenols: optimal conditions and origin of precipitation. *Journal of Agricultural and Food Chemistry*, 44, 599–606.

McKenzie, B.E. & Peterson, C.A. (1995) Root browning in *Pinus banksiana* and *Eucalyptus pilularis* Sm.1. Anatomy and permeability of the white and tannin zones. *Botanica Acta*, 108, 127–137.

McSweeney, C.S., Palmer, B., McNeill, D.M. & Krause, D.O. (2001) Microbial interactions with tannins: nutritional consequences for ruminants. *Animal Feed Science and Technology*, 91, 83–93.

Mellway, R.D., Tran, L.T., Prouse, M.B., *et al.* (2009) The wound-, pathogen-, and ultraviolet B-responsive MYB134 gene encodes an R2R3 MYB transcription factor that regulates proanthocyanidin synthesis in poplar. *Plant Physiology*, 150, 924–941.

Mila, I., Scalbert, A. & Expert, D. (1996) Iron withholding by plant polyphenols and resistance to pathogens and rots. *Phytochemistry*, 42, 1551–1555.

Min, B.R., Attwood, G.T., Reilly, K., *et al.* (2002) *Lotus corniculatus* condensed tannins decrease in vivo populations of proteolytic bacteria and affect nitrogen metabolism in the rumen of sheep. *Canadian Journal of Microbiology*, 48, 911–921.

Min, B.R., Barry, T.N., Attwood, G.T. & McNabb, W.C. (2003) The effect of condensed tannins on the nutrition and health of ruminants fed fresh temperate forages: a review. *Animal Feed Science and Technology*, 106, 3–19.

Mingshu, L., Kai, Y., Qiang, H., & Dongying, J. (2006) Biodegradation of gallotannins and ellagitannins. *Journal of Basic Microbiology*, 46, 68–84.

Miranda, M., Ralph, S.G., Mellway, R., *et al.* (2007) The transcriptional response of hybrid poplar (*Populus trichocarpa x P. deltoides*) to infection by *Melampsora medusae* leaf rust involves induction of flavonoid pathway genes leading to the accumulation of proanthocyanidins. *Molecular Plant-Microbe Interactions*, 20, 816–831.

Mueller-Harvey, I. (2001) Analysis of hydrolysable tannins. *Animal Feed Science and Technology*, 91, 3–20.

Mole, S. (1993) The systematic distribution of tannins in the leaves of angiosperms: A tool for ecological studies. *Biochemical Systematics and Ecology*, 21, 833–846.

Näsholm, T., Ekblad, A., Nordin, A., *et al.* (1998) Boreal forest plants take up organic nitrogen. *Nature*, 392, 914–916.

Nelson, K.E., Pell, A.N., Doane, P.H., *et al.* (1997) Chemical and biological assays to evaluate bacterial inhibition by tannins. *Journal of Chemical Ecology*, 23, 1175–1194.

Nierop, K., Verstraten, J., Tietema, A., *et al.* (2006a) Short- and long-term tannin induced carbon, nitrogen and phosphorus dynamics in Corsican pine litter. *Biogeochemistry*, 79, 275–296.

Nierop, K.G.J., Preston, C.M. & Verstraten, J.M. (2006b) Linking the B ring hydroxylation pattern of condensed tannins to C, N and P mineralization. A case study using four tannins. *Soil Biology and Biochemistry*, 38, 2794–2802.

Norris, C.E., Preston, C.M., Hogg, K.E. & Titus, B.D. (2011) The influence of condensed tannin structure on rate of microbial mineralization and reactivity to chemical assays. *Journal of Chemical Ecology*, 37, 311–319.

Northup, R.R., Yu, Z.S., Dahlgren, R.A. & Vogt, K.A. (1995) Polyphenol control of nitrogen release from pine litter. *Nature*, 377, 227–229.

O'May, C. & Tufenkji, N. (2011) The swarming motility of *Pseudomonas aeruginosa* is blocked by cranberry proanthocyanidins and other tannin-containing materials. *Applied Environmental Microbiology*, 77, 3061–3067.

Oates, J.F., Swain, T. & Zantovska, J. (1977) Secondary compounds and food selection by colobus monkeys. *Biochemical Systematics and Ecology*, 5, 317–321.

Okuda, T., Yoshida, T., Hatano, T. & Ito, H. (2009) Ellagitannins renewed the concept of tannins. In: *Chemistry and Biology of Ellagitannins* (ed. S. Quideau), pp. 1–54. World Scientific, Singapore.

Osawa, H., Endo, I., Hara, Y., *et al.* (2011) Transient proliferation of proanthocyanidin-accumulating cells on the epidermal apex contributes to highly aluminum-resistant root elongation in camphor tree. *Plant Physiology*, 155, 433–46.

Osier, T.L. & Lindroth, R.L. (2001) Effects of genotype, nutrient availability, and defoliation on aspen phytochemistry and insect performance. *Journal of Chemical Ecology*, 27, 1289–1313.

Osier, T.L. & Lindroth, R.L. (2006) Genotype and environment determine allocation to and costs of resistance in quaking aspen. *Oecologia*, 148, 293–303.

Peng, S., Scalbert, A. & Monties, B. (1991) Insoluble ellagitannins in *Castanea sativa* and *Quercus petraea* woods. *Phytochemistry*, 30, 775–778.

Peters, D.J. & Constabel, C.P. (2002) Molecular analysis of herbivore-induced condensed tannin synthesis: cloning and expression of dihydroflavonol reductase from trembling aspen (*Populus tremuloides*). *Plant Journal*, 32, 701–712.

Peterson, C.A., Enstone, D.E. & Taylor, J.H. (1999) Pine root structure and its potential significance for root function. *Plant and Soil*, 217, 205–213.

Porter, L.J. (1988). Flavans and proanthocyanidins. In: *The Flavonoids – Recent Advances in Research Since 1980* (ed. J.B. Harborne), pp. 21–62. Chapman & Hall, London.

Porter, L.J., Hrstich, L.N. & Chan, B.G. (1986) The conversion of procyanidins and prodelphinidins to cyanidin and delphidin. *Phytochemistry*, 25, 223–230.

Pourcel, L., Routaboul, J.-M., Kerhoas, L., *et al.* (2005) TRANSPARENT TESTA10 encodes a laccase-like enzyme involved in oxidative polymerization of flavonoids in Arabidopsis seed coat. *Plant Cell*, 17, 2966–2980.

Prior, R.L. & Gu, L. (2005) Occurrence and biological significance of proanthocyanidins in the American diet. *Phytochemistry*, 66, 2264–80.

Prior, R.L., Lazarus, S.A., Cao, G., *et al.* (2001) Identification of procyanidins and anthocyanins in blueberries and cranberries (*Vaccinium* spp.) using high-performance liquid chromatography/mass spectrometry. *Journal of Agricultural and Food Chemistry*, 49, 1270–1276.
Puupponen-Pimiä, R., Nohynek, L., Hartmann-Schmidlin, S., *et al.* (2005) Berry phenolics selectively inhibit the growth of intestinal pathogens. *Journal of Applied Microbiology*, 98, 991–1000.
Quideau, S., Deffieux, D., Douat-Casassus, C. & Pouysegu, L. (2011) Plant polyphenols: chemical properties, biological activities, and synthesis. *Angewandte Chemie International Edition*, 50, 586–621.
Rasmussen, S.E., Frederiksen, H., Krogholm, K.S. & Poulsen, L. (2005) Dietary proanthocyanidins: occurrence, dietary intake, bioavailability, and protection against cardiovascular disease. *Molecular Nutrition & Food Research*, 49, 159–174.
Rawat, M.S.M., Pant, G., Prasad, D., *et al.* (1998) Plant growth inhibitors (Proanthocyanidins) from *Prunus armeniaca*. *Biochemical Systematics and Ecology*, 26, 13–23.
Rehill, B.J., Whitham, T.G., Martinsen, G.D., *et al.* (2006) Developmental trajectories in cottonwood phytochemistry. *Journal of Chemical Ecology*, 32, 2269–2285.
Sakaguchi, T. & Nakajima, A. (1994) Accumulation of uranium by immobilized persimmon tannin. *Separation Science and Technology*, 29, 205–221.
Salminen, J.-P., Karonen, M. & Sinkkonen, J. (2011) Chemical ecology of tannins: recent developments in tannin chemistry reveal new structures and structure-activity patterns. *Chemistry – A European Journal*, 17, 2806–2816.
Santos-Buelga, C. & Scalbert, A. (2000) Proanthocyanidins and tannin-like compounds – nature, occurrence, dietary intake and effects on nutrition and health. *Journal of the Science of Food and Agriculture*, 80, 1094–1117.
Scalbert, A. (1991) Antimicrobial properties of tannins. *Phytochemistry*, 30, 3875–3883.
Schaefer, H.M., Schmidt, V. & Winkler, H. (2003) Testing the defence trade-off hypothesis: how contents of nutrients and secondary compounds affect fruit removal. *Oikos*, 102, 318–328.
Schimel, J.P., Cleve, K.V., Cates, R.G., *et al.* (1996) Effects of balsam poplar (*Populus balsamifera*) tannins and low molecular weight phenolics on microbial activity in taiga floodplain soil: implications for changes in N cycling during succession. *Canadian Journal of Botany*, 74, 84–90.
Schimel, J.P., Cates, R.G. & Ruess, R. (1998) The role of balsam poplar secondary chemicals in controlling soil nutrient dynamics through succession in the Alaskan taiga. *Biogeochemistry*, 42, 221–234.
Schmidt, M.A., Kreinberg, A.J., Gonzalez, J.M., *et al.* (2013) Soil microbial communities respond differently to three chemically defined polyphenols. *Plant Physiology and Biochemistry*, 72, 190–197.
Schofield, P., Mbugua, D.M. & Pell, A.N. (2001) Analysis of condensed tannins: a review. *Animal Feed Science and Technology*, 91, 21–40.
Schweitzer, J.A., Bailey, J.K., Rehill, B.J., *et al.* (2004) Genetically based trait in a dominant tree affects ecosystem processes. *Ecology Letters*, 7, 127–134.
Schweitzer, J.A., Madritch, M.D., Bailey, J.K., *et al.* (2008) From genes to ecosystems: the genetic basis of condensed tannins and their role in nutrient regulation in a *Populus* model system. *Ecosystems*, 11, 1005–1020.
Schweitzer, J.A., Fischer, D.G., Rehill, B.J., *et al.* (2011) Forest gene diversity is correlated with the composition and function of soil microbial communities. *Population Ecology*, 53, 35–46.
Scioneaux, A.N., Schmidt, M.A., Moore, M.A., *et al.* (2011) Qualitative variation in proanthocyanidin composition of *Populus* species and hybrids: genetics is the key. *Journal of Chemical Ecology*, 37, 57–70.
Shimada, T. (2006) Salivary proteins as a defense against dietary tannins. *Journal of Chemical Ecology*, 32, 1149–1163.
Shimada, T., Saitoh, T., Sasaki, E., *et al.* (2006) Role of tannin-binding salivary proteins and tannase-producing bacteria in the acclimation of the Japanese wood mouse to acorn tannins. *Journal of Chemical Ecology*, 32, 1165–1180.

Sivakumaran, S., Molan, A.L., Meagher, L.P., *et al.* (2004) Variation in antimicrobial action of proanthocyanidins from *Dorycnium rectum* against rumen bacteria. *Phytochemistry*, 65, 2485–2497.

Smith, A.H., Zoetendal, E. & Mackie, R.I. (2005) Bacterial mechanisms to overcome inhibitory effects of dietary tannins. *Microbial Ecology*, 50, 197–205.

Smolander, A., Loponen, J., Suominen, K. & Kitunen, V. (2005) Organic matter characteristics and C and N transformations in the humus layer under two tree species, *Betula pendula* and *Picea abies*. *Soil Biology & Biochemistry*, 37, 1309–1318.

Souquet, J.-M., Cheynier, V., Brossaud, F. & Moutounet, M. (1996) Polymeric proanthocyanidins from grape skins. *Phytochemistry*, 43, 509–512.

Souquet, J.-M., Labarbe, B., Le Guernevé, C., *et al.* (2000) Phenolic composition of grape stems. *Journal of Agricultural and Food Chemistry*, 48, 1076–1080.

Stark, S., Julkunen-Tiitto, R. & Kumpula, J. (2007) Ecological role of reindeer summer browsing in the mountain birch (*Betula pubescens* ssp. *czerepanovii*) forests: effects on plant defense, litter decomposition, and soil nutrient cycling. *Oecologia*, 151, 486–498.

Stoutjesdijk, P.A., Sale, P.W. & Larkin, P.J. (2001) Possible involvement of condensed tannins in aluminium tolerance of *Lotus pedunculatus*. *Australian Journal of Plant Physiology*, 28, 1063–1074.

Strumeyer, D.H. & Malin, M.J. (1975) Condensed tannins in grain sorghum. Isolation, fractionation, and characterization. *Journal of Agricultural and Food Chemistry*, 23, 909–914.

Taganna, J.C., Quanico, J.P., Perono, R.M., *et al.* (2011) Tannin-rich fraction from *Terminalia catappa* inhibits quorum sensing (QS) in *Chromobacterium violaceum* and the QS-controlled biofilm maturation and LasA staphylolytic activity in *Pseudomonas aeruginosa*. *J Ethnopharmacology*, 134, 865–871.

Takahashi, A., Ichihara, Y., Isagi, Y. & Shimada, T. (2010) Effects of acorn tannin content on infection by the fungus *Ciboria batschiana*. *Forest Pathology*, 40, 96–99.

Takos, A.M., Ubi, B.E., Robinson, S.P. & Walker, A.R. (2006) Condensed tannin biosynthesis genes are regulated separately from other flavonoid biosynthesis genes in apple fruit skin. *Plant Science*, 170, 487–499.

Tarascou, I., Souquet, J.M., Mazauric, J.P., *et al.* (2010) The hidden face of food phenolic composition. *Archives of Biochemistry and Biophysics*, 501, 16–22.

Thibault, J.R., Fortin, J.A. & Smirnoff, W.A. (1982) *in vitro* allelopathic inhibition of nitrification by balsam poplar and balsam fir. *American Journal of Botany*, 69, 676–679.

Vander Wall, S.B. (2001) The evolutionary ecology of nut dispersal. *Botanical Review*, 67, 74–117.

Wallis, I.R., Edwards, M.J., Windley, H., *et al.* (2012) Food for folivores: nutritional explanations linking diets to population density. *Oecologia*, 169, 281–291.

Whitham, T.G., Bailey, J.K., Schweitzer, J.A., *et al.* (2006) A framework for community and ecosystem genetics: from genes to ecosystems. *Nature Reviews Genetics*, 7, 510–523.

Wrangham, R.W. & Waterman, P.G. (1983) Condensed tannins fruits eaten by chimpanzees. *Biotropica*, 15, 217–222.

Yoneda, S. & Nakatsubo, F. (1998) Effects of the hydroxylation patterns and degrees of polymerization of condensed tannins on their metal-chelating capacity. *Journal of Wood Chemistry and Technology*, 18, 193–205.

Yoshida, T., Hatano, T., Ito, H. & Okuda, T. (2009) Structural diversity and antimicrobial activities of ellagitannins. In: *Chemistry and Biology of Ellagitannins* (ed. S. Quideau), pp. 55–93. World Scientific, Singapore.

Zhou, H.-C., Lin, Y.-M., Wei, S.-D. & Tam, N.F.Y. (2011) Structural diversity and antioxidant activity of condensed tannins fractionated from mangosteen pericarp. *Food Chemistry*, 129, 1710–1720.

Zhou, H.C., Tam, N.F.Y., Lin, Y.M., *et al.* (2012) Changes of condensed tannins during decomposition of leaves of *Kandelia obovata* in a subtropical mangrove swamp in China. *Soil Biology & Biochemistry*, 44, 113–121.

Zifkin, M., Jin, A., Ozga, J.A., *et al.* (2012) Gene expression and metabolite profiling of developing highbush blueberry (*Vaccinium corymbosum* L.) fruit indicates transcriptional regulation of flavonoid metabolism and activation of abscisic acid metabolism. *Plant Physiology*, 158, 200–224.

# Chapter 6
# Epigenetics, Plant (Poly)phenolics, and Cancer Prevention

Clarissa Gerhauser
*Epigenomics and Cancer Risk Factors, German Cancer Research Center (DKFZ), Heidelberg, Germany*

**Abstract:** Epigenetics describes a phenomenon of modifications in gene expression caused by heritable, but potentially reversible, alterations in the chromatin structure, DNA methylation, and post-transcriptional effects of small noncoding (micro) RNAs (miRNAs), without changes in the DNA sequence. These mechanisms represent novel targets for polyphenols and other natural products in the prevention and treatment of cancer and other diseases. This chapter summarizes the influence of various classes of polyphenol on the enzymatic activities of major enzymes involved in epigenetic gene regulation: DNA and histone methyltransferases (DNMTs and HMTs), histone acetyltransferases (HATs), histone deacetylases (HDACs), and histone demethylases (HDMs). Inhibitory effects mediated by (−)-epigallocatechin 3-gallate (ECGC) from green tea and the isoflavone genistein from soy *in vitro* and *in vivo* are described in detail. Functional consequences on gene expression in cell culture, in rodent models and in human intervention studies, are highlighted.

**Keywords:** gene regulation, DNA methylation, histone modifications, DNA methyltransferase, epigenome, flavonoids

## 6.1 Introduction

Genetic information is stored in the cell nucleus in the form of the deoxyribonucleic acid (DNA) sequence, which is organized as a double-stranded molecule consisting of two complementary strands of DNA. All individual cells in an organism contain an identical DNA sequence. Yet mammals possess over 200 different tissue types, which are characterized by unique gene-expression patterns established during development. These

---

*Recent Advances in Polyphenol Research, Volume 4*, First Edition.
Edited by Annalisa Romani, Vincenzo Lattanzio, and Stéphane Quideau.

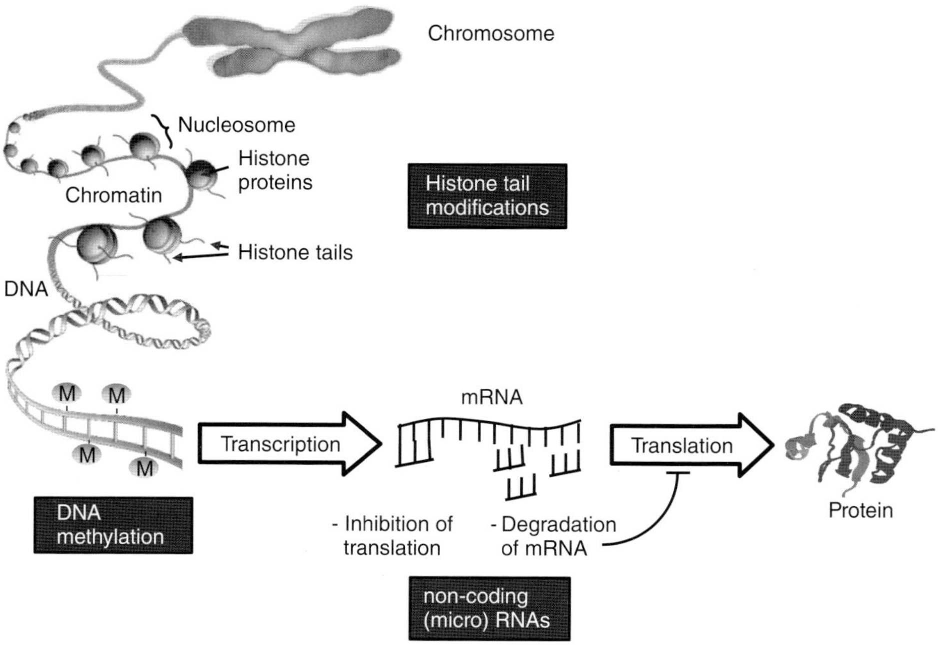


**Fig. 6.1** Overview of chromatin organization and epigenetic mechanisms, including DNA methylation, histone modifications, and noncoding miRNAs, targeting DNA, N-terminal histone tails, and mRNA. (Adapted from Qiu, 2006. Reproduced with permission of Nature Publishing Group.) See text for details. For color details, please see Plate 6.1.

fine-tuned differences in gene-expression patterns are maintained by "epigenetics"; that is, modifications in gene expression caused by heritable, but potentially reversible, alterations in chromatin structure and/or DNA methylation, without changes in the DNA sequence (Henikoff & Matzke, 1997). An important level of regulation is embedded in the temporal and spatial accessibility of genomic regions controlled by the chromatin structure. As summarized in Fig. 6.1, chromatin is a higher-order genomic structure, made up of DNA and proteins. The DNA helix is wrapped around pairs of the core histones H2A, H2B, H3, and H4, forming bead-like structures called nucleosomes. For genomic organization, nucleosomes are arranged on the DNA like "beads on a string", which are then folded to form higher-organized chromatin structures. DNA accessibility is regulated by the degree of chromatin condensation. *Heterochromatin* is highly condensed and contains mainly repetitive sequences, whereas *euchromatin* represents genomic regions that are decondensed and contain actively transcribed genes (Editors, 2003). Epigenetic mechanisms that modulate chromatin accessibility include DNA methylation and post-translational modifications of histone tails, such as acetylation, phosphorylation, and methylation. In addition, small noncoding (micro) RNAs (miRNAs) regulate gene expression at the level of translation by degrading mRNA or inhibiting its translation. These mechanisms represent novel targets for polyphenols and other natural products in the prevention and treatment of cancer and other diseases.

# 6.2 Influence of polyphenols on DNA methylation

## *6.2.1 DNA methylation in normal and tumor cells*

DNA methylation is a post-replication modification of DNA catalyzed by DNA methyltransferases DNMT1, 3a, and 3b that transfer methyl groups from S-adenosyl-*L*-methionine (SAM) as a methyl donor to cytosines, creating 5-methylcytosine (5mC), mainly within the context of CpG dinucleotides (Editors, 2003). DNMT1 is a maintenance methyltransferase that acts on hemimethylated DNA after replication and thus assures transmission of DNA methylation patterns to daughter cells. DNMT3a and DNMT3b are *de novo* methyltransferases that transfer methyl groups to previously unmethylated DNA sequences. DNMT3b is believed to play an important role during tumorigenesis (Beaulieu *et al.*, 2002; Linhart *et al.*, 2007).

In normal cells, CpG-rich sequences (so-called CpG islands, CGIs) in gene-promoter regions are generally unmethylated, with the exception of about 6–8%, which are methylated in a tissue-specific manner (Illingworth *et al.*, 2008). Conversely, the majority of CpG sites in repetitive genomic sequences are heavily methylated, limiting their accessibility and contributing to chromosomal stability (Esteller, 2007). This controlled pattern of DNA methylation is disrupted during aging and the development of chronic diseases. During carcinogenesis, increased methylation (DNA hypermethylation) of promoter CGIs contributes to transcriptional gene silencing, for example of tumor-suppressor genes (Kopelovich *et al.*, 2003; Esteller, 2007; Jones & Baylin, 2007). On the other hand, carcinogenesis is associated with a global loss of DNA methylation at repetitive genomic sequences (DNA hypomethylation), leading to genomic instability and chromosomal aberrations (Gama-Sosa *et al.*, 1983; Goelz *et al.*, 1985) (Fig. 6.2). Importantly, genes silenced by changes in DNA methylation remain intact and can potentially be reactivated by small molecules that affect the activity or expression of DNMTs. Consequently, the development of agents or food components that prevent or reverse cancer-induced DNA methylation and the consequent inactivation of gene expression is a promising new approach in cancer prevention (Huang *et al.*, 2011; Singh *et al.*, 2013).

## *6.2.2 Inhibition of DNMTs* in vitro

Polyphenols are a heterogeneous group of compounds that can be broadly divided into flavonoids, non-flavonoid polyphenols, and phenolic acids (for an overview, see Malireddy *et al.*, 2012). Research on the potential of polyphenols and other natural products to inhibit DNMTs and modulate aberrant DNA methylation was initiated in 2003, when Yang and coworkers first described the DNMT inhibitory activity of the green tea catechin (−)-epigallocatechin 3-gallate (EGCG) (Fang *et al.*, 2003). Since then, various flavonoids and phenolic compounds have been tested in several studies for their potential to inhibit the enzymatic activity of human or bacterial DNMT *in vitro* (Lee *et al.*, 2005; Lee & Zhu, 2006; Fang *et al.*, 2007; Paluszczak *et al.*, 2010; Rajavelu *et al.*, 2011). Table 6.1 gives an overview of the DNMT inhibitory effects of polyphenols, including information on chemical structures and dietary sources. Concentrations tested, percentage of inhibition,

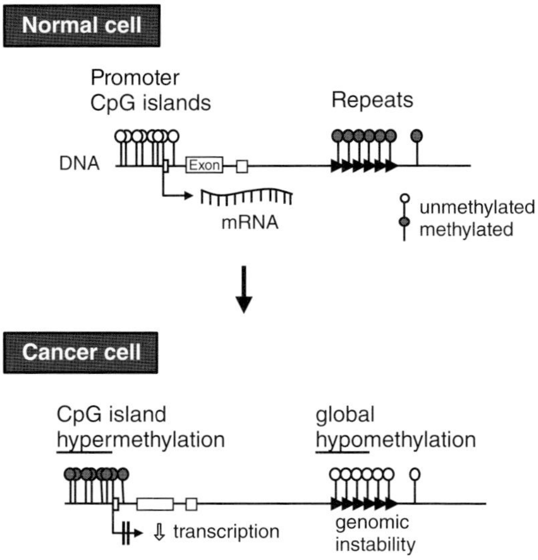


**Fig. 6.2** Overview of DNA methylation changes during carcinogenesis, including gain of methylation (red circles) at CpG islands (CGIs), often located in gene promoter regions (CGI hypermethylation) and associated with epigenetic silencing of gene transcription and global loss of DNA methylation (DNA hypomethylation, open circles) at intergenic regions and repetitive sequences, contributing to genomic instability. For color details, please see Plate 6.2.

and half-maximal inhibitory concentrations ($IC_{50}$) are reported when available. Test systems are based on radioactive detection or an ELISA-based approach, and depending on the assay conditions (human or bacterial enzyme, nuclear extracts or recombinant enzymes as enzyme source, choice of substrate, incubation time, buffer composition, mode of detection), results from these *in vitro* studies may not always be comparable or reproducible (Gilbert & Liu, 2010).

Flavonoid polyphenols tested include flavones, flavonols, flavanones, flavanols, and anthocyanidins. Among the flavones tested, baicalein from *Scrutellaria baicalensis* inhibited DNMT activity by 80% at a concentration of 20 μM and was more potent than apigenin and luteolin. Galangin from propolis or galangal root was identified as the most potent DNMT-inhibitory flavonol, with 67% inhibition at 20 μM. Quercetin, myricetin, and fisetin were less active. Between the two flavanones tested, hesperetin was found to be more active that naringenin, with 49% inhibition at 50 μM. The anthocyanidin cyanidin demonstrated high inhibitory potential, with 94% inhibition at a test concentration of 20 μM.

For polyphenols bearing a catechol moiety, studies by Lee and coworkers (Lee *et al.*, 2005; Lee & Zhu, 2006) suggest an alternative indirect mechanism of DNMT inhibition. These polyphenols may serve as substrates for methylation by catechol-*O*-methyltransferase (COMT), leading to the conversion of S-adenosyl-*L*-methionine (SAM) to S-adenosyl-*L*-homocysteine (SAH), which has been identified as a negative-feedback regulator of methyltransferases, including DNMTs and COMT. On the other hand, catechol-methylation by COMT may lead to depletion of SAM. In the presence of COMT, catechol-polyphenols inhibit bacterial M.Sss1 or human DNMT activity with $IC_{50}$ values in the range 0.75–8.40 μM.

Non-flavonoid polyphenols tested for DNMT inhibitory potential include the dihydrochalcone phloretin, garcinol from *Garcinia indica* (see Table 6.3), ellagic acid, and stilbenes. Rosmarinic acid was identified as the most potent phenolic acid with DNMT inhibitory potential, with 74% inhibition at a test concentration of 20 μM, and 90% at 40 μM.

### 6.2.3 *Inhibition of DNA methylation in cellular systems and* in vivo

As *in vitro* inhibition of enzymatic activities by polyphenols could result from unspecific protein binding, especially at elevated concentrations of inhibitors, demonstration of the functional consequences of DNMT inhibition might be a more reliable approach to the identification of polyphenols targeting the epigenetic mechanisms, including the methylome. As summarized in Table 6.2, quercetin, nordihydroguaiaretic acid (NDGA), resveratrol, and complex extracts of apple and black raspberry polyphenols have been reported to influence DNA methylation in cell culture or *in vivo*.

#### 6.2.3.1 *Quercetin*

Quercetin is a cancer-chemopreventive flavonoid occurring in the form of its glycosides ubiquitously in fruits, vegetables, and beverages (Russo *et al.*, 2012). It has a broad spectrum of cancer-preventive activities: it acts as an antioxidant and modulates enzymes and signaling cascades involved in detoxification, inflammation, proliferation, apoptosis, angiogenesis, autophagy, immune defense, and senescence (for a summary, see Murakami *et al.*, 2008; Vargas & Burd, 2010; Russo *et al.*, 2012). Cancer-preventive efficacy has been demonstrated in several rodent models for colon, mammary-gland, skin, and lung cancer (Murakami *et al.*, 2008). In addition, studies in human subjects indicate that quercetin intervention reduces markers of oxidative stress and inflammation (Russo *et al.*, 2012).

Beside its potential to inhibit DNMT activity *in vitro*, quercetin treatment of a colon-cancer cell line led to dose-dependent demethylation of the cell cycle-regulator $p16^{INK4A}$ (inhibitor of cyclin-dependent kinase CDK4, also known as CDKN2, CDK inhibitor 2). This was associated with p16 up-regulation at the mRNA and protein level and inhibition of cell proliferation (Tan *et al.*, 2009).

#### 6.2.3.2 *Nordihydroguaiaretic acid (NDGA)*

NDGA belongs to the naturally occurring plant phenolic lignans and is found at high concentrations in the Creosote bush (*Larrea tridentata* (Sesse and Moc. ex DC) Coville). It is a potent antioxidant and inhibits lipoxygenases and receptor tyrosine kinases such as insulin-like growth factor receptor 1 (IGF-1R). This is associated with tumor-cell growth inhibition *in vitro* and *in vivo*. Chemopreventive efficacy has been demonstrated in UVB- and chemically-induced rodent models of breast, prostate, lung, esophageal, and skin cancer (Blecha *et al.*, 2007; Lu *et al.*, 2010). NDGA is undergoing early clinical trials for prostate

**Table 6.1** Effects of plant (poly)phenols on DNA methyltransferases activity *in vitro*.

| Group/class | Compound | Structure | Dietary source | DNMT inhibitory activity[a] | References |
|---|---|---|---|---|---|
| Flavones | Apigenin | 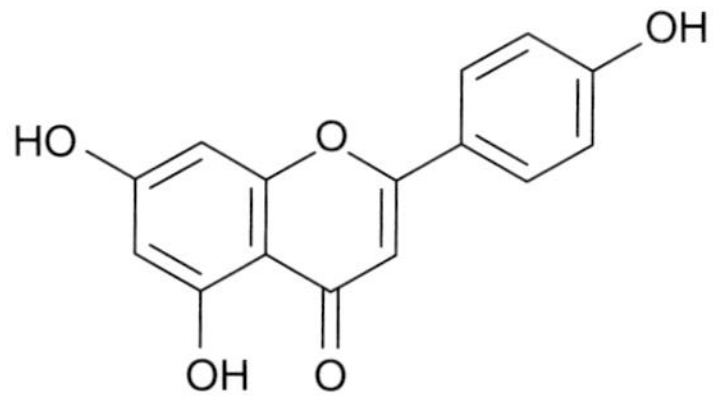 | Celery, chamomile | • Weak DNMT inhibition *in vitro*<br>20 μM: 30%<br>50 μM: 37% | Fang *et al.* (2007) |
| Flavones | Baicalein | 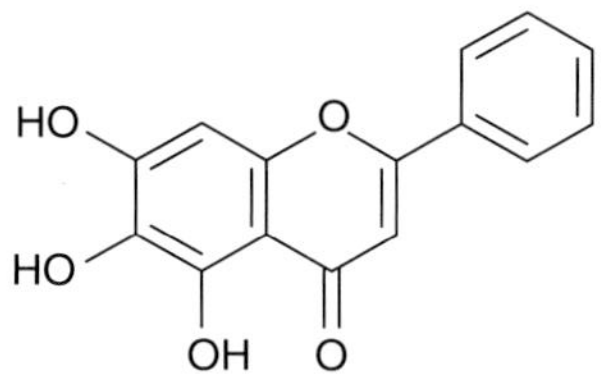 | *Scutellaria baicalensis* | • DNMT inhibition *in vitro*<br>20 μM: 80%<br>40 μM: 80% | Paluszczak *et al.* (2010) |
| Flavones | Luteolin | 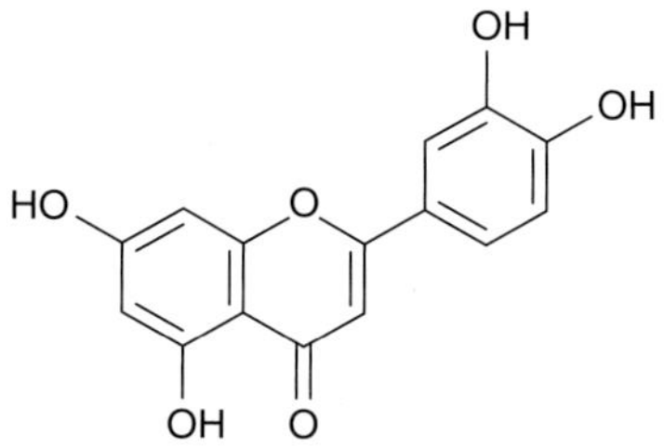 | Parsley, celery | • DNMT inhibition *in vitro*<br>20 μM: 33%<br>50 μM: 55% | Fang *et al.* (2007) |

| | | | | | |
|---|---|---|---|---|---|
| Flavonols | Quercetin | 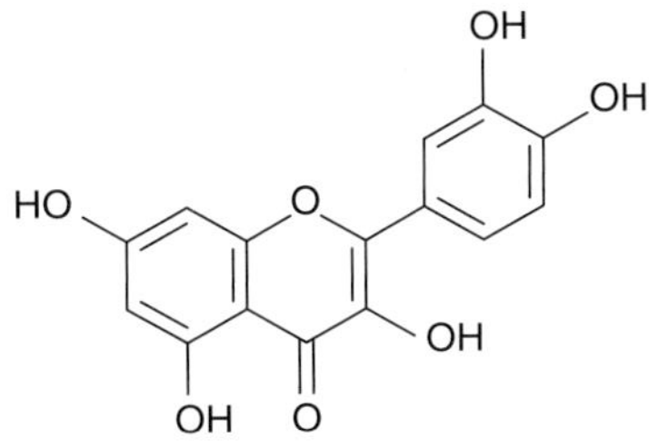 | Onion, broccoli, apples, berries | • DNMT inhibition *in vitro*<br>20 μM: 31%<br>50 μM: 41%<br>• Possible association with COMT-mediated *O*-methylation leading to SAH formation<br>M.Sss1 (+COMT): $IC_{50}$ 1.6 μM<br>DNMT1 (+COMT): $IC_{50}$ 1.6 μM | Lee *et al.* (2005), Fang *et al.* (2007) |
| Flavonols | Myricetin | 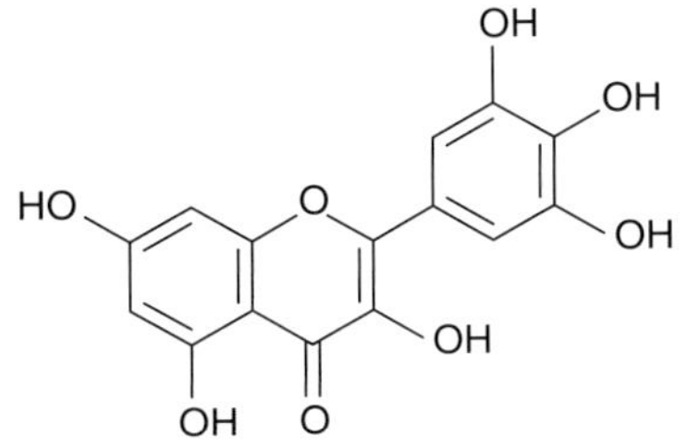 | Fruit, herbs, vegetables | • Weak DNMT inhibition *in vitro*<br>20 μM: 21%, 50 μM: 30%<br>40 μM: 16%<br>• Possible association with COMT-mediated *O*-methylation leading to SAH formation<br>M.Sss1(+COMT): $IC_{50}$ 1.0 μM<br>DNMT1(+COMT): $IC_{50}$ 3.5 μM | Lee *et al.* (2005), Fang *et al.* (2007), Paluszczak *et al.* (2010) |
| Flavonols | Fisetin | 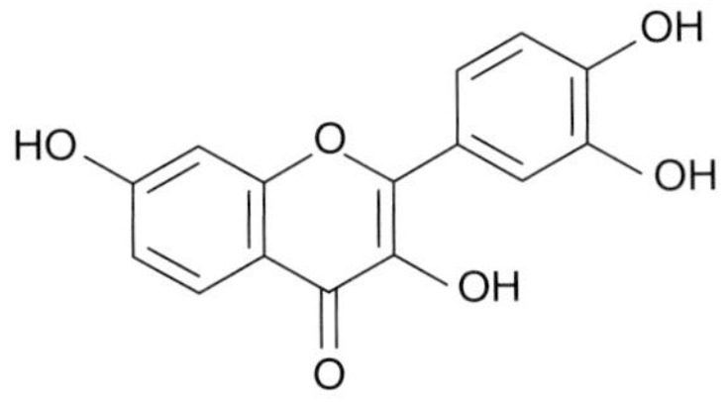 | Strawberries | • DNMT inhibition *in vitro*<br>• Possible association with COMT-mediated *O*-methylation leading to SAH formation<br>M.Sss1: $IC_{50}$ 9.1 μM<br>M.Sss1(+COMT): $IC_{50}$ 0.7 μM<br>DNMT1: $IC_{50}$ 10.7 μM<br>DNMT1(+COMT): $IC_{50}$ 1.2 μM | Lee *et al.* (2005) |

(continued overleaf)

**Table 6.1** (*continued*)

| Group/class | Compound | Structure | Dietary source | DNMT inhibitory activity[a] | References |
|---|---|---|---|---|---|
| Flavonols | Galangin | 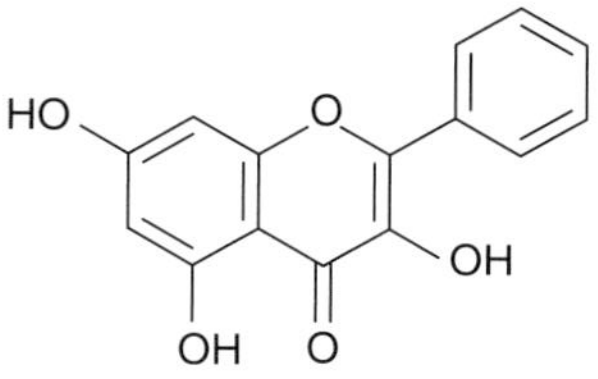 | Propolis, galangal root | • DNMT inhibition *in vitro*<br>20 μM: 67% | Paluszczak *et al.* (2010) |
| Flavanones | Hesperetin | 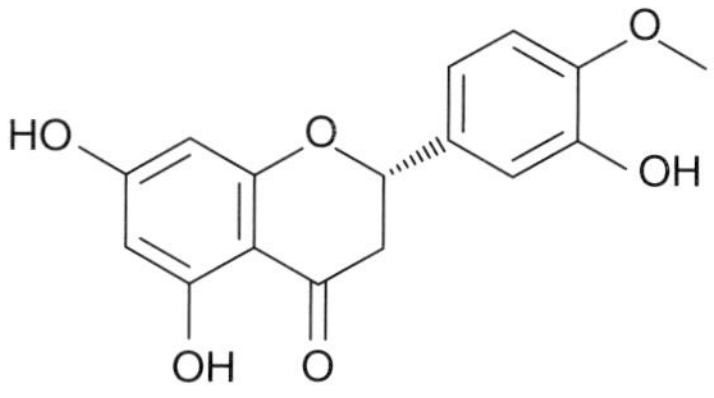 | Citrus fruit | • DNMT inhibition *in vitro*<br>20 μM: 26%<br>50 μM: 49% | Fang *et al.* (2007) |
| Flavanones | Naringenin | 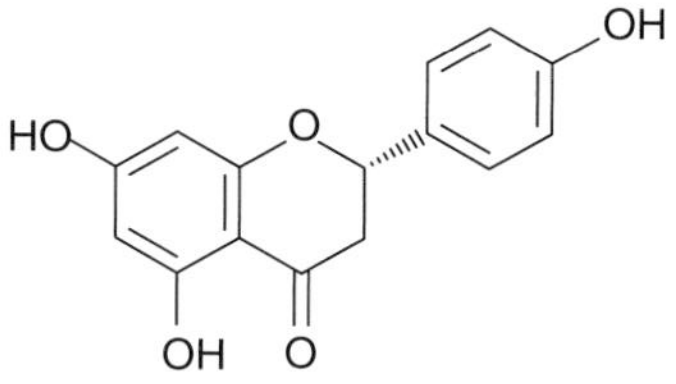 | Orange peel | • Weak DNMT inhibition *in vitro*<br>20 μM: 10%<br>50 μM: 21% | Fang *et al.* (2007) |

| | | | | | |
|---|---|---|---|---|---|
| Flavanols (catechins) | Catechin | 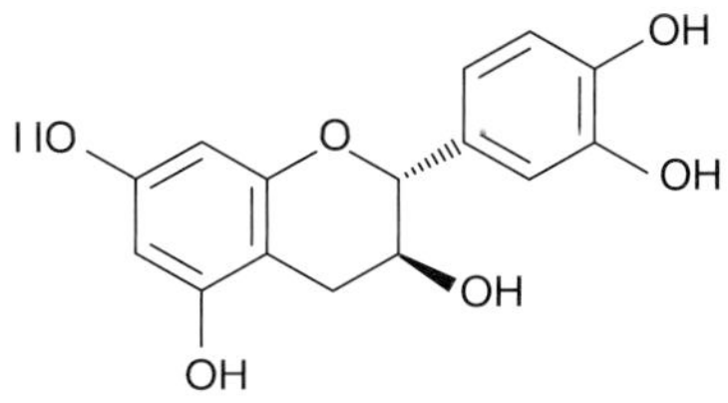 | Green tea | • DNMT inhibition *in vitro*<br>• Possible association with COMT-mediated *O*-methylation leading to SAH formation<br>M.Sss1 (+COMT): $IC_{50}$ 6.5 μM<br>DNMT1 (+COMT): $IC_{50}$ 4.6 μM<br>• ↓ promoter methylation of *RARβ* | Lee *et al.* (2005) |
| Flavanols (catechins) | Epicatechin | 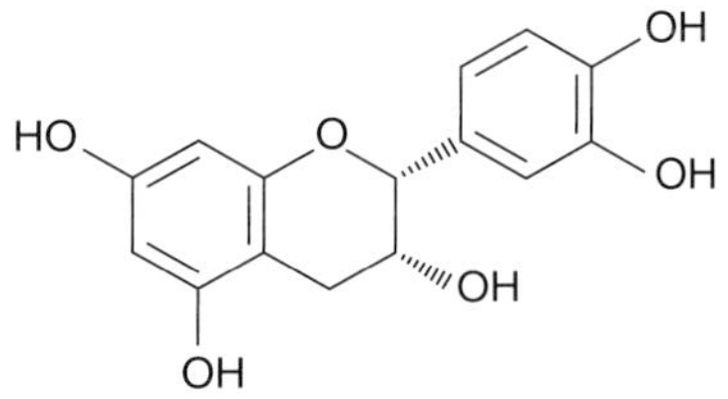 | Green tea | • DNMT inhibition *in vitro*<br>• Possible association with COMT-mediated *O*-methylation leading to SAH formation<br>M.Sss1 (+COMT): $IC_{50}$ 8.1 μM<br>DNMT1 (+COMT): $IC_{50}$ 8.4 μM | Lee *et al.* (2005) |
| Catechin polymer | Theaflavin 3,3′-digallate, thearubigins | 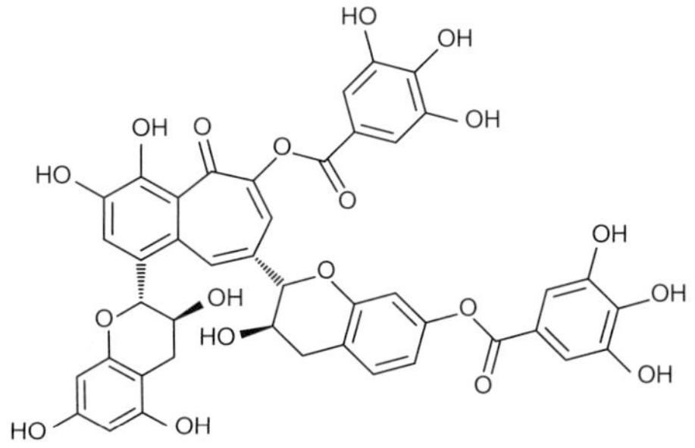 | Black tea | • DNMT3a inhibtion *in vitro*<br>Theaflavin 3,3′-digallate $IC_{50}$ = 44 μM<br>Thearubigins $IC_{50}$ = 40, 28 μM | Rajavelu *et al.* (2011) |

(*continued overleaf*)

**Table 6.1** *(continued)*

| Group/class | Compound | Structure | Dietary source | DNMT inhibitory activity[a] | References |
|---|---|---|---|---|---|
| Anthocyanidins | Cyanidin | 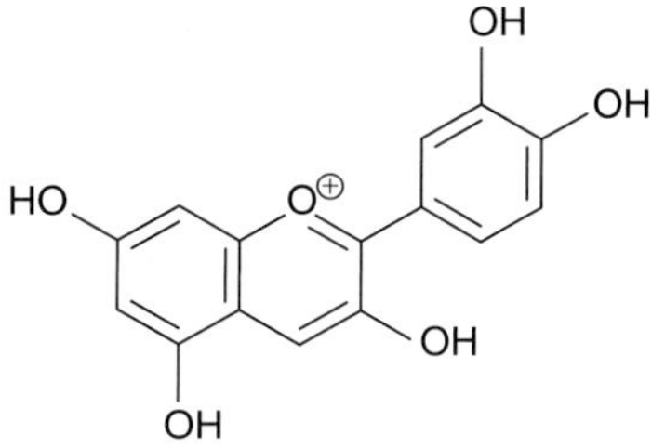 | Cherries, strawberries, blueberries, red cabbage, roses | • DNMT inhibition *in vitro*<br>20 µM: 94%<br>40 µM: 86% | Paluszczak *et al.* (2010) |
| Chalcones | Phloretin (dihydrochalcone) | 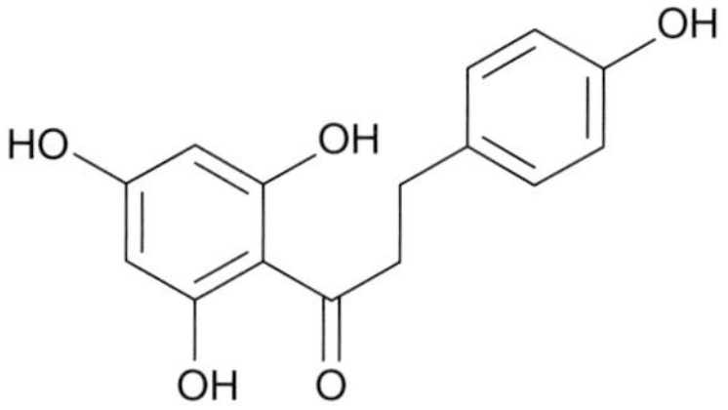 | Apples | • DNMT inhibition *in vitro*<br>20 µM: 81%<br>40 µM: 27% | Paluszczak *et al.* (2010) |
| Non-flavonoid polyhenols | Ellagic acid | 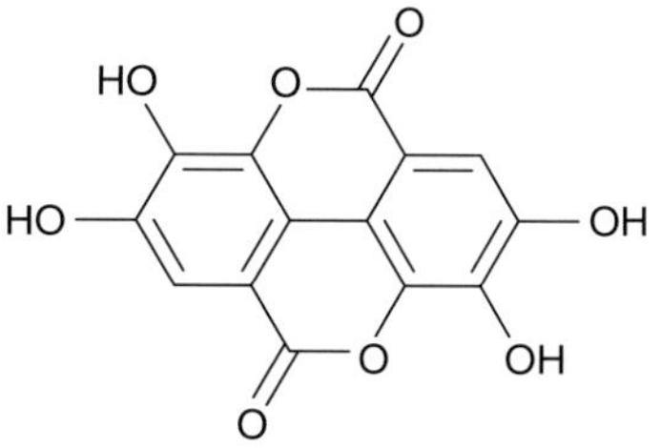 | Berries | • DNMT inhibition *in vitro*<br>20 µM: 53%<br>40 µM: 88% | Paluszczak *et al.* (2010) |

| | | | | | |
|---|---|---|---|---|---|
| Non-flavonoid polyhenols (Stilbenes) | Piceatannol | 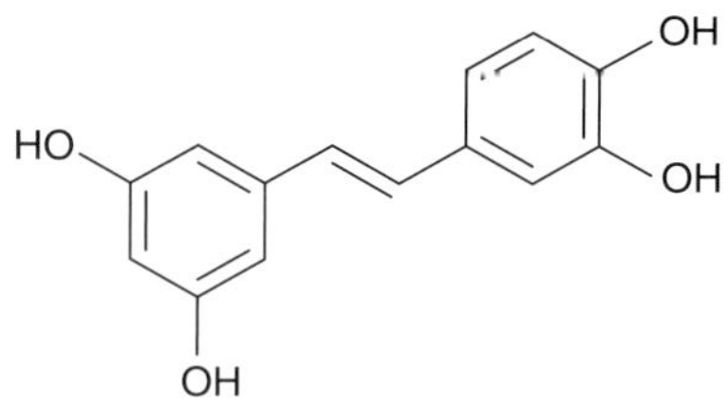 | Grapes | • Weak DNMT inhibition *in vitro*<br>20 μM: 32%<br>40 μM: 31% | Paluszczak *et al.* (2010) |
| Phenolic acids | Caffeic acid | 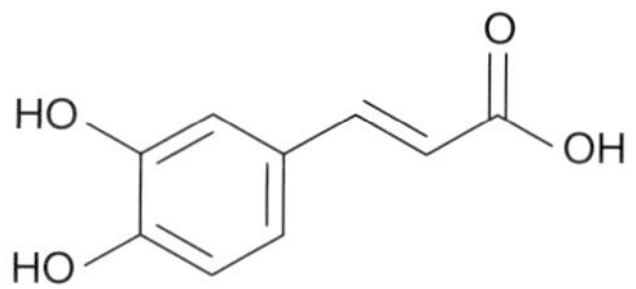 | Coffee | • Non-competitive DNMT1 inhibition *in vitro*, possible association with COMT-mediated *O*-methylation leading to SAH formation<br>M.Sss1 (+COMT): $IC_{50}$ 3.0 μM<br>DNMT1 (+COMT): $IC_{50}$ 2.3 μM<br>• No effect on DNMT3a at 100 μM<br>• Inhibition of HDM[b] KDM4C: $IC_{50}$ 13.7 μM | Lee & Zhu (2006), Rajavelu *et al.* (2011), Nielsen *et al.* (2012) |
| Phenolic acids | Chlorogenic acid | 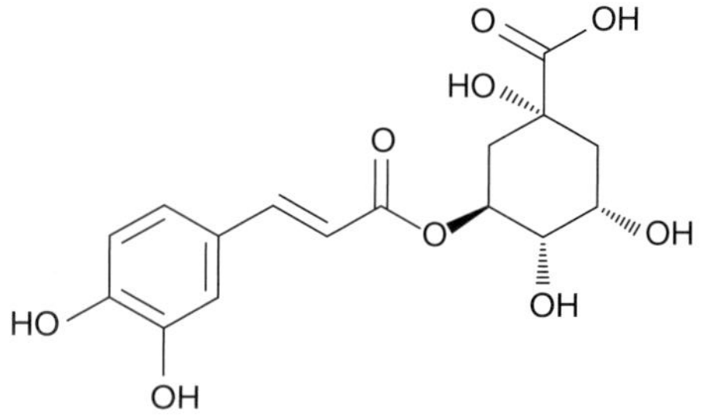 | Coffee | • Non-competitive DNMT1 inhibition *in vitro*, possible association with COMT-mediated *O*-methylation leading to SAH formation<br>M.Sss1 (+COMT): $IC_{50}$ 0.75 μM<br>DNMT1 (+COMT): $IC_{50}$ 0.9 μM | Lee & Zhu (2006) |

(*continued overleaf*)

**Table 6.1** (*continued*)

| Group/class | Compound | Structure | Dietary source | DNMT inhibitory activity[a] | References |
|---|---|---|---|---|---|
| Phenolic acids | 1,3-dicaffeoyl-muco-quinic acid diacetal | 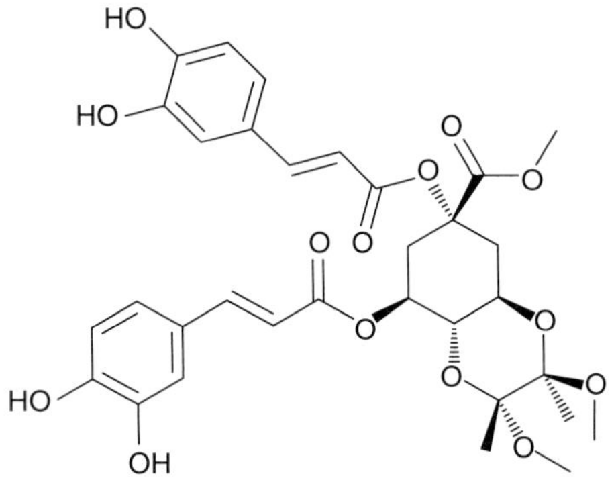 | Synthetic<br>A series of 12 chlorogenic acid derivatives screened | • DNMT3a inhibition *in vitro*<br>$IC_{50} = 35\,\mu M$<br>Only diacyl quinic acids active | Rajavelu *et al.* (2011) |
| Phenolic acids | Rosmarinic acid | 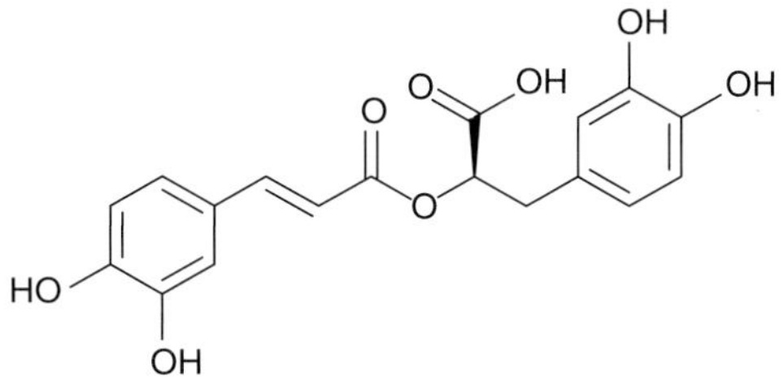 | Rosemary | • DNMT inhibition *in vitro*<br>20 μM: 74%<br>40 μM: 90% | Paluszczak *et al.* (2010) |
| Phenolic acids | Hydroxycinnamic acid | | Fruit | • DNMT inhibition *in vitro*<br>20 μM: 37%<br>50 μM: 74% | Fang *et al.* (2007) |

| | | | | | |
|---|---|---|---|---|---|
| Phenolic acids | Sinapic acid | 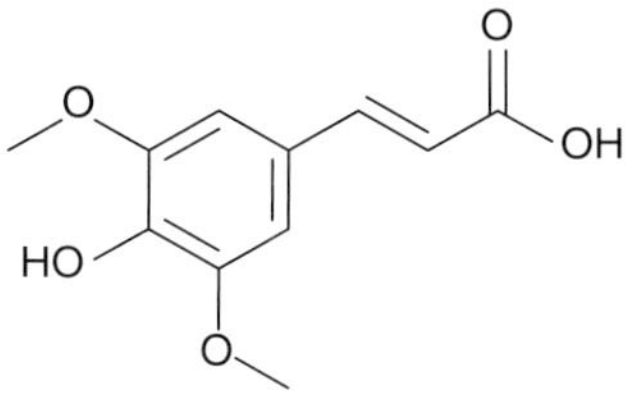 | Rape seed | • DNMT inhibition *in vitro*<br>20 μM: 58%<br>40 μM: 61% | Paluszczak *et al.* (2010) |
| Phenolic acids | Syringic acid | 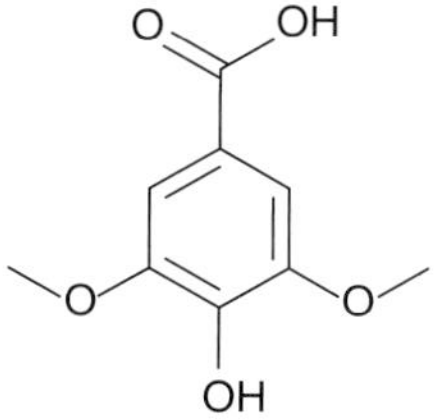 | Acai oil, wine | • DNMT inhibition *in vitro*<br>20 μM: 81%<br>40 μM: 51% | Paluszczak *et al.* (2010) |
| Phenolic acids | Protocatechuic acid | 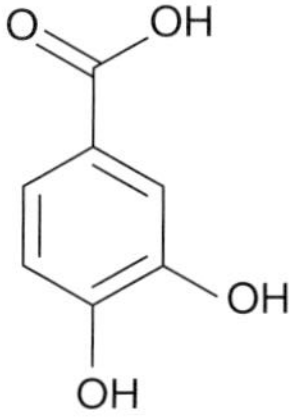 | Acai oil, olives | • DNMT inhibition *in vitro*<br>20 μM: 75%<br>40 μM: 64%<br>• HDM[b] KMD6A inhibition *in vitro*:<br>$IC_{50}$: 12.0 μM | Paluszczak *et al.* (2010), Nielsen *et al.* (2012) |

[a]Fang *et al.* (2007): Reaction based on $^3$H-labeled SAM as a substrate and poly(dI-dC)·poly(dI-dC) as a template. Paluszczak *et al.* (2010): Activity of DNMTs in nuclear extracts assessed based on the methylation of CG-rich oligonucleotides immobilized in wells; methylated cytosines detected by an ELISA assay. Lee *et al.* (2005): Reaction based on $^3$H-labeled SAM as a substrate and poly(dI-dC)·poly(dI-dC) as a template. Rajavelu *et al.* (2011): Inhibition of human Dnmt3a catalytic domain analyzed by Biotin–Avidin methylation kinetics assay (Roth & Jeltsch, 2000) using a biotinylated oligonucleotide substrate and $^3$H-labeled SAM.
[b]HDM, histone lysine demethylase.

**Table 6.2** Effects of plant polyphenols on DNA methyltransferases in cell culture and *in vivo*.

| Group/class | Compound | Structure | Dietary source | Epigenetic mechanism | References |
|---|---|---|---|---|---|
| Flavonols | Quercetin | 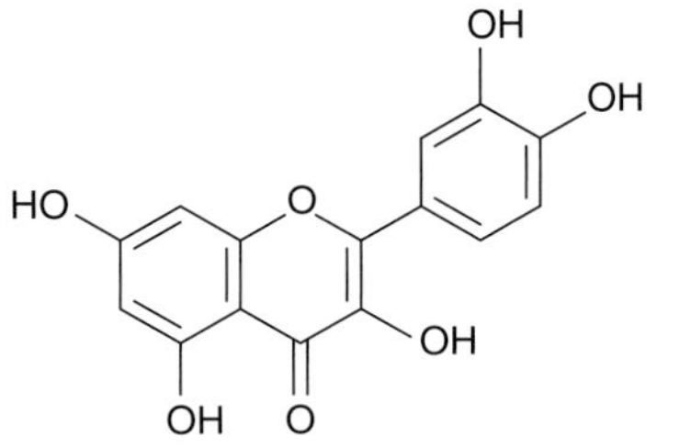  | Onion, broccoli, apples, berries | • ↓ promoter methylation of p16, leading to mRNA and protein re-expression | Tan *et al.* (2009) |
| Lignans | Nordihydroguaiaretic acid (NDGA) | 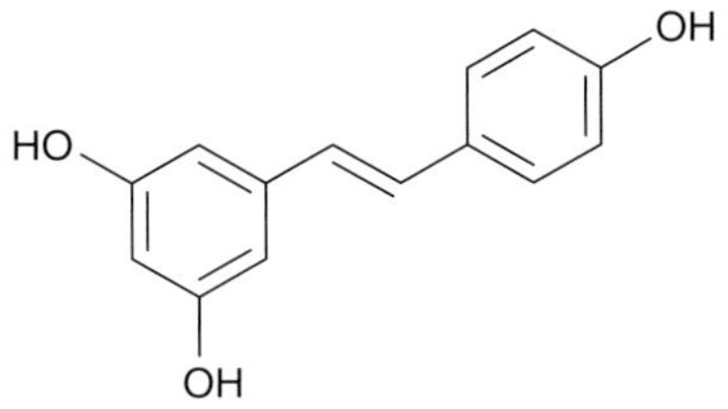  | Creosote bush | • ↓ promoter methylation of p16INK4, leading to mRNA and protein re-expression<br>• ↓ promoter methylation of E-cadherin, leading to mRNA re-expression *in vitro* and *in vivo*<br>• ↔ global (Line1) methylation | Byun *et al.* (2008), Cui *et al.* (2008a,b) |
| Other polyphenolic compounds (Stilbenes) | Resveratrol | 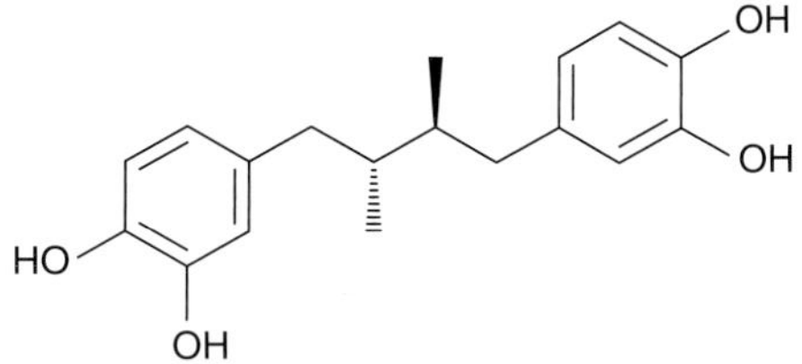  | Grapes | • Weak DNMT inhibition *in vitro*[a]<br>20 μM: 19%, 40 μM: 14%<br>• ↓ MBD2 and DNMT1 recruitment to BRCA1 promoter<br>• ↓ promoter methylation of RAR*β*, PTEN<br>• ↓ DNMT expression | Paluszczak *et al.* (2010), Papoutsis *et al.* (2010), Stefanska *et al.* (2010, 2012) |

| | | | | |
|---|---|---|---|---|
| Mixed polyphenols | Apple polyphenols | *Annurca* apples | • ↓ DNMT expression<br>• ↓ promoter hypermethylation of hMLH1, p14(ARF), and p16(INK4a)<br>• ↓ global DNA hypomethylation in normal mouse colon and polyps<br>• ↓ promoter hypomethylation of Igf2 and P2rx7 in normal mouse colon | Fini *et al.* (2007, 2011) |
| Mixed polyphenols | Black raspberry powder | Black raspberries | • ↓ DNMT expression<br>• ↓ promoter hypermethylation of SFRP2, SFRP5, WIF1, and PAX6a in human colon (cancer) biopsies | Wang *et al.* (2011b) |

[a]Paluszczak *et al.* (2010): Activity of DNMTs in nuclear extracts assessed based on the methylation of CG-rich oligonucleotides immobilized in wells; methylated cytosines detected by ELISA assay.

cancer treatment. However, hepatotoxicity might limit its use in humans (Ryan *et al.*, 2008; Lu *et al.*, 2010).

NDGA was first reported to reduce global methylation in malignant glioma cells (cited in Cui *et al.*, 2008b). Subsequently, NDGA treatment of breast and colon cancer cell lines resulted in p16 promoter demethylation, and p16 reactivation was associated with cell-cycle arrest and increased senescence (Cui *et al.*, 2008b). Similarly, NDGA reduced promoter hypermethylation and reactivated mRNA expression of the epigenetically silenced epithelial cell marker E-cadherin in breast and colon cancer cell lines, induced re-expression of E-cadherin protein in MDA-MB-435 breast-cancer xenografts, and reduced tumor growth (Cui *et al.*, 2008a,b). In contrast, NDGA treatment in HepG2 liver-cancer cells had no effect on global DNA methylation measured via methylation levels of the repetitive element LINE-1 (Byun *et al.*, 2008).

### *6.2.3.3 Resveratrol*

Resveratrol is a plant-derived stilbene derivative found in the skin of red grapes and in other fruits. It was first identified as a chemopreventive agent in 1997 (Pezzuto, 2008). It has a broad spectrum of health-beneficial effects, including antioxidant, cardioprotective, and antitumor activities, which have mechanistically been linked to effects on hormone receptors, drug metabolism, and cell signaling related to inflammation, cell survival and apoptosis, tumor angiogenesis, and metastasis formation (Kundu & Surh, 2008; Pezzuto, 2008). Resveratrol was shown to prevent carcinogenesis in animal models for breast, skin, esophageal, and colon cancer, and reduced xenograft growth of various tumor cell lines. Although plasma levels after oral consumption are extremely low (Baur & Sinclair, 2006), resveratrol is currently undergoing clinical testing in several small intervention trials (Pezzuto, 2008).

Resveratrol only weakly inhibited DNMT enzymatic activity *in vitro* (Paluszczak *et al.*, 2010). Three recent studies investigated its effect in breast-cancer cell lines. Activation of the aryl hydrocarbon receptor (AhR) has been shown to lead to epigenetic silencing of the DNA repair gene BRCA1 in breast cancer (Papoutsis *et al.*, 2010). This is associated with enhanced binding of repressive histone marks and methyl binding-domain protein 2 (MBD2) to the BRCA1 promoter. MBD2 is known to associate with DNMT1. Papoutsis *et al.* (2010) could demonstrate that pre- or cotreatment with resveratrol reduced MBD2 and DNMT1 levels at the BRCA1 promoter and prevented DNA damage induced by the AhR ligand 2,3,7,8 tetrachlorobenzo(p)dioxin (TCDD).

Retinoid acid receptor (RAR) $\beta 2$ and the tumor-suppressor PTEN (phosphatase and tensin homolog) are hypermethylated in breast and various other tumor types. Stefanska *et al.* (2010, 2012) used methylation-sensitive restriction analysis to test whether RAR$\beta 2$ and PTEN silencing could be reversed in breast-cancer cell lines after incubation with resveratrol alone and in combination with nucleoside analogs. Promoters of both genes were about 30% methylated in MCF-7 cells. Incubation with resveratrol reduced methylation levels of both genes; this was associated with down-regulation of DNMT1. However, only PTEN was re-expressed by resveratrol treatment (Stefanska *et al.*, 2010, 2012). This might indicate that alternative indirect mechanisms other than demethylation are involved in the observed re-expression of PTEN.

#### 6.2.3.4 Apple polyphenols

Apples (*Malus* sp., Rosaceae) are a rich source of polyphenols, including hydroxycinnamic acids, dihydrochalcones, flavonols (quercetin-glycosides), catechins, and oligomeric procyanidins, as well as anthocyanins in red apples and triterpenoids in apple peel (Gerhauser, 2008). Apple polyphenol extracts and isolated components have been shown to possess antimutagenic, antioxidant, and anti-inflammatory activities, to modulate carcinogen metabolism and signal transduction pathways, and to induce apoptosis. In animal models, apple components prevented skin, mammary, and colon carcinogenesis. Epidemiological observations indicate that regular consumption of one or more apples a day may reduce the risk for lung and colon cancer (Gerhauser, 2008).

Fini *et al.* (2007, 2011) asked whether some of the chemopreventive potential of apples might be related to epigenetic mechanisms. They could demonstrate that treatment of colon-cancer cell lines with an *Annurca* apple extract reduced protein expression of DNMT1 and DNMT3b. This was associated with reversal of promoter methylation of the cell-cycle regulators p16(INK4a) and p14(ARF) and the DNA repair gene hMLH1 (human mutL homolog 1), associated with reactivation of mRNA or protein expression (Fini *et al.*, 2007). In the $APC^{Min/+}$ mouse model for intestinal cancer (APC for adenomatous polyposis coli, a master regulator of the so-called Wnt signaling pathway, which is often deregulated in colon cancer; Min for multiple intestinal neoplasia, characterized by the occurrence of large numbers of adenoma in the small intestine), application of an apple extract in drinking water reduced the number of small-intestinal polyps by about 40% and prevented global DNA hypomethylation. Also, age-related hypomethylation of the maternally imprinted mitogenic growth factor insulin-like growth factor 2 (Igf2) and of P2rx7 (P2X purinoceptor 7) was prevented in normal colon by the apple extract, whereas age-related hypermethylation of several genes, including Esr1 (estrogen receptor alpha ER$\alpha$), Nkx2–5 (NK2 homeobox 5, a homeobox containing cardiac transcription factor), Gpr37 (G protein-coupled receptor 37, also known as endothelin receptor type B-like), Prdm5 (PR domain containing 5, a transcription factor of the PR-domain protein family, known to be involved in cell differentiation and tumorigenesis), Myod1 (myogenic differentiation 1, a basic helix–loop–helix transcription actor), Nptx2 (neuronal pentraxin II), and Igf2-DMR2 (Igf2 differentially methylated region 2), was not affected (Fini *et al.*, 2011).

#### 6.2.3.5 Black raspberry polyphenols

Black raspberries are rich in ellagic acid, quercetin glycosides, and anthocyanins. For more than 20 years, Stoner *et al.* (2007) investigated the *in vitro* and *in vivo* cancer-preventive potential of freeze-dried black raspberries, mainly for esophageal and colon cancer. They could demonstrate that black raspberries influenced carcinogen metabolism, inhibited cell growth, inflammation, and angiogenesis, and induced apoptosis. In human intervention studies, black raspberries were well tolerated at a dose of 45 g per day for 7 day (Stoner *et al.*, 2007).

In a recent small human phase I pilot study, Wang *et al.* (2011b) provided the first evidence for the modulation of epigenetic markers and downstream effects in human target tissue after chemopreventive intervention. They investigated whether black raspberry intervention (60 g/day for 1–9 weeks) could change the methylation levels of several genes involved in the Wnt signaling pathway, which as already mentioned is often deregulated in colon cancer, including WIF1 (Wnt inhibitory factor 1), sFRP2 and sFRP4 (secreted frizzled-related protein 2 and 4, antagonists of Wnt signaling), the cell-cycle regulator p16, and PAX1 (paired box protein 1, a tissue-specific developmental transcription factor), in normal and cancer tissue of colorectal-cancer patients. After a minimum 4-week intervention, promoter methylation levels of sFRP2 and Pax6 (paired box protein 6, also known as oculorhombin, a transcription factor involved in eye development) were significantly reduced in normal and tumor tissue. Black raspberry uptake also reduced WIF1 promoter methylation in tumor samples, especially after prolonged uptake. These findings were associated with reduced expression of DNMT1 and of downstream Wnt target genes such as $\beta$-catenin (a transcription factor activated by aberrant Wnt signaling, often over-expressed in cancer tissue), E-cadherin (a calcium-dependent adhesion protein with high expression in epithelial cells, regulating the cellular distribution of $\beta$-catenin), and Ki67 as proliferation markers in tumor tissue (Wang *et al.*, 2011b).

In summary, multiple classes of flavonoids, non-flavonoid polyphenols, and phenolic acids have been screened *in vitro* for DNMT inhibitory activity, using nuclear extracts or recombinant human or bacterial proteins as an enzyme source. In general, inhibitory activities are found at an intermediate μM range. In comparison to the more specific inhibition of other enzyme classes of importance for carcinogenesis (e.g. the inhibition of cytochrome P450 1A1 by quercetin at nanomolar concentrations), it is currently unclear whether the observed inhibitory potential has relevance for changes in gene expression and cancer-preventive activity. In cell culture, reduced expression of DNMT proteins after polyphenol treatment might contribute to inhibitory effects on DNA methylation, independent of direct inhibition of DNMT enzyme activity. With few exceptions, the effects of polyphenols on DNA methylation in animal models and in humans are still limited, and a conclusive link between modulation of DNA methylation levels and cancer-preventive efficacy is often still missing.

## 6.3 Influence of polyphenols on histone-modifying enzymes

In addition to DNA methylation, chromatin accessibility and gene expression are controlled by various post-translational modifications of N-terminal histone tails, including acetylation, methylation, phosphorylation, ubiquitinylation, sumoylation, and ADP ribosylation (Kouzarides, 2007; Bannister & Kouzarides, 2011; Fullgrabe *et al.*, 2011). Histone acetylation is controlled by the interplay of histone acetyltransferases (HATs) and histone

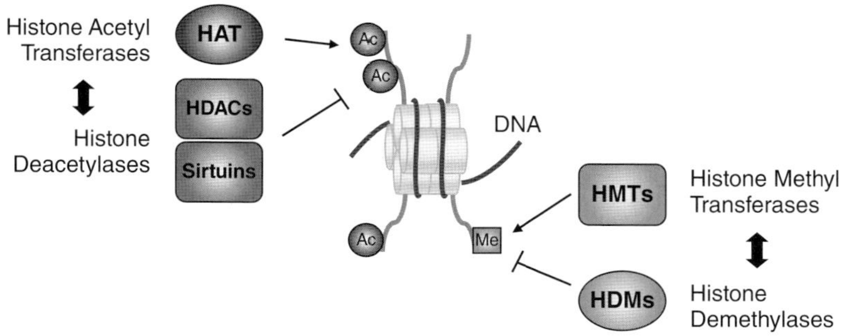


**Fig. 6.3** Simplified overview of histone-modifying enzymes, with a focus on histone deacetylases (HDACs), histone acetyltransferases (HATs), histone methyl transferases (HTMs), and histone demethylases (HDMs) and their influence on chromatin structure. Sirtuins represent a $NAD^+$-dependent subclass of HDACs (class III).

deacetylases (HDACs), whereas histone lysine methylation is controlled by the balance between histone methyltransferases (HMTs) and histone demethylases (HDMs) (Fig. 6.3).

## 6.3.1 *Acetylation of histones and non-histone proteins*

Histone acetylation is catalyzed by HATs, which add acetyl groups from acetyl-coenyzme A as a cofactor to the $\epsilon$-amino group of lysine residues in histone tails (whereas HDACs remove acetyl groups). Histone acetylation leads to an open chromatin structure, allowing transcription factors to access DNA. Consequently, HAT proteins are often transcriptional coactivators, such as p300/CBP (Creb-binding protein) or PCAF (p300/CBP-associated factor). So far, at least 25 proteins with HAT catalytic activity and distinct histone specificity have been identified, subdivided into five subgroups – the GNAT, MYST, p300/CBP, SRC, and TAFII250 families – with varying kinetics of lysine acetylation (Suzuki & Miyata, 2006; Fu & Kurzrock, 2010; Bannister & Kouzarides, 2011; Salvador & Luesch, 2012). Since 2004, several polyphenols have been described as inhibitors of HAT activity, including anacardic acid, curcumin, garcinol, gallic acid, and delphinidin (Table 6.3). The HAT-modulating activities of EGCG and genistein will be described in Section 6.5.

### 6.3.1.1 *Anacardic acid*

Anacardic acid (6-nonadecyl salicylic acid) is a component of cashew nut shell liquid (Hemshekhar *et al.*, 2012) and was identified in an *in vitro* inhibition screen as the first natural-product inhibitor of p300 and PCAF HAT activity, with $IC_{50}$ values of 8.5 and 5.0 μM respectively (Balasubramanyam *et al.*, 2003). Anacardic acid also inhibits the HAT enzyme Tip60 (a member of the MYST family), which is involved in the activation of key enzymes of DNA damage response, with an $IC_{50}$ value of 9 μM (Sun *et al.*, 2006). Recent simulations of molecular docking to the p300 active site indicate that anacardic acid mimics binding of the cofactor acetyl-CoA, suggesting that it nonspecifically inhibits HAT activity (Ghizzoni *et al.*, 2010). Interestingly, anacardic acid also inhibited p300-mediated

**Table 6.3** Effects of plant (poly)phenols on histone acetyl transferases (HAT).

| Group/class | Compound | Structure | Dietary source | Epigenetic mechanism | References |
|---|---|---|---|---|---|
| Phenolic acids | Anacardic acid | 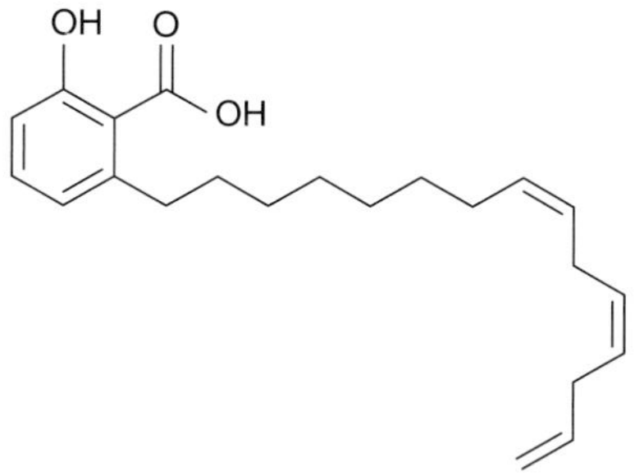 | Cashew nuts | • HAT inhibition *in vitro*<br>$IC_{50}$ p300: 8.5 μM<br>$IC_{50}$ PCAF: 5 μM<br>$IC_{50}$ Tip60: 9 μM<br>• ↓ acetylation and nuclear translocation of NF-κB p65 | Balasubramanyam *et al.* (2003),<br>Sun *et al.* (2006),<br>Sung *et al.* (2008) |
| Phenolic acid derivative | Curcumin (diferuloyl-methane) | 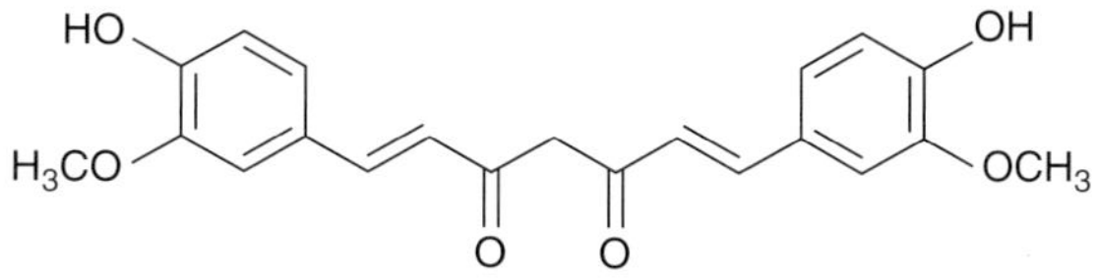 | Turmeric (*Curcuma longa*) | • HAT inhibition *in vitro*<br>$IC_{50}$ p300: ~19.5 μM<br>$IC_{50}$ CBP: ~22.5 μM<br>• ↓ histone and p53 acetylation<br>No effect on PCAF, HDAC, HMT activity *in vitro* | Balasubramanyam *et al.* (2004b),<br>Kang *et al.* (2005) |
| | | | | ↓ HAT p300 expression<br>• ↓ NF-κB p65 acetylation | Chen *et al.* (2007),<br>Ryu *et al.* (2008),<br>Yun *et al.* (2011) |
| | | | | ↓ HDAC1,2,3,4,8 expression<br>• ↑ ac-H4<br>• ↑ tubulin acetylation<br>• ↓ NF-κB activity | Liu *et al.* (2005),<br>Chen *et al.* (2007),<br>Lee *et al.* (2011),<br>Yun *et al.* (2011) |

| | | | | | |
|---|---|---|---|---|---|
| | | | | ↓ HMT EZH2 expression<br>• ↓ H3K27me3 | Hua *et al.* (2010),<br>Shu *et al.* (2011) |
| | | | | ↓ DNMT activity, expression | Liu *et al.* (2009, 2011a,b),<br>Kuck *et al.* (2010),<br>Abusnina *et al.* (2011),<br>Khor *et al.* (2011),<br>Medina-Franco *et al.* (2011),<br>Shu *et al.* (2011) |
| | | | | ↓ promoter methylation of several genes in cell culture | Jha *et al.* (2010),<br>Khor *et al.* (2011),<br>Liu *et al.* (2011a,b),<br>Shu *et al.* (2011),<br>Vilas-Zornoza *et al.* (2011) |
| Non-flavonoid polyphenols | Garcinol | 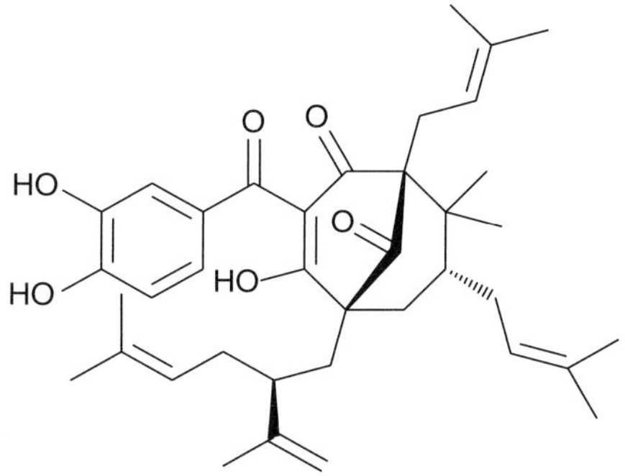 | Mangosteen tree<br>*Garcinia indica* | • HAT (p300, PCAF) inhibition *in vitro*<br>$IC_{50}$ p300: 7 μM<br>$IC_{50}$ PCAF: 5 μM<br>• ↓ H3 and H4 acetylation<br>• ↓ global gene expression in HeLa cells<br>• DNMT inhibition *in vitro*[a]<br>20 μM: 36%<br>50 μM: 65% | Balasubramanyam *et al.* (2004a),<br>Fang *et al.* (2007),<br>Arif *et al.* (2009) |

*(continued overleaf)*

**Table 6.3** *(continued)*

| Group/class | Compound | Structure | Dietary source | Epigenetic mechanism | References |
|---|---|---|---|---|---|
| Phenolic acids | Gallic acid | 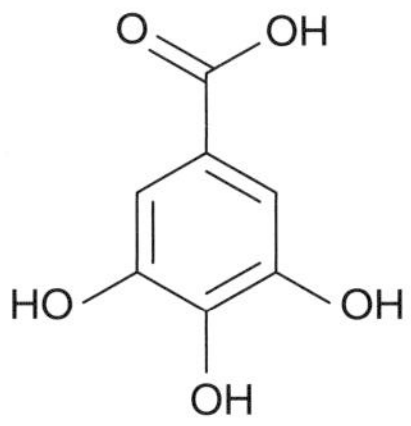 | Fruit, berries, tea, coffee, oak bark, walnuts, witch hazel, sumach | • HAT (p300, CBP, PCAF, Tip60) inhibition *in vitro*<br>$IC_{50}$ p300: 14 μM<br>$IC_{50}$ CBP: 24 μM<br>$IC_{50}$ PCAF: 34 μM<br>$IC_{50}$ Tip60: 25 μM<br>• No effect on SIRT1, HDAC, and HMT activities<br>• No effect on DNMT3a at 100 μM | Choi *et al.* (2009b), Rajavelu *et al.* (2011) |
| Anthocyanidins | Delphinidin | 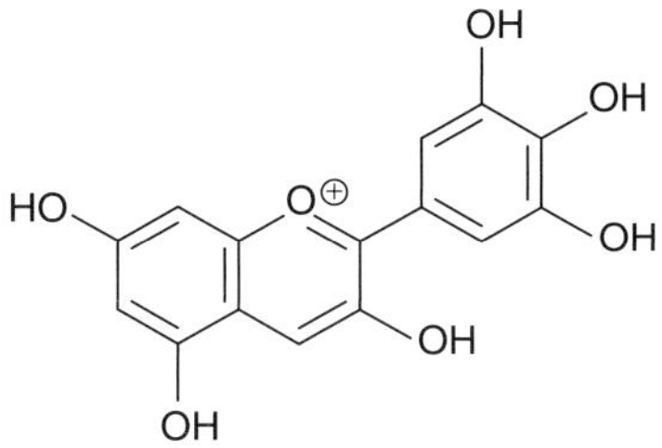 | Dark fruit and berries, e.g. blueberries | • Specific HAT (*p300*) inhibition *in vitro*<br>$IC_{50}$ *p300*: ~30 μM<br>• ↔ no effect on PCAF, SIRT1, HDAC, HMT activities<br>• ↓ acetylation of NF-κB p65 | Seong *et al.* (2011) |

[a]Fang *et al.* (2007): Reaction based on 3H-labeled SAM as a substrate and poly(dI-dC)·poly(dI-dC) as a template.

acetylation of the p65 subunit of NF-κB (nuclear factor "kappa light-chain enhancer" of activated B cells) as a non-histone substrate of HATs, and inhibited NF-κB-mediated signaling involved in cell survival, proliferation, invasion, and inflammation (Sung *et al.*, 2008). A disadvantage of anacardic acid might be its low cellular permeability (Balasubramanyam *et al.*, 2003).

#### *6.3.1.2 Curcumin*

The yellow pigment curcumin (diferuloyl methane) is a component of turmeric (*Curcuma longa*) and a major ingredient of the spice curry. Curcumin belongs to the most actively investigated cancer chemopreventive agents and has a long-term use in traditional medicine in India and other countries (Gupta *et al.*, 2013b). It has a broad spectrum of activities and acts on multiple signaling pathways, particularly NF-κB signaling. Curcumin has been shown to induce apoptosis and block invasion, metastasis, and angiogenesis *in vitro* and to prevent or inhibit tumor growth in rodent models, essentially for all major tumor entities (Surh & Chun, 2007; Kunnumakkara *et al.*, 2008; Ravindran *et al.*, 2009; Gupta *et al.*, 2013a,b). Investigations of curcumin in healthy human subjects and in patients with inflammatory diseases and cancer have shown promising first results and demonstrate that the compound is well tolerated even at elevated concentrations of up to 15 g/day (Kunnumakkara *et al.*, 2008; Gupta *et al.*, 2013a,b).

Curcumin was first identified as an inhibitor of HAT activity in 2004. Balasubramanyam *et al.* (2004b) described a decrease in histone H4 acetylation in Raji cell culture after curcumin treatment and identified curcumin as a specific inhibitor of p300/CBP, whereas PCAF, HDAC, and HMT activities were not affected by curcumin even at elevated concentrations. Curcumin also inhibited acetylation of the tumor suppressor protein p53 as a non-histone protein target of p300/CBP (Balasubramanyam *et al.*, 2004b). Acetylation of p53 was shown to be essential for p53-dependent growth arrest and apoptosis (Tang *et al.*, 2008b). HAT p300 inhibition was confirmed by Kang *et al.* (2005), and exposure of human hepatoma cells to curcumin led to a significant decrease in histone acetylation, as would be expected from HAT inhibition. This histone hypoacetylation was suggested to involve increased production of reactive oxygen species (ROS) by curcumin treatment and was blocked by antioxidants.

In addition to direct inhibition of HAT activity, treatment with curcumin and related compounds reduced the expression of p300. This was associated with downstream effects on Wnt- and NF-κB signaling (Chen *et al.*, 2007; Ryu *et al.*, 2008; Yun *et al.*, 2011). Curcumin treatment also led to a reduced expression of HDACs 1, 2, 3, 4, and 8, and consequently reduced cell proliferation *in vitro* and *in vivo* (Liu *et al.*, 2005; Chen *et al.*, 2007; Lee *et al.*, 2011; Yun *et al.*, 2011). Recent studies point to an effect of curcumin on HMT activities. Expression of EZH2 (enhancer of zeste homolog 2), a member of the polycomb repressive complex 2 (PRC2), and levels of the repressive histone mark H3K27me3 were reduced in breast- and prostate-cancer cell lines after curcumin treatment (Hua *et al.*, 2010; Shu *et al.*, 2011).

Liu *et al.* (2009) performed *in silico* docking studies to the homology domain of DNMT and identified curcumin and related analogs as putative DNMT inhibitors and DNA demethylating agents; this was later confirmed by Medina-Franco *et al.* (2011).

Subsequently, curcumin and its analogs were shown to inhibit DNMT activity *in vitro* (Kuck *et al.*, 2010; Liu *et al.*, 2011a,b) or to reduce DNMT expression (Abusnina *et al.*, 2011). Sequence-specific demethylation at the promoter regions of epigenetically silenced genes was demonstrated in various cancer cell lines (Jha *et al.*, 2010; Khor *et al.*, 2011; Liu *et al.*, 2011a,b; Shu *et al.*, 2011; Vilas-Zornoza *et al.*, 2011).

A recent study by Link *et al.* (2010) performed genome-wide profiling of DNA methylation changes in three colon-cancer cell lines after short- (6 days) and long-term (240 days) exposure to curcumin, using the Illumina 27 k bead array technology. These arrays covered roughly 27 000 CpG sites associated with 14 000 genes. Long-term incubation with curcumin at concentrations leading to about 50% reduction of cell growth led to changes in DNA methylation of 814–3051 individual CpG sites, with about equal distribution of hyper- and hypomethylation events. Sixty-eight loci were hypomethylated in all three cell lines. In contrast to the global demethylation observed with the demethylating agent 5-aza-2′-deoxy-cytidine (decitabine), curcumin mainly targeted CpG sites with intermediate methylation levels. Comparison of methylation with gene-expression changes in long-term treated cell lines indicated that genes involved in cell metabolism, cell signaling, cell proliferation, cell-, tissue-, and cancer development, amongst others, were regulated by curcumin via changes in DNA methylation. Mechanistically, the authors interpreted long-term curcumin effects on DNA methylation as a DNMT-independent event. Instead, they suggested that long-term silencing of important transcription factor-mediated pathways might allow passive gain of methylation at CpG sites that no longer interacted with transcription factors. Overall, short-term curcumin treatment had only negligible effects.

Curcumin and its derivatives have also been identified as potent modulators of miRNAs (Li *et al.*, 2010a; Izzotti *et al.*, 2012; Gupta *et al.*, 2013a). Further details on gene regulation and the activities of curcumin through epigenetic mechanisms have been summarized in several comprehensive review articles (Fu & Kurzrock, 2010; Li *et al.*, 2010a; Huang *et al.*, 2011; Reuter *et al.*, 2011; Khan *et al.*, 2012; Vanden Berghe, 2012).

#### *6.3.1.3 Garcinol*

The polyisoprenylated benzophenone garcinol was isolated from the mangosteen tree *Garcinia indica* Choisy (*Clusiaceae*). A chemopreventive effect was attributed to its antioxidant and anti-inflammatory activities. It induced phase II detoxifying enzymes and inhibited cell proliferation through induction of apoptosis. *In vivo*, it prevented colon and tongue cancer (Padhye *et al.*, 2009).

Garcinol was identified as a cell-permeable HAT inhibitor active against PCAF and p300 with $IC_{50}$ values of 5 and 7 μM, respectively. It was shown to act as a noncompetitive and competitive inhibitor of acetyl-CoA and the histone substrate (Salvador & Luesch, 2012). Molecular-docking studies suggested that the phenol moiety facilitates binding to the acetyl-CoA pocket, whereas the prenyl group substitutions affect specificity and binding to allosteric HAT sites (Arif *et al.*, 2009; Salvador & Luesch, 2012). In HeLa cells, garcinol repressed histone acetylation and global gene expression and induced apoptosis (Balasubramanyam *et al.*, 2004a). A recent study suggests that garcinol also affects miRNAs of the miR-200 and let-7 families and may reduce the metastatic potential of aggressive breast-cancer cells through reversion of epithelial–mesenchymal transition (EMT) (Ahmad *et al.*, 2012).

### 6.3.1.4 Gallic acid

Gallic acid (3,4,5-trihydroxybenzoic acid) is found in various fruit and berries, tea and coffee, oak bark, walnuts, witch hazel (*Hamamelis virginiana*), sumach (*Rhus coriaria*), and other plants, both free and as part of hydrolyzable tannins (gallotannins) (Choi *et al.*, 2009b; Ferk *et al.*, 2011). Gallic acid has good antioxidant activity both *in vitro* and *in vivo*, partly through increasing the activity of antioxidant enzymes (Ferk *et al.*, 2011). It has been shown to reduce oxidative DNA damage (Ferk *et al.*, 2011) and to induce apoptosis preferentially in cancer cells by various mechanisms (Locatelli *et al.*, 2012).

In a recent study by Choi *et al.* (2009b), gallic acid was identified as a specific inhibitor of HAT activity *in vitro*, without affecting other epigenetic enzymes. $IC_{50}$ values for p300, CBP, PCAF, and Tip60 were in the range 14–34 µM. Like anacardic acid, gallic acid inhibited the p300-mediated acetylation of RelA (v-Rel reticuloendotheliosis viral oncogene homolog A), the p65 subunit of NF-$\kappa$B, and subsequently reduced NF-$\kappa$B activation and expression of anti-apoptotic genes in response to pro-inflammatory stimuli (Choi *et al.*, 2009b).

### 6.3.1.5 Delphinidin

Delphinidin is an anthocyanidin that occurs at high concentrations (about 10× higher than other flavonoids such as quercetin) in dark fruit, particularly blueberries. Anthocyanidins have high antioxidant potential and induce detoxifying enzymes. They often possess antiproliferative activity and induce apoptosis and cell differentiation. Inhibition of angiogenesis and invasiveness also contributes to their chemopreventive potential (Wang & Stoner, 2008). Anthocyanidins have been shown to prevent cancer in rodent models for esophageal, lung, skin, and colon cancer. In human studies, intervention with anthocyanidin-rich fruit powders or fruit juices reduced levels of oxidative stress markers, but results on cancer-preventive potential from epidemiological studies remain inconclusive. This might be associated with low absorption and rapid elimination of anthocyanidins or their naturally occurring glycosides (Wang & Stoner, 2008).

Seong *et al.* (2011) identified delphinidin as a HAT inhibitor. The $IC_{50}$ value for inhibition of p300 activity was determined as 30 µM. Delphinidin also inhibited CBP activity, whereas PCAF and other epigenetic enzymes were not affected. Like gallic acid, delphinidin reduced pro-inflammatory signaling by preventing acetylation of the NF-$\kappa$B subunit p65.

Overall, HAT-mediated acetylation of histones and non-histone proteins seems to play an important role in regulating inflammatory signaling, in particular the NF-$\kappa$B pathway, and inhibition of HAT activity might contribute to the anti-inflammatory activity of chemopreventive polyphenols (Ghizzoni *et al.*, 2011).

### 6.3.2 Deacetylation by HDACs and sirtuins

HDACs catalyze the transfer of acetyl groups from acetylated histones or other acetylated proteins to coenzyme A. In general, histone deacetylation leads to chromatin condensation and transcriptional repression. Eighteen proteins with HDAC activity have been classified so far. HDACs 1–11 are subdivided into three classes – I, II, and IV – based on homology, size, subcellular expression, and number of enzymatic domains. Class III comprises sirtuins 1–7, which are structurally unrelated to class I and II HDACs and require $NAD^+$ as a cofactor for activity (Sauve *et al.*, 2006; Mottet & Castronovo, 2008). In addition to histones, important regulatory non-histone proteins and transcription factors such as p53, E2F, NF-$\kappa$B, FOXO (forkhead-box-protein O3), and PGC-1$\alpha$ (peroxisome proliferator-activated receptor (PPAR) gamma, coactivator 1 alpha) involved in inflammation, stress response, apoptosis, and energy metabolism have been shown to be regulated by acetylation (Kim *et al.*, 2006; Sauve *et al.*, 2006; Mottet & Castronovo, 2008; Ghizzoni *et al.*, 2011; Li & Kazgan, 2011).

#### 6.3.2.1 Inhibition of HDAC activity

Structure–activity-relationship analyses and molecular-docking studies have defined structural requirements for HDAC inhibition. The active center of an HDAC is characterized by the presence of a $Zn^{2+}$ ion and lies within a pocket that accommodates the lysine residue to be acetylated. Consequently, HDAC inhibitors generally possess (i) a metal-binding functional group that interacts with the $Zn^{2+}$ ion at the active site, (ii) a hydrocarbon linker that mimics the lysine residue, and (iii) a cap group that interacts with amino acids at the entrance of the active-site pocket (Bora-Tatar *et al.*, 2009) to stabilize the interaction. Spatial constraints might explain why, so far, complex multiring structured flavonoid-type polyphenols have not been identified as inhibitors of HDAC enzymatic activity. However, as described in Section 6.3.1.2 for curcumin, and as outlined in later sections for compounds such as EGCG and genistein, polyphenols have been reported to modulate HDAC activity by affecting the expression of various HDAC isoenzymes.

In a screen of 33 carboxylic-acid derivatives for inhibition of HDAC activity, chlorogenic acid and curcumin were identified as HDAC inhibitors, with $IC_{50}$ values of 375 and 115 μM, respectively. This was equivalent to HDAC inhibition mediated by the short-chain fatty acid (SCFA) butyric acid ($IC_{50}$ 116 μM) and its salt, sodium butyrate ($IC_{50}$ 365 μM), well-known natural HDAC inhibitors. Inhibitory potential was substantiated by molecular docking to the HDAC8 binding cavity (Bora-Tatar *et al.*, 2009). It should be noted that inhibitory concentrations are relatively high and might not be reached *in vivo* after consumption of dietary sources of chlorogenic acid or curcumin.

#### 6.3.2.2 Modulation of sirtuin activity

Silent information regulator 1 (SIRT1, sirtuin 1) has been shown to participate in various cellular processes, from fat and glucose metabolism to aging and cancer (Lavu *et al.*, 2008), but its role in carcinogenesis is still poorly understood (Brooks & Gu, 2009; Stunkel & Campbell, 2011). SIRT1 negatively affects stress response, but is also involved in recovery from DNA damage. Interestingly, over-expression of SIRT1 is regarded as oncogenic in the presence of wild-type (wt) p53, but is tumor-suppressive in tumors with mutant p53 (Huang *et al.*, 2011).

Interest in SIRT1 was raised when Howitz *et al.* (2003) identified the chemopreventive agent resveratrol (see Section 6.2.3.3) as an activator of SIRT1 and demonstrated that resveratrol increased lifespan in yeast and *C. elegans*, mimicking the effects of calorie restriction (Cohen *et al.*, 2004; for a recent review, see Farghali *et al.*, 2013). Since then, other polyphenols have been identified as activators (silymarin) or inhibitors (3,2′,3′,4′-tetrahydroxy-chalcone) of sirtuins (Table 6.4).

***SIRT1 activator: flavolignan silymarin from milk thistle***

Milk thistle (*Silybum marianum*) is traditionally used to protect against various liver diseases. Silymarin derived from milk-thistle seeds contains at least seven flavolignans and additional compounds. The most abundant component is silybinin (or silibinin), which constitutes a mixture of two isomers, silybin A and B. Although often used synonymously, silymarin and silybinin show distinct activity profiles (Kroll *et al.*, 2007). Cancer-preventive potential *in vitro* and *in vivo* has been attributed to the inhibition of inflammation, cell growth, angiogenesis, tumor invasion, and metastasis (Ramasamy & Agarwal, 2008; Loguercio & Festi, 2011). In clinical trials, milk-thistle extracts were shown to delay progression of prostate cancer (Tamayo & Diamond, 2007).

In a recent study, silymarin was identified as an activator of SIRT1 in human melanoma cells (Li *et al.*, 2007). SIRT1 activation was associated with cell survival and protected the cells from UV light-induced apoptosis. On the other hand, Ciu *et al.* (2009) reported that silybinin treatment reduced the growth of human liver cancer xenografts through induction of apoptosis, and this was associated, besides effects on PTEN/AKT (v-akt murine thymoma viral oncogene homolog 1, also known as protein kinase B, PKB) and ERK (extracellular signal-regulated kinase) signaling, with an increase in histone H3 and H4 acetylation. The underlying mechanism for histone hyperacetylation (e.g. HDAC inhibition versus HAT activation) was not elucidated.

***SIRT inhibitor 3,2′,3′,4′-tetrahydroxychalcone***

In an attempt to identify novel activators or inhibitors of SIRT1, Kahyo *et al.* (2008) tested a series of polyphenols, including 10 stilbenes, 8 chalcones, 10 flavanones, and 8 flavonols, using purified recombinant SIRT1 enzyme and an acetylated peptide of p53 as a substrate for the deacetylation reaction. At a test concentration of 100 μM, three hydroxy-chalcones inhibited SIRT1 deacetylase activity by more than 90%. Further testing using crude cell lysates indicated that only one of the chalcones, synthetic 3,2′,3′,4′-tetrahydroxychalcone, efficiently inhibited deacetylation of full-length p53. $IC_{50}$ values for SIRT1 and 3 inhibition were calculated as 40.3 and 48.0 μM, respectively. In human embryonic kidney cells, the chalcone dose-dependently induced hyperacetylation of endogenous p53 and increased protein expression of the p53 target p21. This was associated with suppression of cell growth (Kahyo *et al.*, 2008). Since the chalcone was a more effective inhibitor than the known SITR1 inhibitor sirtinol, further testing of its cancer-preventive efficacy is warranted.

**Table 6.4** Effects of plant polyphenols on sirtuins.

| Group/class | Compound | Structure | Dietary source | Epigenetic mechanism | References |
|---|---|---|---|---|---|
| Non-flavonoid polyphenols (Stilbene) | Resveratrol | 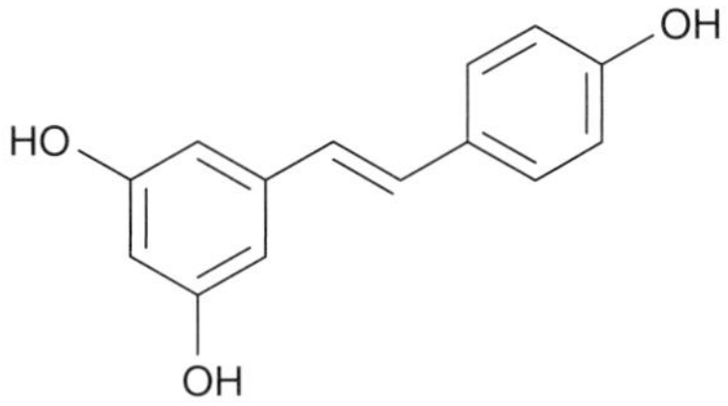 | Grapes | • Activation of SIRT1 | Howitz *et al.* (2003) |
| Flavonolignan | Silymarin | 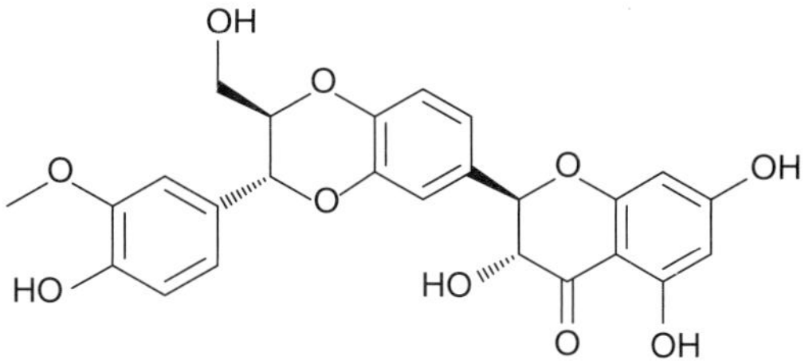 | Milk thistle | • SIRT activation<br>• ↑ ac-H3 and ac-H4 | Li *et al.* (2007), Cui *et al.* (2009) |
| Chalcones | Chalcone derivative 3,2′,3′,4′-tetrahydroxy-chalcone | 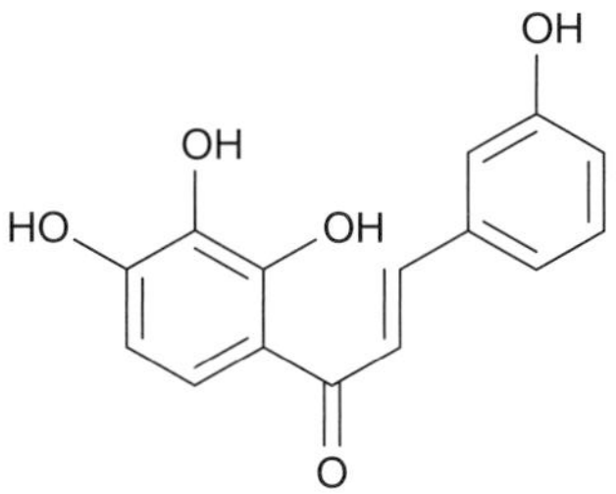 | synthetic | • SIRT inhibition *in vitro*<br>• $IC_{50}$ SIRT1: 40.3 μM<br>• $IC_{50}$ SIRT3: 48 μM<br>• $IC_{50}$ HDACs: 445 μM<br>• ↑ p53 acetylation | Kahyo *et al.* (2008) |

### 6.3.3 Histone methylation marks

Besides histone acetylation, histone methylation is an important epigenetic marker involved in the regulation of chromatin structure. Histone methylation takes place at lysine and arginine residues. Histone lysine methylation has activating or repressive effects on gene expression, dependent on the lysine residue that is methylated (e.g. K4, K9, K27, K36, K79 in H3), the methylation status (mono-, di-, or trimethylation), and the location (interaction with promoter versus gene-coding regions) (Kouzarides, 2007; Bannister & Kouzarides, 2011; Brait & Sidransky, 2011). Methylation at H3K4, H3K36, and H3K79 is generally associated with transcriptional active chromatin (*euchromatin*), whereas methylation at H3K9, H3K27, and H4K20 is frequently associated with transcriptional inactive heterochromatin (Suzuki & Miyata, 2006; Upadhyay & Cheng, 2011).

#### 6.3.3.1 Histone lysine methylation

Histone lysine methylation is mediated by histone lysine methyltransferases (HMTs), which transfer a methyl group from SAM to the lysine residue. HMTs can be classified as belonging to the Dot1 (disruptor of telomeric silencing) protein family and as proteins containing a so-called SET domain, based on sequence similarity with the *Drosophila* proteins SUV (suppressor of variegation), EZH (enhancer of zeste), and TRX (homeobox gene regulator trithorax). So far, more than 50 SET-domain family members, which can be grouped in six subfamilies, have been identified in humans (Upadhyay & Cheng, 2011).

Synthetic HMT inhibitors are developed based on structural similarity with the substrate SAM. So far, only a few natural compounds have been described as affecting histone lysine methylation (Huang *et al.*, 2011). The alkaloid chaetocin was one of the first selective inhibitors of the SUV39 class of HMTs targeting H3K9 to be identified (Ellis *et al.*, 2009). Trimethylation of H3K9 is generally associated with repressed chromatin. n-3 polyunsaturated fatty acids (PUFAs) were shown to target EZH. EZH2 is over-expressed in various cancer types, including breast and prostate cancer, and promotes H3K27 trimethylation and thereby maintains the transcriptional repressive states of genes. n-3 PUFAs were shown to reduce the expression of EZH2 by increasing its proteasomal degradation (Dimri *et al.*, 2010). As described in Section 6.5.1.2, the tea polyphenol EGCG also reduces the expression of several HMTs in various cancer cell lines (Balasubramanian *et al.*, 2010).

#### 6.3.3.2 Histone lysine demethylation

Several types of histone lysine demethylase (HDM) have been identified so far, including lysine-specific demethylase 1 (LSD1, also known as KDM1A and B) and a family of about 20 Jumonji domain-containing (JmjC) HDMs (Kouzarides, 2007; Bannister & Kouzarides, 2011; Upadhyay & Cheng, 2011). LSD1 is a FAD (flavin adenine dinucleotide, a coenzyme and electron carrier)-dependent amine oxidase, which removes methyl groups from mono- or dimethylated H3K4 or H3K9 through a FAD-dependent oxidation reaction that leads to the formation of $H_2O_2$ and formaldehyde. This demethylation alters the chromatin structure and results in context-dependent activation or repression of transcription (Teperino *et al.*, 2010). Like lysine acetylation, lysine methylation is not limited to histone proteins, and several non-histone protein substrates, including p53, retinoblastoma protein (RB), the NF-$\kappa$B subunit RelA, and estrogen receptor $\alpha$ (ER$\alpha$), have been identified (Yang *et al.*, 2009b; Munro *et al.*, 2010; West & Gozani, 2011).

In a screen to detect LSD1 inhibitors, two polyamine analogs, PG11144 and PG11150, were identified as potent competitive inhibitors with $IC_{50}$ values of 5 μM (Huang *et al.*, 2009). In a study by Nielsen *et al.* (2012) to discover novel HDM inhibitors, a natural product library representing 640 compounds was screened against the HDM KDM4C (lysine (K)-specific demethylase 4C, also known as GASC1/JmjD2C), a putative oncogene; a subset was further tested against two additional HDMs, KDM6A (lysine (K)-specific demethylase 6A, involved in cellular differentiation), and PHF8 (PHD finger protein 8, another putative oncogene over-expressed in prostate cancer). A short biotin-labeled histone tail fragment with trimethylation at lysine 9 was used as a substrate, and the resulting H3K9me2 fragment was detected with a specific antibody. Caffeic acid (Table 6.1) was identified as an inhibitor of both KDM4C and KDM6A, with $IC_{50}$ values of 13.7 and 5.5 μM, respectively. Protocatechuic acid (PCA, Table 6.1) was more potent in inhibiting KDM6A than KDM4C, with $IC_{50}$ values of 12.0 and 52.0 μM. None of the compounds inhibited PHF-8 activity, showing that the effects were not a result of general unspecific binding or inhibition. Mechanistic analyses of the inhibitory potential of PCA indicated a mixed mode of inhibition with respect to the peptide substrate, and a noncompetitive mode with respect to the 2-oxoglutarate cofactor. The authors concluded that the compound might interact with the peptide instead of binding to the cofactor pocket. In human osteosarcoma cells over-expressing KDM4C, PCA treatment at concentrations in the range 0.156–2.500 mM indeed inhibited the decrease in H3K9 trimethylation levels, although it should be noted that the concentrations used for testing were extremely high.

## 6.4 Influence of noncoding miRNAs on gene expression

MiRNAs are small, noncoding RNAs with a length of 20–22 nucleotides that regulate gene expression at the post-transcriptional level by controlling the translation of mRNA into proteins. They either bind to the 3′-untranslated region of mRNA in order to repress protein synthesis or affect mRNA stability. Each miRNA is estimated to control several hundred genes involved in key biological processes, including development, differentiation, apoptosis, and proliferation (Calin & Croce, 2006). MiRNAs are produced from RNA precursor structures by a complex multistep process involving proteins of the Argonaute family, polymerase II-dependent transcription, and the ribonucleases Drosha and Dicer (Winter *et al.*, 2009). Recent research indicates that miRNA expression is altered in various chronic degenerative diseases, including cancer (Calin & Croce, 2006). During carcinogenesis, miRNAs can be aberrantly up-regulated (oncomiRs) or down-regulated (tumor-suppressor miRs). Major mechanisms contributing to their deregulation include genetic and epigenetic alterations such as DNA methylation and defects in the miRNA processing machinery (Brait & Sidransky, 2011).

Several chemopreventive polyphenols have been shown to modulate gene expression via miRNAs. Since a number of recent reviews comprehensively summarize the action of dietary chemopreventive agents on miRNAs (Davis & Ross, 2008; Li *et al.*, 2010a; Link *et al.*, 2010; Saini *et al.*, 2010; Che & Huang, 2012; Izzotti *et al.*, 2012; Karius *et al.*, 2012; Neelakandan *et al.*, 2012; Parasramka *et al.*, 2012; Shah *et al.*, 2012; Lancon *et al.*, 2013; Sethi *et al.*, 2013; Wang *et al.*, 2013; Yi *et al.*, 2013), this section will not cover the activities of polyphenols on miRNAs and miRNA-mediated gene expression. Exceptions are EGCG and genistein, which are described in more detail in the following section.

## 6.5 Chemopreventive polyphenols affecting the epigenome via multiple mechanisms

Several polyphenols with chemopreventive efficacy *in vitro* and *in vivo* affect the epigenome by multiple and eventually interacting mechanisms. Two compounds will be highlighted in this review: the catechin EGCG from green tea (*Camellia sinensis* L. Kuntze) and the isoflavone genistein from soy (*Glycine max* L. Merr.) (Fig. 6.4).

### *6.5.1 (−)-epigallocatechin 3-gallate (EGCG) and green-tea polyphenols (GTPs)*

Green–tea polyphenols (GTPs) constitute a mixture of flavan-3-ols containing a catechol moiety. The main GTPs are EGCG, epigallocatechin (EGC), epicatechin gallate (ECG), and epicatechin (EC). EGCG has a broad spectrum of chemopreventive activities *in vitro*, including activity as a pro- and antioxidant, effects on various signal-transduction pathways and receptor-dependent signaling cascades (MAPK, EGFR, NF-$\kappa$B, IGF), inhibition of enzyme activities involved in drug metabolism and inflammation, and inhibition of cell growth and angiogenesis (Khan & Mukhtar, 2008). *In vivo*, EGCG and GTP prevented carcinogenesis in animal carcinogenesis models of all major organ sites, including lung,

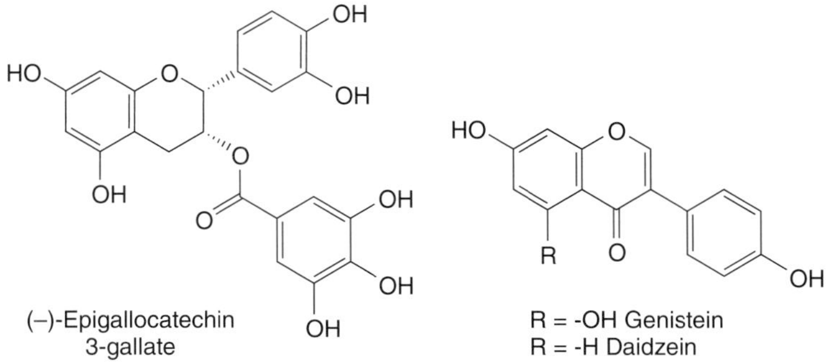


**Fig. 6.4** Chemical structures of (−)-epigallocatechin 3-gallate (ECGC), genistein, and daidzein.

colon, breast, prostate, skin, stomach, and liver cancer and tumors of the oral cavity. So far, human epidemiological studies have provided less conclusive results, most likely due to the low quantities of tea consumed (Yang *et al.*, 2009a).

#### *6.5.1.1 DNA methylation*

In a landmark study, Fang *et al.* (2003) described the DNA demethylating activity of EGCG (Table 6.5). Using nuclear extracts as an enzyme source, they suggested that EGCG is a competitive inhibitor of DNMT activity. Based on *in silico* modeling using a homology model of the catalytic domain of human DNMT1, EGCG was postulated to directly bind to the catalytic pocket, stabilized by hydrogen bonding with key amino acids. This was supported by two subsequent studies (Lee *et al.*, 2005; Yoo & Medina-Franco, 2011). Lee *et al.* (2005) demonstrated that $Mg^{2+}$ enhances the potential of EGCG to inhibit human DNMT by stabilizing the interaction between Glu1265 and the gallic acid moiety of EGCG. In several studies, DNMT activity was reduced in cultured cells treated with EGCG (Pandey *et al.*, 2009; Meeran *et al.*, 2011, 2012; Nandakumar *et al.*, 2011); this was partly associated with a reduced expression of DNMT mRNA or protein. EGCG or GTP intervention *reduced* genomic 5-methyl cytosine levels in A431 and Jurkat cell lines (Nandakumar *et al.*, 2011; Wong *et al.*, 2011) but was not effective at reducing global methylation levels or methylation of repetitive elements in two other studies (Chuang *et al.*, 2005; Stresemann *et al.*, 2006). On the other hand, GTP *enhanced* methylation of LINE-1 repetitive elements in LNCaP prostate-cancer cells (Pandey *et al.*, 2009). Since loss of global methylation has been associated with enhanced genomic instability, an increase of methylation at repetitive elements might indicate a preventive effect.

Several studies investigated whether EGCG treatment of cultured cells would lead to promoter methylation and subsequent re-expression of selected candidate genes. In KYSE510 esophageal and HT29 colon cancer cells, incubation with EGCG led to promoter demethylation and re-expression of the cell-cycle inhibitor p16 (Fang *et al.*, 2003, 2007). These results could be confirmed in A431 epidermoid carcinoma cells (Nandakumar *et al.*, 2011), but not in T24, HT29, or PC3 cells (Chuang *et al.*, 2005). EGCG treatment resulted in reduced promoter methylation and mRNA/protein re-expression of the DNA repair genes MGMT (*O*-6-methylguanine-DNA methyltransferase) and hMLH1 and of the retinoic acid receptor RAR$\beta$2 in human esophageal carcinoma cells (Fang *et al.*, 2003, 2007). In breast-cancer cell lines, RAR$\beta$2 promoter methylation was slightly reduced by EGCG intervention (Lee *et al.*, 2005).

Human telomerase reverse transcriptase (hTERT) is a catalytic subunit of the enzyme telomerase and is often upregulated in cancer cells. In normal cells, hTERT transcription is generally repressed through binding of the repressor E2F to its promoter region, whereas in cancer cells binding of E2F is often prohibited by methylation in the hTERT promoter. In two studies with breast-cancer cell lines, EGCG or an EGCG prodrug with enhanced stability and bioavailability reduced promoter methylation at selected CpG sites in the hTERT promoter. This facilitated E2F binding, as measured by chromatin immunoprecipitation (ChIP), and led to reduced hTERT mRNA expression and reduced cell proliferation (Berletch *et al.*, 2008; Meeran *et al.*, 2011).

**Table 6.5** Influence of (−)-epigallocatechin gallate (EGCG) and green-tea polyphenols (GTPs) on DNA methylation.

| Mechanism/target | Effect | Cell line/tissue | Genes/factors analyzed | References |
|---|---|---|---|---|
| **DNA methylation *in vitro*** | | | | |
| *In silico* modeling | | Homology model of the catalytic domain of hDNMT1 | Hydrogen bonding with Glu1265, Ser1229, and Arg1309 in the catalytic pocket of DNMT | Fang *et al.* (2003), Lee *et al.* (2005), Yoo & Medina-Franco (2011) |
| DNMT activity *in vitro* | Inhibition | KYSE 510 nuclear extracts | $IC_{50}$: 20 µM | Fang *et al.* (2003), Lee *et al.* (2005) |
| | | *M.Sss1* bacterial enzyme | $IC_{50}$: 0.2 µM | |
| | | rec. DNMT1 | $IC_{50}$: 0.5 µM | |
| | Inhibition | rec. DNMT1 | Considered an artifact due to unspecific enzyme inhibition | Medina-Franco *et al.* (2011) |
| DNMT activity in cell culture | Reduction | LNCaP nuclear extracts | | Pandey *et al.* (2009) |
| | Reduction | A431 | | Nandakumar *et al.* (2011) |
| | Reduction | MCF7, MDA-MB-231 | | Meeran *et al.* (2011, 2012) |
| DNMT mRNA/protein expression | Reduction (GTP) | LNCaP | DNMT1 | Pandey *et al.* (2009) |
| | Reduction | A431 | DNMT1, DNMT3a, DNMT3b | Nandakumar *et al.* (2011) |
| | Reduction | Jurkat | DNMT1, DNMT3a, DNMT3b | Wong *et al.* (2011) |
| | No effect | KYSE 510 | DNMT1, DNMT3a, DNMT3b, MBD2 | Fang *et al.* (2003) |
| Global methylation | Reduction | A431, Jurkat | Genomic 5meC content | Nandakumar *et al.* (2011), Wong *et al.* (2011) |
| | No effect | T24, HT29, PC3 TK6, HCT116 | MAGE-A1, Alu, LINE repetitive elements, genomic 5meC content | Chuang *et al.* (2005), Stresemann *et al.* (2006) |
| | Increase (GTP) | LNCaP | LINE-1 repetitive elements | Pandey *et al.* (2009) |
| Promoter hypermethylation | Reduction | KYSE 510 | p16, RAR$\beta$, MGMT, hMLH1, associated with gene re-expression | Fang *et al.* (2003, 2007) |
| | Reduction | HT29, Caco-2 | p16 (via folate metabolism) | Fang *et al.* (2003), Navarro-Peran *et al.* (2007) |

*(continued overleaf)*

**Table 6.5** (*continued*)

| Mechanism/target | Effect | Cell line/tissue | Genes/factors analyzed | References |
|---|---|---|---|---|
| | No effect | T24, HT29, PC3 | p16, also no effect on gene expression | Chuang *et al.* (2005) |
| | Reduction | PC3, KYSE 150, MCF7, MDA-MB-231 | RAR$\beta$ | Fang *et al.* (2003), Lee *et al.* (2005) |
| | Reduction | MCF7, MDA-MB-231 | hTERT, associated with reduced gene expression | Berletch *et al.* (2008), Meeran *et al.* (2011) |
| | Reduction | HSC3, HSC4, SCC9, SCC25 | RECK, associated with gene re-expression, ↓ invasive potential | Kato *et al.* (2008) |
| | Reduction | H460, A549 | WIF-1, associated with gene re-expression | Gao *et al.* (2009) |
| | Reduction (GTP) | LNCaP | GSTP1 | Pandey *et al.* (2009) |
| mRNA/protein re-expression | Induction | KYSE 510, CL13 | p16, RAR$\beta$, DAPK | Fang *et al.* (2007) |
| | Induction | A431 | p16, p21 | Nandakumar *et al.* (2011) |
| | Induction (GTP) | LNCaP | GSTP1 | Pandey *et al.* (2009) |
| | Induction (EGCG, GTP) | Jurkat | Foxp3 | Wong *et al.* (2011) |
| **DNA methylation *in vivo*** | | | | |
| SAM levels; SAM:SAH ratio | Moderate reduction | Plasma, small intestine, liver in healthy mice | | Fang *et al.* (2007) |
| SAH, methionine, homocysteine level | No effect | Plasma, small intestine, liver in healthy mice | | Fang *et al.* (2007) |
| Global methylation | No effect | Prostate, gut, liver in TRAMP mice | Global 5mC level; B1 repetitive elements, MAGE-a8 | Morey Kinney *et al.* (2009) |
| Promoter methylation | Reduction | Colon, small intestine, AOM-treated $APC^{Min/+}$ mice | RXR$\alpha$ | Volate *et al.* (2009) |
| | No effect | Prostate, gut, liver in TRAMP mice | IRX3, CACNA1A, CDKN2A, NRX2 | Morey Kinney *et al.* (2009) |
| | Reduction | Gastric-cancer patients | CDX2,BMP2 | Yuasa *et al.* (2009) |
| | No effect | Gastric-cancer patients | p16, CACNA2D3, GATA5, ER | Yuasa *et al.* (2009) |

Reversion-inducing cysteine-rich protein with Kazal motifs (RECK) is a novel tumor-suppressor gene that inhibits matrix-metalloproteases. Kato *et al.* (2008) treated oral squamous cell carcinoma cell lines with EGCG and detected reduced promoter methylation, accompanied by mRNA re-expression of RECK. This led to reduced levels and activities of metalloproteases, and invasive potential through collagen matrices was inhibited.

Disruption of the Wnt signaling pathway and aberrant activation of the transcriptional coactivator $\beta$-catenin play an important role in cancer development in various organs, including colon, skin, liver, ovary, breast, and lung (Barker & Clevers, 2006). WIF1 is an endogenous antagonist of the Wnt signaling pathway that is frequently disrupted by DNA methylation (Klaus & Birchmeier, 2008). In lung-cancer cell lines, EGCG treatment enhanced WIF1 mRNA expression by lowering promoter methylation levels. This was associated with reduced $\beta$-catenin expression and inhibition of Wnt signaling in a reporter assay (Gao *et al.*, 2009).

Glutathione S-transferase P1 (GSTP1) is frequently silenced in prostate cancer by promoter hypermethylation. In a study by Pandey *et al.* (2009), inhibition of DNMT activity in LNCaP prostate cancer cells by GTP reduced GSTP1 promoter methylation and led to its re-expression. ChIP experiments indicated reduced MBD2 (methyl-CpG binding domain protein 2) binding to transcription factor Sp1 (specificity protein 1) binding sites and induced transcriptional activation of the GSTP1 promoter.

Wong *et al.* (2011) investigated the effect of EGCG and GTP on FoxP3 methlyation and expression. FoxP3 controls the development and function of regulatory T cells (Treg). Both interventions reduced FoxP3 promoter methylation and led to significant up-regulation of FoxP3 mRNA expression.

Taken together, these data demonstrate that EGCG and GTP not only inhibit DNMT activity *in vitro* but also reduce promoter methylation in cultured cells, leading often to gene re-expression. So far, these studies have focused on selected candidate genes. Consequently, it is presently not known whether EGCG affects only selected genes or has a global reactivating effect on genes silenced by promoter methylation. One question that also needs to be answered is whether EGCG-mediated gene reactivation is relevant for its cancer-preventive efficacy.

Studies investigating the influence of EGCG on DNA methylation in rodent models and in human epidemiological studies have so far produced inconsistent results (Table 6.5). Fang *et al.* (2007) reported a moderate reduction of SAM levels and of the SAM : SAH ratio in various tissues and plasma of healthy mice, without an effect on SAH, methionine, or homocysteine levels. Morey Kinney *et al.* (2009) analyzed the prostate cancer preventive potential of GTP in "TRansgenic Adenocarcinoma of Mouse Prostate" (TRAMP) mice and used a genome-wide approach to identify genes with altered DNA methylation in prostate, gut, and liver. Although earlier studies had repeatedly reported good prostate-cancer preventive activity of GTP in this model (Gupta *et al.*, 2001; Caporali *et al.*, 2004; Adhami *et al.*, 2009), Morey Kinney *et al.* (2009) could not demonstrate any change in global me5C levels, any locus-specific or genome-wide DNA methylation, or any preventive effect on prostate cancer development.

$APC^{Min/+}$ mice develop multiple adenomas in the small intestine due to a genetic defect in the Wnt signaling pathway. Additional treatment with the carcinogen azoxymethane induces development of colonic tumors, which are generally rare in the $APC^{Min/+}$ model. Volate *et al.* (2009) treated $APC^{Min/+}$ mice and wt controls with GTP in drinking water for 4–8 weeks. Numbers of both small-intestinal adenomas and colon tumors were significantly reduced: by 28 and 50%, respectively. Also, GTP prevented AOM-induced promoter methylation and silencing of RXR$\alpha$ in colonic mucosa. RXR$\alpha$ expression is often reduced in human colon cancer. Therefore, these data might be of relevance for human colon cancer prevention.

In a human intervention study with 106 gastric-cancer patients, high consumption of green tea ($\geq$7 cups per day) significantly lowered the frequency of promoter methylation of the homeobox transcription factor CDX2 (caudal type homeobox 2) and bone morphogenetic protein BMP2 in primary tumor samples, whereas methylation of p16 (INK4A), calcium channel-related CACNA2D3 (calcium channel, voltage-dependent, alpha 2/delta subunit 3), transcription factor GATA-5 (GATA binding protein 5, binding to the DNA sequence GATA), and estrogen receptor (ER) was not changed (Yuasa *et al.*, 2009).

So far, the reason why EGCG inhibits or prevents methylation at certain promoter regions while others genes are not affected has not been investigated in detail. In principle, a sequence-specific effect on DNA methylation is feasible. It is well documented that transcription factors recruit activating or repressing complexes containing DNMTs to the DNA (Robertson *et al.*, 2000; Rountree *et al.*, 2000; Di Croce *et al.*, 2002). This is additionally regulated in a cell type-specific manner by the abundance of coactivators and corepressors. A further layer of regulation is the CpG density and the level of DNA methylation, as well as the regulation of DNA accessibility by the chromatin substructure at a certain DNA region. As long as genome-wide DNA methylation information is not available and interpretations are limited to a number of candidate genes, this question cannot be conclusively answered.

#### *6.5.1.2 Histone-modifying enzymes (HATs, HDACs, HMTs)*

Since 2009, several studies have revealed that the epigenetic effects of EGCG and GTP are not limited to influencing DNA methylation. As summarized in Table 6.6, EGCG has been shown to enhance histone acetylation by both *enhancing* p300 HAT expression and activity and *reducing* HDAC expression and activity in cultured cells. This leads to enhanced expression of ER$\alpha$ and reduced expression of hTERT in breast-cancer cells (Li *et al.*, 2010c; Meeran *et al.*, 2011, 2012), and to induction of the cell-cycle regulators p16 and p21 in lung-cancer cells (Nandakumar *et al.*, 2011). In LNCaP cells, reduced expression of HDAC isoforms enhanced acetylation and activation of p53 as a non-histone HDAC substrate. As a consequence, expression of p21 and the pro-apoptotic Bax were induced, leading to cell-cycle arrest and induction of apoptosis (Thakur *et al.*, 2012a,b).

Several studies have reported an *inhibitory effect* of EGCG on HAT activity, mainly p300, associated with reduced acetylation levels of histone H3 and H4, as well as of the non-histone HAT substrates NF-$\kappa$B p65 and androgen receptor (AR). As a consequence, NF-$\kappa$B- and AR-mediated cell signaling were reduced by EGCG treatment. The activities

**Table 6.6** Influence of (−)-epigallocatechin gallate (EGCG) and green-tea polyphenols (GTPs) on histone-modifying enzymes.

| Mechanism/target | Effect | Cell line/tissue | Genes/factors analyzed | References |
|---|---|---|---|---|
| HAT activity<br>HDAC activity | Induction<br>Reduction | MDA-MB-231 | ↑ p300 expression<br>↓ HDAC1 expression<br>↑ ac-H3, H3K9ac, ac-H4<br>↓ HMT SUV39H1 expression<br>↑ H3K4me2; ↓ H3K9me3<br>↑ ERα reactivation | Li *et al.* (2010c),<br>Meeran *et al.* (2012) |
| HDAC activity | Reduction | A431 | ↑ ac-H3, ac-H4<br>↑ p16, p21 protein expression | Nandakumar *et al.* (2011) |
| | Reduction (prodrug) | MCF7,<br>MDA-MB-231 | ↑ ac-H3, H3K9ac<br>↑ binding of MAD1, E2F1 to hTERT promoter;<br>↓ binding of c-Myc to hTERT promoter<br>↓ hTERT expression | Meeran *et al.* (2011) |
| | Reduction | LNCaP, PC3 | ↓ HDAC1, 2, 3, 8 protein expression<br>↑ p53 acetylation and activation<br>↑ p21, Bax expression<br>↑ cell cycle arrest, apoptosis | Thakur *et al.* (2012a,b) |
| HAT activity:<br>p300, PCAF, Tip60 | Reduction | HeLa, Hek293 | ↔ SIRT1, HDAC, HMT activity<br>↓ ac-H3, ac-H4<br>↓ NF-κB p65 acetylation<br>↑ cytosolic IκBα<br>↓ p65-binding to IL6 promoter,<br>↓ NF-κB-target gene expression<br>↓ EBV-mediated IL-6, IL-12 expr<br>↓ EBV-mediated cell transformation | Choi *et al.* (2009a) |

(*continued overleaf*)

**Table 6.6** (*continued*)

| Mechanism/target | Effect | Cell line/tissue | Genes/factors analyzed | References |
|---|---|---|---|---|
| | | TNF$\alpha$-stimulated mice | ↓ serum IL-6<br>↓ p65 acetylation in macrophages<br>↓ cytokine expression | Choi *et al.* (2009a) |
| HAT activity: p300<br>SIRT1, HDAC, HMT activity | Reduction<br>No effect | LNCaP | ↓ ac-H3<br>↓ AR acetylation, nuclear translocation<br>↓ AR-mediated PSA, NKX3.1 mRNA expression<br>↓ cell growth | Lee *et al.* (2012) |
| HMT protein expression: BMI-1, SUZ12, EZH2, EED, Mel18 | reduction | SCC-13, HaCaT, A431 | ↓ H3K27me3<br>↑ p21 and p27 expression<br>↑ cell cycle arrest<br>↑ apoptosis<br>BMI over-expression reverses the effects of EGCG | Balasubramanian *et al.* (2010), Choudhury *et al.* (2011) |

of other histone-modifying enzymes (SIRT1, HDAC, HMT) were not inhibited by the polyphenol (Choi *et al.*, 2009a; Lee *et al.*, 2012).

A recent study suggests that EGCG also targets histone methylation (Balasubramanian *et al.*, 2010). The polycomb group (PcG) proteins BMI-1 (B lymphoma Mo-MLV insertion region 1 homolog) and EZH2 regulate gene expression by increasing the repressive H3K27 methylation mark and reducing histone acetylation. EGCG reduced the expression of BMI-1 and EZH2 in skin-cancer cells, which was consistent with reduced H3K27me3 levels. Also, EGCG enhanced the expression of p21 and p27 cell-cycle regulators and induced apoptosis. Over-expression of BMI-1 reversed the effects of EGCG, indicating that inhibition of BMI-1-mediated H3K27 methylation contributes to the cell-growth inhibitory effect of EGCG (Balasubramanian *et al.*, 2010; Choudhury *et al.*, 2011).

#### *6.5.1.3 miRNAs*

The development of microarray-based approaches has facilitated investigations of natural product-mediated effects on miRNAs expression, similar to gene-expression studies. To avoid over-interpretation, the results of microarray-based studies should be verified by RT-PCR or by over-expression/antagonization of altered miRNAs and measurement of the expression of their target genes. The present review is therefore limited to those miRNA screening results obtained with EGCG, for which alterations in miRNA expression were confirmed by alternative methods (Table 6.7).

Tsang & Kwok (2010) investigated the effect of EGCG on miRNA expression in HepG2 human hepatocellular carcinoma cells. miR-16, let-7a, and miR-221 were up-regulated, whereas miR-18a, miR34b, miR-193, miR-222, and miR-342 were down-regulated. One of the targets of miR-16 is the anti-apoptotic protein Bcl-2. The authors could confirm that EGCG treatment reduced expression of Bcl-2 and induced apoptosis in HepG2 cells. Data were verified by transfection with a miR-16 inhibitor (Tsang & Kwok, 2010). In MCF7 breast-cancer cells, intervention with a GTP extract, polyphenon-60, up-regulated miR-548m and miR-720 expression. These data were confirmed by RT-PCR analyses. Concomitantly, miR-27a was down-regulated, but this finding could not be verified (Fix *et al.*, 2010). The influence of polyphenon-60 on miRNA target genes was not investigated.

Wang *et al.* (2011a) analyzed the effect of EGCG on proliferation of mouse and human lung-cancer cells and investigated regulation of miRNAs as an underlying mechanism. HIF-1$\alpha$ (hypoxia-inducible factor 1-alpha), a major transcription factor involved in angiogenesis, is unstable under normoxic conditions and is stabilized during hypoxia. Over-expression of miR-210, induced by EGCG, was found to be associated with stabilization of HIF-1$\alpha$, which was shown to regulate miR-210 expression (Wang *et al.*, 2011a).

In a prostate-cancer xenograft model, treatment of mice with EGCG reduced tumor growth and resulted in up-regulation of the tumor suppressor miR-330 and down-regulation of androgen-responsive miR-21 (Siddiqui *et al.*, 2011). Molecular modeling and functional studies indicate that EGCG binds to the AR and disrupts androgen signaling.

In a recent study, EGCG in combination with a retinol derivative inhibited the growth of human malignant neuroblastoma cells. Expression of the tumor-suppressor miRs miR-7-1, miR-34a, and miR-99a was increased by the combination, whereas expression of oncogenic

**Table 6.7** Influence of (−)-epigallocatechin gallate (EGCG) and green-tea polyphenols (GTPs) on miRNA expression (verified by RT-PCR).

| Mechanism/target | Effect | Cell line/tissue | Genes/factors analyzed | References |
|---|---|---|---|---|
| miR-16, let-7a, miR-221 | Up-regulation | HepG2 | ↓ Bcl-2 by miR-16 confirmed by precursor and inhibitor transfection | Tsang & Kwok (2010) |
| miR548m, miR-720 | Up-regulation (GTP) | MCF7 | | Fix *et al.* (2010) |
| miR-210 | Up-regulation | H1299<br>CL13 (mouse lung) | Mediated via HIF-1 stabilization<br>↓ cell growth | Wang *et al.* (2011a) |
| miR-330 | Up-regulation | C4-2 and 22Rν1prostate xenografts | Mediated via reduced AR signaling | Siddiqui *et al.* (2011) |
| miR-7-1, miR-34a, miR-99a | Up-regulation | SK-N-BE2<br>IMR-32 | Over-expression of miR-7-1, augmenting induction of apoptosis | Chakrabarti *et al.* (2012) |
| miR-18a, miR-34b, miR-193b, miR-222, miR-342 | Reduction | HepG2 | | Tsang & Kwok (2010) |
| miR-27a | Reduction (GTP) | MCF7 | | Fix *et al.* (2010) |
| miR-21 | Reduction | C4-2 and 22Rν1prostate xenografts | Mediated via reduced AR signaling | Siddiqui *et al.* (2011) |
| miR-92, miR-93, miR-106b | Reduction | SK-N-BE2<br>IMR-32 | Over-expression of miR-93, counteracting induction of apoptosis | Chakrabarti *et al.* (2012) |

miR-92, miR-93, and miR-106b was down-regulated. Over-expression of miR-7-1 and miR-93 augmented and counteracted the effects of the EGCG and retinamide combination.

Overall, these data indicate that alterations in the expression of miRNAs need to be considered when EGCG treatment modulates expression of miRNA target genes. The mechanism by which specific miRNAs are up-regulated or repressed by EGCG treatment is presently not known.

### *6.5.2 Genistein and soy isoflavones*

Isoflavones are a class of flavonoids abundantly found in plants of the *Fabaceae* family and characterized by phytoestrogenic properties. The isoflavones genistein and daidzein are contained in high concentrations in soybean (*Glycine max* L.) and are enriched in soy products. Epidemiological evidence indicates an inverse correlation between a traditional soy-rich, low-fat Asian diet and the risk of developing breast and prostate cancer (Messina *et al.*, 2006; Jian, 2009; Magee & Rowland, 2012). As phytoestrogens, soy isoflavones bind to the estrogen receptor and modulate ER signaling. Genistein has been shown to affect several additional chemopreventive mechanisms, including activation of carcinogens, inhibition of oxidative stress, inflammation, cell signaling, angiogenesis, modulation of cell-cycle regulation, and induction of apoptosis (Steiner *et al.*, 2008; Molinie & Georgel, 2009; Magee & Rowland, 2012). Preventive efficacy against breast, ovarian, prostate, skin, stomach, and colon cancer has been demonstrated in various rodent models (Banerjee *et al.*, 2008). Recent observations of a growth-promoting effect of genistein in ER-positive breast-cancer cell lines and xenograft models have indicated a potential risk of genistein for human health (Helferich *et al.*, 2008; Rietjens *et al.*, 2013). A recent review does not support these concerns (Magee & Rowland, 2012). Genistein is currently being tested in various clinical trials for the treatment and prevention of prostate, bladder, kidney, breast, and endometrial cancer (Taylor *et al.*, 2009).

#### *6.5.2.1 DNA methylation*

Several studies have identified genistein as a polyphenol that affects epigenetic mechanisms at the level of DNA methylation (Table 6.8), modulation of histone marks, and miRNA expression. Fang *et al.* (2005) identified genistein as a dose-dependent inhibitor of DNMT activity with both nuclear extracts derived from human esophageal cells and human recombinant DNMT1. Genistein was also found to reduce DNMT activity in various cell lines; this was at least partly due to reduced expression of all three DNMTs: 1, 3a, and 3b (Li *et al.*, 2009a; Majid *et al.*, 2009). Fang *et al.* (2005, 2007) further reported that genistein reduced promoter methylation and led to re-expression of RAR$\beta$, the cell-cycle inhibitor p16, and the DNA repair gene MGMT in KYSE 510 esophageal cancer cells. RAR$\beta$ demethylation associated with re-expression was also seen in the KSYE 150 esophageal and two prostate cancer cell lines in a time- and dose-dependent manner (Fang *et al.*, 2005), as well as in HCT116 colon and SiHa squamous cervical-cancer cells (Spurling *et al.*, 2008), but not in two breast-cancer cell lines (King-Batoon *et al.*, 2008).

**Table 6.8** Effects mediated by genistein and soy isoflavones on DNA methylation.

| Mechanism/target | Effect | Cell line/tissue | Genes/factors analyzed | References |
|---|---|---|---|---|
| **DNA methylation *in vivo*** | | | | |
| DNMT activity *in vitro* | Inhibition | KYSE 510 nuclear extracts, human recombinant DNMT1 | | Fang *et al.* (2005) |
| DNMT activity in cell culture | Reduction | A498, ACHN, HEK-293 nuclear extracts | | Majid *et al.* (2009) |
| DNMT mRNA/protein expression | Weak reduction | A498, ACHN, HEK-293 | DNMT1, DNMT3a, DNMT3b | Majid *et al.* (2009) |
| | Reduction | MCF7, (MCF10AT) | DNMT1, DNMT3a, DNMT3b | Li *et al.* (2009a) |
| Promoter hypermethylation | Reduction | KYSE 510 | RAR$\beta$, p16, MGMT, associated with gene re-expression | Fang *et al.* (2005, 2007) |
| | Reduction | PC3, LNCaP, KYSE 150 | RAR$\beta$, associated with gene re-expression | Fang *et al.* (2005) |
| | Reduction | HCT116, SiHa | RAR$\beta$2, associated with gene re-expression | Spurling *et al.* (2008), Jha *et al.* (2010) |
| | No effect | MCF7, MDA-MB-468 | RAR$\beta$ | King-Batoon *et al.* (2008) |
| | Reduction | MDA-MB-468 | GSTP1, associated with gene re-expression | King-Batoon *et al.* (2008) |
| | No effect | MCF7 | GSTP1 | King-Batoon *et al.* (2008) |
| | Reduction | DU-145, PC-3, LNCaP | GSTP1, EPHB2 | Vardi *et al.* (2010) |
| | No effect | DU-145, PC-3, LNCaP | RASSF1A, BRCA1 | Vardi *et al.* (2010) |
| | Reduction | DU-145, PC-3 | GSTP1, EPHB2, BRCA, associated with gene re-expression | Adjakly *et al.* (2011) |
| | Reduction | SW1116 | Wnt5a, associated with gene re-expression | Wang & Chen (2010) |
| | Reduction | MCF10AT, MCF7 | hTERT (E2F-1 binding site), associated with reduced expression | Li *et al.* (2009a) |
| | Reduction | A498, ACHN, HEK-293, LNCaP, PC-3, RWPE-1 | BTG, associated with gene re-expression in cancer cell lines | Majid *et al.* (2009, 2010b) |
| | Reduction | Mouse embryonic stem cells | Ucp1, Sytl1 | Sato *et al.* (2011) |
| | Reduction | AGS | PCDH17, associated with gene ge-expression | Yang *et al.* (2012) |
| | Reduction | MCF7, MDA-MB-231, MCF10a | BRCA1, BRAC2, associated with weak protein re-expression, also seen with the daidzein metabolite equol | Bosviel *et al.* (2012a,b) |

| | | | | |
|---|---|---|---|---|
| **DNA methylation *in vivo*** | Increase or no effect | Pancreas, rats treated neonatally with equol (daidzein metabolite) | Oncogenic Ha-Ras, no effect on c-Myc, c-Fos | Lyn-Cook *et al.* (1995) |
| | Increase | prostate, healthy male mice, up to 4 weeks' intervention | Specific clones (no gene specified) | Day *et al.* (2002) |
| | Increase | *In utero* exposure, $A^{vy}$ (Agouti) mice | $A^{vy}$ intracisternal A particle (IAP) murine retrotransposon | Dolinoy *et al.* (2006) |
| | No effect | *In utero* exposure, Agouti mice | Coat color | Badger *et al.* (2008) |
| | No effect | *In utero* exposure, Agouti mice | Coat color | Rosenfeld *et al.* (2013) |
| | Increase | *In utero* exposure, mice | Repetitive elements in bone marrow, down-regulation of estrogen-responsive genes | Vanhees *et al.* (2011) |
| | Increase | PBMC of heavy smokers, 4 weeks' intervention with soy and green tea | Line1 repetitive elements | Scoccianti *et al.* (2011) |
| | Reduction | Uterus in healthy mice, intact and ovarextomized (OVX) | Nsbp1 | Tang *et al.* (2008a) |
| | Reduction | Pancreas, liver in healthy mice | ActaI | Guerrero-Bosagna *et al.* (2008) |
| | Reduction | Endometrium, ovariectomized (OVX) mice | SF-1, associated with re-expression | Matsukura *et al.* (2011) |
| | Reduction or increase | Cynomolgus monkeys | Fat tissue: ↓ABCG5, ↑TBX5, ↑HoxB1<br>Muscle: ↑HoxA5, ↑HoxA11, ↓NTRK3 | Howard *et al.* (2011) |
| | Reduction | Healthy premenopausal women, intraductal breast specimens | RAR$\beta$2, CCDN2 (in women with low circulating serum genistein levels) | Qin *et al.* (2009) |
| | No effect | Healthy premenopausal women, intraductal breast specimens | ER, p16, RASSF1A | Qin *et al.* (2009) |
| | No effect | PBMC of heavy smokers, 4 weeks' intervention with soy and green tea | RASSF1A, ARF, CDKN2, MLH1, MTHFR | Scoccianti *et al.* (2011) |

Methylation at the GSTP1 promoter was reduced and the gene was weakly re-expressed after repetitive treatments with genistein only in ER-negative, Her2-overexpressing MDA-MB-468 breast-cancer cells, whereas similar treatment was ineffective in reactivating GSTP1 in ER-positive MCF-7 cells (King-Batoon *et al.*, 2008). However, genistein and daidzein intervention significantly reduced GSTP1 promoter methylation in the prostate-cancer cell lines LNCaP, PC3, and DU-145, leading to re-expression of GSTP1 protein, as determined by immunocytochemistry and Western blotting (Vardi *et al.*, 2010; Adjakly *et al.*, 2011).

In a study by Vardi *et al.* (2010), methylation of two other genes often silenced by promoter hypermethylation in cancer, the candidate tumor suppressor gene RASSF1A (Ras association domain family 1, isoform A) and the DNA repair gene BRCA1 (breast cancer 1, early-onset), was not reduced in prostate-cancer cell lines; Adjakly *et al.* (2011) could demonstrate BRCA and EPHB2 (ephrin receptor B2) demethylation and re-expression in DU-145 and PC-3 prostate cancer cells, on the other hand. Consistently, genistein and daidzein treatment, as well as the daidzein metabolite equol, led to demethylation and re-expression of BRCA1 and 2 protein in three breast (cancer) cell lines: MCF-7, MDA-MB-231 and MCF 10a (Bosviel *et al.*, 2012a,b).

Wnt5a is a ligand of the Wnt receptor that is silenced by promoter methylation in colon-cancer cell lines. Genistein treatment of SW1116 colon-cancer cells reduced cell viability by 80%. Concomitantly, Wnt5a promoter methylation decreased by about 10% in association with an increase of Wnt5a mRNA levels (Wang & Chen, 2010).

Hypomethylation at the hTERT core promoter facilitates binding of the repressor E2F-1 and reduces hTERT expression. In the breast-cancer cell lines MCF7 and MCF10AT, genistein-mediated down-regulation of DNMT expression was associated with reduced hTERT methylation and reduced expression (Li *et al.*, 2009a).

The tumor suppressor BTG3 (B cell translocation gene 3) is a negative regulator of cell signaling and inhibits E2F. Genistein intervention in renal- and prostate-cancer cell lines reduced promoter methylation and increased BTG3 mRNA levels. This was further facilitated by genistein-mediated effects on HDAC and HAT activity, in association with enhanced binding of acetylated and methylated histones to the BGT promoter, indicating increased transcription (Majid *et al.*, 2009, 2010b).

Using a mouse embryonic stem (ES) cell differentiation system as a model, Sato *et al.* (2011) analyzed the influence of genistein on epigenetic reprogramming during fetal development. Genome-wide methylation analyses indicated a moderate effect on DNA methylation in a subset of genes. Ucp1 and Sytl1 promoters were selected for focused analyses of methylation dynamics. Methylation at selected CpG sites was reduced by genistein treatment after initial *de novo* methylation, which was not affected by genistein treatment. Also, differentiation of ES cells to contracting cardiomyocytes was not affected by genistein (Sato *et al.*, 2011).

PCDH17 (protocadherin 17) belongs to the cadherin protein family (Yang *et al.*, 2012). In the gastric-cancer cell line AGS, expression of PCDH17 was found to be reduced by promoter methylation. Treatment with genistein lowered methylation levels and reactivated PCDH17 expression to a similar extent as seen with the demethylating agent

5-aza-2′-deoxycytidine. It also induced cell-cycle arrest and reduced cell proliferation (Yang *et al.*, 2012).

As indicated before, analyses of the demethylating effects of genistein in cell culture are not totally conclusive. This may be a result of the varying dose range, cell culture-, and compound-treatment conditions, as well as of differences in the methodology used to determine DNA methylation. Although metabolism of isoflavones differs between rodents and humans, modulation of epigenetic effects in rodent models might be a more reliable indicator for activity in humans than the results of *in vitro* cell-culture studies.

An early study published by Lyn-Cook *et al.* (1995) used methylation-sensitive restriction digestion to compare fragmentation patters of the c-Ha-Ras oncogene in rat pancreas after neonatal intervention with the daidzein metabolite equol (100 µg) and another phytoestrogen, coumestrol (10 µg), for 10 days. Southern blotting indicated increased methylation at the c-Ha-Ras gene, but not at c-Myc of Fos proto-oncogenes.

Day *et al.* (2002) applied genistein at a dose of 300 mg/kg diet to male mice for up to 4 weeks. Changes in DNA CGI methylation in the prostate were analyzed by differential methylation hybridization after digestion of DNA with methylation-sensitive restriction enzymes. Methylation of three DNA fragments increased after 4 weeks' genistein intervention; no methylation changes were detected in liver DNA. Dolinoy *et al.* (2006) investigated the effect of *in utero* exposure to genistein in the Agouti mouse model. Expression of the Agouti gene responsible for coat color was epigenetically regulated by a retrotransposon upstream of its transcription start site. An increase in methylation at six CpG sites in the retrotransposon was detected in the offspring of female mice exposed to dietary genistein (250 mg/kg diet) during gestation, and coat color was shifted towards pseudoagouti, indicative of reduced expression of the Agouti gene. Two later studies did not observe coat-color changes following *in utero* genistein exposure (Badger *et al.*, 2008; Rosenfeld *et al.*, 2013). A study by Vanhees *et al.* (2011) indicated that prenatal maternal exposure of mice to genistein (270 mg/kg diet) was associated with hypermethylation at repetitive elements in bone marrow and down-regulated ER-responsive genes in hematopoeietic cells of the offspring.

Methylation at repetitive elements can be interpreted as a sign of genomic stability. Scoccianti *et al.* (2011) performed a 4-week intervention study in heavy smokers with diets enriched in cruciferous vegetables or supplemented with flavonoids in the form of soy products and green tea. They compared methylation of the LINE-1 (long interspersed nuclear element) repetitive element and of a series of genes known as tumor-suppressor genes, involved in DNA repair and folate metabolism in DNA derived from peripheral white blood cells (PBMCs) collected before and after intervention. Both diets slightly but significantly increased methylation of LINE-1, by 0.8 and 2.1% respectively, whereas methylation of selected candidate genes such as RassF1A, the cell-cycle regulators ARF (alternate reading frame of the INK4a/ARF locus CDKN2A) and CDKN2A (cyclin-dependent kinase inhibitor 2A, also known as p16), the DNA repair gene MLH, and MTHRF (methylenetetrahydrofolate reductase) with a major role in folate metabolism did not change significantly (Scoccianti *et al.*, 2011). There is an ongoing debate over whether PBMC can be used as surrogate tissue to identify physiologically relevant changes in DNA methylation (Wang *et al.*, 2010; Brennan *et al.*, 2012). Notably, analyses can be confounded by variation in white blood cell composition (Adalsteinsson *et al.*, 2012; Houseman *et al.*, 2012; Reinius *et al.*, 2012).

The influence of genistein or daidzein exposure on gene-promoter methylation was investigated in several rodent models (Guerrero-Bosagna *et al.*, 2008; Tang *et al.*, 2008a; Matsukura *et al.*, 2011), a study on macaques (Howard *et al.*, 2011), and in breast specimens of healthy women (Qin *et al.*, 2009). In a study by Tang *et al.* (2008a), mice were neonatally exposed to genistein (50 mg/kg bodyweight, b.w.) or diethylstilbestrol (DES) at three doses by subcutaneous injection at days 1–5. At the age of 6 months, uterine DNA was extracted and differentially methylated regions were identified by methylation-sensitive restriction fingerprinting. Five hyper- and five hypomethylated regions mostly localized in intronic CGIs were identified after genistein treatment; four additional regions showed a modulation of DNA methylation after treatment with DES at the highest dose (1 mg/kg b.w.). A region associated with a promoter CGI of nucleosomal binding protein 1 (Nsbp1) that was hypomethylated after DES treatment was selected for further analysis. Nsbp1 plays an important role in chromatin remodeling. Neonatal DES/genistein exposure prevented hypermethylation of the Nsbp1 promoter during puberty and aging in intact animals, but not in ovarectomized mice, and led to its continuous expression (Tang *et al.*, 2008a).

In a mouse study by Guerrero-Bosagna *et al.* (2008), dietary soy exposure *in utero* (2% in the parental diet, starting 2 weeks before mating) led to accelerated vaginal opening in female offspring, indicating physiological activity. The intervention did not modulate DNA methylation levels of skeletal $\alpha$-actin (Acta1) in the liver and pancreas of male and female offspring, but reduced gender differences in methylation levels. Exposure of ovarectomized mice to high- (200 kg/kg diet) and low-dose (60 mg/kg diet) genistein for 1 week reduced methylation and induced expression of steroidogenic factor 1 (SF-1, gene name NR5A1) in endometrial cells. SF-1 is an orphan nuclear receptor and transcription factor for key enzymes involved in steroidogenesis (Matsukura *et al.*, 2011).

Howard *et al.* (2011) were interested in whether typical commercial animal chows based on soy protein and rich in soy isoflavones would modulate DNA methylation patterns in various tissues of non-human primates. A total of eight cynomolgus macaques were assigned to two high-fat diets based on either casein or soy (equivalent to 180 mg soy isoflavones for a human adult per day) as the protein source, with similar macronutrient contents, for 8 weeks in a cross-over design. HumanMethylation 27k arrays (Illumina) were used to determine genome-wide methylation changes in subcutaneous adipose tissue, muscle, and liver. A switch from a soy-based to a casein-based diet resulted in a general increase in methylation levels in liver and muscle tissue. Specifically, in muscle tissue two homeobox genes, HOXA5 and HOXA11, with functions as developmental transcription factors were demethylated during the switch from the soy- to the casein-based diet, whereas methylation at the promoter of the NTRK3 gene (neurotrophic tyrosine kinase, receptor, type 3) increased. In fat tissue, the switch from the casein- to the soy-based diet resulted in reduced methylation at the ABC-transporter G5 (ABCG5) and increased methylation at the promoters of HOXB1 and T-box protein 5 (TBX5). T-box proteins share a T-box-binding domain; they are transcription factors with important functions in the regulation of developmental processes. Methylation changes for these genes in fat tissue were confirmed by pyrosequencing (Howard *et al.*, 2011). In addition to changes in methylation, the diet switch from casein to soy resulted in reduced fasting insulin levels and improved insulin sensitivity. The same group had shown earlier that maintaining the animals on a soy-based diet for up to 2 years led to reduced body weight and improved blood lipids compared to those on the casein-based diet. Qin *et al.* (2009) performed a 4-week dietary intervention in 34 healthy premenopausal women with daily doses of 40 or 140 mg isoflavones. They reported dose-dependent changes in RAR$\beta$2 and CCND2 (cyclin D2) promoter methylation in mammary tissue in women with low circulating serum genistein levels, whereas the intervention did not influence Rassf1A, ER, or p16 methylation in intraductal breast specimens (Qin *et al.*, 2009).

Although these results are promising, so far a direct relation between modulation of DNA methylation by genistein and chemopreventive potential *in vivo* has not been demonstrated.

#### 6.5.2.2 *Influence on histone acetylation and methylation*

In addition to its effects on DNA methylation, genistein has been reported to affect acetylation and methylation of histones and non-histone proteins, and as a consequence to alter gene expression (Table 6.9). Several studies have focused on effects related to breast-cancer prevention. Genistein and other phytoestrogens are known to interact with ER-$\alpha$ and -$\beta$. After ligand binding, the ER dimer recruits nuclear coactivators and the complex interacts with estrogen-responsive elements in the promoter regions of estrogen-regulated genes. In contrast, anti-estrogen binding to the ER results in the recruitment of corepressors. Coactivators often have HAT activity, whereas corepressors possess HDAC activity (Powles, 2002). In a test-tube experiment with isolated ER-$\beta$, coactivators, chromatin, and [$^3$H]-labeled acetyl-CoA as a substrate, genistein, daidzein, and equol stimulated ER-$\beta$-mediated histone acetylation (Hong *et al.*, 2004). In ER-positive MCF-7 breast-cancer cells, long-term treatment for 40–60 days with low-dose genistein at a concentration of 10 nM reduced basal histone H3 acetylation and desensitized the cells towards the HDAC inhibitors TSA and apicidin. Also, cell proliferation and the mitogenic response to estradiol and epidermal growth factor were reduced, suggesting that long-term genistein treatment might lead to epigenetic changes supporting cancer-preventive potential (Jawaid *et al.*, 2010).

A recent study by Li *et al.* (2013b) indicates that genistein can reactivate ER-$\alpha$ expression in ER-negative cell lines such as MDA-MB-231 and resensitize these cells to treatment with anti-estrogens such as tamoxifen via a histone acetylation-related mechanism, especially in combination with TSA. Accordingly, enhanced tumor-growth inhibition by dietary genistein (250 mg/kg) in combination with tamoxifen was demonstrated *in vivo* in a MDA-MB-231 xenograft model and in the C3(1) SV40 TAg transgenic mouse model for basal breast cancer. This combined treatment may have implications for the clinical management of hormone receptor-negative breast cancer (Li *et al.*, 2013b). Li *et al.* (2013a) also used human mammary epithelial cells (HMEC) transformed with SV40 and hTERT (SH) to a precancerous state and then with H-Ras to a cancerous state (SHR) to analyze genistein effects. Reactivation of the cell-cycle regulators p16 and p21 by genistein was associated with enhanced binding of the activating acetylated histone H3 and reduced binding of the repressive histone markers H3K9me3 and H3K27me3 to their promoter regions.

**Table 6.9** Effects mediated by genistein and soy isoflavones on enzymes modifying histones and non-histone proteins.

| Mechanism/target | Effect | Cell line/tissue | Genes/factors analyzed | References |
|---|---|---|---|---|
| Histone acetylation | Induction | *In vitro* | ↑ ac-histones, mediated via ER activation and recruitment of coactivator HATs | Hong *et al.* (2004) |
| | Reduction | MCF7 after long-term genistein exposure | ↓ H3 expression and acetylation response after HDACi treatment<br>↓ growth rate, ↓ response to mitogens (EGF), TSA, and apicidin | Jawaid *et al.* (2010) |
| | Induction | MDA-MB-231 *in vitro* and as xenografts, C3(1) SV40 TAg mice | ↑ acH3, H3K9ac, acH4 at the ER-$\alpha$ promoter, especially in combination with TSA<br>↓ cell growth, ↓ growth of MDA-MB-231 xenografts and mammary tumors in c3(1) mice | Li *et al.* (2013b) |
| | Induction | HMEC transformed with SV40, hTERT (SH) and H-Ras (SHR) *in vitro* and in xenografts | ↑ acH3, ↓ H3K27me3, H3K9me3 at p21 and p16 promoter<br>expression ↑ p21, p16, ↓ Bmi-1, c-Myc<br>↓ xenograft growth | Li *et al.* (2013a) |
| | Induction | SW480, HCT15 | ↑ acH3 at the DKK1 promoter, ↓ cell growth | Wang *et al.* (2012) |
| *HDAC6* expression | Reduction | LNCaP | ↑ acetylation of HSP90, promoting dissociation and degradation of AR | Basak *et al.* (2008) |
| *SIRT1* activity/expression | Reduction | LNCaP, PC-3 | ↑ ac-H3K9, ↓ H3K9me2, ↓ Akt signaling through PTEN, CYLD, p53, FOXO3a | Kikuno *et al.* (2008) |
| HAT expression: *p300, PCAF, CREBBP, HAT1* | Induction | LNCaP, DuPro | ↑ ac-H3, ac-H4, ↑ p21, $p16^{INK4a}$ expression | Majid *et al.* (2008) |
| Histone methylation: H3K27me3 | Reduction | Eker rats, uterine tissue | Mediated via Akt activation and ↑EZH2 phosphorylation, ↑ hypersensitivity of ER-responsive genes<br>↑ uterine tumor incidence and multiplicity | Greathouse *et al.* (2012) |

In a xenograft model with the SHR cell line, xenograft growth was strongly inhibited by dietary genistein (250 mg/kg).

Dickkopf 1 (DKK1) is an important regulator of Wnt signaling, but it is often lost during colon carcinogenesis (e.g. by DNA methylation). Wang *et al.* (2012) demonstrated that genistein treatment of colon-cancer cell lines leads to reactivation of DKK1 expression and inhibition of cell growth. Induction of DKK1 expression was less pronounced in colon-cancer cell lines with a highly methylated DKK1 promoter, and genistein treatment did not reduce methylation levels. Rather, in cell lines with unmethylated DKK1, such as SW480 and HCT15, genistein treatment led to enhanced binding of acetylated H3 to the DKK1 promoter and coding region. Induction of DKK1 following TSA treatment confirmed that enhanced histone acetylation was causally linked to DKK1 re-expression (Wang *et al.*, 2012).

Several studies have also investigated whether epigenetic mechanisms might be involved in the prostate-cancer chemopreventive activity of genistein. AR signaling and androgen-mediated DNA damage is strongly involved in prostate carcinogenesis (Weischenfeldt *et al.*, 2013). Under basal conditions, AR is stabilized by interaction with the protein chaperone heat-shock protein 90 (HSP90). Basak *et al.* (2008) observed a dose-dependent reduction of AR protein levels in genistein-treated LNCaP prostate-cancer cells. AR mRNA was only reduced at the highest concentration of genistein (50 μM). Instead, genistein treatment prevented the AR-HSP90 interaction via enhanced acetylation of HSP90 and increased ubiquitination and subsequent degradation of the AR. Mechanistically, the anti-estrogenic activity of genistein led to down-regulation of HDAC6, which is important for HSP90 deacetylation. The genistein-mediated effects on AR expression were mimicked by knock down of HDAC6. These data demonstrate that the prostate cancer-preventive potential of genistein may be mediated by an indirect anti-androgen mechanism (Basak *et al.*, 2008).

In prostate cancer, AKT signaling is often aberrantly activated through loss of the tumor-suppressor gene (TSG) PTEN. Other TSGs downstream of PTEN, such as p53, the deubiquitination enzyme CYLD (cylindromatosis), and FoxA3, are often also lost by mechanisms independent of DNA methylation. Kikuno *et al.* (2008) showed that genistein treatment of LNCaP and PC-3 prostate-cancer cell lines reactivated PTEN expression, inactivated AKT, and induced the downstream effectors p53 and FOXO3a. Genistein treatment also led to the up-regulation of CYLD and inhibition of NF-$\kappa$B. These effects were linked to demethylation (PTEN and CYLD) and increased acetylation (PTEN, CYLD, p53, FOXO3a) of histone H3K9. The authors could demonstrate that H3K9 hyperacetylation was mediated by the down-regulation and reduced nuclear localization of SIRT1, a class III HDAC, highlighting the importance of epigenetic mechanisms for the regulation of AKT and NF-$\kappa$B signaling by genistein (Kikuno *et al.*, 2008). In another study with LNCaP prostate-cancer cells, genistein treatment led to up-regulation of HATs including p300, PCAF, CBP, and HAT1. As a consequence, acetylation of the histones H3 and H4 increased and the cell-cycle inhibitors p21 and p16 were up-regulated by enhanced association of acetylated H3K4 with their promoters (Majid *et al.*, 2008).

Information on the influence of natural products on histone methylation is still scant. In a study to determine whether exposure to genistein and the estrogenic plasticizer bisphenol A

(BPA) would increase the risk for uterine tumorigenesis, Greathouse *et al.* (2012) identified the HMT EZH2 as a target of genistein-induced PI3K/AKT-mediated nongenomic ER signaling. For developmental reprogramming analyses, Eker rats were postnatally exposed to BPA (50 mg/kg b.w.), estradiol (1 mg/kg), or genistein (50 mg/kg) by subcutaneous injection on days 10–12, and sacrificed at 3 month of age. For acute exposure studies, Eker rats were sacrificed 6 hours after a single injection to the estrogenic compounds on postnatal day 12. Exposure to genistein, but not BPA, led to uterine activation of AKT, enhanced phosphorylation of EZH2, reduced levels of the H3K27me3 repressive methyl marker, and consequently reactivated gene expression. Developmental reprogramming by genistein also promoted uterine tumorigenesis in adult Eker rats. These data indicate that differential epigenetic effects mediated by nongenomic ER signaling can contribute to differences in developmental reprogramming and tumorigenic response to environmental estrogens such as genistein and BPA.

### *6.5.2.3 miRNAs affected by isoflavones*

Genistein and soy isoflavones have been reported to modulate gene expression via a miRNA-related mechanism (Table 6.10). Pancreatic cancer is a cancer type with very poor prognosis, due to late detection and development of drug resistance. EMT is a phenomenon observed during development of drug resistance and cancer-cell metastasis and has been related to altered miRNA expression, such as down-regulation of members of the miR-200 and let-7 families (Li *et al.*, 2009b). Li *et al.* (2009b) addressed the question of whether an isoflavone mixture (containing 70.5% genistein, 26.3% daidzein, and 0.3% glycitein) might affect EMT in gemcitabine-resistant pancreatic-cancer cells. miRNA profiling identified 67 out of 711 miRNAs as differentially expressed after treatment with 25 μM isoflavone mixture (calculated as genistein) for 48 hours. Up-regulation of members of the let-7 (let-7b, let-7e) and miR-200 families (miR-200b, miR-200c) was associated with a change in morphology to a more epithelial-like phenotype and confirmed at the miRNA level by quantitative RT-PCR. After treatment of pancreatic cancer cells with low-dose isoflavones (10 μM) for 3 weeks, the epithelial marker E-cadherin was up-regulated at the mRNA and protein level, whereas the mesenchymal markers ZEB1, vimentin, and slug were down-regulated (Li *et al.*, 2009b), indicating isoflavone-induced reversal of the EMT phenotype. Similar effects were observed in FoxM1 and Notch-1 over-expressing cell lines also characterized by an EMT phenotype (Bao *et al.*, 2011a,b).

Pancreatic cancer cells are characterized by over-expression of EGFR (epidermal growth factor receptor) and activation of the NF-$\kappa$B signaling pathway. Li *et al.* (2010b) related these observations to lower expression of miR-146 in pancreatic cancer cells than in normal pancreatic-duct epithelial cells. Up-regulation of miR-146 reduced expression of EGFR as well as NF-$\kappa$B, I$\kappa$B, IRAK-1 (interleukin 1 receptor-associated kinase 1), and metastasis-associated protein 2 (MTA-2) and reduced the invasive capacity of the cells. Interestingly, all these effects could be mimicked by up-regulation of miR-146 after incubation with 25 μM isoflavones (Li *et al.*, 2010b).

Minichromosome maintenance (MCM) genes play a role in DNA replication during cell division and are often up-regulated in cancer cells, including prostate cancer. Conversely, miR-1296, which has been shown to regulate the expression of MCM2, is down-regulated

**Table 6.10** Effects mediated by genistein and soy isoflavones on miRNA expression (verified by RT-PCR).

| Mechanism/target | Effect | Cell line/tissue | Genes/factors analyzed | References |
|---|---|---|---|---|
| let-7b, let-7e, miR-200b, miR-200c | Up-regulation | MiaPaCa-2, L3.6pl, Panc-1 | ↑ E-cadherin as epithelial marker<br>↓ vimentin, Slug, Zeb1, expression<br>↓ reversal of EMT | Li *et al.* (2009b), Bao *et al.* (2011a,b) |
| miR-146 | Up-regulation | Colo357, Panc-1 | ↓ expression of EGFR, IRAK-1, NF-κB, MTA-2<br>↓ invasive capacity | Li *et al.* (2010b) |
| miR-1296 | Up-regulation | LNCaP, PC3 | ↓ MCM genes and CDK2, CDK7, CDT1 | Majid *et al.* (2010a) |
| miR-29a, miR-1256 | Up-regulation | C4-2B, LNCaP | ↓ TRIM68, PGK-1 mRNA expression<br>↓ cell growth and invasion | Li *et al.* (2012) |
| miR-221, miR-222 | Reduction | PC3 | ↑ ARHI expression<br>(↓ cell proliferation, colony formation, invasion) | Chen *et al.* (2011) |
| miR-27a | Reduction | C918 *in vitro* and *in vivo* | ↑ ZBTB10 expression<br>↓ xenograft growth<br>(i.p. application 25–100 mg/kg/day) | Sun *et al.* (2009) |
| miR-21 | Reduction | A-498 *in vitro* and *in vivo* | ↓ xenograft growth (pretreatment with 25 μM genistein for 4 days) | Zaman *et al.* (2012) |

in prostate tumor tissue compared with samples of benign prostatic hyperplasia (BPH) (Majid *et al.*, 2010a). Over-expression of miR-1296 reduced MCM2 mRNA expression and induced cell-cycle arrest in S phase. Similarly, treatment of prostate-cancer cell lines with genistein concentrations of 25 and 50 μM induced miR-1296 up to fivefold. Therefore, the antiproliferative activity of genistein may involve a miRNA-mediated effect on cell-cycle progression (Majid *et al.*, 2010a). Li *et al.* (2012) recently observed the low expression of miR-1256 and miR-29 in prostate-cancer tissue and cell lines and associated it with partial promoter methylation. They identified TRIM68 (tripartite motif containing 68; involved in AR-mediated transactivation) and phosphoglycerate kinase 2 (PGK-1; has important functions in glycolysis), both often up-regulated in prostate cancer, as novel targets of miR1256 and miR-29a. The treatment of prostate-cancer cell lines with 20 μM isoflavone mixture for 5 days reduced the methylation of both miRNA promoters and led to re-expression of both target genes. This was associated with reduced cell growth and invasion. On the other hand, isoflavone treatment reduced the expression of oncomiRs miR-155 and miR-421 (Li *et al.*, 2012).

Chen *et al.* (2011) also investigated the activity of genistein on miRNA expression in prostate cancer. ARHI (age-related hearing impairment 1) is an imprinted tumor-suppressor gene that is down-regulated in prostate cancer. Over-expression studies confirmed its role in the regulation of cell proliferation, colony formation, invasion, and induction of apoptosis (Chen *et al.*, 2011). ARHI was identified as a novel target of the miRNAs miR-221 and -222, which are both up-regulated in prostate cancer. In PC3 prostate-cancer cells treated with 50 μM genistein for 4 days, genistein up-regulated ARHI levels by regulating miR-221 and -222 (Chen *et al.*, 2011).

In another study, Sun *et al.* (2009) demonstrated that genistein treatment at concentrations of 10–200 μM dose-dependently reduced proliferation of uveal melanoma cells *in vitro* and in a xenograft model when applied intraperitoneally at a dose of 25, 50, and 100 mg/kg b.w./day for 1 month. They additionally investigated the effect of genistein on expression of miR-27a, which is often over-expressed in cancer cells, and its target gene ZBTB10 (zinc finger and BTB domain containing 10). In the uveal melanoma cell line, genistein at the highest dose reduced miR-27a levels by about 50% and led to about a 1.3-fold increased expression of ZBTB10 (Sun *et al.*, 2009). The effect on miR-27a and ZBTB10 expression was not analyzed in the xenograft model. Since high doses of genistein were used in this study, and reversal of the effects by transfection of miR-27a was not performed, the observed inhibition of cell growth might be caused by alternative mechanisms unrelated to the modulation of miR-27a expression.

MiR-21 is an oncomiR that is frequently over-expressed in various types of cancer (Zaman *et al.*, 2012). Zaman *et al.* (2012) were interested in determining miR-21 expression and the effect of genistein intervention in renal-cell carcinoma. In this study, kidney-cancer patients with high miR-21 expression had a significantly lower 5-year survival than patients with low miR-21 expression. MiR-21 was shown to be involved in the regulation of apoptosis, migration, and invasion, and regulated expression of the cell-cycle inhibitor p21. Treatment of A-498 renal cancer cells with 25 μM genistein reduced miR-21 expression by about 30%. Pretreatment of A-498 cells with genistein also reduced tumor growth in a xenograft model. Under these conditions, miR-21 expression was about 40% lower than in control tumors.

So far it is not clear how genistein treatment influences miRNA expression. Demethylation of miRNAs silenced by promoter methylation might be relevant to the re-expression of tumor-suppressor miRNAs. Further studies are needed to explore how natural products such as genistein might lead to down-regulation of oncomiRs.

## 6.6 Conclusions

This chapter has summarized current knowledge of the potential of chemopreventive polyphenols to affect several epigenetic mechanisms, focusing on DNA methylation and histone modifications, including acetylation and methylation. The effects of miRNAs on gene expression were not broadly covered, with the exception of EGCG and genistein.

### *6.6.1 DNA methylation*

Multiple polyphenols of various structural classes have been shown to inhibit DNMT enzyme activity *in vitro*, with more limited and less consistent reports on demethylating activity in cultured cell lines. At present, there is no conclusive understanding as to why some of the data reported by one group are not easily reproducible by another. One reason might be the methodology chosen to detect DNA methylation levels. In earlier studies, proof of demethylating activity in cell culture was often obtained by methylation-specific PCR (MSP), based on the design of selective primer pairs for exclusive amplification of methylated or unmethylated DNA sequences. This method was initially developed for the ultra-sensitive detection of methylated DNA as a cancer biomarker in a pool of unmethylated DNA (Herman *et al.*, 1996), but it might not be suitable for the quantitative detection of a reduction in DNA methylation, as is expected when treating cancer cells with demethylating agents. More reliable state-of-the-art methodology for the quantification of sequence-specific DNA methylation levels includes bisulfite sequencing, pyrosequencing, and the mass spectrometry-based Sequenom MassArray technology (for a summary of these methods, see supplement in Huang *et al.*, 2011). These tools for the sequence-specific quantification of DNA methylation should be preferentially used to quantify demethylating activity.

With the exception of EGCG and genistein, among a few others, the effects of polyphenols on DNA methylation *in vivo* are still limited, and a conclusive link between modulation of DNA methylation levels and cancer preventive efficacy is often still missing. Also, most studies have focused on only a few selected candidate genes, and effects on genome-wide DNA methylation are largely unknown. With the recent advancement of array- and sequencing-based methods for the assessment DNA methylation at a genome-wide scale (Rizzo & Buck, 2012), these analyses are now feasible and affordable, and should be included in future chemoprevention studies in animal models and pilot studies in humans.

### *6.6.2 Histone-modifying enzymes*

The development of HDAC inhibitors has been a focus of medicinal chemistry and the pharmaceutical industry for more than 2 decades. Structural requirements for the inhibition of HDACs are well defined. So far, no flavonoid-type HDAC inhibitor has been identified, most likely due to the spatial restriction of the active binding pocket of HDACs. In contrast, several sulfur-containing chemopreventive agents, such as metabolites of isothiocyanates from cruciferous vegetables, have been identified as potent HDAC inhibitors *in vitro* and *in vivo*. Some polyphenolic compounds, such as curcumin, EGCG, and genistein, however, have been reported to modulate HDAC activity, histone acetylation, and chromatin accessibility by affecting the expression of HDAC isoenzymes. Interestingly, various polyphenols have been identified as inhibitors of histone acetyl transferases (HATs). This might seem contradictory to the development of HDAC inhibitors for cancer treatment and prevention. However, HAT inhibition should lead to reduced histone acetylation and more condensed chromatin, making it less susceptible to DNA-damaging agents. Furthermore, HATs acetylate not only histones but also a series of important signaling molecules, such as p53, the NF-$\kappa$B subunit p65, and the AR. Since both NF-$\kappa$B and AR signaling become deregulated during carcinogenesis, HAT inhibition may indirectly contribute to the anti-inflammatory and anti-androgenic activities of some chemopreventive polyphenols.

### 6.6.3 miRNAs

Analysis of the influence of miRNAs and other noncoding RNAs on gene regulation is an emerging topic in current epigenetic research. Numerous recent reports based on microarray detection of miRNAs have described multiple changes in miRNA levels following treatment of cancer-cell lines with chemopreventive polyphenols. Interpretation of the data is often hampered by the fact that information on the validated gene targets of miRNAs is limited. Also, most studies do not include mechanistic insights into how chemopreventive polyphenols modulate the expression of miRNAs. These questions should be addressed in future studies.

### 6.6.4 Summary

In conclusion, 10 years of research have provided a wealth of information on how polyphenols and other chemopreventive agents influence key epigenetic mechanisms mainly in *in vitro* systems. Epigenetic research is highly dynamic and rapidly evolving, both in terms of characterization of epigenetic enzymes in basic research and as drug targets and in terms of technological developments in the detection of DNA methylation and chromatin modifications. Future studies are challenged to make use of these novel technologies in order to translate the *in vitro* findings of chemopreventive polyphenols on epigenetic mechanisms to the *in vivo* situation and to demonstrate the significance of epigenetic gene regulation for chemopreventive efficacy.

## References

Abusnina, A., Keravis, T., Yougbare, I., *et al.* (2011) Anti-proliferative effect of curcumin on melanoma cells is mediated by PDE1A inhibition that regulates the epigenetic integrator UHRF1. *Molecular Nutrition and Food Research*, 55, 1677–1689.

Adalsteinsson, B.T., Gudnason, H., Aspelund, T., *et al.* (2012) Heterogeneity in white blood cells has potential to confound DNA methylation measurements. *PloS One*, 7, e46705.

Adhami, V.M., Siddiqui, I.A., Sarfaraz, S., *et al.* (2009) Effective prostate cancer chemopreventive intervention with green tea polyphenols in the TRAMP model depends on the stage of the disease. *Clinical Cancer Research*, 15, 1947–1953.

Adjakly, M., Bosviel, R., Rabiau, N., *et al.* (2011) DNA methylation and soy phytoestrogens: quantitative study in DU-145 and PC-3 human prostate cancer cell lines. *Epigenomics*, 3, 795–803.

Ahmad, A., Sarkar, S.H., Bitar, B., *et al.* (2012) Garcinol regulates EMT and Wnt signaling pathways in vitro and in vivo, leading to anticancer activity against breast cancer cells. *Molecular Cancer Therapeutics*, 11, 2193–2201.

Arif, M., Pradhan, S.K., Thanuja, G.R., *et al.* (2009) Mechanism of p300 specific histone acetyltransferase inhibition by small molecules. *Journal of Medicinal Chemistry*, 52, 267–277.

Badger, T.M., Ronis, M.J., Wolff, G., *et al.* (2008) Soy protein isolate reduces hepatosteatosis in yellow Avy/a mice without altering coat color phenotype. *Experimental Biology and Medicine (Maywood, N.J.)*, 233, 1242–1254.

Balasubramanian, S., Adhikary, G. & Eckert, R.L. (2010) The Bmi-1 polycomb protein antagonizes the (−)-epigallocatechin-3-gallate-dependent suppression of skin cancer cell survival. *Carcinogenesis*, 31, 496–503.

Balasubramanyam, K., Swaminathan, V., Ranganathan, A. & Kundu, T.K. (2003) Small molecule modulators of histone acetyltransferase p300. *Journal of Biological Chemistry*, 278, 19134–19140.

Balasubramanyam, K., Altaf, M., Varier, R.A., *et al.* (2004a) Polyisoprenylated benzophenone, garcinol, a natural histone acetyltransferase inhibitor, represses chromatin transcription and alters global gene expression. *Journal of Biological Chemistry*, 279, 33716–33726.

Balasubramanyam, K., Varier, R.A., Altaf, M., *et al.* (2004b) Curcumin, a novel p300/CREB-binding protein-specific inhibitor of acetyltransferase, represses the acetylation of histone/nonhistone proteins and histone acetyltransferase-dependent chromatin transcription. *Journal of Biological Chemistry*, 279, 51163–51171.

Banerjee, S., Li, Y., Wang, Z. & Sarkar, F.H. (2008) Multi-targeted therapy of cancer by genistein. *Cancer Letters*, 269, 226–242.

Bannister, A.J. & Kouzarides, T. (2011) Regulation of chromatin by histone modifications. *Cell Research*, 21, 381–395.

Bao, B., Wang, Z., Ali, S., *et al.* (2011a) Over-expression of FoxM1 leads to epithelial-mesenchymal transition and cancer stem cell phenotype in pancreatic cancer cells. *Journal of Cellular Biochemistry*, 112, 2296–2306.

Bao, B., Wang, Z., Ali, S., *et al.* (2011b) Notch-1 induces epithelial-mesenchymal transition consistent with cancer stem cell phenotype in pancreatic cancer cells. *Cancer Letters*, 307, 26–36.

Barker, N. & Clevers, H. (2006) Mining the Wnt pathway for cancer therapeutics. *Nature Reviews: Drug Discovery*, 5, 997–1014.

Basak, S., Pookot, D., Noonan, E.J. & Dahiya, R. (2008) Genistein down-regulates androgen receptor by modulating HDAC6–Hsp90 chaperone function. *Molecular Cancer Therapeutics*, 7, 3195–3202.

Baur, J.A. & Sinclair, D.A. (2006) Therapeutic potential of resveratrol: the in vivo evidence. *Nature Reviews: Drug Discovery*, 5, 493–506.

Beaulieu, N., Morin, S., Chute, I.C., *et al.* (2002) An essential role for DNA methyltransferase DNMT3B in cancer cell survival. *Journal of Biological Chemistry*, 277, 28176–28181.

Berletch, J.B., Liu, C., Love, W.K., *et al.* (2008) Epigenetic and genetic mechanisms contribute to telomerase inhibition by EGCG. *Journal of Cellular Biochemistry*, 103, 509–519.

Blecha, J.E., Anderson, M.O., Chow, J.M., *et al.* (2007) Inhibition of IGF-1R and lipoxygenase by nordihydroguaiaretic acid (NDGA) analogs. *Bioorganic and Medicinal Chemistry Letters*, 17, 4026–4029.

Bora-Tatar, G., Dayangac-Erden, D., Demir, A.S., *et al.* (2009) Molecular modifications on carboxylic acid derivatives as potent histone deacetylase inhibitors: activity and docking studies. *Bioorganic and Medicinal Chemistry*, 17, 5219–5228.

Bosviel, R., Dumollard, E., Dechelotte, P., *et al.* (2012a) Can soy phytoestrogens decrease DNA methylation in BRCA1 and BRCA2 oncosuppressor genes in breast cancer? *OMICS: A Journal of Integrative Biology*, 16, 235–244.

Bosviel, R., Durif, J., Dechelotte, P., *et al.* (2012b) Epigenetic modulation of BRCA1 and BRCA2 gene expression by equol in breast cancer cell lines. *British Journal of Nutrition*, 108, 1187–1193.

Brait, M. & Sidransky, D. (2011) Cancer epigenetics: above and beyond. *Toxicology Mechanisms and Methods*, 21, 275–288.

Brennan, K., Garcia-Closas, M., Orr, N., *et al.* (2012) Intragenic ATM methylation in peripheral blood DNA as a biomarker of breast cancer risk. *Cancer Research*, 72, 2304–2313.

Brooks, C.L. & Gu, W. (2009) How does SIRT1 affect metabolism, senescence and cancer? *Nature Reviews: Cancer*, 9, 123–128.

Byun, H.M., Choi, S.H., Laird, P.W., *et al.* (2008) 2′-deoxy-N4–[2–(4–nitrophenyl)ethoxycarbonyl]-5–azacytidine: a novel inhibitor of DNA methyltransferase that requires activation by human carboxylesterase 1. *Cancer Letters*, 266, 238–248.

Calin, G.A. & Croce, C.M. (2006) MicroRNA signatures in human cancers. *Nature Reviews: Cancer*, 6, 857–866.

Caporali, A., Davalli, P., Astancolle, S., *et al.* (2004) The chemopreventive action of catechins in the TRAMP mouse model of prostate carcinogenesis is accompanied by clusterin over-expression. *Carcinogenesis*, 25, 2217–2224.

Chakrabarti, M., Khandkar, M., Banik, N.L. & Ray, S.K. (2012) Alterations in expression of specific microRNAs by combination of 4-HPR and EGCG inhibited growth of human malignant neuroblastoma cells. *Brain Research*, 1454, 1–13.

Che, X. & Huang, C. (2012) MicroRNA, cancer and cancer chemoprevention. *Current Molecular Pharmacology*, 5, 362–371.

Chen, Y., Shu, W., Chen, W., *et al.* (2007) Curcumin, both histone deacetylase and p300/CBP-specific inhibitor, represses the activity of nuclear factor kappa B and Notch 1 in Raji cells. *Basic & Clinical Pharmacology & Toxicology*, 101, 427–433.

Chen, Y., Zaman, M.S., Deng, G., *et al.* (2011) MicroRNAs 221/222 and genistein-mediated regulation of ARHI tumor suppressor gene in prostate cancer. *Cancer Prevention Research (Philadelphia, Pa.)*, 4, 76–86.

Choi, K.C., Jung, M.G., Lee, Y.H., *et al.* (2009a) Epigallocatechin-3-gallate, a histone acetyltransferase inhibitor, inhibits EBV-induced B lymphocyte transformation via suppression of RelA acetylation. *Cancer Research*, 69, 583–592.

Choi, K.C., Lee, Y.H., Jung, M.G., *et al.* (2009b) Gallic acid suppresses lipopolysaccharide-induced nuclear factor-kappaB signaling by preventing RelA acetylation in A549 lung cancer cells. *Molecular Cancer Research*, 7, 2011–2021.

Choudhury, S.R., Balasubramanian, S., Chew, Y.C., *et al.* (2011) (−)-epigallocatechin-3-gallate and DZNep reduce polycomb protein level via a proteasome-dependent mechanism in skin cancer cells. *Carcinogenesis*, 32, 1525–1532.

Chuang, J.C., Yoo, C.B., Kwan, J.M., *et al.* (2005) Comparison of biological effects of non-nucleoside DNA methylation inhibitors versus 5-aza-2′-deoxycytidine. *Molecular Cancer Therapeutics*, 4, 1515–1520.

Cohen, H.Y., Miller, C., Bitterman, K.J., *et al.* (2004) Calorie restriction promotes mammalian cell survival by inducing the SIRT1 deacetylase. *Science*, 305, 390–392.

Cui, Y., Lu, C., Kang, A., *et al.* (2008a) Nordihydroguaiaretic acid restores expression of silenced E-cadherin gene in human breast cancer cell lines and xenografts. *Anti-Cancer Drugs*, 19, 487–494.

Cui, Y., Lu, C., Liu, L., *et al.* (2008b) Reactivation of methylation-silenced tumor suppressor gene p16INK4a by nordihydroguaiaretic acid and its implication in G1 cell cycle arrest. *Life Sciences*, 82, 247–255.

Cui, W., Gu, F. & Hu, K.Q. (2009) Effects and mechanisms of silibinin on human hepatocellular carcinoma xenografts in nude mice. *World Journal of Gastroenterology*, 15, 1943–1950.

Davis, C.D. & Ross, S.A. (2008) Evidence for dietary regulation of microRNA expression in cancer cells. *Nutrition Reviews*, 66, 477–482.

Day, J.K., Bauer, A.M., DesBordes, C., *et al.* (2002) Genistein alters methylation patterns in mice. *Journal of Nutrition*, 132, 2419S–2423S.

Di Croce, L., Raker, V.A., Corsaro, M., *et al.* (2002) Methyltransferase recruitment and DNA hypermethylation of target promoters by an oncogenic transcription factor. *Science*, 295, 1079–1082.

Dimri, M., Bommi, P., Sahasrabuddhe, A.A., *et al.* (2010) Dietary omega-3 polyunsaturated fatty acids suppress expression of EZH2 in breast cancer cells. *Carcinogenesis*, 31, 489–495.

Dolinoy, D.C., Weidman, J.R., Waterland, R.A. & Jirtle, R.L. (2006) Maternal genistein alters coat color and protects Avy mouse offspring from obesity by modifying the fetal epigenome. *Environmental Health Perspectives*, 114, 567–572.

Editors (2003) Definitions in epigenetics. *Annals of the New York Academy of Sciences*, 983, 321–328.

Ellis, L., Atadja, P.W. & Johnstone, R.W. (2009) Epigenetics in cancer: targeting chromatin modifications. *Molecular Cancer Therapeutics*, 8, 1409–1420.

Esteller, M. (2007) Cancer epigenomics: DNA methylomes and histone-modification maps. *Nature Reviews: Genetics*, 8, 286–298.

Fang, M.Z., Wang, Y., Ai, N., *et al.* (2003) Tea polyphenol (−)-epigallocatechin-3-gallate inhibits DNA methyltransferase and reactivates methylation-silenced genes in cancer cell lines. *Cancer Research*, 63, 7563–7570.

Fang, M.Z., Chen, D., Sun, Y., *et al.* (2005) Reversal of hypermethylation and reactivation of p16INK4a, RARbeta, and MGMT genes by genistein and other isoflavones from soy. *Clinical Cancer Research*, 11, 7033–7041.

Fang, M., Chen, D. & Yang, C.S. (2007) Dietary polyphenols may affect DNA methylation. *Journal of Nutrition*, 137, 223S–228S.

Farghali, H., Kutinova Canova, N. & Lekic, N. (2013) Resveratrol and related compounds as antioxidants with an allosteric mechanism of action in epigenetic drug targets. *Physiological Research*, 62, 1–13.

Ferk, F., Chakraborty, A., Jager, W., *et al.* (2011) Potent protection of gallic acid against DNA oxidation: results of human and animal experiments. *Mutation Research*, 715, 61–71.

Fini, L., Selgrad, M., Fogliano, V., *et al.* (2007) Annurca apple polyphenols have potent demethylating activity and can reactivate silenced tumor suppressor genes in colorectal cancer cells. *Journal of Nutrition*, 137, 2622–2628.

Fini, L., Piazzi, G., Daoud, Y., *et al.* (2011) Chemoprevention of intestinal polyps in ApcMin/+ mice fed western or balanced diets by drinking Annurca apple polyphenol extract. *Cancer Prevention Research (Philadelphia, Pa.)*, 4, 907–915.

Fix, L.N., Shah, M., Efferth, T., *et al.* (2010) MicroRNA expression profile of MCF-7 human breast cancer cells and the effect of green tea polyphenon-60. *Cancer Genomics & Proteomics*, 7, 261–277.

Fu, S. & Kurzrock, R. (2010) Development of curcumin as an epigenetic agent. *Cancer*, 116, 4670–4676.

Fullgrabe, J., Kavanagh, E. & Joseph, B. (2011) Histone onco-modifications. *Oncogene*, 30, 3391–3403.

Gama-Sosa, M.A., Slagel, V.A., Trewyn, R.W., *et al.* (1983) The 5-methylcytosine content of DNA from human tumors. *Nucleic Acids Research*, 11, 6883–6894.

Gao, Z., Xu, Z., Hung, M.S., *et al.* (2009) Promoter demethylation of WIF-1 by epigallocatechin-3-gallate in lung cancer cells. *Anticancer Research*, 29, 2025–2030.

Gerhauser, C. (2008) Cancer chemopreventive potential of apples, apple juice, and apple components. *Planta Medica*, 74, 1608–1624.

Ghizzoni, M., Boltjes, A., Graaf, C., *et al.* (2010) Improved inhibition of the histone acetyltransferase PCAF by an anacardic acid derivative. *Bioorganic and Medicinal Chemistry*, 18, 5826–5834.

Ghizzoni, M., Haisma, H.J., Maarsingh, H. & Dekker, F.J. (2011) Histone acetyltransferases are crucial regulators in NF-kappaB mediated inflammation. *Drug Discovery Today*, 16, 504–511.

Gilbert, E.R. & Liu, D. (2010) Flavonoids influence epigenetic-modifying enzyme activity: structure - function relationships and the therapeutic potential for cancer. *Current Medicinal Chemistry*, 17, 1756–1768.

Goelz, S.E., Vogelstein, B., Hamilton, S.R. & Feinberg, A.P. (1985) Hypomethylation of DNA from benign and malignant human colon neoplasms. *Science*, 228, 187–190.

Greathouse, K.L., Bredfeldt, T., Everitt, J.I., *et al.* (2012) Environmental estrogens differentially engage the histone methyltransferase EZH2 to increase risk of uterine tumorigenesis. *Molecular Cancer Research*, 10, 546–557.

Guerrero-Bosagna, C.M., Sabat, P., Valdovinos, F.S., *et al.* (2008) Epigenetic and phenotypic changes result from a continuous pre and post natal dietary exposure to phytoestrogens in an experimental population of mice. *BMC Physiology*, 8, 17.

Gupta, S., Hastak, K., Ahmad, N., *et al.* (2001) Inhibition of prostate carcinogenesis in TRAMP mice by oral infusion of green tea polyphenols. *Proceedings of the National Academy of Sciences USA*, 98, 10350–10355.

Gupta, S.C., Kismali, G. & Aggarwal, B.B. (2013a) Curcumin, a component of turmeric: from farm to pharmacy. *Biofactors*, 39, 2–13.

Gupta, S.C., Sung, B., Kim, J.H., *et al.* (2013b) Multitargeting by turmeric, the golden spice: from kitchen to clinic. *Molecular Nutrition and Food Research*, 57, 1510–1528.

Helferich, W.G., Andrade, J.E. & Hoagland, M.S. (2008) Phytoestrogens and breast cancer: a complex story. *Inflammopharmacology*, 16, 219–226.

Hemshekhar, M., Sebastin Santhosh, M., Kemparaju, K. & Girish, K.S. (2012) Emerging roles of anacardic acid and its derivatives: a pharmacological overview. *Basic & Clinical Pharmacology & Toxicology*, 110, 122–132.

Henikoff, S. & Matzke, M.A. (1997) Exploring and explaining epigenetic effects. *Trends in Genetics*, 13, 293–295.

Herman, J.G., Graff, J.R., Myohanen, S., *et al.* (1996) Methylation-specific PCR: a novel PCR assay for methylation status of CpG islands. *Proceedings of the National Academy of Sciences USA*, 93, 9821–9826.

Hong, T., Nakagawa, T., Pan, W., *et al.* (2004) Isoflavones stimulate estrogen receptor-mediated core histone acetylation. *Biochemical and Biophysical Research Communications*, 317, 259–264.

Houseman, E.A., Accomando, W.P., Koestler, D.C., *et al.* (2012) DNA methylation arrays as surrogate measures of cell mixture distribution. *BMC Bioinformatics*, 13, 86.

Howard, T.D., Ho, S.M., Zhang, L., *et al.* (2011) Epigenetic changes with dietary soy in cynomolgus monkeys. *PloS One*, 6, e26791.

Howitz, K.T., Bitterman, K.J., Cohen, H.Y., *et al.* (2003) Small molecule activators of sirtuins extend Saccharomyces cerevisiae lifespan. *Nature*, 425, 191–196.

Hua, W.F., Fu, Y.S., Liao, Y.J., *et al.* (2010) Curcumin induces down-regulation of EZH2 expression through the MAPK pathway in MDA-MB-435 human breast cancer cells. *European Journal of Pharmacology*, 637, 16–21.

Huang, Y., Stewart, T.M., Wu, Y., *et al.* (2009) Novel oligoamine analogues inhibit lysine-specific demethylase 1 and induce reexpression of epigenetically silenced genes. *Clinical Cancer Research*, 15, 7217–7228.

Huang, J., Plass, C. & Gerhauser, C. (2011) Cancer chemoprevention by targeting the epigenome. *Current Drug Targets*, 12, 1925–1956.

Illingworth, R., Kerr, A., Desousa, D., *et al.* (2008) A novel CpG island set identifies tissue-specific methylation at developmental gene loci. *PLoS Biology*, 6, e22.

Izzotti, A., Cartiglia, C., Steele, V.E. & De Flora, S. (2012) MicroRNAs as targets for dietary and pharmacological inhibitors of mutagenesis and carcinogenesis. *Mutation Research*, 751, 287–303.

Jawaid, K., Crane, S.R., Nowers, J.L., *et al.* (2010) Long-term genistein treatment of MCF-7 cells decreases acetylated histone 3 expression and alters growth responses to mitogens and histone deacetylase inhibitors. *Journal of Steroid Biochemistry and Molecular Biology*, 120, 164–171.

Jha, A.K., Nikbakht, M., Parashar, G., *et al.* (2010) Reversal of hypermethylation and reactivation of the RARbeta2 gene by natural compounds in cervical cancer cell lines. *Folia Biologica (Praha)*, 56, 195–200.

Jian, L. (2009) Soy, isoflavones, and prostate cancer. *Molecular Nutrition and Food Research*, 53, 217–226.

Jones, P.A. & Baylin, S.B. (2007) The epigenomics of cancer. *Cell*, 128, 683–692.

Kahyo, T., Ichikawa, S., Hatanaka, T., *et al.* (2008) A novel chalcone polyphenol inhibits the deacetylase activity of SIRT1 and cell growth in HEK293T cells. *Journal of Pharmacological Sciences*, 108, 364–371.

Kang, J., Chen, J., Shi, Y., *et al.* (2005) Curcumin-induced histone hypoacetylation: the role of reactive oxygen species. *Biochemical Pharmacology*, 69, 1205–1213.

Karius, T., Schnekenburger, M., Dicato, M. & Diederich, M. (2012) MicroRNAs in cancer management and their modulation by dietary agents. *Biochemical Pharmacology*, 83, 1591–1601.

Kato, K., Long, N.K., Makita, H., *et al.* (2008) Effects of green tea polyphenol on methylation status of RECK gene and cancer cell invasion in oral squamous cell carcinoma cells. *British Journal of Cancer*, 99, 647–654.

Khan, N. & Mukhtar, H. (2008) Multitargeted therapy of cancer by green tea polyphenols. *Cancer Letters*, 269, 269–280.

Khan, S.I., Aumsuwan, P., Khan, I.A., *et al.* (2012) Epigenetic events associated with breast cancer and their prevention by dietary components targeting the epigenome. *Chemical Research in Toxicology*, 25, 61–73.

Khor, T.O., Huang, Y., Wu, T.Y., *et al.* (2011) Pharmacodynamics of curcumin as DNA hypomethylation agent in restoring the expression of Nrf2 via promoter CpGs demethylation. *Biochemical Pharmacology*, 82, 1073–1078.

Kikuno, N., Shiina, H., Urakami, S., *et al.* (2008) Genistein mediated histone acetylation and demethylation activates tumor suppressor genes in prostate cancer cells. *International Journal of Cancer*, 123, 552–560.

Kim, S.C., Sprung, R., Chen, Y., *et al.* (2006) Substrate and functional diversity of lysine acetylation revealed by a proteomics survey. *Molecular Cell*, 23, 607–618.

King-Batoon, A., Leszczynska, J.M. & Klein, C.B. (2008) Modulation of gene methylation by genistein or lycopene in breast cancer cells. *Environmental and Molecular Mutagenesis*, 49, 36–45.

Klaus, A. & Birchmeier, W. (2008) Wnt signalling and its impact on development and cancer. *Nature Reviews: Cancer*, 8, 387–398.

Kopelovich, L., Crowell, J.A. & Fay, J.R. (2003) The epigenome as a target for cancer chemoprevention. *Journal of the National Cancer Institute*, 95, 1747–1757.

Kouzarides, T. (2007) Chromatin modifications and their function. *Cell*, 128, 693–705.

Kroll, D.J., Shaw, H.S. & Oberlies, N.H. (2007) Milk thistle nomenclature: why it matters in cancer research and pharmacokinetic studies. *Integrative Cancer Therapies*, 6, 110–119.

Kuck, D., Singh, N., Lyko, F. & Medina-Franco, J.L. (2010) Novel and selective DNA methyltransferase inhibitors: Docking-based virtual screening and experimental evaluation. *Bioorganic and Medicinal Chemistry*, 18, 822–829.

Kundu, J.K. & Surh, Y.J. (2008) Cancer chemopreventive and therapeutic potential of resveratrol: mechanistic perspectives. *Cancer Letters*, 269, 243–261.

Kunnumakkara, A.B., Anand, P. & Aggarwal, B.B. (2008) Curcumin inhibits proliferation, invasion, angiogenesis and metastasis of different cancers through interaction with multiple cell signaling proteins. *Cancer Letters*, 269, 199–225.

Lancon, A., Michaille, J.J. & Latruffe, N. (2013) Effects of dietary phytophenols on the expression of microRNAs involved in mammalian cell homeostasis. *Journal of the Science of Food and Agriculture*, 93, 3155–3164.

Lavu, S., Boss, O., Elliott, P.J. & Lambert, P.D. (2008) Sirtuins – novel therapeutic targets to treat age-associated diseases. *Nature Reviews: Drug Discovery*, 7, 841–853.

Lee, W.J. & Zhu, B.T. (2006) Inhibition of DNA methylation by caffeic acid and chlorogenic acid, two common catechol-containing coffee polyphenols. *Carcinogenesis*, 27, 269–277.

Lee, W.J., Shim, J.Y. & Zhu, B.T. (2005) Mechanisms for the inhibition of DNA methyltransferases by tea catechins and bioflavonoids. *Molecular Pharmacology*, 68, 1018–1030.

Lee, S.J., Krauthauser, C., Maduskuie, V., *et al.* (2011) Curcumin-induced HDAC inhibition and attenuation of medulloblastoma growth in vitro and in vivo. *BMC Cancer*, 11, 144.

Lee, Y.H., Kwak, J., Choi, H.K., *et al.* (2012) EGCG suppresses prostate cancer cell growth modulating acetylation of androgen receptor by anti-histone acetyltransferase activity. *International Journal of Molecular Medicine*, 30, 69–74.

Li, X. & Kazgan, N. (2011) Mammalian sirtuins and energy metabolism. *International Journal of Biological Sciences*, 7, 575–587.

Li, L.-H., Wu, L.-J., Tashiro, S.-I., *et al.* (2007) Activation of SIRT1 pathway and modulation of the cell cycle were involved in silimarin's protection against UV-induced A375-S2 cell apoptosis. *Journal of Asian Natural Products Research*, 9, 245–252.

Li, Y., Liu, L., Andrews, L.G. & Tollefsbol, T.O. (2009a) Genistein depletes telomerase activity through cross-talk between genetic and epigenetic mechanisms. *International Journal of Cancer*, 125, 286–296.

Li, Y., VandenBoom, T.G. 2nd,, Kong, D., *et al.* (2009b) Up-regulation of miR-200 and let-7 by natural agents leads to the reversal of epithelial-to-mesenchymal transition in gemcitabine-resistant pancreatic cancer cells. *Cancer Research*, 69, 6704–6712.

Li, Y., Kong, D., Wang, Z. & Sarkar, F.H. (2010a) Regulation of microRNAs by natural agents: an emerging field in chemoprevention and chemotherapy research. *Pharmaceutical Research*, 27, 1027–1041.

Li, Y., Vandenboom, T.G. 2nd,, Wang, Z., *et al.* (2010b) miR-146a suppresses invasion of pancreatic cancer cells. *Cancer Research*, 70, 1486–1495.

Li, Y., Yuan, Y.Y., Meeran, S.M. & Tollefsbol, T.O. (2010c) Synergistic epigenetic reactivation of estrogen receptor-alpha (ERalpha) by combined green tea polyphenol and histone deacetylase inhibitor in ERalpha-negative breast cancer cells. *Molecular Cancer*, 9, 274.

Li, Y., Kong, D., Ahmad, A., *et al.* (2012) Epigenetic deregulation of miR-29a and miR-1256 by isoflavone contributes to the inhibition of prostate cancer cell growth and invasion. *Epigenetics*, 7, 940–949.

Li, Y., Chen, H., Hardy, T.M. & Tollefsbol, T.O. (2013a) Epigenetic regulation of multiple tumor-related genes leads to suppression of breast tumorigenesis by dietary genistein. *PloS One*, 8, e54369.

Li, Y., Meeran, S.M., Patel, S.N., *et al.* (2013b) Epigenetic reactivation of estrogen receptor-alpha (ERalpha) by genistein enhances hormonal therapy sensitivity in ERalpha-negative breast cancer. *Molecular Cancer*, 12, 9.

Linhart, H.G., Lin, H., Yamada, Y., *et al.* (2007) Dnmt3b promotes tumorigenesis in vivo by gene-specific de novo methylation and transcriptional silencing. *Genes and Development*, 21, 3110–3122.

Link, A., Balaguer, F. & Goel, A. (2010) Cancer chemoprevention by dietary polyphenols: promising role for epigenetics. *Biochemical Pharmacology*, 80, 1771–1792.

Liu, H.L., Chen, Y., Cui, G.H. & Zhou, J.F. (2005) Curcumin, a potent anti-tumor reagent, is a novel histone deacetylase inhibitor regulating B-NHL cell line Raji proliferation. *Acta Pharmacologica Sinica*, 26, 603–609.

Liu, Z., Xie, Z., Jones, W., *et al.* (2009) Curcumin is a potent DNA hypomethylation agent. *Bioorganic and Medicinal Chemistry Letters*, 19, 706–709.

Liu, Y.L., Yang, H.P., Gong, L., *et al.* (2011a) Hypomethylation effects of curcumin, demethoxycurcumin and bisdemethoxycurcumin on WIF-1 promoter in non-small cell lung cancer cell lines. *Molecular Medicine Reports*, 4, 675–679.

Liu, Y.L., Yang, H.P., Zhou, X.D., *et al.* (2011b) The hypomethylation agent bisdemethoxycurcumin acts on the WIF-1 promoter, inhibits the canonical Wnt pathway and induces apoptosis in human non-small-cell lung cancer. *Current Cancer Drug Targets*, 11, 1098–1110.

Locatelli, C., Filippin-Monteiro, F.B. & Creczynski-Pasa, T.B. (2012) Alkyl esters of gallic acid as anticancer agents: a review. *European Journal of Medicinal Chemistry*, 60C, 233–239.

Loguercio, C. & Festi, D. (2011) Silybin and the liver: from basic research to clinical practice. *World Journal of Gastroenterology*, 17, 2288–2301.

Lu, J.M., Nurko, J., Weakley, S.M., *et al.* (2010) Molecular mechanisms and clinical applications of nordihydroguaiaretic acid (NDGA) and its derivatives: an update. *Medical Science Monitor*, 16, RA93–100.

Lyn-Cook, B.D., Blann, E., Payne, P.W., *et al.* (1995) Methylation profile and amplification of proto-oncogenes in rat pancreas induced with phytoestrogens. *Proceedings of the Society for Experimental Biology and Medicine*, 208, 116–119.

Magee, P.J. & Rowland, I. (2012) Soy products in the management of breast cancer. *Current Opinion in Clinical Nutrition and Metabolic Care*, 15, 586–591.

Majid, S., Kikuno, N., Nelles, J., *et al.* (2008) Genistein induces the p21WAF1/CIP1 and p16INK4a tumor suppressor genes in prostate cancer cells by epigenetic mechanisms involving active chromatin modification. *Cancer Research*, 68, 2736–2744.

Majid, S., Dar, A.A., Ahmad, A.E., *et al.* (2009) BTG3 tumor suppressor gene promoter demethylation, histone modification and cell cycle arrest by genistein in renal cancer. *Carcinogenesis*, 30, 662–670.

Majid, S., Dar, A.A., Saini, S., *et al.* (2010a) Regulation of minichromosome maintenance gene family by microRNA-1296 and genistein in prostate cancer. *Cancer Research*, 70, 2809–2818.

Majid, S., Dar, A.A., Shahryari, V., *et al.* (2010b) Genistein reverses hypermethylation and induces active histone modifications in tumor suppressor gene B-Cell translocation gene 3 in prostate cancer. *Cancer*, 116, 66–76.

Malireddy, S., Kotha, S.R., Secor, J.D., *et al.* (2012) Phytochemical antioxidants modulate mammalian cellular epigenome: implications in health and disease. *Antioxidants & Redox Signaling*, 17, 327–339.

Matsukura, H., Aisaki, K., Igarashi, K., *et al.* (2011) Genistein promotes DNA demethylation of the steroidogenic factor 1 (SF-1) promoter in endometrial stromal cells. *Biochemical and Biophysical Research Communications*, 412, 366–372.

Medina-Franco, J.L., Lopez-Vallejo, F., Kuck, D. & Lyko, F. (2011) Natural products as DNA methyltransferase inhibitors: a computer-aided discovery approach. *Molecular Diversity*, 15, 293–304.

Meeran, S.M., Patel, S.N., Chan, T.H. & Tollefsbol, T.O. (2011) A novel prodrug of epigallocatechin-3-gallate: differential epigenetic hTERT repression in human breast cancer cells. *Cancer Prevention Research (Philadelphia, Pa.)*, 4, 1243–1254.

Meeran, S.M., Patel, S.N., Li, Y., *et al.* (2012) Bioactive dietary supplements reactivate ER expression in ER-negative breast cancer cells by active chromatin modifications. *PloS One*, 7, e37748.

Messina, M., McCaskill-Stevens, W. & Lampe, J.W. (2006) Addressing the soy and breast cancer relationship: review, commentary, and workshop proceedings. *Journal of the National Cancer Institute*, 98, 1275–1284.

Molinie, B. & Georgel, P. (2009) Genetic and epigenetic regulations of prostate cancer by genistein. *Drug News & Perspectives*, 22, 247–254.

Morey Kinney, S.R., Zhang, W., Pascual, M., *et al.* (2009) Lack of evidence for green tea polyphenols as DNA methylation inhibitors in murine prostate. *Cancer Prevention Research (Philadelphia, Pa.)*, 2, 1065–1075.

Mottet, D. & Castronovo, V. (2008) Histone deacetylases: target enzymes for cancer therapy. *Clinical and Experimental Metastasis*, 25, 183–189.

Munro, S., Khaire, N., Inche, A., *et al.* (2010) Lysine methylation regulates the pRb tumour suppressor protein. *Oncogene*, 29, 2357–2367.

Murakami, A., Ashida, H. & Terao, J. (2008) Multitargeted cancer prevention by quercetin. *Cancer Letters*, 269, 315–325.

Nandakumar, V., Vaid, M. & Katiyar, S.K. (2011) (−)-epigallocatechin-3-gallate reactivates silenced tumor suppressor genes, Cip1/p21 and p16INK4a, by reducing DNA methylation and increasing histones acetylation in human skin cancer cells. *Carcinogenesis*, 32, 537–544.

Navarro-Peran, E., Cabezas-Herrera, J., Campo, L.S. & Rodriguez-Lopez, J.N. (2007) Effects of folate cycle disruption by the green tea polyphenol epigallocatechin-3–gallate. *International Journal of Biochemistry and Cell Biology*, 39, 2215–2225.

Neelakandan, K., Babu, P. & Nair, S. (2012) Emerging roles for modulation of microRNA signatures in cancer chemoprevention. *Current Cancer Drug Targets*, 12, 716–740.

Nielsen, A.L., Kristensen, L.H., Stephansen, K.B., *et al.* (2012) Identification of catechols as histone-lysine demethylase inhibitors. *FEBS Letters*, 586, 1190–1194.

Padhye, S., Ahmad, A., Oswal, N. & Sarkar, F.H. (2009) Emerging role of Garcinol, the antioxidant chalcone from Garcinia indica Choisy and its synthetic analogs. *Journal of Hematology & Oncology*, 2, 38.

Paluszczak, J., Krajka-Kuzniak, V. & Baer-Dubowska, W. (2010) The effect of dietary polyphenols on the epigenetic regulation of gene expression in MCF7 breast cancer cells. *Toxicology Letters*, 192, 119–125.

Pandey, M., Shukla, S. & Gupta, S. (2009) Promoter demethylation and chromatin remodeling by green tea polyphenols leads to re-expression of GSTP1 in human prostate cancer cells. *International Journal of Cancer*, 126, 2520–2533.

Papoutsis, A.J., Lamore, S.D., Wondrak, G.T., *et al.* (2010) Resveratrol prevents epigenetic silencing of BRCA-1 by the aromatic hydrocarbon receptor in human breast cancer cells. *Journal of Nutrition*, 140, 1607–1614.

Parasramka, M.A., Ho, E., Williams, D.E. & Dashwood, R.H. (2012) MicroRNAs, diet, and cancer: new mechanistic insights on the epigenetic actions of phytochemicals. *Molecular Carcinogenesis*, 51, 213–230.

Pezzuto, J.M. (2008) Resveratrol as an inhibitor of carcinogenensis. *Pharmaceutical Biology*, 46, 443–573.

Powles, T.J. (2002) Anti-oestrogenic prevention of breast cancer – the make or break point. *Nature Reviews: Cancer*, 2, 787–794.

Qin, W., Zhu, W., Shi, H., *et al.* (2009) Soy isoflavones have an antiestrogenic effect and alter mammary promoter hypermethylation in healthy premenopausal women. *Nutrition and Cancer*, 61, 238–244.

Qiu, J. (2006) Epigenetics: unfinished symphony. *Nature*, 441, 143–145.

Rajavelu, A., Tulyasheva, Z., Jaiswal, R., *et al.* (2011) The inhibition of the mammalian DNA methyltransferase 3a (Dnmt3a) by dietary black tea and coffee polyphenols. *BMC Biochemistry*, 12, 16.

Ramasamy, K. & Agarwal, R. (2008) Multitargeted therapy of cancer by silymarin. *Cancer Letters*, 269, 352–362.

Ravindran, J., Prasad, S. & Aggarwal, B.B. (2009) Curcumin and cancer cells: how many ways can curry kill tumor cells selectively? *American Association of Pharmaceutical Scientists (AAPS) Journal*, 11, 495–510.

Reinius, L.E., Acevedo, N., Joerink, M., *et al.* (2012) Differential DNA methylation in purified human blood cells: implications for cell lineage and studies on disease susceptibility. *PloS One*, 7, e41361.

Reuter, S., Gupta, S.C., Park, B., *et al.* (2011) Epigenetic changes induced by curcumin and other natural compounds. *Genes & Nutrition*, 6, 93–108.

Rietjens, I.M., Sotoca, A.M., Vervoort, J. & Louisse, J. (2013) Mechanisms underlying the dualistic mode of action of major soy isoflavones in relation to cell proliferation and cancer risks. *Molecular Nutrition and Food Research*, 57, 100–113.

Rizzo, J.M. & Buck, M.J. (2012) Key principles and clinical applications of "next-generation" *DNA sequencing. Cancer Prevention Research (Philadelphia, Pa.)*, 5, 887–900.

Robertson, K.D., Ait-Si-Ali, S., Yokochi, T., *et al.* (2000) DNMT1 forms a complex with Rb, E2F1 and HDAC1 and represses transcription from E2F-responsive promoters. *Nature Genetics*, 25, 338–342.

Rosenfeld, C.S., Sieli, P.T., Warzak, D.A., *et al.* (2013) Maternal exposure to bisphenol A and genistein has minimal effect on Avy/a offspring coat color but favors birth of agouti over nonagouti mice. *Proceedings of the National Academy of Sciences USA*, 110, 537–542.
Roth, M. & Jeltsch, A. (2000) Biotin-avidin microplate assay for the quantitative analysis of enzymatic methylation of DNA by DNA methyltransferases. *Biological Chemistry*, 381, 269–272.
Rountree, M.R., Bachman, K.E. & Baylin, S.B. (2000) DNMT1 binds HDAC2 and a new co-repressor, DMAP1, to form a complex at replication foci. *Nature Genetics*, 25, 269–277.
Russo, M., Spagnuolo, C., Tedesco, I., *et al.* (2012) The flavonoid quercetin in disease prevention and therapy: facts and fancies. *Biochemical Pharmacology*, 83, 6–15.
Ryan, C.J., Harzstark, A.H., Rosenberg, J., *et al.* (2008) A pilot dose-escalation study of the effects of nordihydroguareacetic acid on hormone and prostate specific antigen levels in patients with relapsed prostate cancer. *BJU International*, 101, 436–439.
Ryu, M.J., Cho, M., Song, J.Y., *et al.* (2008) Natural derivatives of curcumin attenuate the Wnt/beta-catenin pathway through down-regulation of the transcriptional coactivator p300. *Biochemical and Biophysical Research Communications*, 377, 1304–1308.
Saini, S., Majid, S. & Dahiya, R. (2010) Diet, microRNAs and prostate cancer. *Pharmaceutical Research*, 27, 1014–1026.
Salvador, L.A. & Luesch, H. (2012) Discovery and mechanism of natural products as modulators of histone acetylation. *Current Drug Targets*, 13, 1029–1047.
Sato, N., Yamakawa, N., Masuda, M., *et al.* (2011) Genome-wide DNA methylation analysis reveals phytoestrogen modification of promoter methylation patterns during embryonic stem cell differentiation. *PloS One*, 6, e19278.
Sauve, A.A., Wolberger, C., Schramm, V.L. & Boeke, J.D. (2006) The biochemistry of sirtuins. *Annual Review of Biochemistry*, 75, 435–465.
Scoccianti, C., Ricceri, F., Ferrari, P., *et al.* (2011) Methylation patterns in sentinel genes in peripheral blood cells of heavy smokers: influence of cruciferous vegetables in an intervention study. *Epigenetics*, 6, 1114–1119.
Seong, A.R., Yoo, J.Y., Choi, K., *et al.* (2011) Delphinidin, a specific inhibitor of histone acetyltransferase, suppresses inflammatory signaling via prevention of NF-kappaB acetylation in fibroblast-like synoviocyte MH7A cells. *Biochemical and Biophysical Research Communications*, 410, 581–586.
Sethi, S., Li, Y. & Sarkar, F.H. (2013) Regulating MiRNA by natural agents as a new strategy for cancer treatment. *Current Drug Targets*, 14, 1167–1174.
Shah, M.S., Davidson, L.A. & Chapkin, R.S. (2012) Mechanistic insights into the role of microRNAs in cancer: influence of nutrient crosstalk. *Frontiers in Genetics*, 3, 305.
Shu, L., Khor, T.O., Lee, J.H., *et al.* (2011) Epigenetic CpG demethylation of the promoter and reactivation of the expression of neurog1 by curcumin in prostate LNCaP cells. *American Association of Pharmaceutical Scientists (AAPS) Journal*, 13, 606–614.
Siddiqui, I.A., Asim, M., Hafeez, B.B., *et al.* (2011) Green tea polyphenol EGCG blunts androgen receptor function in prostate cancer. *FASEB Journal*, 25, 1198–1207.
Singh, V., Sharma, P. & Capalash, N. (2013) DNA methyltransferase-1 inhibitors as epigenetic therapy for cancer. *Current Cancer Drug Targets*, 13, 379–399.
Spurling, C.C., Suhl, J.A., Boucher, N., *et al.* (2008) The short chain fatty acid butyrate induces promoter demethylation and reactivation of RARbeta2 in colon cancer cells. *Nutrition and Cancer*, 60, 692–702.
Stefanska, B., Rudnicka, K., Bednarek, A. & Fabianowska-Majewska, K. (2010) Hypomethylation and induction of retinoic acid receptor beta 2 by concurrent action of adenosine analogues and natural compounds in breast cancer cells. *European Journal of Pharmacology*, 638, 47–53.
Stefanska, B., Salame, P., Bednarek, A. & Fabianowska-Majewska, K. (2012) Comparative effects of retinoic acid, vitamin D and resveratrol alone and in combination with adenosine analogues on methylation and expression of phosphatase and tensin homologue tumour suppressor gene in breast cancer cells. *British Journal of Nutrition*, 107, 781–790.

Steiner, C., Arnould, S., Scalbert, A. & Manach, C. (2008) Isoflavones and the prevention of breast and prostate cancer: new perspectives opened by nutrigenomics. *British Journal of Nutrition*, 99E(Suppl. 1), ES78–ES108.

Stoner, G.D., Wang, L.S., Zikri, N., *et al.* (2007) Cancer prevention with freeze-dried berries and berry components. *Seminars in Cancer Biology*, 17, 403–410.

Stresemann, C., Brueckner, B., Musch, T., *et al.* (2006) Functional diversity of DNA methyltransferase inhibitors in human cancer cell lines. *Cancer Research*, 66, 2794–2800.

Stunkel, W. & Campbell, R.M. (2011) Sirtuin 1 (SIRT1): the misunderstood HDAC. *Journal of Biomolecular Screening*, 16, 1153–1169.

Sun, Y., Jiang, X., Chen, S. & Price, B.D. (2006) Inhibition of histone acetyltransferase activity by anacardic acid sensitizes tumor cells to ionizing radiation. *FEBS Letters*, 580, 4353–4356.

Sun, Q., Cong, R., Yan, H., *et al.* (2009) Genistein inhibits growth of human uveal melanoma cells and affects microRNA-27a and target gene expression. *Oncology Reports*, 22, 563–567.

Sung, B., Pandey, M.K., Ahn, K.S., *et al.* (2008) Anacardic acid (6–nonadecyl salicylic acid), an inhibitor of histone acetyltransferase, suppresses expression of nuclear factor-kappaB-regulated gene products involved in cell survival, proliferation, invasion, and inflammation through inhibition of the inhibitory subunit of nuclear factor-kappaBalpha kinase, leading to potentiation of apoptosis. *Blood*, 111, 4880–4891.

Surh, Y.J. & Chun, K.S. (2007) Cancer chemopreventive effects of curcumin. *Advances in Experimental Medicine and Biology*, 595, 149–172.

Suzuki, T. & Miyata, N. (2006) Epigenetic control using natural products and synthetic molecules. *Current Medicinal Chemistry*, 13, 935–958.

Tamayo, C. & Diamond, S. (2007) Review of clinical trials evaluating safety and efficacy of milk thistle (Silybum marianum [L.] Gaertn.). *Integrative Cancer Therapies*, 6, 146–157.

Tan, S., Wang, C., Lu, C., *et al.* (2009) Quercetin is able to demethylate the p16INK4a gene promoter. *Chemotherapy*, 55, 6–10.

Tang, W.Y., Newbold, R., Mardilovich, K., *et al.* (2008a) Persistent hypomethylation in the promoter of nucleosomal binding protein 1 (Nsbp1) correlates with overexpression of Nsbp1 in mouse uteri neonatally exposed to diethylstilbestrol or genistein. *Endocrinology*, 149, 5922–5931.

Tang, Y., Zhao, W., Chen, Y., *et al.* (2008b) Acetylation is indispensable for p53 activation. *Cell*, 133, 612–626.

Taylor, C.K., Levy, R.M., Elliott, J.C. & Burnett, B.P. (2009) The effect of genistein aglycone on cancer and cancer risk: a review of in vitro, preclinical, and clinical studies. *Nutrition Reviews*, 67, 398–415.

Teperino, R., Schoonjans, K. & Auwerx, J. (2010) Histone methyl transferases and demethylases: can they link metabolism and transcription? *Cell Metabolism*, 12, 321–327.

Thakur, V.S., Gupta, K. & Gupta, S. (2012a) Green tea polyphenols causes cell cycle arrest and apoptosis in prostate cancer cells by suppressing class I histone deacetylases. *Carcinogenesis*, 33, 377–384.

Thakur, V.S., Gupta, K. & Gupta, S. (2012b) Green tea polyphenols increase p53 transcriptional activity and acetylation by suppressing class I histone deacetylases. *International Journal of Oncology*, 41, 353–361.

Tsang, W.P. & Kwok, T.T. (2010) Epigallocatechin gallate up-regulation of miR-16 and induction of apoptosis in human cancer cells. *Journal of Nutritional Biochemistry*, 21, 140–146.

Upadhyay, A.K. & Cheng, X. (2011) Dynamics of histone lysine methylation: structures of methyl writers and erasers. *Progress in Drug Research*, 67, 107–124.

Vanden Berghe, W. (2012) Epigenetic impact of dietary polyphenols in cancer chemoprevention: Lifelong remodeling of our epigenomes. *Pharmacological Research*, 65, 565–576.

Vanhees, K., Coort, S., Ruijters, E.J., *et al.* (2011) Epigenetics: prenatal exposure to genistein leaves a permanent signature on the hematopoietic lineage. *FASEB Journal*, 25, 797–807.

Vardi, A., Bosviel, R., Rabiau, N., *et al.* (2010) Soy phytoestrogens modify DNA methylation of GSTP1, RASSF1A, EPH2 and BRCA1 promoter in prostate cancer cells. *In Vivo*, 24, 393–400.

Vargas, A.J. & Burd, R. (2010) Hormesis and synergy: pathways and mechanisms of quercetin in cancer prevention and management. *Nutrition Reviews*, 68, 418–428.

Vilas-Zornoza, A., Agirre, X., Martin-Palanco, V., *et al.* (2011) Frequent and simultaneous epigenetic inactivation of TP53 pathway genes in acute lymphoblastic leukemia. *PloS One*, 6, e17012.

Volate, S.R., Muga, S.J., Issa, A.Y., *et al.* (2009) Epigenetic modulation of the retinoid X receptor alpha by green tea in the azoxymethane-Apc Min/+ mouse model of intestinal cancer. *Molecular Carcinogenesis*, 48, 920–933.

Wang, Z. & Chen, H. (2010) Genistein increases gene expression by demethylation of WNT5a promoter in colon cancer cell line SW1116. *Anticancer Research*, 30, 4537–4545.

Wang, L.S. & Stoner, G.D. (2008) Anthocyanins and their role in cancer prevention. *Cancer Letters*, 269, 281–290.

Wang, X., Zhu, H., Snieder, H., *et al.* (2010) Obesity related methylation changes in DNA of peripheral blood leukocytes. *BMC Medicine*, 8, 87.

Wang, H., Bian, S. & Yang, C.S. (2011a) Green tea polyphenol EGCG suppresses lung cancer cell growth through upregulating miR-210 expression caused by stabilizing HIF-1alpha. *Carcinogenesis*, 32, 1881–1889.

Wang, L.S., Arnold, M., Huang, Y.W., *et al.* (2011b) Modulation of genetic and epigenetic biomarkers of colorectal cancer in humans by black raspberries: a phase I pilot study. *Clinical Cancer Research*, 17, 598–610.

Wang, H., Li, Q. & Chen, H. (2012) Genistein affects histone modifications on Dickkopf-related protein 1 (DKK1) gene in SW480 human colon cancer cell line. *PloS One*, 7, e40955.

Wang, Y., Li, Y., Liu, X. & Cho, W.C. (2013) Genetic and epigenetic studies for determining molecular targets of natural product anticancer agents. *Current Cancer Drug Targets*, 13, 506–518.

Weischenfeldt, J., Simon, R., Feuerbach, L., *et al.* (2013) Integrative genomic analyses reveal an androgen-driven somatic alteration landscape in early-onset prostate cancer. *Cancer Cell*, 23, 159–170.

West, L.E. & Gozani, O. (2011) Regulation of p53 function by lysine methylation. *Epigenomics*, 3, 361–369.

Winter, J., Jung, S., Keller, S., *et al.* (2009) Many roads to maturity: microRNA biogenesis pathways and their regulation. *Nature Cell Biology*, 11, 228–234.

Wong, C.P., Nguyen, L.P., Noh, S.K., *et al.* (2011) Induction of regulatory T cells by green tea polyphenol EGCG. *Immunology Letters*, 139, 7–13.

Yang, C.S., Wang, X., Lu, G. & Picinich, S.C. (2009a) Cancer prevention by tea: animal studies, molecular mechanisms and human relevance. *Nature Reviews: Cancer*, 9, 429–439.

Yang, X.D., Lamb, A. & Chen, L.F. (2009b) Methylation, a new epigenetic mark for protein stability. *Epigenetics*, 4, 429–433.

Yang, Y., Liu, J., Li, X. & Li, J.C. (2012) PCDH17 gene promoter demethylation and cell cycle arrest by genistein in gastric cancer. *Histology and Histopathology*, 27, 217–224.

Yi, B., Piazza, G.A., Su, X. & Xi, Y. (2013) MicroRNA and cancer chemoprevention. *Cancer Prevention Research (Philadelphia, Pa.)*, 6, 401–409.

Yoo, J. & Medina-Franco, J.L. (2011) Homology modeling, docking and structure-based pharmacophore of inhibitors of DNA methyltransferase. *Journal of Computer-Aided Molecular Design*, 25, 555–567.

Yuasa, Y., Nagasaki, H., Akiyama, Y., *et al.* (2009) DNA methylation status is inversely correlated with green tea intake and physical activity in gastric cancer patients. *International Journal of Cancer*, 124, 2677–2682.

Yun, J.M., Jialal, I. & Devaraj, S. (2011) Epigenetic regulation of high glucose-induced proinflammatory cytokine production in monocytes by curcumin. *Journal of Nutritional Biochemistry*, 22, 450–458.

Zaman, M.S., Shahryari, V., Deng, G., *et al.* (2012) Up-regulation of microRNA-21 correlates with lower kidney cancer survival. *PloS One*, 7, e31060.

Chapter 7

# Discovery of Polyphenol-Based Drugs for Cancer Prevention and Treatment: The Tumor Proteasome as a Novel Target

Fathima R. Kona[1], Min Shen[1,2], Di Chen[1], Tak Hang Chan[3], and Q. Ping Dou[4]

[1]*Department of Oncology and Barbara Ann Karmanos Cancer Institute, School of Medicine, Wayne State University, Detroit, MI, USA*

[2]*Department of Pharmacology and Barbara Ann Karmanos Cancer Institute, School of Medicine, Wayne State University, Detroit, MI, USA*

[3]*Department of Chemistry, McGill University, Montreal, QC, Canada*

[4]*Departments of Oncology, Pharmacology, and Pathology and Barbara Ann Karmanos Cancer Institute, School of Medicine, Wayne State University, Detroit, MI, USA*

**Abstract:** Epidemiological evidence suggests that diets rich in vegetables and fruits have the potential to decrease the risk of premature mortality from many clinical conditions, including cancer. However, it is not clearly understood what components and combinations of these dietary products are protective and what their mechanisms of action might be. Plant polyphenols have recently gained much attention as they may play a role in the prevention of chronic diseases. The health effects of green tea are attributed to various polyphenolic compounds, among which EGCG has been scientifically scrutinized and is a promising target of ongoing anticancer research. This chapter summarizes the importance of these polyphenols and discusses the proteasome complex as an important target of polyphenols in cancer prevention and treatment, as well as in overcoming drug resistance.

**Keywords:** polyphenols, epigallocatechin gallate, proteasome, curcumin, genestein, tumor drug resistance

## 7.1 Introduction

Plant polyphenols are secondary metabolites of plants that possess aromatic ring structures with hydroxyl substituents and are essential to plant physiology. The health benefits of plant

*Recent Advances in Polyphenol Research, Volume 4*, First Edition.
Edited by Annalisa Romani, Vincenzo Lattanzio, and Stéphane Quideau.

polyphenols are widely supported by results from epidemiological, cell-culture, animal, and clinical studies, although the exact molecular mechanisms of action have not been well characterized.

This chapter will summarize recent advances in the field of plant polyphenols, represented mainly by green-tea polyphenols (GTPs), in terms of biological effects, cancer-related studies, and drug development. One of the major molecular targets for GTPs is the tumor proteasome, whose activity can be inhibited potently and specifically by the formation of an ester bond with a GTP (e.g. (−)-epigallocatechin gallate (EGCG) and its synthetic analogs). However, methylation of GTPs under physiological conditions decreases their proteasome inhibitory activity, leading to decreased cancer-preventive effects of tea consumption. Because EGCG is unstable under physiological conditions, its prodrug has been developed to improve its stability. The peracetate-protected, or prodrug, form of EGCG (a patented product) increases the bioavailability, stability, and proteasomal inhibitory and anticancer activities of EGCG in human breast, prostate, and colon cancer cultures and mouse xenografts, suggesting its potential for cancer prevention and treatment. Other plant polyphenols with proteasomal inhibitory activity will also be discussed in this chapter.

Tumor drug resistance, which can occur through different mechanisms, is one of the major limiting factors of current chemotherapies. Up-regulated proteasomal activity in tumor cells has been suggested as a major factor in the development of tumor drug resistance through several different mechanisms, including evading apoptosis, increasing drug efflux, relieving drug-induced oxidative stress, and enhancing DNA repair. In this chapter, we will discuss how plant polyphenols can aid in overcoming drug resistance *in vitro* and *in vivo*.

Since most plant polyphenols can be administrated orally, which provides advantages in clinical trials, the efficacy of plant polyphenols in chemoprevention and chemotherapy has been widely investigated. Results from epidemiological and clinical studies have suggested that green tea consumption has promising effects against the development of some types of cancer, such as prostate cancer, without inducing major toxicities. However, data obtained from the chemotherapeutic application of green-tea extracts as a treatment for established solid tumors are not impressive. On the other hand, several clinical trials combining chemotherapeutics with plant polyphenols as a sensitizer have resulted in encouraging outcomes and warrant further investigation.

In summary, current evidence from laboratories and clinics suggests that plant polyphenols and their analogs can be used in the future for chemoprevention and chemosensitization/radiosensitization, but not necessarily for chemotherapy.

## 7.2 Secondary metabolites of plants

Plant metabolites are a large class of compounds comprising many diverse chemicals. In the pool of plant metabolites, only a few are involved in "primary" metabolic pathways; the remainder are termed "secondary" metabolites. Initially, the secondary metabolites were considered non-essential compounds, with levels being either environmentally induced or genetically controlled in plants. While primary metabolites are essential for the basic catabolic and anabolic processes required for cellular maintenance and proliferation, secondary metabolites are the compounds that are required for plant survival and are an integral part of plant development (Fraenkel, 1959).

The ability to produce secondary metabolites has been conserved in plant evolution, as such metabolites are essential for their specific needs. Secondary metabolites are essential for defense against microbes and viruses. Some protect against UV radiation, some act as signal compounds to attract seed-dispersing animals, and some are essential for pollination. Thhrough natural selection, plants have thus adopted many of them for their diverse biological activities (Hattenschwiler & Vitousek, 2000). Of the estimated 60 000 genes in the plant genome, 15–20% encode enzymes essential for secondary metabolism. While animals have developed immune and nervous systems to detect and respond to danger, plants have developed a rich repertoire of complex natural products that hold adaptive significance for their survival. By producing and conserving these secondary metabolites, plants are able to kill pathogens and protect themselves.

The secondary metabolites of plants are classified into several groups. Plant polyphenols are the most widely distributed and ubiquitously present in the plant kingdom. The term "polyphenol" can be defined chemically as a substance that possesses at least two aromatic rings with hydroxyl substituents. A large number of these plant phenolic metabolites are small molecules with no tanning action. As a general rule, the term "plant polyphenols" should refer to secondary natural metabolites arising biogenetically from the shikimate or phenylpropanoid pathway or the "polyketide" acetate/malonate pathway, or both, producing monomeric and polymeric phenols (Quideau *et al.*, 2011). Such plant phenolics with only few hydroxyl groups are soluble in ether, chloroform, ethyl acetate, methanol, and ethanol. All of these compounds exhibit intense absorption in the UV region. In recent years, the view that plant polyphenols are one of nature's unwanted products has changed, and it is now clear that they play an important role in plant physiology and ecology. Recent biomedical research has also revealed that dietary phenolics play an important role as antioxidants with free-radical scavenging properties, as well as in the prevention of many chronic diseases (Kutchan, 2001).

## 7.3 Plant polyphenols and their analogs

### *7.3.1 Classification and bioavailability of plant polyphenols*

As secondary metabolites of plants, polyphenols protect them from UV radiation and pathogen assault (Ross & Kasum, 2002; Manach *et al.*, 2004). These compounds constitute an important component of our daily diet, and are found in fruits, vegetables, and beverages. Plant polyphenols can be divided into three classes based on their chemical structure: phenolic acids, flavonoids, and other non-flavonoid polyphenols (Rossi *et al.*, 2008), and their physiological action is strongly influenced by their chemical structure. For example, many isoflavones have pseudo-hormonal properties due to their structural similarity to estradiol, and they are consequently classified as phyto-estrogens (Ross & Kasum, 2002; Manach *et al.*, 2004). Phenolic acids are a class of substances containing one or more hydroxyl functions and a carboxylic acid function at the benzene ring. Two subclasses of phenolic acid can be distinguished as derivatives of benzoic acid, whose polymer forms hydrolyzable tannins (either gallotannins or ellagitannins), and as derivatives of cinnamic acid, whose polymer forms lignins (Manach *et al.*, 2004). Flavonoids are the largest class of polyphenols, with a common structure of two aromatic rings linked through three carbons ($C_6$–$C_3$–$C_6$) (Ross & Kasum, 2002). They can be further divided into six subclasses: flavones, flavonones, flavonols, flavanonols, flavanols, and isoflavones (Narayana *et al.*, 2001). The two most common non-flavonoid polyphenols are curcumin and resveratrol (Rossi *et al.*, 2008). The plant polyphenol structures, representative compounds, and dietary sources of each class are summarized in Table 7.1.

Plant polyphenols are mainly absorbed in the small intestine or colon. Many of these substances are present in the form of esters, glycosides, or polymers, which cannot be absorbed directly; they must be hydrolyzed by intestinal enzymes or colonic microflora before absorption (Manach *et al.*, 2004; D'Archivio *et al.*, 2007). Once absorbed, polyphenols are subject to intestinal and hepatic conjugations, mainly methylation, sulfation, and glucuronidation. An enzyme like *O*-methyltransferase (OMT) catalyzes the transfer of a methyl group from S-adenosyl-L-methionine (SAM) to polyphenols that have an *O*-phenolic moiety. The methylation occurs predominantly in the $3'$ position of the polyphenols, with a minor proportion in the $4'$ position. Sulfotransferase (SULT) catalyzes the transfer of a sulfate moiety from $3'$-phosphoadenosine-$5'$-phosphosulfate to a hydroxyl group on various substrates, including polyphenols, while UDP-glucuronosyltransferase (UGT) catalyzes the transfer of a glucuronic acid from UDP-glucuronic acid to its substrates, including polyphenols (Manach *et al.*, 2004; D'Archivio *et al.*, 2007). Not only do plant polyphenols protect plants from UV radiation and pathogen assault, they also protect humans against a variety of diseases, particularly cardiovascular disease and neoplastic diseases. Reported pharmacological effects include antioxidant, antibacterial, antifungal, antiviral, anti-inflammatory, and antineoplastic (Manach *et al.*, 2004; D'Archivio *et al.*, 2007). These versatile bioactivities are achieved through modulation (by either activation or inhibition) of multiple intracellular targets and signaling pathways, including reactive oxygen species, nuclear receptors, transcription factors, NF-$\kappa$B and PI3K/Akt pathways, and the ubiquitin–proteasome system (Aggarwal *et al.*, 2004; Garg *et al.*, 2005). There are many plants that are specifically studied for their polyphenol constituents, and tea is one of the most important.

### 7.3.2 Tea and tea polyphenols

Tea has been consumed for thousands of years and is the most consumed beverage in the world, after water, with one-third of the world drinking it regularly. Tea was used for medicinal purposes as a drink in ancient China. There are three different types of tea, which are made by processing the *Camellia sinensis* plant to different extents of fermentation: green, black, and oolong. The approximate composition of the world's tea consumption is 78% black tea (in Western countries), 20% green tea (in Asian countries), and 2% oolong tea (mainly in China). The oxidation of simple polyphenols occurs through fermentation, forming more complex condensed polyphenols (converting catechins to theaflavins and thearubigins). Green tea is the least fermented type and has been most investigated for

**Table 7.1** Classes and subclasses of plant polyphenols.

| Class | Backbone | Representative compounds | Dietary sources |
|---|---|---|---|
| **Phenolic acid** | | | |
| | 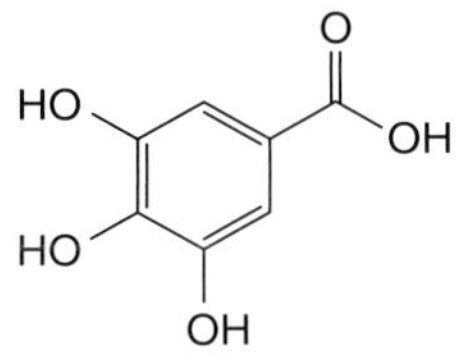 | Gallic acid | Blackberry |
| | 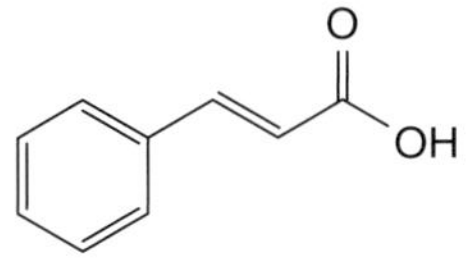 | Cinnamic acid | Cinnamon |
| **Flavonoids** | | | |
| Flavones | 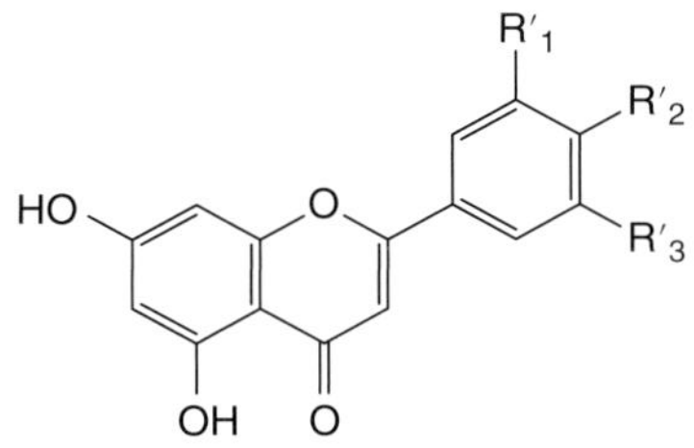 | Luteolin ($R'_1$=$R'_2$=OH, $R'_3$=H)<br>Apigenin ($R'_1$=$R'_3$=H, $R'_2$=OH)<br>Chrysin ($R'_1$=$R'_2$=$R'_3$=H) | Parsley |

(continued overleaf)

**Table 7.1** (*continued*)

| Class | Backbone | Representative compounds | Dietary sources |
|---|---|---|---|
| Flavonones | 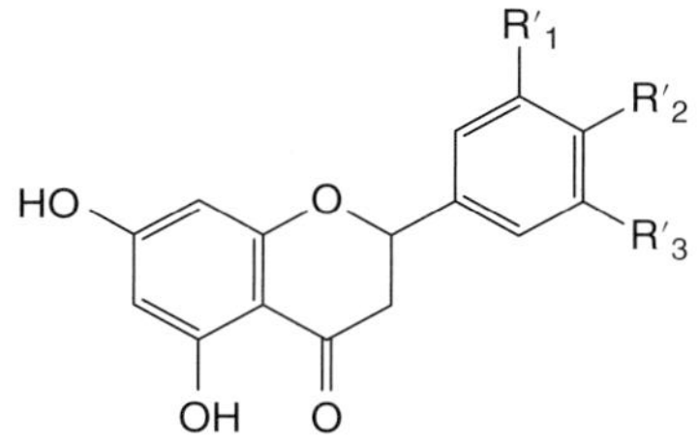 | Eriodictyol ($R'_1$=$R'_2$=OH, $R'_3$=H)<br>Naringenin ($R'_1$=$R'_3$=H, $R'_2$=OH) | Orange juice, grapefruit juice |
| Flavonols | 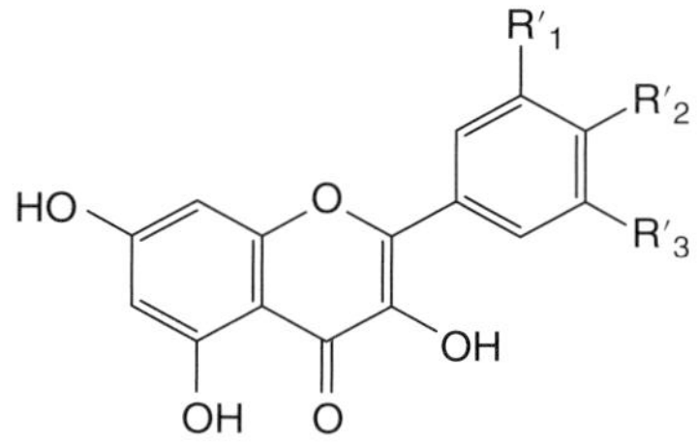 | Quercetin ($R'_1$=$R'_2$=OH, $R'_3$=H)<br>Kaempferol ($R'_1$=$R'_3$=H, $R'_2$=OH)<br>Myricetin ($R'_1$=$R'_2$=$R'_3$=OH) | Black and green tea, red onion |
| Flavanonols | 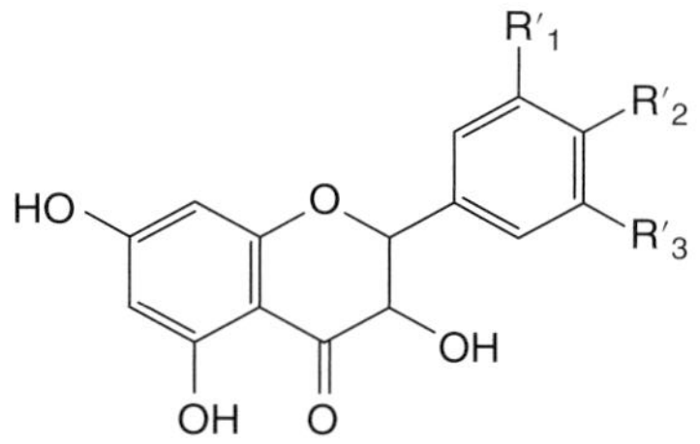 | | |

| | | | |
|---|---|---|---|
| Flavanols | 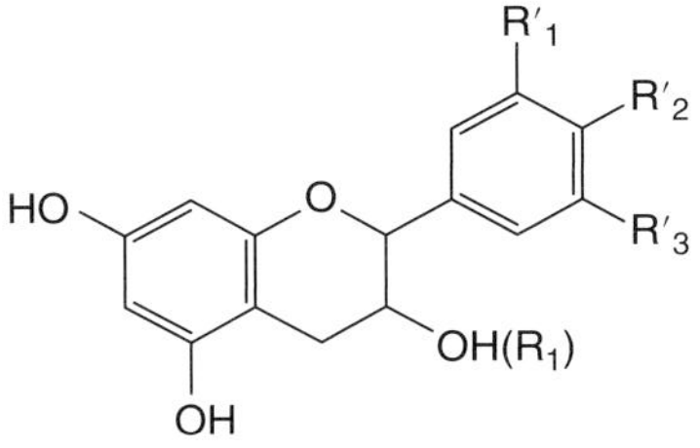 | EC ($R'_1$=$R'_2$=OH, $R'_3$=H; $R_1$=H)<br>ECG ($R'_1$=$R'_2$=OH, $R'_3$=H; $R_1$=gallate)<br>EGC ($R'_1$=$R'_2$= $R'_3$=OH; $R_1$=H)<br>EGCG ($R'_1$=$R'_2$= $R'_3$=OH; $R_1$=gallate) | Green tea |
| Isoflavones | 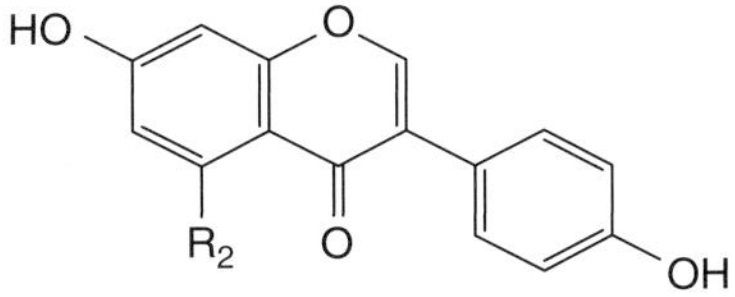 | Genistein ($R_2$=OH)<br>Daidzin ($R_2$=H) | Soybeans |
| **Non-flavonoid polyphenols** | | | |
| | 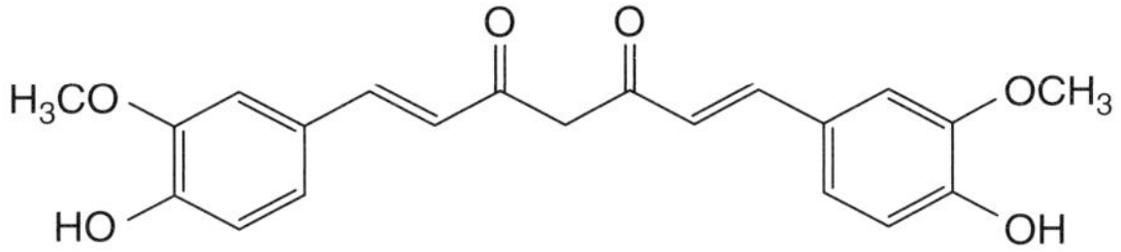 | Curcumin | Turmeric |
| | 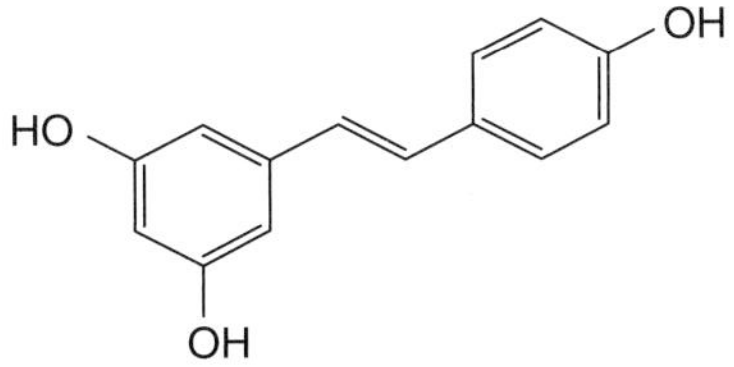 | Resveratrol | Red wine, grapes |

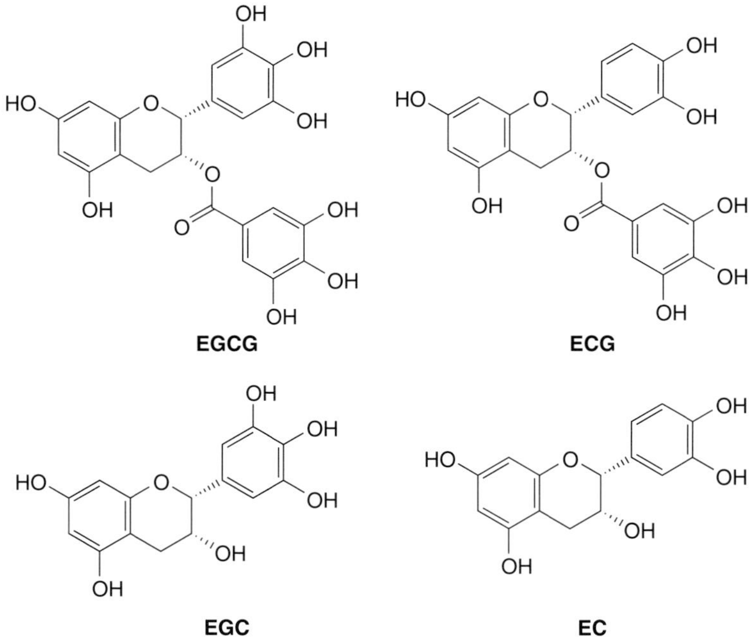


**Fig. 7.1** Chemical structures of (−)-epigallocatechin-3-gallate (EGCG), (−)-epigallocatechin (EGC), (−)-epicatechin-3-gallate (ECG), and (−)-epicatechin (EC).

its benefits to health. Although clinical evidence is still not clear, epidemiological and laboratory data suggest that drinking green tea may reduce the incidence of many types of cancer. Cohort and case-control studies have revealed the benefits of green tea consumption in breast, colon, pancreas, lung, and ovarian cancers, as well as in preventing recurrence in stage I–II breast-cancer patients. The major polyphenols in green tea, referred to as catechins, are EGCG, (−)-epigallocatechin (EGC), (−)-epicatechin-3-gallate (ECG), and (−)-epicatechin (EC) (Dou *et al.*, 2008) (Fig. 7.1). Of these, EGCG is the most abundant and biologically active, and it has been extensively studied for its biological activities and cellular targets. The others are much less abundant and are pharmacologically inactive.

Laboratory studies have shown that EGCG exerts antioxidant and anti-inflammatory activities and induces apoptosis in prostate cancer. It has also been shown to inhibit metallo- and serine proteases involved in tumor invasion and metastasis. EGCG can alter the endocrine system, reducing both 17 $\beta$-estradiol and insulin-like growth factor 1. Despite abundant research, the biological activities of green tea and EGCG are still unclear, but one major mechanism of action has been shown to occur through inhibition of the tumor proteasome.

## *7.3.3 Targeting of the tumor proteasome by tea polyphenols*

One potentially important cellular target of EGCG is the proteasome (Dou *et al.*, 2008). More than 80% of intracellular protein degradation is performed by the ubiquitin–proteasome system. Proteins marked for degradation with polyubiquitin chains are hydrolyzed by the proteasome in an ATP-dependent manner. In contrast to the lysosome, which is mainly responsible for degrading extracellular and transmembrane proteins, the proteasome is in charge of degrading intracellular proteins that are aberrantly folded or normally short-lived. A large and growing body of evidence indicates that the proteasome affects cell-cycle progression partially by regulating the turnover of cyclins and cyclin-dependent kinase inhibitors (CIP/KIP family). Inhibition of proteasome function could result in cell-cycle arrest (Adams, 2004). In addition, the ubiquitin–proteasome system can regulate apoptotic activity by affecting the Bcl-2 family, NF-$\kappa$B family, and p53-MDM2 complexes, among others (Jesenberger & Jentsch, 2002; Adams, 2004). Therefore, the proteasome is crucial for cell survival and proliferation.

The proteasome is a 26S multiprotein complex consisting of a 20S core particle associated with one or two 19S regulatory particles. The 20S particle of the proteasome is a cylindrical structure consisting of four stacked heptameric rings composed of two different types of subunit, $\alpha$ and $\beta$. The two $\alpha$ rings are structural in nature, whereas the two $\beta$ rings are predominantly catalytic (Fig. 7.2). In each $\beta$ ring, the $\beta 1$, $\beta 2$, and $\beta 5$ subunits are catalytic and have three distinct specificities for substrates, although they share a common mechanism. The three catalytic activities are chymotrypsin-like, trypsin-like, and peptidyl-glutamyl peptide-hydrolyzing (PHGH)-like (Fig. 7.2). Proteasome activity can be either up-regulated or down-regulated under various *in vivo* conditions; in many cancers, for example, proteasome activity is up-regulated by cellular oncogenic factors. Enhanced proteasome activity in turn promotes the degradation of tumor-suppressor proteins, resulting in cancer cell survival and proliferation, as well as the development of drug resistance (Adams, 2004; Voorhees & Orlowski, 2006). On the other hand, proteasome activity can be suppressed by several endogenous inhibitors, as well as various exogenous inhibitors, including synthetic compounds such as bortezomib (the first therapeutic proteasome inhibitor) and many natural products, including plant polyphenols (Adams, 2004; Voorhees & Orlowski, 2006). Other inhibitors of the proteasome are epoxy-ketones such as carfilzomib and natural products such as lactacystin, which is related to NP1-0052 (Fig. 7.2). It has been shown that plant polyphenols modulate proteasome activity in drug-sensitive and drug-resistant cancer cells (Yang *et al.*, 2008, 2009).

### 7.3.4 EGCG analogs as proteasome inhibitors

Both naturally occurring (−)-EGCG and its synthetic enantiomer (+)-EGCG are able to potently, specifically, and irreversibly inhibit the chymotrypsin-like activity of proteasomal $\beta 5$ subunits *in vitro* ($IC_{50}$ = 86–194 nM) and *in vivo* (1–10 $\mu$M) (Nam *et al.*, 2001a; Smith *et al.*, 2002). EGCG is able to interact with not only the $\beta 5$ subunit in the constitutive proteasome but also the $\beta 5$i subunit in the interferon-$\gamma$-inducible immune-proteasome, with even higher affinity (Mozzicafreddo *et al.*, 2009). Like another proteasome inhibitor, $\beta$-lactone, molecular-docking and structure–activity relationship (SAR) analyses suggest that the ester carbon linking the C and D rings of EGCG is important in mediating its proteasome-inhibitory activity, which could be subjected to nucleophilic attack by the hydroxy group of the catalytic Thr1 residue in the $\beta 5$ subunit (Nam *et al.*, 2001a). In

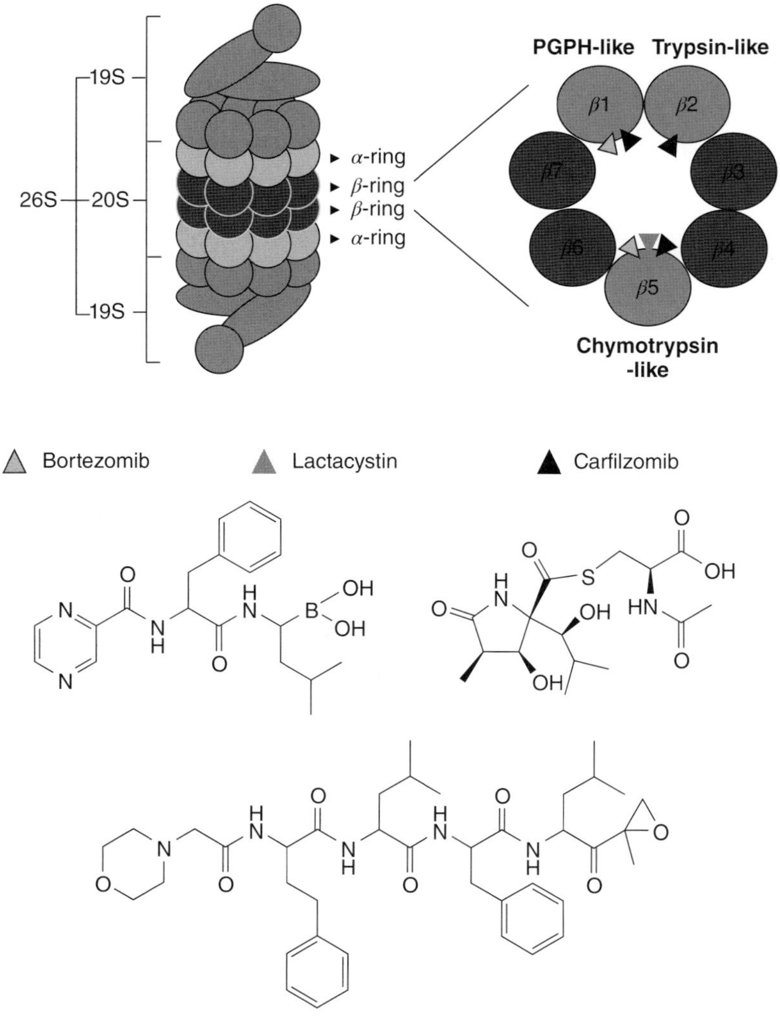


**Fig. 7.2** Schematic structure of the 26S proteasome and representative proteasome inhibitors. The 26S proteasome is composed of a 20S core proteasome and 19S regulatory particles. The $\alpha$- and $\beta$-ring subunits form the 20S core, among which $\beta 1$, $\beta 2$, and $\beta 5$ are mainly responsible for the catalytic activity of the proteasome. The drugs bortezomib, lactacystin, and carfilzomib bind to different $\beta$ subunits to inhibit proteasomal activity.

addition, the A ring of EGCG mimics the Tyr residue and binds to the hydrophobic S1 pocket of the $\beta 5$ subunit (Smith *et al.*, 2004). The B and D rings are oriented towards the aqueous phase, and the number of hydroxy groups on them is positively correlated with proteasome-inhibitory activity (Wan *et al.*, 2005). However, the vicinal trihydroxy groups of EGCG not only contribute to its proteasome-inhibitory activity but also cause its susceptibility to auto-oxidation, even under physiological conditions (Sang *et al.*, 2006). Besides the auto-oxidation issue, methylation also impedes the application of EGCG. As mentioned previously, once absorbed, EGCG is readily methylated by catechol-*O*-methyltransferase (COMT) at the 4′ and 4″ positions (Lu *et al.*, 2003). COMT is a mammalian OMT mainly involved in the inactivation of catecholamine neurotransmitters. Methylation of EGCG significantly affects its binding to the proteasomal $\beta 5$ subunit; as the number of methyl groups increases, the inhibitory potency decreases (Landis-Piwowar *et al.*, 2007b). Thus, cancer cells with lower COMT activity are more vulnerable to EGCG-mediated proteasome inhibition and apoptosis induction than those with higher COMT activity. In an effort to increase the stability of EGCG, a series of analogs with various replacements of the hydroxyl groups on benzene rings were synthesized and examined.

Sanchez-del-Campo *et al.* (2008) developed a new EGCG analog, 3-*O*-(3, 4, 5-trimethoxybenzoyl)-(−)-epicatechin (TMECG), which showed significant antiproliferative activity against several cancer cell lines, especially melanoma. They reported that TMECG binds effectively to human dihydrofolate reductase and down-regulates folate-cycle gene expression in melanoma cells. It was found that the 3-galloyl moiety was essential for its biological activity. It was suggested that the observed biological effects were a result of disruption of the folate cycle by TMECG and thus that TMECG could be of clinical value in cancer therapy (Table 7.2).

### *7.3.4.1 Peracetate and other prodrugs of EGCG*

In our laboratory, we synthesized several (−)-EGCG analogs with acetyl protected hydroxy groups as prodrugs (Lam *et al.*, 2004). We reported that these acetylated synthetic analogs were much more potent than natural (−)-EGCG in inhibiting the proteasome in cultured tumor cells (Kuhn *et al.*, 2005). In sharp contrast, these protected analogs had little to no effect on human normal and nontransformed cells (Kuhn *et al.*, 2005).

In our follow-up study, we synthesized (−)-EGCG analogs with one or more hydroxy groups eliminated from the B and/or D rings, as well as their putative prodrugs with hydroxy groups protected by peracetate, which can be hydrolyzed by cellular cytosolic esterases (Landis-Piwowar *et al.*, 2005). Structure–activity-relationship analysis of these unprotected and protected compounds indicated that removing hydroxy groups from either the B or the D ring led to diminished proteasome-inhibitory activity *in vitro*, whereas in cultured tumor cells only the protected analogs could potently inhibit the proteasome activity. Furthermore, these protected analogs were tumor cell-specific in inducing apoptosis. The superior efficacy of the protected (−)-EGCG analogs indicated that they act as prodrugs and can be converted into natural active polyphenols in intact tumor cells. These data also suggest that the B-ring/D-ring peracetate-protected (−)-EGCG analogs have great potential to be developed into novel anticancer and cancer-preventive agents (Landis-Piwowar *et al.*, 2005).

These prodrugs lacked inhibitory activity towards purified 20S proteasome but displayed increased proteasome-inhibitory activity in intact leukemic cells compared to the parent compound (Landis-Piwowar *et al.*, 2007b). Additionally, pro-EGCG (Table 7.2) inhibited the proteasome, suppressed tumor growth, and induced apoptosis in cultured MDA-MB-231 breast cancer cells more effectively than natural (−)-EGCG (Landis-Piwowar *et al.*, 2007a). An *in vivo* study using triple-negative breast cancer MDA-MB-231 mouse xenografts treated with pro-EGCG or (−)-EGCG showed significant tumor growth ablation (56%), proteasome inhibition, and apoptosis induction from pro-EGCG as compared to (−)-EGCG (Landis-Piwowar *et al.*, 2007a).

**Table 7.2** EGCG analogs.

| EGCG analog | Chemical structure | Mode of action |
|---|---|---|
| EGCG-peracetate (pro-EGCG) | 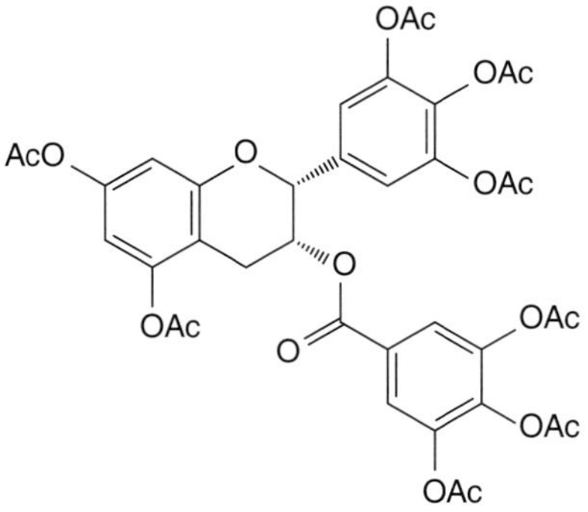  | Significant tumor-growth ablation (56%), proteasome inhibition, and apoptosis induction compared to (−)-EGCG (Landis-Piwowar *et al.*, 2007a) |
| Fluoro substituent of pro-EGCG (pro-F-EGCG) | 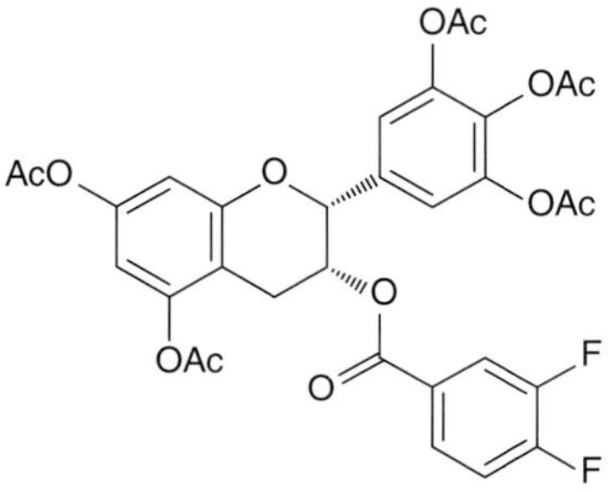  | Proteasome inhibition and apoptotic cell death in MDA-MB-231 tumors (Yang *et al.*, 2010) |
| *Para*-amino substituent of EGCG | 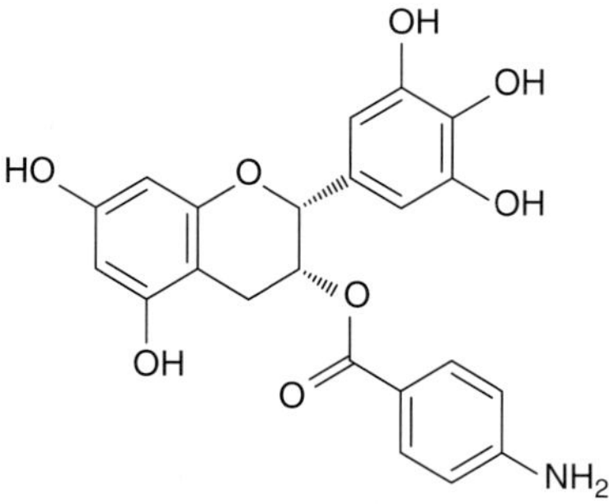  | Inhibition of breast-cancer cell growth *in vitro*, similar to (−)-EGCG and hypoxia- and serum starvation-induced production of VEGF mRNA in breast-cancer cells (Osanai *et al.*, 2007) |
| Trimethoxy substituent of ECG (TMECG) | 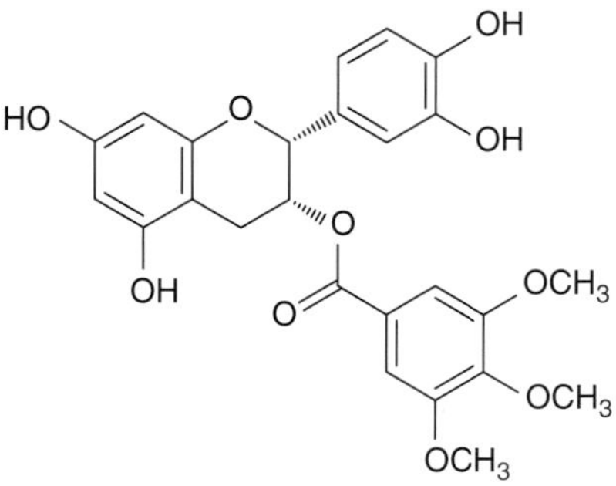  | Decreased proteasome inhibition compared to (−)-EGCG (Sanchez-del-Campo *et al.*, 2008) |

**Table 7.2** (*continued*)

| EGCG analog | Chemical structure | Mode of action |
|---|---|---|
| Bis-gallate derivative A | 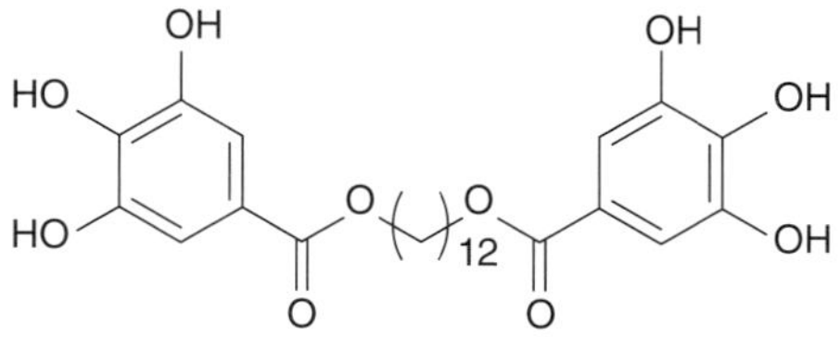 | Similar potency against the proteasome as that of (−)-EGCG (Dodo *et al.*, 2009) |
| Bis-gallate derivative B | 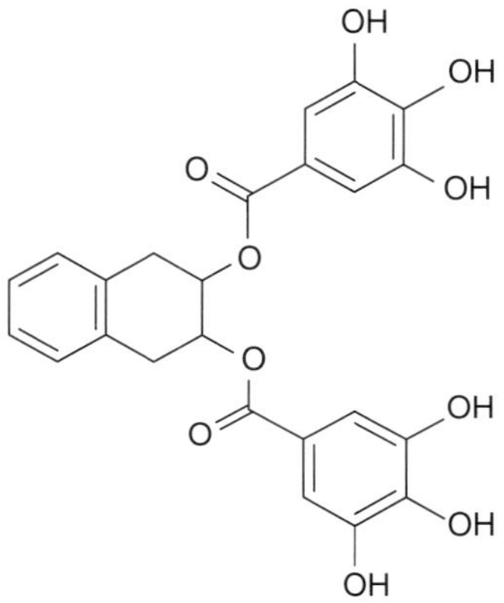 | Similar potency against the proteasome as that of (−)-EGCG (Dodo *et al.*, 2009) |
| COMT-resistant EGCG analogs | 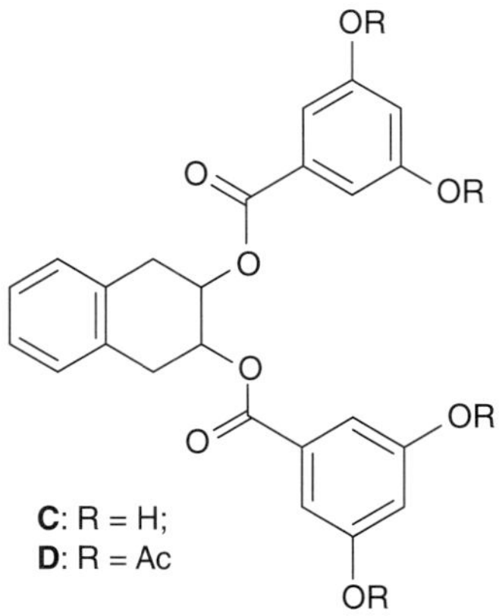 **C**: R = H; **D**: R = Ac | More active than pro-EGCG in MDA-MB-231 cells (Dou *et al.*, 2008) |
| AMPK activation by EGCG analogs | 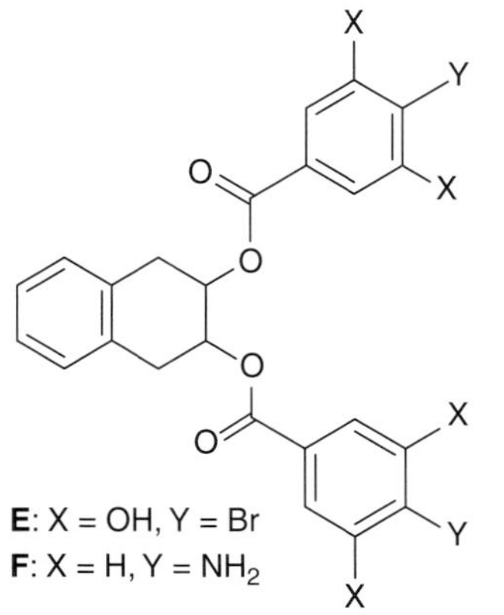 **E**: X = OH, Y = Br **F**: X = H, Y = $NH_2$ | More potent than EGCG in AMPK activation (Chen *et al.*, 2012) |

It has been established that in cultured tumor cells the acetate-protected analogs were capable of potently inhibiting proteasomal chymotrypsin-like activity by as much as 97% (Landis-Piwowar *et al.*, 2005). Furthermore, when compared to (−)-EGCG, protected analogs exhibited greater potency towards inhibiting cell proliferation and inducing apoptosis in human leukemic, prostate, breast, and simian virus 40-transformed cells (Kuhn *et al.*, 2005). Importantly, the protected analogs were nontoxic towards human normal and nontransformed cells (Kuhn *et al.*, 2005). Lee *et al.* (2008) investigated the inhibitory effects of (−)-EGCG and a synthetic derivative, consisting of the peracetate of (−)-EGCG [pro-(−)-EGCG], towards the growth of androgen-independent prostate cancer *in vivo*. They also reported that (−)-EGCG and pro-(−)-EGCG compounds suppressed the growth of CWR22R xenografts without causing any detectable side effects in nude mice. Suppression of tumor growth correlated with decreased serum prostrate-specific antigen (PSA) levels reduced angiogenesis and increased apoptosis *in vivo*. These results suggest that pro-(−)-EGCG may be a more stable and useful analog for increasing therapeutic efficacy in androgen-independent prostate cancer.

Matsumura *et al.* (2008) used a lipase-catalyzed transesterification method to synthesize EGCG monoesters modified with butanoyl (EGCG-C4), octanoyl (EGCG-C8), and palmitoyl groups (EGCG-C16) and subsequently investigated their antitumor activities *in vitro* and *in vivo*. The *in vitro* antitumor activities of EGCG-monoester derivatives increased in an alkyl chain length-dependent manner. EGCG-C16 inhibited cell proliferation and induced apoptosis in the presence of catalase by inhibiting the phosphorylation of the epidermal growth factor receptor (EGFR). EGCG-C16 was consistently found to suppress tumor growth *in vivo* in colorectal tumor-bearing mice in comparison to untreated control, vector control (DMSO), and EGCG.

### *7.3.4.2 Fluoro-substituted EGCG analogs*

In order to improve upon the stability and potency of (−)-EGCG analogs, we synthesized fluoro- analogs by replacing the hydroxy groups from the D ring with one or two fluorine atoms. Prodrugs of fluoro-substituted (−)-EGCG with fluorine at the meta-position on the phenyl ring or difluoro-substituted (−)-EGCG at both meta- and para-positions on the phenyl ring (pro-F-EGCG; Table 7.2) were synthesized (Yang *et al.*, 2010). We reported that the two novel fluoro-substituted (−)-EGCG analogs inhibited tumor growth with potencies similar to that of pro-EGCG, which has improved potency over natural (−)-EGCG in human breast-cancer MDA-MB-231 xenografts. Fluoro-substituted (−)-EGCG analogs exhibited proteasome inhibition and apoptotic cell death in MDA-MB-231 tumors, suggesting that the proteasome might be one of the key cellular targets and that its inhibition may be partially responsible for the observed antitumor activity (Yang *et al.*, 2010).

### *7.3.4.3 Para-amino substituent on the D ring*

(−)-EGCG analogs containing a para-amino group on the D ring (*p*-$NH_2$-EGCG; Table 7.2) in place of the hydroxyl groups were synthesized and tested for their proteasome-inhibitory activities (Osanai *et al.*, 2007). The *O*-acetylated (−)-EGCG analogs possessing a *p*-$NH_2$

or *p*-NHBoc (Boc; tert-butoxycarbonyl) in the D ring could inhibit the tumor-cellular proteasomal activity with potency similar to natural (−)-EGCG and (−)-EGCG peracetate. According to this study, the acetylated amino analogs of GTPs have potential to be developed into novel anticancer agents, since the presence of the para-amino substituent can limit biotransformation reactions and increase the bioavailability of the compound.

Waleh *et al.* (2005) studied other synthetic D-ring analogs of (−)-EGCG and showed their ability to inhibit breast-cancer cell growth *in vitro* with potencies equivalent to that of (−)-EGCG. In addition, the synthetic analogs were found to inhibit hypoxia- and serum starvation-induced production of VEGF mRNA in breast-cancer cells. Such synthetic analogs are valuable for elucidating the structure–function relationship of (−)-EGCG and identifying relevant mechanisms of the chemopreventive action of (−)-EGCG.

### *7.3.4.4 Bis-galloyl derivatives of EGCG*

The D-ring galloyl group in (−)-EGCG is essential for its biological activity. Bis-galloyl compound A (Table 7.2) has shown significant antiproliferative activity in PC-9 human lung-cancer cells (Dodo *et al.*, 2009). The hydroxyl groups of the galloyl moiety were also found to be essential for cell-growth inhibition, indicating that the galloyl group greatly contributes to its bioactivities. There was a strong antiproliferative effect of alkyl gallates and gallamide derivatives towards human leukemia HL-60 cells through induction of apoptosis (Dodo *et al.*, 2009). Various bisgallate and bisgallamide derivatives were synthesized, and their antiproliferative activity was tested against HL-60 cells. In gallate ester derivatives, the addition of another galloyl group showed little to no effect towards antiproliferative activity (Dodo *et al.*, 2009).

The fact that both enantiomers of EGCG are nearly equipotent in their inhibition of the proteasome suggests that the active site may be pseudo-symmetric and bind equally well to both enantiomers (Smith *et al.*, 2002). We have synthesized the bis-gallate of tetrahydronaphth-2, 3-diol (compound B; Table 7.2) and its analogs as the symmetrical analogs of EGCG and examined their activities (Kazi *et al.*, 2004). Compound B was found to be nearly as active as (−)-EGCG, suggesting that the benzene A ring is critical for binding to the active site of the proteasome.

### *7.3.4.5 Methylation-resistant (−)-EGCG analogs*

Biotransformation reactions such as *O*-methylation by COMT, an enzyme ubiquitously present in humans, may reduce the cancer-preventive effects of catechins (Wu *et al.*, 2003). In MDA-MB-231 triple-negative breast-cancer cells containing high COMT activity, the methylation-resistant prodrug D of analog C (Table 7.2) was more active than the prodrugs of (−)-EGCG and its analogs, due to the fact that analog C is not a substrate of COMT (Huo *et al.*, 2010). These results suggest that high COMT activity could methylate and inactivate (−)-EGCG, resulting in an attenuation of its health benefits in patients with high COMT levels (Landis-Piwowar *et al.*, 2007b). In the course of our studies, we synthesized nine different methylated catechins that are potential biomethylated metabolites of tea catechins (Wan *et al.*, 2006). Upon investigation, we found that the addition of a methyl group on the B or D rings of (−)-EGCG or (−)-ECG led to decreased proteasome inhibition. Furthermore, as the number of methyl groups attached increased, a compensatory decrease in proteasome inhibitory potency was observed (Dou *et al.*, 2008).

## 7.3.5 Other molecular targets of tea polyphenols

(−)-EGCG also has been found to induce cell-cycle arrest and apoptosis in several cancer cell lines, including lung, colon, prostate, and skin cancer, with minimal effect on normal cells (Ahmad *et al.*, 1997; Khan *et al.*, 2006). (−)-EGCG can exist in various chemical conformations which allow it to act as a chemical chaperone that can interact with various biological molecules, such as DNA, RNA, lipids, and proteins. (−)-EGCG has been shown to inhibit the expression of the anti-apoptotic proteins Bcl-2 and Bcl-$X_L$, while increasing the expression of pro-apoptotic Bax and Bak proteins (Qin *et al.*, 2007). Recent studies using NMR spectroscopy have revealed that GTPs can directly bind to the BH3 domain of anti-apoptotic Bcl-2-family proteins and mediate their cancer-preventive effects through induction of apoptosis (Leone *et al.*, 2003). *In vitro* studies have shown that (−)-EGCG can induce apoptotic cell death in patients with chronic lymphocytic leukemia (CLL) (Shanafelt *et al.*, 2006). Molecular-modeling studies have shown that (−)-EGCG can target biological molecules, such as the tumor proteasome (Nam *et al.*, 2001a) and urokinase plasminogen activator (uPA), a matrix-remodeling protein, and thereby prevent tumor metastasis (Jankun & Skrzypczak-Jankun, 2001).

Britschgi *et al.* (2010) reported in acute promyelocytic leukemia (APL) that (−)-EGCG treatment resulted in a significant increase in death-associated protein kinase 2 (DAPK2) levels and 67 kDa laminin receptor expression, associated with increased cell death. We previously reported that (−)-EGCG potently and specifically inhibited the chymotrypsin-like activity of the proteasome under *in vitro* and *in vivo* conditions. We found that (−)-EGCG induced tumor cell-growth arrest in the $G_1$ phase of the cell cycle, while having minimal effect on normal, nontransformed cells (Smith *et al.*, 2002). We were the first to report, using *in silico* docking methods, that the ester bond within (−)-EGCG played a critical role in its inhibition of the proteasome. We also reported that the GTP analogs (+)-EGCG and (+)-GCG potently and specifically inhibited the chymotrypsin-like activity of purified 20S proteasome and 26S proteasome in tumor-cell lysates, while Bn-(+)-EGCG failed to do so (Smith *et al.*, 2002). Treatment of leukemic Jurkat T or prostate-cancer LNCaP cells with either (+)-EGCG or (+)-GCG resulted in accumulation of p27 and I$\kappa$B-$\alpha$ proteins, associated with an increased $G_1$ population. (+)-EGCG treatment also caused accumulation of pro-apoptotic Bax protein and induced apoptosis in LNCaP cells expressing high basal levels of Bax, but not in DU145 prostate-cancer cells with low Bax expression. Finally, synthetic GTPs were able to significantly inhibit colony formation in LNCaP cancer cells.

### 7.3.5.1 AMPK activation

AMP-activated protein kinase (AMPK) controls processes related to tumor development, including cell-cycle progression, protein synthesis, cell growth, and survival. It has received intensive attention recently as a possible anticancer target. It has been reported that some flavonoids, such as quercetin, genistein, and the GTP epigallocatechin gallate (EGCG), can activate AMPK and inhibit cancer cell growth. We recently demonstrated that the synthetic EGCG analogs E and F (Table 7.2) were more potent AMPK activators than EGCG (Chen *et al.*, 2012). Activation of AMPK by these EGCG analogs resulted in inhibition of cell proliferation, up-regulation of the cyclin-dependent kinase inhibitor p21, down-regulation of the mTOR pathway, and suppression of stem-cell populations in human breast-cancer cells. Our findings suggest that novel potent and specific AMPK activators that have potential to be used clinically for anticancer therapy can be discovered from tea polyphenols and their analogs.

### 7.3.6 Proteasome inhibitory action of other plant polyphenols

After the discovery of the proteasome-inhibitory activity of EGCG, several other dietary flavonoids, such as genistein (Kazi *et al.*, 2003), luteolin (Chen *et al.*, 2007a), apigenin (Chen *et al.*, 2005a, 2007a,b), chrysin (Chen *et al.*, 2007a), and quercetin (Chen *et al.*, 2005a), were subsequently found to possess proteasome-inhibitory activity as well. The median inhibitory concentration ($IC_{50}$) of these flavonoids for the inhibition of chymotrypsin-like activity in both purified 20S and intracellular 26S proteasome can be found in the original reports. Docking studies suggested that the carbonyl carbon on the C ring might be the site of nucleophilic attack by the hydroxy group of Thr1 in the proteasomal $\beta5$ subunit (Chen *et al.*, 2005a). SAR analysis found that flavonoids with a hydroxylated B ring and/or unsaturated C ring tended to be potent proteasome inhibitors and tumor-cell apoptosis inducers (Chen *et al.*, 2007a). Flavonoids with a pyrocatechol structure such as quercetin could also be methylated by COMT, but less efficiently than EGCG (van Duursen *et al.*, 2004; Chen *et al.*, 2005b), and although methylated flavonoids have been suggested to be more stable metabolically than unmethylated ones, their proteasome-inhibitory activity and cytotoxicity against tumor cells are largely compromised (Landis-Piwowar *et al.*, 2008). Like EGCG, those flavonoids with a vicinal diol structure could also antagonize the anticancer effect of bortezomib. Thus, all flavonoids with pyrocatechol or pyrogallol structures are potentially able to chemically react with the boronic acid group in bortezomib and other boronic acid-based proteasome inhibitors, rendering them ineffectual (Liu *et al.*, 2008).

Curcumin is a non-flavonoid polyphenol extracted from turmeric, which is commonly used as a spice. Due to the observation that curcumin-induced cell death is very similar to that of proteasome inhibitors, the proteasome-inhibitory ability of curcumin was proposed and verified in proliferating neuroblastoma cells (Jana *et al.*, 2004). Proteasome inhibition, as one of the mechanisms responsible for curcumin-induced cell death, was further supported by local and metastatic colon-cancer models both *in vitro* and *in vivo* (Milacic *et al.*, 2008). Consistent with previous findings that the ester-bond carbon of $\beta$-lactone and EGCG and the carbonyl carbon of flavonoids confer their proteasome-inhibitory activities, the two carbonyl carbons on curcumin were found to be susceptible to nucleophilic attack by the hydroxy group of Thr1 in the proteasomal $\beta5$ subunit (Milacic *et al.*, 2008). Although most polyphenols tend to be water-soluble, curcumin itself has poor solubility in water, leading to its poor bioavailability in tissues remote from the gastrointestinal tract, which largely compromises its anticancer effect (Wan *et al.*, 2010). Among the different analogs synthesized to improve its water solubility, amino-acid conjugates of curcumin were found to be soluble in water and showed a potent antiproliferative effect in cell models, indicating their potential as curcumin prodrugs (Wan *et al.*, 2010). In contrast to the discussed flavonoids containing pyrocatechol or pyrogallol structures, curcumin was reported to potentiate the effect of bortezomib against human multiple myeloma in a nude mouse model (Sung *et al.*, 2009).

Tannic acid (TA), which is a hydrolyzable tannin, is composed of either gallotannins or ellagitannins. Upon hydrolysis, gallotannin yields a D glucose and six to nine gallate groups, which link either to each other or the glucose core through ester bonds. Since ester bond-containing tea polyphenols are potent proteasome inhibitors, the ability of TA to inhibit proteasome activity was tested and verified in purified 20S proteasome (Nam *et al.*, 2001b) and cellular 26S proteasome in different cell types (Nam *et al.*, 2001b; Marienfeld *et al.*, 2003), as well as in mouse xenograft models (Marienfeld *et al.*, 2003). TA inhibits proteasome function, resulting in increased p27 and Bax expression and impaired cell-cycle progression (Nam *et al.*, 2001b).

## 7.4 Natural polyphenols in reversal of drug resistance

### *7.4.1 Mechanisms of tumor drug resistance*

Tumor drug resistance is one of the major limiting factors of current chemotherapies and can occur through different mechanisms, some of which are intrinsic, while others are acquired after chemotherapy. Acquired resistance is particularly critical, as tumors not only become resistant to the drugs originally used to treat them but may also become cross-resistant to other drugs with different mechanisms of action (Longley & Johnston, 2005). In recent years, proteasome inhibition has become an important topic in cancer treatment and drug resistance. The molecular mechanisms usually involved in drug resistance include evasion of apoptosis, increase in drug efflux, increase in drug detoxification, alterations in drug targets, loss of dependence on drug targets, and enhanced processing of drug-induced damage (Longley & Johnston, 2005). Research on the ubiquitin–proteasome system has revealed that this system participates extensively in the development of drug resistance (Landis-Piwowar *et al.*, 2006), and inhibition of this pathway has been shown to result in accumulation of tumor-suppressor and pro-apoptotic proteins, making it a viable target for cancer therapy.

### *7.4.2 The ubiquitin–proteasome pathway in drug resistance*

Many chemotherapeutic drugs exert antitumor effects via induction of apoptosis in tumor cells. Drug-resistant tumor cells evade apoptosis by enhancing survival signals and/or suppressing apoptotic signals. Up-regulated proteasome activity in tumor cells helps them evade apoptosis on at least two levels. First, increased proteasome activity can facilitate the NF-$\kappa$B survival pathway in tumor cells by accelerating I-$\kappa$B turnover. Proteasome inhibitors have consistently been reported to enhance tumor response to conventional chemotherapy in a colon-cancer model (Cusack *et al.*, 2006). Increased proteasome activity in cancer cells can also interfere with the balance between pro- and anti-apoptotic Bcl-2 family members by accelerating Bax degradation and increasing Bcl-2 expression. Therefore, proteasome inhibition not only stabilizes/activates Bax but also down-regulates Bcl-2 expression (Li & Dou, 2000; Fennell *et al.*, 2008; Zhang *et al.*, 2008). Second, drug efflux is mediated by ATP-binding cassette (ABC) transporters, including ABCB1 (MDR1/P-glycoprotein/P-gp), ABCC subfamily members (MRPs), and ABCG2 (BCRP) (Choudhuri & Klaassen, 2006; Dean, 2009). Newly synthesized P-gp is immature (only core-glycosylated, associated with chaperones) and trapped in the endoplasmic reticulum (ER) (Loo *et al.*, 1999). The proteasome is believed to be involved in its maturation (full glycosylation), as proteasome inhibition significantly suppresses P-gp maturation and exportation from the ER to the plasma membrane (Loo *et al.*, 1999). In agreement with this, proteasome inhibitors successfully reversed the multidrug-resistance phenotype in P-gp over-expressing cancer cells through down-regulation of P-gp expression (Fujita *et al.*, 2005; Zhang *et al.*, 2008).

In addition to involvement in these major mechanisms of drug resistance, increased proteasome activity also facilitates the elimination of damaged proteins due to oxidation caused by cytotoxic drugs, which relieves drug-induced oxidative stress (Lee *et al.*, 2001). Proteasome inhibition leads to considerable accumulation of oxidative proteins (Lee *et al.*, 2001) and hyper-sensitizes tumor cells to oxidative stress inducers (Lev *et al.*, 2006).

The ubiquitin–proteasome system also plays a crucial role in regulating DNA repair. The ubiquitin–proteasome system is involved in at least four DNA-repair pathways (Motegi *et al.*, 2009), namely nucleotide excision repair (Wang *et al.*, 2005), homologous recombination (Murakawa *et al.*, 2007), post-replication repair (Podlaska *et al.*, 2003), and the Fanconi anemia pathway (Jacquemont *et al.*, 2007). Proteasome inhibitors significantly potentiated the cytotoxicity of cisplatin, a DNA damaging agent that causes the formation of cisplatin-DNA adducts, by preventing nucleotide excision repair (Mimnaugh *et al.*, 2000).

### *7.4.3 EGCG and overcoming drug resistance*

Studies have demonstrated that EGCG has chemosensitizing effects *in vitro* and *in vivo* through inhibition of the drug-efflux transporter P-gp. For example, Kitagawa *et al.* (2004) reported that catechins increased cellular accumulation of P-gp substrates in the order EGCG > ECG > EGC, while EC had no effect on substrate accumulation in P-gp over-expressing KB-C2 human cervical epidermal carcinoma cells. Chemosensitization by EGCG through inhibition of P-gp expression/activity has also been validated in Caco-2 human intestinal adenocarcinoma cells, MCF-7 human breast-cancer cells (Farabegoli *et al.*, 2010), and xenograft mouse models using either doxorubicin-resistant KB-A1 cells (Mei *et al.*, 2004; Qian *et al.*, 2005) or doxorubicin-resistant BEL-7404/DOX hepatocarcinoma cells (Liang *et al.*, 2010). A study with tamoxifen-resistant MCF-7 cells also found that EGCG was able to inhibit the activity of the BCRP transporter (Farabegoli *et al.*, 2010). EGCG has been reported to enhance the sensitivity of glioblastoma to temozolomide

(Chen *et al.*, 2011), of prostate carcinoma to doxorubicin (Stearns *et al.*, 2010), and of breast carcinoma to paclitaxel (Luo *et al.*, 2010) in corresponding ectopic or orthotopic xenograft mouse models. Furthermore, the peracetate-protected prodrug of EGCG also exhibits a chemosensitizing effect in the treatment of leukemia cells, augmenting the efficacy of the conventional chemotherapeutics daunorubicin and cytosine arabinoside (Davenport *et al.*, 2010). A recent study reported that polymer-based nanoparticles of the polyphenols EGCG and theaflavin retained biological effectiveness with an over 20-fold dose advantage compared to EGCG/theaflavin in exerting anticancer effects, as well as enhancing the potency of cisplatin in several different tumor cell types (Singh *et al.*, 2011).

### 7.4.4 Genistein and overcoming drug resistance

Genistein has been studied as a promising agent for chemoprevention, chemotherapy, and chemosensitization. Genistein enhances the efficacy of chemotherapy by inhibiting survival signals and/or enhancing apoptotic signals. Dr Sarkar's group reported that genistein pre-treatment inactivated NF-$\kappa$B and contributed to increased growth inhibition and apoptosis induced by cisplatin, docetaxel, doxorubicin, or gemcitabine in various cancer cells, including prostate, breast, lung, pancreatic, and ovarian cancers (Banerjee *et al.*, 2005; Li *et al.*, 2005; Solomon *et al.*, 2008). Inactivation of Akt/NF-$\kappa$B by genistein not only potentiated the proliferation-inhibiting and apoptosis-inducing effects of arsenic trioxide on human hepatocellular carcinoma cells *in vitro* but also dramatically augmented its suppressive effect on tumor growth and angiogenesis in tumor-bearing nude mice (Ma *et al.*, 2011). Genistein also restored the sensitivity of platinum-resistant epithelial ovarian cancer cells to cisplatin, taxotere, and gemcitabine via the inhibition of NF-$\kappa$B and its target genes (c-IAP1, Bcl-2, Bcl-xL, and survivin) (Solomon *et al.*, 2008). Similarly, genistein also restored sensitivity to oxaliplatin in gemcitabine-resistant pancreatic cancer cells both *in vitro* and *in vivo* (Banerjee *et al.*, 2011). The combination of gemcitabine and genistein enhanced antitumor efficacy in osteosarcoma cells through the same mechanism (abrogating the Akt/NF-$\kappa$B pathway) (Zhang *et al.*, 2010). Genistein also enhanced the cytotoxicity of cisplatin and, to a lesser extent, vincristine in medulloblastoma cells (Khoshyomn *et al.*, 2000). Moreover, in acute myeloid leukemia cells genistein exhibited a synergistic antileukemia effect with cytosine arabinoside in both *in vitro* and *in vivo* models by affecting the MAPK signaling pathway (Shen *et al.*, 2007).

### 7.4.5 Curcumin and overcoming drug resistance

Research with curcumin revealed that it can sensitize tumors to different chemotherapeutic drugs, including doxorubicin, 5-FU, paclitaxel, vincristine, vinorelbine, melphalan butyrate, cisplatin, oxaliplatin, celecoxib, gemcitabine, etoposide, sulfinosine, and thalidomide (Goel & Aggarwal, 2010). Chemosensitization by curcumin has been observed in solid tumors of the lung, liver, pancreas, stomach, colon, head and neck, brain, breast, prostate, bladder, cervix, and ovary, as well as in hematological tumors of multiple myeloma, leukemia, and lymphoma (Goel & Aggarwal, 2010). Curcumin remarkably increased sensitivity to vinblastine/vincristine and reversed multidrug resistance via modulation of P-gp expression and function in vincristine-selected P-gp over-expressing KB-V1 cervical epidermal carcinoma cells (Anuchapreeda *et al.*, 2002) and SGC7901/VCR gastric carcinoma cells (Tang *et al.*, 2005). Curcumin was also shown to sensitize BRCP-expressing human breast-cancer cells to different chemotherapeutic drugs, including mitoxantrone, topotecan, SN-38, and doxorubicin, by down-regulating the function of the BRCP transporter (Chearwae *et al.*, 2006); this was further confirmed in a transgenic mouse model (Shukla *et al.*, 2009). Both curcumin and its major metabolite tetrahydrocurcumin were able to inhibit the efflux function of P-gp, MRP1, and BRCP, and thereby reverse multidrug resistance (Limtrakul *et al.*, 2007).

It has also been found that curcumin is able to improve the sensitivity of drug-resistant cancer cells by suppressing NF-$\kappa$B signaling. Dhandapani *et al.* (2007) reported that curcumin sensitized glioma cells to clinically used chemotherapeutic agents (cisplatin, etoposide, camptothecin, and doxorubicin) via inhibition of AP-1 and NF-$\kappa$B, which correlated with reduced expression of Bcl-2 and IAP family members and some DNA repair enzymes. By blocking the NF-$\kappa$B pathway, curcumin can also potentiate the effect of paclitaxel in breast-cancer xenografts (Kang *et al.*, 2009) and the effect of cisplatin in head and neck squamous-cell carcinoma xenografts (Duarte *et al.*, 2010). Also, in a recent study, Yoon *et al.* (2010) showed for the first time that curcumin effectively induced paraptosis, a new type of nonapoptotic cell death, in malignant breast-cancer cell lines. Proteasomal dysfunction and superoxide anion contributed to those paraptotic changes characterized by vacuolation that begins with physical enlargement of mitochondria and ER. These findings suggest that curcumin could be beneficial in the treatment of apoptosis-resistant tumors (Yoon *et al.*, 2010).

### 7.4.6 Clinical trials using polyphenols and chemotherapy

Studies have been conducted using green-tea extract as monotherapy in cancer patients. Although the oral consumption of green-tea extract was well tolerated in these studies, its anticancer activity was unsatisfactory. For example, a phase I trial in patients with advanced lung cancer suggested that while it is relatively nontoxic at a dose of 3 g/m$^2$ per day, green-tea extract likely has limited activity as a cytotoxic agent (Laurie *et al.*, 2005); a phase II trial suggested that oral administration of green-tea powder (6 g/day) carried limited antineoplastic activity, as defined by a decline in PSA level, among patients with androgen-independent prostate cancer (Jatoi *et al.*, 2003). Therefore, a prospective clinical trial using green tea as a complementary/alternative medicine was carried out in patients with hormone-refractory prostate cancer. However, the results suggested that green tea had minimal clinical activity against hormone-refractory prostate cancer (Choan *et al.*, 2005). In view of the greater efficacy of the prodrugs of EGCG (peracetate-protected or fluoro-substituted) than of EGCG itself in xenograft mouse models of both estrogen-independent human breast cancer and androgen-independent human prostate cancer (Landis-Piwowar *et al.*, 2007a; Lee *et al.*, 2008; Yang *et al.*, 2010), these EGCG prodrugs may be useful in clinical studies.

Despite its excellent safety record, in 2006 the US Food and Drug Administration (FDA) temporarily suspended all human trials of EGCG for additional review of toxicity data, and later allowed trials of Polyphenon E. In a phase I trial in patients with asymptomatic Rai stage 0–II CLL, daily oral consumption of Polyphenon E was well tolerated, and declines in absolute lymphocyte count (ALC) and/or lymphadenopathy were observed in the majority of patients. Based on the encouraging phase I result, a phase II trial using 2 g Polyphenon E twice a day to evaluate its efficacy was recently concluded (Shanafelt *et al.*, 2013). Overall, 29 out of a total of 42 patients (69%) fulfilled the criteria for a biologic response, with either a sustained decline ≥20% in ALC or a reduction ≥30% in the sum of the products of all lymph-node areas at some point during the 6 months of active treatment (Shanafelt *et al.*, 2013).

Genistein has been investigated as a sensitizer to gemcitabine and erlotinib in patients with breast or pancreatic cancer. In a phase II study in the United States, 20 patients with locally advanced or metastatic pancreatic cancer received gemcitabine- and erlotinib-based chemotherapy, along with soy isoflavones (Novasoy 0.531 g twice daily; genistein is the major component). According to the RECIST criteria, one patient had a partial response (PR) and six had stable disease (SD); the 6-month survival rate was 50% and the median survival time was 5.2 months. Since these results were comparable to the outcome observed in gemcitabine/erlotinib bitherapy without genistein, the investigators concluded that the antitumor activity of gemcitabine and erlotinib was not improved by soy isoflavones (El-Rayes *et al.*, 2010).

The results from the prototype genistein are unfavorable, but phenoxodiol, a synthetic derivative of genistein, has been granted "fast track" status by the FDA for development as a chemosensitizer for platinum and taxane drugs used in the treatment of recurrent ovarian cancer (Silasi *et al.*, 2009). A phase III study treating recurrent platinum-resistant ovarian-cancer patients with phenoxodiol (400 mg/8 hours) and carboplatin (weekly) is currently ongoing. Notably, phenoxodiol could be administered not only orally but also via intravenous infusion, which greatly enhanced the bioavailability of the drug. In a phase II study, patients with platinum/taxane-refractory/resistant ovarian cancer received phenoxodiol (intravenously 3 mg/kg on days 1 and 2) and cisplatin or paclitaxel (on day 2) each cycle (7 days per cycle). Among 16 platinum-resistant patients, 3 had PR and 9 had SD. Among 15 taxane-resistant patients, 1 had a complete response, 2 had PR, and 8 had SD. These results, particularly that from the cisplatin–phenoxodiol combination, are quite encouraging and warrant further investigation (Kelly *et al.*, 2011).

Clinical studies using the combination of chemotherapy and curcumin have been conducted around the world. In a phase I/II trial in Japan, 21 patients with gemcitabine-resistant pancreatic cancer received curcumin (8 g/day) in combination with gemcitabine-based chemotherapy (gemcitabine/S-1 bitherapy or gemcitabine monotherapy). No dose-limiting toxicity was observed in the phase I study and oral curcumin 8 g/day was selected for the phase II study. Among 18 evaluable patients, 0 had PR and 5 had SD. These results warrant further investigation into the efficacy of curcumin in combination with gemcitabine-based chemotherapy (Kanai *et al.*, 2011).

In a phase I trial in France, 14 patients with advanced and metastatic breast cancer were treated with docetaxel along with curcumin (dose started from 0.5 g/day for 7 consecutive days and escalated until a dose-limiting toxicity occurred). The maximum tolerated dose of curcumin was 8 g/day, and a dose of 6 g/day in combination with docetaxel was recommended. Among 8 evaluable patients, 5 had PR and 3 had SD. Based on these encouraging efficacy results, a comparative phase II trial is ongoing (Bayet-Robert *et al.*, 2010). Furthermore, a combined regimen of gemcitabine, Celebrex, and curcumin in the treatment of colon and pancreatic cancer has entered phase III trial in Israel.

All of these clinical-trial studies related to plant-based polyphenols and their analogs in the cancer field reveal that they can be used as drugs for chemoprevention and chemosensitization/radiosensitization, but not necessarily for chemotherapy.

## 7.5 Conclusions

The health-beneficial effects of plant polyphenols are widely supported by results from epidemiological, cell-culture, animal, and clinical studies. The bioactivities of plant polyphenols are achieved through modulation of multiple intracellular targets and signaling pathways, the proteasome among them. EGCG, as a representative of plant polyphenols, exhibits potent proteasomal inhibitory and anticancer activities *in vitro* and *in vivo*. Analogs of EGCG resulting from chemical modification improve bioavailability, stability, and biological activity. However, data obtained from the chemotherapeutic application of tea polyphenols as treatment for established solid tumors are not impressive. Since the tumor proteasome has also been found to participate extensively in the development of tumor drug resistance, much effort has been directed towards evaluating the effect of plant polyphenols on overcoming drug resistance. Several clinical trials combining chemotherapies with plant polyphenols as a sensitizer have received encouraging results and warrant further investigation. Therefore, given the evidence from laboratories and clinics, plant polyphenols and their analogs may likely be used for chemoprevention and chemosensitization/ radiosensitization, but not necessarily for chemotherapy, in the future.

## Acknowledgments

This work was partially supported by National Cancer Institute grants (1R01CA20009, 3R01CA120009-04S1 and 5R01CA127258-05) given to Q.P. Dou. We thank Sara Schmitt and Daniela Buac for critical reading of the manuscript.

## References

Adams, J. (2004) The proteasome: a suitable antineoplastic target. *Nature Review Cancer*, 4, 349–360.

Aggarwal, B.B., Takada, Y., Oommen, O.V. (2004) From chemoprevention to chemotherapy: common targets and common goals. *Expert Opinion on Investigational Drugs*, 13, 1327–1338.

Ahmad, N., Feyes, D.K., Nieminen, A.L., *et al.* (1997) Green tea constituent epigallocatechin-3-gallate and induction of apoptosis and cell cycle arrest in human carcinoma cells. *Journal of National Cancer Institute*, 89, 1881–1886.

Anuchapreeda, S., Leechanachai, P., Smith, M.M., *et al.* (2002) Modulation of P-glycoprotein expression and function by curcumin in multidrug-resistant human KB cells. *Biochemical Pharmacology*, 64, 573–582.

Banerjee, S., Zhang, Y., Ali, S., *et al.* (2005) Molecular evidence for increased antitumor activity of gemcitabine by genistein in vitro and in vivo using an orthotopic model of pancreatic cancer. *Cancer Research*, 65, 9064–9072.

Banerjee, S., Kong, D., Azmi, A.S., *et al.* (2011) Restoring sensitivity to oxaliplatin by a novel approach in gemcitabine-resistant pancreatic cancer cells in vitro and in vivo. *International Journal of Cancer*, 128, 1240–1250.

Bayet-Robert, M., Kwiatkowski, F., Leheurteur, M., *et al.* (2010) Phase I dose escalation trial of docetaxel plus curcumin in patients with advanced and metastatic breast cancer. *Cancer Biological Therapy*, 9, 8–14.

Britschgi, A., Simon, H.U., Tobler, A., *et al.* (2010) Epigallocatechin-3-gallate induces cell death in acute myeloid leukaemia cells and supports all-trans retinoic acid-induced neutrophil differentiation via death-associated protein kinase 2. *British Journal of Haematology*, 149, 55–64.

Chearwae, W., Shukla, S., Limtrakul, P., *et al.* (2006) Modulation of the function of the multidrug resistance-linked ATP-binding cassette transporter ABCG2 by the cancer chemopreventive agent curcumin. *Molecular Cancer Therapy*, 5, 1995–2006.

Chen, D., Daniel, K.G., Chen, M.S., *et al.* (2005a) Dietary flavonoids as proteasome inhibitors and apoptosis inducers in human leukemia cells. *Biochemical Pharmacology*, 69, 1421–1432.

Chen, D., Wang, C.Y., Lambert, J.D., *et al.* (2005b) Inhibition of human liver catechol-O-methyltransferase by tea catechins and their metabolites: structure-activity relationship and molecular-modeling studies. *Biochemical Pharmacology*, 69, 1523–1531.

Chen, D., Chen, M.S., Cui, Q.C., *et al.* (2007a) Structure-proteasome-inhibitory activity relationships of dietary flavonoids in human cancer cells. *Front Biosciences*, 12, 1935–1945.

Chen, D., Landis-Piwowar, K.R., Chen, M.S., *et al.* (2007b) Inhibition of proteasome activity by the dietary flavonoid apigenin is associated with growth inhibition in cultured breast cancer cells and xenografts. *Breast Cancer Research*, 9, R80.

Chen, T.C., Wang, W., Golden, E.B., *et al.* (2011) Green tea epigallocatechin gallate enhances therapeutic efficacy of temozolomide in orthotopic mouse glioblastoma models. *Cancer Letters*, 302, 100–108.

Chen, D., Pamu, S., Cui, Q. C., *et al.* (2012) Novel epigallocatechin gallate (EGCG) analogs activate AMP-activated protein kinase pathway and target cancer stem cells. *Bioorganic &. Medicinal Chemistry*, 20, 3031–3037.

Choan, E., Segal, R., Jonker, D., *et al.* (2005) A prospective clinical trial of green tea for hormone refractory prostate cancer: an evaluation of the complementary/alternative therapy approach. *Urologic Oncology*, 23, 108–113.

Choudhuri, S. & Klaassen, C.D. (2006) Structure, function, expression, genomic organization, and single nucleotide polymorphisms of human ABCB1 (MDR1), ABCC (MRP), and ABCG2 (BCRP) efflux transporters. *International Journal of Toxicology*, 25, 231–259.

Cusack, J.C., Jr.,, Liu, R., Xia, L., *et al.* (2006) NPI-0052 enhances tumoricidal response to conventional cancer therapy in a colon cancer model. *Clinical Cancer Research,* 12, 6758–6764.

D'Archivio, M., Filesi, C., Di Benedetto, R., *et al.* (2007) Polyphenols, dietary sources and bioavailability. *Annali dell'Istituto Superiore di Sanita*, 43, 348–361.

Davenport, A., Frezza, M., Shen, M., *et al.* (2010) Celastrol and an EGCG pro-drug exhibit potent chemosensitizing activity in human leukemia cells. *International Journal of Molecular Medicine*, 25, 465–470.

Dean, M. (2009) ABC transporters, drug resistance, and cancer stem cells. *Journal of Mammary Gland Biology and Neoplasia*, 14, 3–9.

Dhandapani, K.M., Mahesh, V.B. & Brann, D.W. (2007) Curcumin suppresses growth and chemoresistance of human glioblastoma cells via AP-1 and NFkappaB transcription factors. *Journal of Neurochemistry*, 102, 522–538.

Dodo, K., Minato, T., Hashimoto, Y. (2009) Structure-activity relationship of bis-galloyl derivatives related to (−)-epigallocatechin gallate. *Chemical and Pharmaceutical Bulletin (Tokyo)*, 57, 190–194.

Dou, Q.P., Landis-Piwowar, K.R., Chen, D., *et al.* (2008) Green tea polyphenols as a natural tumour cell proteasome inhibitor. *Inflammopharmacology*, 16, 208–212.

Duarte, V.M., Han, E., Veena, M.S., *et al.* (2010) Curcumin enhances the effect of cisplatin in suppression of head and neck squamous cell carcinoma via inhibition of IKKbeta protein of the NFkappaB pathway. *Molecular Cancer Therapy*, 9, 2665–2675.

El-Rayes, B., Philip, P., Sarkar, F., *et al.* (2010) A phase II study of isoflavones, erlotinib, and gemcitabine in advanced pancreatic cancer. *Investigational New Drugs*, 1–6.

Farabegoli, F., Papi, A., Bartolini, G., *et al.* (2010) (−)-epigallocatechin-3-gallate downregulates Pg-P and BCRP in a tamoxifen resistant MCF-7 cell line. *Phytomedicine*, 17, 356–362.

Fennell, D.A., Chacko, A. & Mutti, L. (2008) BCL-2 family regulation by the 20S proteasome inhibitor bortezomib. *Oncogene*, 27, 1189–1197.

Fraenkel, G.S. (1959) The raison d'être of secondary plant substances. *Science*, 129, 1466–1470.

Fujita, T., Washio, K., Takabatake, D., *et al.* (2005) Proteasome inhibitors can alter the signaling pathways and attenuate the P-glycoprotein-mediated multidrug resistance. *International Journal of Cancer*, 117, 670–682.

Garg, A.K., Buchholz, T.A. & Aggarwal, B.B. (2005) Chemosensitization and radiosensitization of tumors by plant polyphenols. *Antioxidants & Redox Signaling*, 7, 1630–1647.

Goel, A. & Aggarwal, B.B. (2010) Curcumin, the golden spice from Indian saffron, is a chemosensitizer and radiosensitizer for tumors and chemoprotector and radioprotector for normal organs. *Nutrition and Cancer*, 62, 919–930.

Hattenschwiler, S. & Vitousek, P.M. (2000) The role of polyphenols in terrestrial ecosystem nutrient cycling. *Trends in Ecology& Evolution*, 15, 238–243.

Huo, C., Yang, H., Cui, Q. C., *et al.* (2010) Proteasome inhibition in human breast cancer cells with high catechol-O-methyltransferase activity by green tea polyphenol EGCG analogs. *Bioorganic & Medicinal Chemistry*, 18, 1252–1258.

Jacquemont, C. & Taniguchi, T. (2007) Proteasome function is required for DNA damage response and fanconi anemia pathway activation. *Cancer Research*, 67, 7395–7405.

Jana, N.R., Dikshit, P., Goswami, A., *et al.* (2004) Inhibition of proteasomal function by curcumin induces apoptosis through mitochondrial pathway. *Journal of Biological Chemistry*, 279, 11680–11685.

Jankun, J. & Skrzypczak-Jankun, E. (2001) Binding site of amiloride to urokinase plasminogen activator depends on species. *International Journal of Molecular Medicine*, 8, 365–371.

Jatoi, A., Ellison, N., Burch, P.A., *et al.* (2003) A phase II trial of green tea in the treatment of patients with androgen independent metastatic prostate carcinoma. *Cancer*, 97, 1442–1446.

Jesenberger, V. & Jentsch, S. (2002) Deadly encounter: ubiquitin meets apoptosis. *Nature Review Molecular Cell Biology*, 3, 112–121.

Kanai, M., Yoshimura, K., Asada, M., *et al.* (2011) A phase I/II study of gemcitabine-based chemotherapy plus curcumin for patients with gemcitabine-resistant pancreatic cancer. *Cancer Chemotherapy and Pharmacology*, 68, 157–164.

Kang, H.J., Lee, S.H., Price, J.E., *et al.* (2009) Curcumin suppresses the paclitaxel-induced nuclear factor-kappaB in breast cancer cells and potentiates the growth inhibitory effect of paclitaxel in a breast cancer nude mice model. *Breast Journal*, 15, 223–229.

Kazi, A., Daniel, K.G., Smith, D.M., *et al.* (2003) Inhibition of the proteasome activity, a novel mechanism associated with the tumor cell apoptosis-inducing ability of genistein. *Biochemical Pharmacology*, 66, 965–976.

Kazi, A., Wang, Z., Kumar, N., *et al.* (2004) Structure-activity relationships of synthetic analogs of (−)-epigallocatechin-3-gallate as proteasome inhibitors. *Anti-Cancer Research*, 24, 943–954.

Kelly, M.G., Mor, G., Husband, A., *et al.* (2011) Phase II evaluation of phenoxodiol in combination with cisplatin or paclitaxel in women with platinum/taxane-refractory/resistant epithelial ovarian,

fallopian tube, or primary peritoneal cancers. *International Journal of Gynecological Cancer*, 21, 633–639.
Khan, N., Afaq, F., Saleem, M., *et al.* (2006) Targeting multiple signaling pathways by green tea polyphenol (−)-epigallocatechin-3-gallate. *Cancer Research*, 66, 2500–2505.
Khoshyomn, S., Manske, G.C., Lew, S.M., *et al.* (2000) Synergistic action of genistein and cisplatin on growth inhibition and cytotoxicity of human medulloblastoma cells. *Pediatric Neurosurgery*, 33, 123–131.
Kitagawa, S., Nabekura, T. & Kamiyama, S. (2004) Inhibition of P-glycoprotein function by tea catechins in KB-C2 cells. *Journal of Pharmacy and Pharmacology*, 56, 1001–1005.
Kuhn, D., Lam, W.H., Kazi, A., *et al.* (2005) Synthetic peracetate tea polyphenols as potent proteasome inhibitors and apoptosis inducers in human cancer cells. *Front Biosciences*, 10, 1010–1023.
Kutchan, T.M. (2001) Ecological arsenal and developmental dispatcher. The paradigm of secondary metabolism. *Plant Physiology*, 125, 58–60.
Lam, W.H., Kazi, A., Kuhn, D.J., *et al.* (2004) A potential prodrug for a green tea polyphenol proteasome inhibitor: evaluation of the peracetate ester of (−)-epigallocatechin gallate [(−)-EGCG]. *Bioorganic & Medicinal Chemistry*, 12, 5587–5593.
Landis-Piwowar, K.R., Kuhn, D.J., Wan, S.B., *et al.* (2005) Evaluation of proteasome-inhibitory and apoptosis-inducing potencies of novel (−)-EGCG analogs and their prodrugs. *International Journal of Molecular Medicine*, 15, 735–742.
Landis-Piwowar, K.R., Milacic, V., Chen, D., *et al.* (2006) The proteasome as a potential target for novel anticancer drugs and chemosensitizers. *Drug Resistance Updates*, 9, 263–273.
Landis-Piwowar, K.R., Huo, C., Chen, D., *et al.* (2007a) A novel prodrug of the green tea polyphenol (−)-epigallocatechin-3-gallate as a potential anticancer agent. *Cancer Research*, 67, 4303–4310.
Landis-Piwowar, K.R., Wan, S.B., Wiegand, R.A., *et al.* (2007b) Methylation suppresses the proteasome-inhibitory function of green tea polyphenols. *Journal of Cellular Physiology*, 213, 252–260.
Landis-Piwowar, K.R., Milacic, V. & Dou, Q.P. (2008) Relationship between the methylation status of dietary flavonoids and their growth-inhibitory and apoptosis-inducing activities in human cancer cells. *Journal Cellular Biochemistry*, 105, 514–523.
Laurie, S.A., Miller, V.A., Grant, S.C., *et al.* (2005) Phase I study of green tea extract in patients with advanced lung cancer. *Cancer Chemotherapy and Pharmacology*, 55, 33–38.
Lee, M.H., Hyun, D.H., Jenner, P., *et al.* (2001) Effect of proteasome inhibition on cellular oxidative damage, antioxidant defences and nitric oxide production. *Journal of Neurochemistry*, 78, 32–41.
Lee, S.C., Chan, W.K., Lee, T.W., *et al.* (2008) Effect of a prodrug of the green tea polyphenol (−)-epigallocatechin-3-gallate on the growth of androgen-independent prostate cancer in vivo. *Nutrition and Cancer*, 60, 483–491.
Leone, M., Zhai, D., Sareth, S., *et al.* (2003) Cancer prevention by tea polyphenols is linked to their direct inhibition of antiapoptotic Bcl-2-family proteins. *Cancer Research*, 63, 8118–8121.
Lev, N., Melamed, E. & Offen, D. (2006) Proteasomal inhibition hypersensitizes differentiated neuroblastoma cells to oxidative damage. *Neuroscience Letters*, 399, 27–32.
Li, B. & Dou, Q.P. (2000) Bax degradation by the ubiquitin/proteasome-dependent pathway: involvement in tumor survival and progression. *Proceedings of the National Academy of Sciences USA*, 97, 3850–3855.
Li, Y., Ahmed, F., Ali, S., *et al.* (2005) Inactivation of nuclear factor kappaB by soy isoflavone genistein contributes to increased apoptosis induced by chemotherapeutic agents in human cancer cells. *Cancer Research*, 65, 6934–6942.
Liang, G., Tang, A., Lin, X., *et al.* (2010) Green tea catechins augment the antitumor activity of doxorubicin in an in vivo mouse model for chemoresistant liver cancer. *International Journal of Oncology*, 37, 111–123.
Limtrakul, P., Chearwae, W., Shukla, S., *et al.* (2007) Modulation of function of three ABC drug transporters, P-glycoprotein (ABCB1), mitoxantrone resistance protein (ABCG2) and multidrug resistance protein 1 (ABCC1) by tetrahydrocurcumin, a major metabolite of curcumin. *Molecular and Cellular Biochemistry*, 296, 85–95.

Liu, F.T., Agrawal, S.G., Movasaghi, Z., *et al.* (2008) Dietary flavonoids inhibit the anticancer effects of the proteasome inhibitor bortezomib. *Blood*, 112, 3835–3846.

Longley, D.B. & Johnston, P.G. (2005) Molecular mechanisms of drug resistance. *Journal of Pathology*, 205, 275–292.

Loo, T.W. & Clarke, D.M. (1999) The human multidrug resistance P-glycoprotein is inactive when its maturation is inhibited: potential for a role in cancer chemotherapy. *Federation of American Societies for Experimental Biology (FASEB) Journal*, 13, 1724–1732.

Lu, H., Meng, X. & Yang, C.S. (2003) Enzymology of methylation of tea catechins and inhibition of catechol-O-methyltransferase by (−)-epigallocatechin gallate. *Drug Metabolism and Disposition*, 31, 572–579.

Luo, T., Wang, J., Yin, Y., *et al.* (2010) (−)-epigallocatechin gallate sensitizes breast cancer cells to paclitaxel in a murine model of breast carcinoma. *Breast Cancer Research*, 12, R8.

Ma, Y., Wang, J., Liu, L., *et al.* (2011) Genistein potentiates the effect of arsenic trioxide against human hepatocellular carcinoma: role of Akt and nuclear factor-kappaB. *Cancer Letters*, 301, 75–84.

Manach, C., Scalbert, A., Morand, C., *et al.* (2004) Polyphenols: food sources and bioavailability. *Americal Journal of Clinical Nutritions*, 79, 727–747.

Marienfeld, C., Tadlock, L., Yamagiwa, Y., *et al.* (2003) Inhibition of cholangiocarcinoma growth by tannic acid. *Hepatology*, 37, 1097–1104.

Matsumura, K., Kaihatsu, K., Mori, S., *et al.* (2008) Enhanced antitumor activities of (−)-epigallocatechin-3-O-gallate fatty acid monoester derivatives in vitro and in vivo. *Biochemical Biophysical Research Communications*, 377, 1118–1122.

Mei, Y., Qian, F., Wei, D., *et al.* (2004) Reversal of cancer multidrug resistance by green tea polyphenols. *Journal of Pharmacy and Pharmacology*, 56, 1307–1314.

Milacic, V., Banerjee, S., Landis-Piwowar, K.R., *et al.* (2008) Curcumin inhibits the proteasome activity in human colon cancer cells in vitro and in vivo. *Cancer Research*, 68, 7283–7292.

Mimnaugh, E.G., Yunmbam, M.K., Li, Q., *et al.* (2000) Prevention of cisplatin-DNA adduct repair and potentiation of cisplatin-induced apoptosis in ovarian carcinoma cells by proteasome inhibitors. *Biochemical Pharmacology*, 60, 1343–1354.

Motegi, A., Murakawa, Y. & Takeda, S. (2009) The vital link between the ubiquitin-proteasome pathway and DNA repair: impact on cancer therapy. *Cancer Letters*, 283, 1–9.

Mozzicafreddo, M., Cuccioloni, M., Cecarini, V., *et al.* (2009) Homology modeling and docking analysis of the interaction between polyphenols and mammalian 20S proteasomes. *Journal of Chemical Information & Modeling*, 49, 401–409.

Murakawa, Y., Sonoda, E., Barber, L.J., *et al.* (2007) Inhibitors of the proteasome suppress homologous DNA recombination in mammalian cells. *Cancer Research*, 67, 8536–8543.

Nam, S., Smith, D.M. & Dou, Q.P. (2001a) Ester bond-containing tea polyphenols potently inhibit proteasome activity in vitro and in vivo. *Journal of Biological Chemistry*, 276, 13322–13330.

Nam, S., Smith, D.M. & Dou, Q.P. (2001b) Tannic acid potently inhibits tumor cell proteasome activity, increases p27 and Bax expression, and induces G1 arrest and apoptosis. *Cancer Epidemiol Biomarkers Prevention*, 10, 1083–1088.

Narayana, K.R., Reddy, M.S., Chaluvadi, M., *et al.* (2001) Bioflavonoids classification, pharmacological, biochemical effects and therapeutic potential. *Indian Journal of Pharmacology*, 33, 2–16.

Osanai, K., Landis-Piwowar, K.R., Dou, Q.P., *et al.* (2007) A para-amino substituent on the D-ring of green tea polyphenol epigallocatechin-3-gallate as a novel proteasome inhibitor and cancer cell apoptosis inducer. *Bioorganic Medicinal Chemistry*, 15, 5076–5082.

Podlaska, A., McIntyre, J., Skoneczna, A., *et al.* (2003) The link between 20S proteasome activity and post-replication DNA repair in Saccharomyces cerevisiae. *Molecular Microbiology*, 49, 1321–1332.

Qian, F., Wei, D., Zhang, Q., *et al.* (2005) Modulation of P-glycoprotein function and reversal of multidrug resistance by (−)-epigallocatechin gallate in human cancer cells. *Biomedicine & Pharmacothererapy*, 59, 64–69.

Qin, J., Xie, L.P., Zheng, X.Y., *et al.* (2007) A component of green tea, (−)-epigallocatechin-3-gallate, promotes apoptosis in T24 human bladder cancer cells via modulation of the PI3K/Akt pathway and Bcl-2 family proteins. *Biochemical Biophysical Research Communications*, 354, 852–857.

Quideau, S., Deffieux, D., Douat-Casassus , *et al.* (2011) Plant polyphenols: chemical properties, biological cctivities, and synthesis. *Angewandte Chemie International Edition*, 50, 586–621.

Ross, J.A. & Kasum, C.M. (2002) Dietary flavonoids: bioavailability, metabolic effects, and safety. *Annual Review of Nutrition*, 22, 19–34.

Rossi, L., Mazzitelli, S., Arciello, M., *et al.* (2008) Benefits from dietary polyphenols for brain aging and Alzheimer's disease. *Neurochemical Research*, 33, 2390–2400.

Sanchez-del-Campo, L., Oton, F., Tarraga, A., *et al.* (2008) Synthesis and biological activity of a 3,4,5-trimethoxybenzoyl ester analogue of epicatechin-3-gallate. *Journal of Medicinal Chemistry*, 51, 2018–2026.

Sang, S., Lambert, J.D. & Yang, C.S. (2006) Bioavailability and stability issues in understanding the cancer preventive effects of tea polyphenols. *Journal of the Science of Food and Agriculture*, 86, 2256–2265.

Shanafelt, T.D., Lee, Y.K., Call, T.G., *et al.* (2006) Clinical effects of oral green tea extracts in four patients with low grade B-cell malignancies. *Leukemia Research*, 30, 707–712.

Shanafelt, T.D., Call, T.G., Zent, C.S., *et al.* (2009) Phase I trial of daily oral Polyphenon E in patients with asymptomatic Rai stage 0 to II chronic lymphocytic leukemia. *Journal of Clinical Oncology*, 27, 3808–3814.

Shanafelt, T.D., Call, T.G., Zent, C.S., *et al.* (2013) Phase 2 trial of daily oral Polyphenon E in patients with asymptomatic Rai stage 0 to II chronic lymphocytic leukemia. *Cancer*, 119, 363–370.

Shen, J., Tai, Y.C., Zhou, J., *et al.* (2007) Synergistic antileukemia effect of genistein and chemotherapy in mouse xenograft model and potential mechanism through MAPK signaling. *Experimental Hematology*, 35, 75–83.

Shukla, S., Zaher, H., Hartz, A., *et al.* (2009) Curcumin inhibits the activity of ABCG2/BCRP1, a multidrug resistance-linked ABC drug transporter in mice. *Pharmaceutical Research*, 26, 480–487.

Silasi, D.A., Alvero, A.B., Rutherford, T.J., *et al.* (2009) Phenoxodiol: pharmacology and clinical experience in cancer monotherapy and in combination with chemotherapeutic drugs. *Expert Opinion on Pharmacotherapy*, 10, 1059–1067.

Singh, M., Bhatnagar, P., Srivastava, A.K., *et al.* (2011) Enhancement of cancer chemosensitization potential of cisplatin by tea polyphenols poly(lactide-co-glycolide) nanoparticles. *Journal of Biomedical Nanotechnology*, 7, 202.

Smith, D.M., Wang, Z., Kazi, A., *et al.* (2002) Synthetic analogs of green tea polyphenols as proteasome inhibitors. *Molecular Medicine*, 8, 382–392.

Smith, D.M., Daniel, K.G., Wang, Z., *et al.* (2004) Docking studies and model development of tea polyphenol proteasome inhibitors: applications to rational drug design. *Proteins*, 54, 58–70.

Solomon, L.A., Ali, S., Banerjee, S., *et al.* (2008) Sensitization of ovarian cancer cells to cisplatin by genistein: the role of NF-kappaB. *Journal of Ovarian Research*, 1, 9.

Stearns, M.E., Amatangelo, M.D., Varma, D., *et al.* (2010) Combination therapy with epigallocatechin-3-gallate and doxorubicin in human prostate tumor modeling studies: inhibition of metastatic tumor growth in severe combined immunodeficiency mice. *American Journal of Pathology*, 177, 3169–3179.

Sung, B., Kunnumakkara, A.B., Sethi, G., *et al.* (2009) Curcumin circumvents chemoresistance in vitro and potentiates the effect of thalidomide and bortezomib against human multiple myeloma in nude mice model. *Molecular Cancer Thererapy*, 8, 959–970.

Tang, X.Q., Bi, H., Feng, J.Q., *et al.* (2005) Effect of curcumin on multidrug resistance in resistant human gastric carcinoma cell line SGC7901/VCR. *Acta Pharmacol Sinica*, 26, 1009–1016.

Van Duursen, M.B., Sanderson, J.T., de Jong, P.C., *et al.* (2004) Phytochemicals inhibit catechol-O-methyltransferase activity in cytosolic fractions from healthy human mammary tissues: implications for catechol estrogen-induced DNA damage. *Toxicol Sciences*, 81, 316–324.

Voorhees, P.M. & Orlowski, R.Z. (2006) The proteasome and proteasome inhibitors in cancer therapy. *Annual Review of Pharmacology and Toxicology*, 46, 189–213.

Waleh, N.S., Chao, W.R., Bensari, A., *et al.* (2005) Novel D-ring analog of epigallocatechin-3-gallate inhibits tumor growth and VEGF expression in breast carcinoma cells. *Anticancer Research*, 25, 397–402.

Wan, S.B., Landis-Piwowar, K.R., Kuhn, D.J., *et al.* (2005) Structure-activity study of epi-gallocatechin gallate (EGCG) analogs as proteasome inhibitors. *Bioorganic and Medicinal Chemistry*, 13, 2177–2185.

Wan, S.B., Dou, Q.P. & Chan TH. (2006) Regiospecific and enantioselective synthesis of methylated metabolites of tea catechins. *Tetrahedron*, 62, 5897–5904.

Wan, S.B., Yang, H., Zhou, Z., *et al.* (2010) Evaluation of curcumin acetates and amino acid conjugates as proteasome inhibitors. *International Journal of Molecular Medicine*, 26, 447–455.

Wang, Q.E., Wani, M.A., Chen, J., *et al.* (2005) Cellular ubiquitination and proteasomal functions positively modulate mammalian nucleotide excision repair. *Molecular Carcinogenesis*, 42, 53–64.

Wu, A.H., Tseng, C.C., Van Den Berg, D., *et al.* (2003) Tea intake, COMT genotype, and breast cancer in Asian-American women. *Cancer Research*, 63, 7526–7529.

Yang, H., Landis-Piwowar, K.R., Chen, D., *et al.* (2008) Natural compounds with proteasome inhibitory activity for cancer prevention and treatment. *Current Protein Peptide Sciences*, 9, 227–239.

Yang, H., Zonder, J.A. & Dou, Q.P. (2009) Clinical development of novel proteasome inhibitors for cancer treatment. *Expert Opinion Investigational Drugs*, 18, 957–971.

Yang, H., Sun, D.K., Chen, D., *et al.* (2010) Antitumor activity of novel fluoro-substituted (−)-epigallocatechin-3-gallate analogs. *Cancer Letters*, 292, 48–53.

Yoon, M.J., Kim, E.H., Lim, J.H., *et al.* (2010) Superoxide anion and proteasomal dysfunction contribute to curcumin-induced paraptosis of malignant breast cancer cells. *Free Radical Biology and Medicine*, 48, 713–726.

Zhang, Y., Shi, Y., Li, X., *et al.* (2008) Proteasome inhibitor MG132 reverses multidrug resistance of gastric cancer through enhancing apoptosis and inhibiting P-gp. *Cancer Biology and Therapy*, 7, 540–546.

Zhang, B., Shi, Z.L., Liu, B., *et al.* (2010) Enhanced anticancer effect of gemcitabine by genistein in osteosarcoma: the role of Akt and nuclear factor-kappaB. *Anticancer Drugs*, 21, 288–296.

Chapter 8

# Flavonoid Occurrence, Bioavailability, Metabolism, and Protective Effects in Humans: Focus on Flavan-3-ols and Flavonols

Luca Calani, Margherita Dall'Asta, Renato Bruni, and Daniele Del Rio
*The $\varphi^2$ Laboratory of Phytochemicals in Physiology, LS9 Bioactives and Health Interlab Group, Department of Food Sciences, University of Parma, Parma, Italy*

**Abstract:** Flavonoids are profusely abundant in the human diet and might be responsible for the advantages ascribed to dietary regimens rich in plant foods. Notwithstanding the evidence from epidemiological and cohort studies, an unambiguous and exhaustive pattern has not yet emerged regarding the mechanisms underlying their food-related health properties. It is now increasingly evident that the sole direct action of substances available *in planta* is not sufficient to fully explain the benefits of polyphenol-rich dietary habits. To elucidate this conundrum, a better comprehension of the interplay between human and microbial metabolism and dietary polyphenols is needed. This chapter provides a detailed overview of the current knowledge on bioavailability and *in vivo* biotransformation of two widespread classes of flavonoids: flavonols and flavan-3-ols. According to data collected, these substances are not always endowed with the highest bioactivities *in vivo*, and a deeper knowledge of their metabolites is needed to properly assess their overall contribution to human health and to single out substances to be monitored for the exhaustive profiling of flavonoid bioavailability.

**Keywords:** nutrition, flavan-3-ols, flavonols, metabolism, colonic microbiota

---

*Recent Advances in Polyphenol Research, Volume 4*, First Edition.
Edited by Annalisa Romani, Vincenzo Lattanzio, and Stéphane Quideau.

## 8.1 Introduction

The health effects of a high consumption of fruits and vegetables are attributed to the presence of fiber and antioxidant vitamins (C, E, and $\beta$-carotene), as well as to other organic compounds known as phytochemicals. Among these, antioxidant compounds known as (poly)phenols play a relevant role, as they are found widespread in plant foods and their dietary intake exceeds that of all other classes of phytochemicals and known dietary antioxidants (Scalbert *et al.*, 2005). Polyphenols can be classified into several groups according to their phenolic and polyphenolic skeletons, but an easier classification divides them into two main groups: the flavonoids and the non-flavonoids (Crozier *et al.*, 2009). The first group comprises a large family of compounds, further divided into several subclasses, of which six have a particularly large dietary relevance: flavonols, flavanones, flavones, isoflavones, anthocyanidins, and flavan-3-ols (Crozier *et al.*, 2009). However, only in the last 2 decades have researchers and food manufacturers focused the public's attention on flavonoids as potential candidates for the prevention of several chronic degenerative diseases. This has led to several epidemiological studies evaluating whether the protective effects of vegetables and fruits against several aging-related disorders can be explained by regular intake of flavonoids (Hertog *et al.*, 1993a, 1995; Arts & Hollman, 2005). Among the flavonoids, flavonols and flavan-3-ols are the most investigated subclasses because they are widespread in edible plants, making their dietary intake particularly relevant, along with that of flavanones (Bettuzzi *et al.*, 2006; Knaze *et al.*, 2011; Zamora-Ros *et al.*, 2011). However, in seeking to fully understand the biological and putative beneficial effects of these plant-derived compounds, one cannot ignore their bioavailability and metabolism. Therefore, research into flavonoids began to focus on these factors within the human organism, revealing that these phytochemicals can be extensively metabolized by humans, so that the *in planta* molecules are often not apparent or bioavailable in amounts not congruent with their putative bioactivity. These human modifications also occurred for flavonols and flavan-3-ols, as reported by *in vitro* and *in vivo* studies (Crozier *et al.*, 2010). This chapter will focus on flavonols and flavan-3-ols, their metabolism and absorption in humans, and the evidence for their protective effect on human health and the mechanisms that might underlie them.

## 8.2 Focus on flavan-3-ols and flavonols: chemical structures and dietary sources

### *8.2.1 Flavan-3-ols*

Flavan-3-ols are the most complex subclass of flavonoids, ranging from the simple monomers (also known as catechins) to oligomeric and polymeric compounds (better known as proanthocyanidins or condensed tannins). The monomeric flavan-3-ol molecular structure have two chiral centers in position C2 and C3, forming four diastereoisomers for each level of B-ring hydroxylation. The two isomers (+)-catechin and (−)-epicatechin (EC) are widespread in nature, unlike their hydroxylated counterparts, namely the (epi)gallocatechins. There is a more limited distribution of (−)-epiafzelechin and its epimer (+)-afzelechin (Crozier *et al.*, 2009; Aron & Kennedy, 2008). Catechins also occur as C-3 esters, commonly with gallic acid, where the main forms are (−)-epicatechin-3-*O*-gallate (ECG) and (−)-epigallocatechin-3-*O*-gallate (EGCG).

Oligomeric and polymeric proanthocyanidins are formed through the C–C link of the flavan-3-ol units, mainly through the C4→C8 bond, although the C4→C6 bond also exists (both are called "B type"). The monomeric units can also be doubly linked by an additional ether bond in C2→O7 to form the A-type proanthocyanidins. Three common flavan-3-ols, which differ in their hydroxylation patterns, are found in proanthocyanidins. The condensed tannins exclusively formed by (epi)catechin units are called procyanidins. Proanthocyanidins containing at least one unit of (epi)gallocatechin or (epi)afzelechin are known as prodelphinidins or propelargonidins, respectively. However, the latter two types of proanthocyanidin are less common in fruits and vegetables than are procyanidins, and B-type procyanidins are more frequently encountered in common foods than A-type (Gu *et al.*, 2004).

The intake of flavan-3-ols in humans can be guaranteed by consumption of several foodstuffs, due to their almost ubiquitous distribution in the plant kingdom. For example, these flavonoids were quantified in 41 kinds of food typically consumed by the US population in one study (Gu *et al.*, 2004). Catechins only contributed to 7.1% of total flavan-3-ol intake, while dimers and trimers accounted for 11.2 and 7.8% of total flavan-3-ols, respectively. The great majority of flavan-3-ol intake was related to consumption of proanthocyanidins above the third degree of polymerization (DP), contributing 73.9% of total intake. The mean daily intake of total flavan-3-ols was estimated to reach almost 58 mg/day, with the major food contributors being apples, chocolate, and grapes. More recently, between 1999 and 2002, the National Health and Nutrition Examination Survey (NHANES) estimated that the intake of flavan-3-ols by the US population was between 143 and 179 mg/day according to age, with higher values for people aged 51–70 (Chun *et al.*, 2007). Apple and chocolate provide only B-type procyanidins, while green and red grapes also contain prodelphinidins (Gu *et al.*, 2004). In Europe, the intake of flavonols and flavones in the Dutch diet was found to be 23 mg/day (Hertog *et al.*, 1993b), while that in the Danish population was estimated at 28 mg/day (Leth & Justesen, 1998). However, it should be noted that the actual content of these substances in foods was often underestimated in the past due to the limited detectability of proanthocyanidin oligomers and polymers, which may be quantitatively relevant and contribute to the overall effects once metabolized by the colonic microflora. Recent improvements in the analytical approach have allowed more detailed studies, and the intake of flavan-3-ols in several European countries has been estimated, subdivided for gender. In the UK, health-conscious men consume 453.6 mg/day total flavan-3-ols, while women have an intake of 377.6 mg/day. Greece, on the other hand, shows the lowest intake of these flavonoids, at 160.5 mg/day for men and 124.8 mg/day for women (Zamora-Ros *et al.*, 2011). In the same study, the intake was also divided for Mediterranean and non-Mediterranean countries, while UK data were kept apart. The total intake of flavan-3-ols was 268.8 mg/day in Mediterranean countries, 274.7 mg/day in non-Mediterranean countries, and more than 400 mg/day in the UK. The proanthocyanidins were the highest contributor to the total flavan-3-ol intake in all three European zones, especially the polymers with DP $>$ 10. The catechins were the second highest contributors to the total intake, providing 18.6% in the Mediterranean region and close to 45% in the UK. This difference was probably related to the high intake of tea in the UK, also highlighted by the higher intake of EGCG than of other catechins.

Theaflavins – catechin-derived compounds formed by the oxidation of green tea leaves in black tea production – contributed less than 1% of total flavan-3-ol intake in the Mediterranean region, while in the UK they contributed more than 5% (Zamora-Ros *et al.*, 2011). Tea is a good source of catechins. However, green tea is a far richer dietary source than black tea, being the main subclass present in fresh tea leaves, contributing 77% of the overall phenolic level. In fact, an infusion of green tea contains up to 4.5 g/L catechins, mainly (−)-epigallocatechin (EGC) and EGCG (Del Rio *et al.*, 2004). Although monomers are the most represented form of polyphenol in green tea, oligomeric flavan-3-ols are also present. In fact, EC, EGC, and their gallate esters can be linked together to form compounds with high molecular weights, up to DP 7, and isomers of proanthocyanidins containing epiafzelechin-gallate units have been identified in green tea (Lakenbrink *et al.*, 1999; Kalili & de Villiers, 2010).

The apple is another rich source of B-type procyanidins, with oligomers representing the major fraction in this class. The oligomers can in fact reach almost 20 g/kg dry weight (d.w.), representing 78% of the total flavan-3-ol fraction. However, dimers and trimers are also present in substantial amounts, mainly as procyanidin B2 (69–2000 mg/kg d.w.) and procyanidin C1 (58–970 mg/kg d.w.), followed by procyanidin B1 and the catechins. Interestingly, the difference in DP measured by thiolysis ranged from 3 to 29 among the cultivars analyzed, with vintage and cider cultivars providing higher amounts than modern, sweeter cultivars, leading to the suggestion that a wide variability in the contribution to the dietary intake may be encountered. Therefore, an accurate estimation of the apple contribution should be determined according to the prevalent cultivar eaten (Wojdylo *et al.*, 2008). Moreover, the uneven distribution of flavan-3-ols in the apple fruit may have some relapses on the actual intake of these substances, as they can be consumed both with and without peels, which contain three to five times more procyanidins than the flesh (Tsao *et al.*, 2003). Other fruits, such as plums, blueberries, and grapes, provide a rich source of flavan-3-ols, mainly proanthocyanidins. The total flavan-3-ol content in US black plums is 257 mg/100 g for black diamond fruits and 238 mg/100 g for other black plums. Both plum types contain A-type proanthocyanidins (Gu *et al.*, 2004). The latter data are higher than those in a Finnish study, where the total flavan-3-ol content in plums was 105 mg/100 g, although the cultivar of the plum analyzed was not specified (Hellström *et al.*, 2009). The flavan-3-ol content in blueberries ranges from a minimum of 180 to a maximum of 332 mg/100 g, mainly as procyanidin polymers. Green and red grapes provide 54 and 33 mg/100 g of total flavan-3-ols, which present as procyanidin- and prodelphinidin-containing monomer gallates (Hellström *et al.*, 2009).

Cocoa beans are also rich in flavan-3-ols, mainly EC and oligomeric procyanidins beyond DP 10. In cocoa-derived products, the cocoa powder contains the highest amount of EC, equal to 185 mg/100 g. The EC content in baking chocolate is 114 mg/100 g, which is higher than that of dark or milk chocolate, with an average content of 34 and 10 mg/100 g, respectively (Katz *et al.*, 2011). In a study aimed at reporting the presence of proanthocyanidins in a wide range of plant foods, cocoa powder had an astonishing content of proanthocyanidins (almost 1.5 g/100 g), as much as 40% of which was quantified in the unextractable fraction (Hellström *et al.*, 2009). A high content of extractable proanthocyanidins in cocoa powder is also reported in the USDA database (http://www.ars.usda.gov/SP2UserFiles/Place/12354500/Data/PA/PA.pdf; last accessed April 2, 2014), albeit with a huge standard deviation ($1.4 \pm 2.1$ g/100 g), probably due to different processing conditions.

A high contribution to human flavan-3-ol intake, especially of flavan-3-ol polymers, is also made by nut consumption. Hazelnuts provide about 500 mg/100 g, as much as 64% of which is recovered as polymers with DP > 10. This flavan-3-ol content is similar to that of pecans (494 mg/100 g), which, like hazelnuts, contain procyanidins and prodelphinidins. A lower amount of flavan-3-ols is found in pistachios and almonds: 237 and 184 mg/100 g, respectively. In walnuts the flavan-3-ol content is about sevenfold lower than that of hazelnut (Gu *et al.*, 2004), but the large fraction of these flavonoids is located in the skin and not the kernel. In hazelnut skins, the monomeric and oligomeric flavan-3-ols account for more than 95% of total polyphenols, as quantified by LC-MS/MS. Depending on the growing region, total flavan-3-ols can exceed 850 mg/100 g in the pellicle (up to DP 3). Among monomers, catechin alone constitutes 80% of the total content on average, with the remaining quantity provided by EC and ECG. Among proanthocyanidins, the B-type procyanidins are the most abundant flavan-3-ols; they are also detected as dimer gallates. However, B-type prodelphinidins can reach more than 110 mg/100 g in hazelnut skin (mainly as dimers) (Del Rio *et al.*, 2011). This number is probably under-estimated, due to the polymers (up to DP 9) previously identified in the tegument surrounding hazelnut seeds (Monagas *et al.*, 2009).

Some spices, despite the limited amounts of their overall consumption, may offer a relevant contribution to the dietary intake of flavan-3-ols if assumed regularly. This is the case with cinnamon, whose total proanthocyanidin content (mainly trimers and above) is almost 40 times that of dark chocolate (Gu *et al.*, 2004).

### 8.2.2 Flavonols

Flavonols are a class of flavonoids that have a 3-hydroxyflavone backbone. Their diversity stems from the different positions of the phenolic OH groups. They are present in a wide variety of fruits and vegetables. Flavonols – in particular kaempferol, quercetin, isorhamnetin, and myricetin – are ubiquitous in edible plants. These flavonoids are typically found as sugar esters, where a glycosidic bond joins one or more phenolic OH groups to the sugar moiety. Several *O*-glycosides are formed involving the 3-position of the C-ring, although other conjugations occur at the 5, 7, 3′, 4′, and 5′ positions (Crozier *et al.*, 2009). The intake of flavonols by the US population estimated by a NHANES survey during 1999–2002 was between 12 and 14 mg/day according to age, with higher values for people aged 51–70 (Chun *et al.*, 2007). Quercetin is the major contributor to the dietary intake of flavonols and is present at high concentrations in onions, apples, tea, broccoli, and red wine (Williamson & Manach, 2005).

Onion bulbs are among the richest sources of dietary flavonols, with at least 25 different flavonols characterized from them. Yellow onions contain 270–1187 mg/kg flavonols, whereas red onions contain 415–1917 mg/kg, with a large variability in flavonol content depending on the cultivar. Unlike yellow and red bulbs, white onions usually have a small flavonol content. The most representative flavonols in onions are quercetin and isorhamnetin, although kaempferol and myricetin are also detected (Slimestad *et al.*, 2007). Seven quercetin glucosides were quantified in southern Italian red onions, with quercetin 3,4′-diglucoside and quercetin 4′-glucoside being the most abundant flavonols. The 3,4′-diglucoside form ranged from 254 to 274 mg/kg; slightly higher than 4′-monoglucoside (208–230 mg/kg). Among the five minor flavonols, the content of isorhamnetin-4′-glucoside was 41–49 mg/kg. Less relevant were quercetin-3-glucoside, quercetin-7,4′-diglucoside, quercetin-3,7,4′-triglucoside, and isorhamnetin-3,4′-diglucoside (Bonaccorsi *et al.*, 2005). Other fruit sources of quercetin include apples (21–72 mg/kg) and blackcurrants (44 mg/kg) (Erlund, 2004).

Tea, both black and green, is a good source of flavonols: in a 2004 study the total flavonol concentration was close to 780 mg/L in green tea and 570 mg/L in black. Quercetin was the highest flavonol quantified (mainly present as the 3-glucoside form), followed by quercetin-3-rutinoside and quercetin-3-galactoside. Kaempferol accounted for smaller but still substantial amounts, in the same form as quercetin (Del Rio *et al.*, 2004). Another edible food that provides a high amount of quercetin and kaempferol is caper, widely used in Mediterranean cooking. A serving of capers (10 g) ensures an intake of 65 mg flavonoid glycosides or the equivalent 40 mg quercetin as aglycone (Inocencio *et al.*, 2000).

A substantial amount of flavonols is also present in other fruits and vegetables, such as broccoli, tomatoes, leeks, black grapes, blueberries, blackcurrants, apples, and apricots (Manach *et al.*, 2004). As these foods are regularly consumed in large amounts, their contribution to the diet is relevant. For example, a single apple (~250 g), eaten with skin, ensures a flavan-3-ol intake of approximately 150–500 mg, according to the cultivar (Vrhovsek *et al.*, 2004). As already mentioned, the intrinsic variability in the abundance of these polyphenols in plant food sources is not the sole constraint on the evaluation of their purported health benefits. Many of these substances, in fact, may undergo a wide array of chemical transformations and rearrangements once they enter the human digestive tract. Such modifications can affect, positively or negatively, their bioavailability, their physicochemical behavior, and ultimately their efficacy in delaying the onset of chronic diseases and disorders, as described in the next section.

## 8.3 Metabolism and bioavailability of flavonoids in humans

Flavonoids exist *in planta* mainly in glycosidic forms, in order to improve their storage in the aqueous medium of cell vacuoles. Their physiological role in the plant kingdom is to provide relief against a plethora of biotic and abiotic stresses, including exposure to noxious free radicals and oxidant molecules produced during excessive exposure to light and heat. On many occasions, such behavior is the rationale for their beneficial effects in humans. However, after ingestion these substances may be extensively modified, altering their functional effects in the human body. After ingestion, flavonoid glycosides can be modified immediately in the oral cavity by the hydrolyzing activity of saliva. Although glycosides can also be hydrolyzed prior to reaching the small intestine, absorption at this site is associated with the most hydrolyzing activity towards flavonoid glycosides, operating through a cascade of enzymes. The release of aglycone is mediated by the action of lactase phloridzin hydrolase (LPH) in the brush-border of the small-intestine epithelial cells (Crozier *et al.*, 2010). LPH exhibits broad specificity for flavonoid-*O*-$\beta$-D-glucosides. The released aglycone may then enter the epithelial cells by passive diffusion, due to its increased lipophilicity (Day *et al.*, 2000). An alternative hydrolytic step is mediated by a cytosolic $\beta$-glucosidase (CBG). For this process, the active sodium-dependent glucose transporter, SGLT-1, seems to be involved in the transport of flavonoid glycosides into epithelial cells, a necessary step in permitting the action of $\beta$-glucosidase (Crozier *et al.*, 2010).

Once absorbed, flavonoids follow the common metabolic pathway of exogenous organic substances and, like drugs and most xenobiotics, undergo phase II enzymatic detoxification. They can be conjugated with glucuronic acid, sulfate, and methyl groups through UDP-glucuronosyltransferases (UGTs), sulfotransferases (SULTs), and catechol-*O*-methyltransferase (COMT), respectively (Manach *et al.*, 2004). This conjugation first occurs in the small intestine, and metabolites then undergo further conversions in the liver before entering the systemic circulation, recycling back to the small intestine through bile excretion, and finally being excreted. However, the fraction of flavonoids unabsorbed in the small intestine will travel further down the gastrointestinal tract to reach the large intestine, where they are exposed to the resident microflora and are rapidly degraded (Manach *et al.*, 2004). In some cases, an enterohepatic circulation may ensure a longer residence time in plasma for some metabolites, thus extending their bioavailability compared to that of their parent compounds. Gut bacteria can hydrolyze glycosides, glucuronides, sulfates, amides, esters, and lactones and are also able to break down the flavonoidic skeleton, as well as to carry out reactions of reduction, decarboxylation, demethylation, and dehydroxylation (Selma *et al.*, 2009). These complex modifications generate several low-molecular-weight metabolites that can be efficiently absorbed *in situ*, subsequently undergoing further phase II metabolism locally and/or in the liver before entering the blood compartment and finally being excreted in substantial amounts by kidneys. Their presence in the urine largely exceed the excretion of flavonoid metabolites formed in the upper gastrointestinal tract (Crozier *et al.*, 2010). These extensive modifications carried out by human and microbial enzymes affect the bioavailability of flavonoids. Palafox-Carlos *et al.* (2011) defined bioavailability for these phytochemicals as the proportion of plant-derived substance that is digested, absorbed, and utilized in normal metabolism. In many works on flavonoids, their bioavailability is assessed by quantification of phytochemicals excreted in urine with respect to the orally consumed dose; it is assumed that a phytochemical appearing in the urine must have access to the systemic circulation and then be distributed at the tissue level (Prasain & Barnes, 2007).

A graphical scheme of the human gastrointestinal metabolism of flavonoids described in this section is shown in Fig. 8.1.

### *8.3.1 Flavan-3-ols*

Dietary flavan-3-ols, unlike flavonols, exist *in planta* mainly as aglycones rather than glycosides, but share with them vacuolar segregation and physiological roles. This characteristic

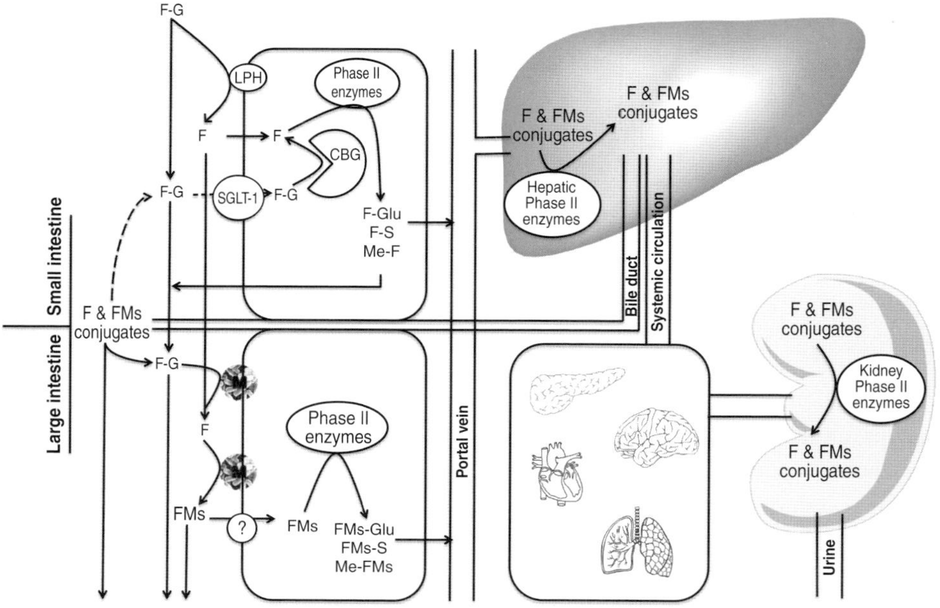


**Fig. 8.1** Metabolic fate of flavonoids in the human gastrointestinal tract. F-G, flavonoid glycosides; F, flavonoid aglycones; Glu, glucuronide; S, sulfate; Me, methyl; FMs, flavonoid microbial metabolites.

affects the mechanism of absorption of this flavonoid subclass, which for some (but not all) components in the circulatory system occurs in the small intestine. The aglycones may enter the epithelial cells without preliminary hydrolysis by passive diffusion as a result of their relative lipophilicity and proximity to the cellular membrane (Day *et al.*, 2000). It should be noted that there are large differences in the metabolic fates, rates, and degrees of bioavailability of flavan-3-ols, related to the molecular weights and chemical structures of catechins, proanthocyanidins, and theaflavins. Several studies have focused on the metabolism and absorption of catechin-rich products, mainly apples, green tea, and cocoa-based products, as will be discussed in this section.

It was finally proved in 2007 that, after apple consumption, almost 90% of the consumed procyanidins reach the colon under physiologic circumstances, confirming that the polymerized forms of these compounds are actually available for colonic biotransformation – previously described only *in vitro* – with fecal starters (Kahle *et al.*, 2007). Apple-derived proanthocyanidins are thus degraded by the colonic microflora into low-molecular-weight organic acids – including 2-(*m*-hydroxyphenyl)-acetic acid, 2-(*m*-hydroxyphenyl)-propionic acid, 5-(*m*-hydroxyphenyl)-valeric acid, phenylacetic acid, 2-(*p*-hydroxyphenyl)-acetic acid, and phenylpropionic acid – which are subsequently absorbed (Déprez *et al.*, 2000). The reported variability in bioavailability of apple proanthocyanidins is also a consequence of other dietary factors, including the overall fruit phytocomplex. In particular, the extent of enzymatic biotransformation of flavan-3-ols is dependent on the chain length of co-ingested proanthocyanidins, and the chain length of proanthocyanidins has a more crucial role in their microbial metabolism than the dose, as it can inhibit some luminal enzymes (Bazzocco *et al.*, 2008).

In an acute feeding study, healthy humans consumed 500 mL of bottled green tea containing 648 µmol of catechins, after which plasma and urine were collected over a 24-hour period and analyzed by HPLC-$MS^2$ (Stalmach *et al.*, 2009). The plasma contained a total of 12 metabolites, in the form of *O*-methylated, sulfated, and glucuronide conjugates of (epi)catechin and (epi)gallocatechin, along with EGCG and ECG, the only catechins recovered in unmetabolized form. The peak plasma concentrations of catechin conjugates ranged from 29 nmol/L of (epi)catechin-glucuronide to 126 nM of (epi)gallocatechin-glucuronide. The maximum plasma concentration of metabolites ranged from 1.6 to 2.3 hours. Urine collected after up to 24 hours contained many of the flavan-3-ol metabolites detected in plasma, except for EGCG and ECG, as well as minor amounts of three additional (epi)gallocatechin-*O*-sulfates (Stalmach *et al.*, 2009). The major (epi)gallocatechin metabolite excreted in urine was 4′-*O*-methyl-(epi)gallocatechin-*O*-sulfate, three times higher than the major plasma metabolite of (epi)gallocatechin ((epi)gallocatechin-glucuronide). However, there was a notable difference between the excretions of (epi)catechin and (epi)gallocatechin metabolites. The recovery of (epi)gallocatechin metabolites was 11%, lower than that of urinary (epi)catechin metabolites, which reached 29% of the total ingested amount. These high excretion levels are similar to those of earlier studies with green tea extract (Auger *et al.*, 2008), confirming that (epi)catechin is more bioavailable than (epi)gallocatechin. Similar results were obtained by the same group when they fed green tea to a group of ileostomy volunteers (Stalmach *et al.*, 2010), where the total excreted catechin in urine was close to 7%. The total excretion of (epi)catechin metabolites was equivalent to 27% of ingested (+)-catechin and EC. However, in this study the flavan-3-ol content in ileal fluid was 70% of the total ingested amount, revealing that a substantial fraction reaches the colon in the form of parent compounds (33%) and 23 phase II metabolites (37%) (Stalmach *et al.*, 2010). Once in the large intestine, the parent compounds and metabolites get in contact with an even wider array of enzymes belonging to microbial cells. Two catechin metabolites had been previously detected in the urine and plasma of human volunteers after ingestion of green tea. These were identified by LC/ESI-MS and NMR as (−)-5-(3′,4′,5′-trihydroxyphenyl)-$\gamma$-valerolactone and (−)-5-(3′,4′-dihydroxyphenyl)-$\gamma$-valerolactone (Li *et al.*, 2000). A third metabolite, (−)-5-(3′,5′-dihydroxyphenyl)-$\gamma$-valerolactone, was identified 2 years later (Meng *et al.*, 2002), and the cumulative excretion of these three valerolactones over a 24-hour period accounted for up to 16% of the ingested catechins. These molecules derived from ring-fission reactions carried out by the colonic microbiota and were excreted in higher amounts than their human enzyme-linked counterparts. Sang *et al.* (2008) provided a detailed description of the metabolite profile of tea polyphenols in human urine using liquid chromatography with ESI-MS/MS with data-dependent acquisition, but without a bioavailability study. Nevertheless, the study is remarkable in that it describes each catabolite and the chemical reactions involved in its generation *in vivo*. The study is also relevant for describing $\gamma$-valerolactones in their conjugated forms, as sulfated, methylated, and glucuronidated derivatives.

A second study, carried out by Roowi *et al.* (2010), investigated specifically the fate of flavan-3-ols in the colon both *in vitro* (with batch fecal fermentations) and in a feeding study. With this double approach, and with the help of a different analytical technique (GC-MS), they showed that monomeric flavan-3-ol undergoes degradation by fecal microflora. Recently, two human colonic bacteria were shown to be involved in the formation of some catechin metabolites. *Eggerthella lenta* and *Flavonifractor plantii* were able to cleave the C ring of both EC and (+)-catechin, producing 1-(3′,4′-dihydroxyphenyl)-3-(2″,4″,6″-trihydroxyphenyl)propan-2-ol). *Flavonifractor plautii* further converted 1-(3′,4′-dihydroxyphenyl)-3-(2″,4″,6″-trihydroxyphenyl)propan-2-ol to 5-(3′,4′-dihydroxyphenyl)-$\gamma$-valerolactone and 4-hydroxy-5-(3′,4′-dihydroxyphenyl)valeric acid (Kutschera *et al.*, 2011). Phenolic acid catabolites, after being absorbed and transferred to the circulatory system, are excreted in urine in quantities corresponding to almost 27% of the flavan-3-ol intake (Roowi *et al.*, 2010). However, in this case the work did not consider flavan-3-ol conjugates and failed to identify one of the three $\gamma$-valerolactones described earlier. The GC-MS sample pretreatment made it impossible to identify conjugations, due to the chemical derivatization step employed for GC analysis, which is able to replace the originally conjugated form of every catabolite. Nevertheless, this study is one of the more detailed in the literature characterizing colonic catabolites derived from green tea catechins.

A feeding study with 400 mL of a ready-to-drink green tea was carried out to investigate green-tea flavan-3-ol catabolism, plasma pharmacokinetic, and urinary excretion by HPLC-MS/MS (Del Rio *et al.*, 2010a). Tea consumed by 20 volunteers contained approximately 400 µmoles of catechins. Plasma and urine were collected for 4 and 24 hours, respectively, and 39 relevant flavan-3-ol catabolites were identified in biological samples. In plasma, EGCG was the only unmetabolized compound, and the highest in absolute concentration compared to (epi)gallocatechin and (epi)catechin conjugates. The phase II metabolites of the ring-fission products 5-(3′,4′,5′-trihydroxyphenyl)-$\gamma$-valerolactone, 5-(3′,4′-dihydroxyphenyl)-$\gamma$-valerolactone, and 5-(3′, 5′-dihydroxyphenyl)-$\gamma$-valerolactone were by far the main urinary catabolites, averaging 10-fold more concentrated than flavan-3-ol conjugates. The calculated bioavailability was 39%. It is interesting to note the variability in urinary excretion of colonic metabolites among participants, which was probably related to differences in their individual colonic microflora. The conjugates deriving from human enzyme activity shared a very fast kinetics in human plasma and were mainly excreted during the first 4 hours, accounting for less than 4% of the final bioavailability value. The only exception was methyl-(epi)gallocatechin-sulfated-glucuronide, whose appearance was delayed in excretion, leading to the hypothesis of a major involvement of hepatic enzymes in its formation. In contrast, the methyl, sulfate, and glucuronide forms of valerolactones showed a completely different excretive kinetics. They were generated in the colon, later than the catechin conjugates, and were therefore absorbed and excreted with a delay of several hours, reaching peak urinary excretion at 24 hours. However, collection of urine for only up to 1 day does not allow a complete excretion of these colon metabolites to be obtained. Indeed, the long-term excretion of these colonic metabolites was tested in one subject for up to 54 hours (Del Rio *et al.*, 2010a).

Based on these findings, a similar study investigating green tea flavan-3-ol metabolism and bioavailability was set up, extending the urine collection period to up to 48 hours from ingestion, without plasma samples (Calani *et al.*, 2012). Twenty healthy subjects consumed 500 mL of ready-to-drink green tea containing 639 µmol of catechins and a total content of proanthocyanidins equal to 88 μmol. No plasma samples were collected. Forty-one different urinary flavan-3-ol metabolites were identified by means of MS/MS detection. Three novel ring-fission metabolites of green-tea flavan-3-ols were identified. Two were a sulfate and a glucuronide form of hydroxyphenyl-γ-valerolactone and a methyl-glucuronide form of 5-(3′,4′-dihydroxyphenyl)-γ-valerolactone, previously identified in human urine after almond skin consumption (Llorach *et al.*, 2010). In this study, it was not possible to determine the exact position of the hydroxyl group of hydroxyphenyl-γ-valerolactone, even though it could have been the 3′-hydroxyl group, previously identified by *in vitro* biotransformation of flavan-3-ols (Stoupi *et al.*, 2010). In contrast, the methyl-glucuronide form of 5-(3′,4′,5′-trihydroxyphenyl)-γ-valerolactone, besides disulfate forms of 3′,4′- and 3′,5′-dihydroxyphenyl-γ-valerolactones, identified in our study, had never previously been identified. Regarding the 24-hour study, it was possible to distinguish the 3′,4′-dihydroxylated valerolactone- from the 3′,5′-dihydroxylated valerolactone-conjugates that share the same molecular weight. This was done by analysis of urine after consumption of sources of (epi)catechin and procyanidins but free of (epi)gallogatechin or prodelphinidins, as 5-(3′,5′-dihydroxyphenyl)-γ-valerolactone can derive only from (epi)gallocatechins. Most of the (epi)catechin and (epi)gallocatechin conjugates reached their excretive peak during the first or second collection time. The valerolactone conjugates, on the other hand, reached their excretive peak at 24 hours after tea ingestion. Some of these molecules showed a different excretive kinetics. In fact, 5-(3′,4′-dihydroxyphenyl)-γ-valerolactone conjugates are generally excreted in larger amounts in the first collection hours than are 5-(3′,5′-dihydroxyphenyl)-γ-valerolactone conjugates, which largely exceed the 3′,4′-dihydroxylated valerolactone conjugates only after the $10^{th}$ hour from tea intake. Their generation pathway in the colon could explain this, as 3′,4′-dihydroxyphenyl-γ-valerolactone has a 3′- and a 4′-hydroxyl group and is derived from a breakdown of the (epi)catechin skeleton, whereas an additional dehydroxylation step at the 4′-hydroxyl group on the B ring is necessary to form 3′,5′-dihydroxyphenyl-γ-valerolactone after the breakdown of (epi)gallocatechin units (Calani *et al.*, 2012). The bioavailability of tea flavan-3-ols estimated in this study was close to 62%, ranging from 18 to 100% among the recruited volunteers. The main reason for this increment was the urinary collection period, which lasted for 2 days after tea ingestion, allowing an almost complete excretion of flavan-3-ol metabolites and the identification and quantification of novel molecules derived by microbial modification of flavan-3-ols. In fact, the excretion of previously identified metabolites during the first 24 hours accounted for 44% of the ingested dose of tea catechins, a value that is only marginally higher than that previously calculated by Del Rio *et al.* (2010a). Moreover, this is the first study among investigations concerning green tea flavan-3-ol absorption to take into account proanthocyanidins, although these molecules were already identified in green tea (Kalili & de Villiers, 2010). In fact, dimers of procyanidin may be converted by human fecal microflora to several low-molecular-weight metabolites, including phenyl-γ-valerolactones (Appeldoorn *et al.*, 2009). For this reason, the exclusion of proanthocyanidins from the total flavan-3-ol quantification would have strongly affected the bioavailability calculation. However, a huge variability in the absorption values occurred among volunteers, ranging from a minimum of 18 to an astounding 100% absorption. These data show that some subjects were unable to efficiently generate, absorb, and excrete this class of flavonoid, allowing us to estimate how much colonic ring-fission metabolism might be relevant in the putative bioactivity of flavan-3-ols (Calani *et al.*, 2012).

Green tea provides mainly monomeric flavan-3-ols, while cocoa-based products also contain flavan-3-ols in oligomeric/polymeric forms. To our knowledge, in only one study has the cocoa procyanidins been detected and quantified in human plasma. The procyanidin B2 reached its peak plasma at 2 hours after the intake by five volunteers of 323 mg monomers and 256 mg dimers of procyanidins, and was also quantifiable at 6 hours post-ingestion (Holt *et al.*, 2002). However, a large fraction of oligomeric/polymeric flavan-3-ols, including the dimers, are not absorbed in the gastrointestinal tract via the small intestine, and so they reach the colon and are then extensively metabolized by host microbiota. Eleven subjects consumed 80 g chocolate containing 439 mg proanthocyanidins and 147 mg catechin monomers, mainly EC. Urinary samples were collected during the 24 hours before chocolate consumption and for up to 48 hours afterwards. After 2 days of urinary monitoring, there was an increase in the excretion of *m*-hydroxyphenylpropionic, ferulic, 3′,4′-dihydroxyphenylacetic, *m*-hydroxyphenylacetic, vanillic, and *m*-hydroxybenzoic acids (Rios *et al.*, 2003).

In a study by Mullen *et al.* (2009) aimed at evaluating the effect of milk on the bioavailability of cocoa flavan-3-ol metabolites, nine healthy humans consumed 250 mL of a cocoa beverage, made with water or milk, containing 45 μmol of EC and (−)-catechin. Plasma and urinary samples were collected for 24 hours and analyzed by HPLC with photodiode array and mass-spectrometric detection. Two catechin metabolites were detected in plasma and in urine. Milk had only minor effects on the plasma pharmacokinetics of an (epi)catechin-sulfate and no effect on its *O*-methyl counterpart. Both conjugate forms of (epi)catechin reached peak plasma concentration (<90 nM) within 1.4 hours of cocoa ingestion. However, at the same time milk significantly lowered the excretion of four urinary flavan-3-ol metabolites, from 18.3 to 10.5% of the ingested dose (Mullen *et al.*, 2009). Another study evaluated the effect of milk at the colonic level, studying the urinary excretion of microbial-derived phenolic acids after ingestion of cocoa powder with either water or milk during acute feeding (Urpi-Sarda *et al.*, 2010). In detail, 21 healthy subjects consumed 40 g of cocoa powder dissolved in 250 mL of water or milk and their urine was collected for up to 24 hours after intake. Using LC-MS/MS analysis, 15 phenolic acids were identified, of which 9 were affected by milk intake. The urinary concentration of caffeic, ferulic, protocatechuic, 4-hydroxybenzoic, 3′,4′-dihydroxyphenylacetic, 4-hydroxyhippuric, and hippuric acids diminished after the intake of cocoa with milk, unlike those of phenylacetic and vanillic acids, which were increased by milk ingestion. A long-period feeding with cocoa was set up with 42 subjects, who received either 20 g of cocoa powder with 500 mL of skimmed milk or 500 mL of skimmed milk only per day for two 4-week periods. Several flavan-3-ol metabolites increased in urine after cocoa ingestion, including glucuronides, sulfate, and methyl conjugates of (epi)catechin and 3′,4′-dihydroxyphenyl-$\gamma$-valerolactone (Urpi-Sarda *et al.*, 2009). However, in addition to these phase II metabolites, cocoa consumption for 4 weeks produced a significant increase in the urinary excretion of colonic microbial-derived phenolic acids, including vanillic, 3′-hydroxyphenylacetic, and 3′,4′-dihydroxyphenylacetic acids. Specifically,

3′-hydroxyphenylacetic acid and 3′,4′-dihydroxyphenyl-$\gamma$-valerolactone were the two microbial metabolites that were increased compared to fasting plasma (Urpi-Sarda *et al.*, 2009). In a recent study, 3′,4′-dihydroxyphenyl-$\gamma$-valerolactone was one of the main catabolites produced *in vitro* by human fecal fermentation of extra-dark chocolate (Dall'Asta *et al.*, 2012a). Thus, it has been confirmed that this metabolite can be considered a good biomarker of cocoa consumption, along with other foodstuffs rich in flavan-3-ols.

Fig. 8.2 shows the chemical structures of the main phase II metabolites of flavan-3-ols (Fig. 8.2a) and their colon-derived metabolites (Fig. 8.2b).

### 8.3.2 Flavonols

In a study by Walle *et al.* (2005), incubation of quercetin-4′-glucoside with human saliva led to the efficient formation of free quercetin, while for quercetin-3-glucoside, aglycone formation was about half as efficient. Unlike these quercetin glucosides, only small amounts of quercetin-3-rutinoside were hydrolyzed into its quercetin counterpart, while quercetin-3-rhamnoside (also known as quercitrin) was not hydrolyzed at all. However, the limitation of this study was the incubation time of up to 2 hours, which does not reflect the much shorter contact time between ingested polyphenols and saliva in the mouth. The mechanism of hydrolysis was probably related to the action of $\beta$-glucosidases, but their source was not determined. The microflora of the oral cavity seemed to play a role, as deduced by using antibacterial agents to inhibit salivary hydrolysis or exposing flavan-3-ols to subcultured oral bacterial colonies. However, cytosols from cultured oral epithelial cells also produced free quercetin through the hydrolysis of quercetin-4′-glucoside (Walle *et al.*, 2005). The hydrolyses of quercetin-4′-glucoside and quercetin-3,4′-diglucoside were further confirmed after consumption of onion soup by humans (Hirota *et al.*, 2001).

However, the modifications towards flavonols in the oral cavity can be considered a minor step in the biotransformation cycle of these flavonoids by humans when compared to those that occur in the gastrointestinal tract. Several studies have focused on flavonol metabolism and bioavailability in human subjects that consumed flavonol-rich foods, such as onions. In volunteers that had consumed these vegetables, the main circulating metabolites in the blood compartment were quercetin-3-*O*-glucuronide (Q3GA), 3′-*O*-methyl-quercetin-3-*O*-glucuronide (isorhamnetin-3-glucuronide), and quercetin-3′-*O*-sulfate (Q3′S). These phase II metabolites of quercetin were already present 1.5 hours after ingestion, showing the absence of quercetin glucosides (Day *et al.*, 2001). In contrast, following consumption of fried onions, five different glucuronides of quercetin were identified in plasma samples by means of HPLC-UV-MS/MS. Neither the free flavonol nor the genuine glycosides were detected in plasma (Wittig *et al.*, 2001). In a feeding study of six volunteers, their plasma and urinary samples were collected after intake of 270 g lightly fried onions. Using HPLC with photodiode array (PDA) and $MS^2$ detection, it was possible to identify 23 quercetin-based compounds in plasma and urine collected 1 and 0–4 hours, respectively, after ingestion. The availability of the reference metabolites allowed the positions of some glucuronide forms to be determined, including the 3-, 3′,- and 4′-OH groups, as well as the 3′-sulfate (Mullen *et al.*, 2004). A study exploring the urinary metabolite profiles of six volunteers after consumption of 200 g cooked onions identified 21

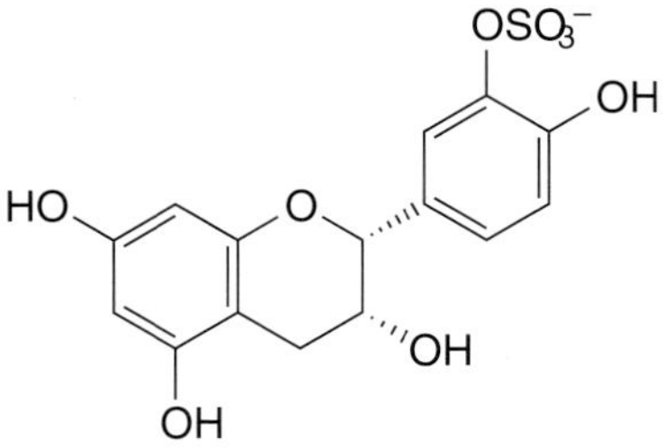

(–)-Epicatechin-3'-*O*-sulfate

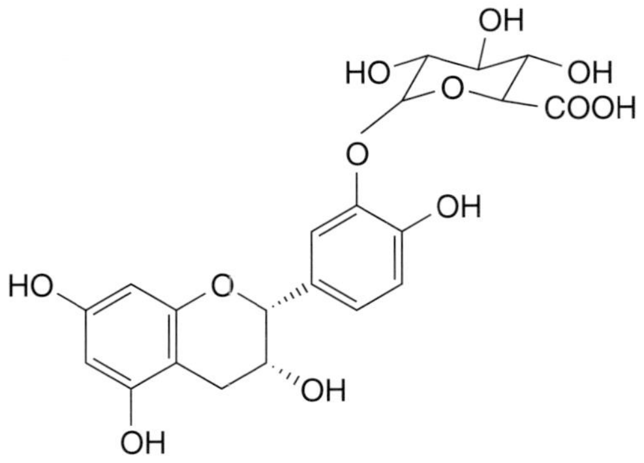

(–)-Epicatechin-3'-*O*-glucuronide

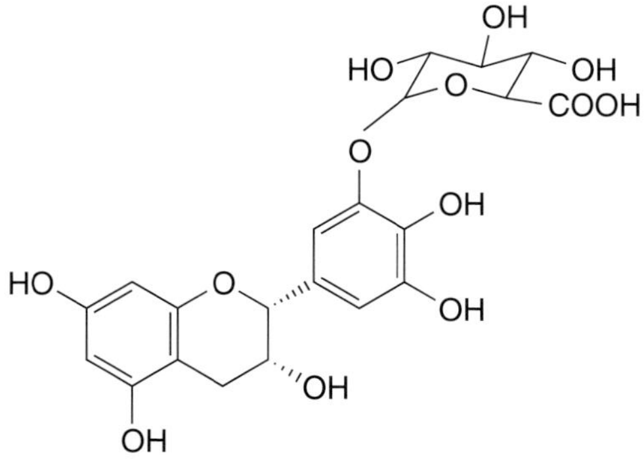

(–)-Epicatechin-3'-*O*-glucuronide

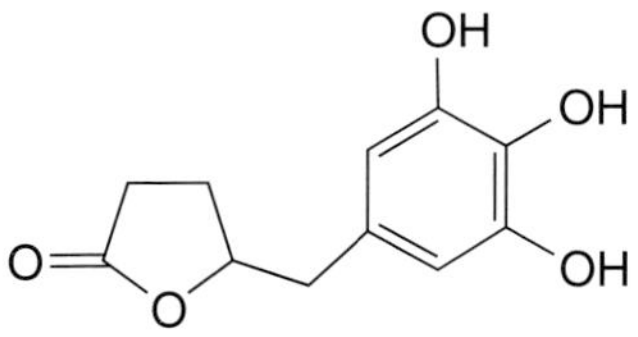

5-(3',4',5'-Trihydroxyphenyl)-γ-valerolactone

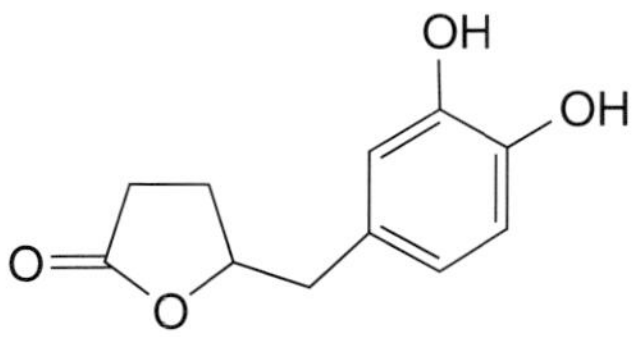

5-(3',4'-Dihydroxyphenyl)-γ-valerolactone

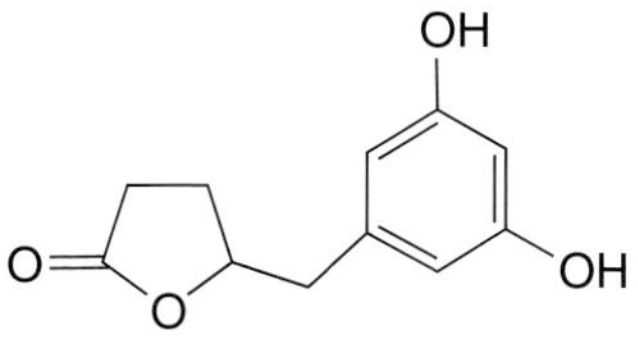

5-(3',5'-Dihydroxyphenyl)-γ-valerolactone

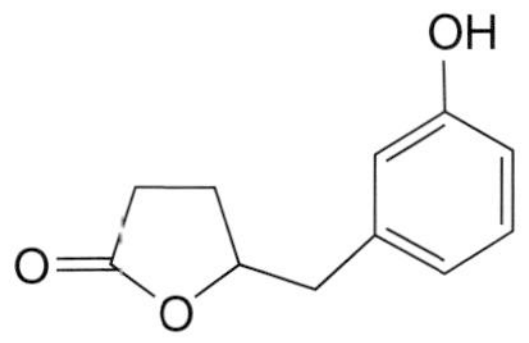

5-(3'-hydroxyphenyl)-γ-valerolactone

**Fig. 8.2** Chemical structures of the main phase II metabolites of flavan-3-ols (Fig. 8.2a) and their colon-derived metabolites (Fig. 8.2b).

flavonol metabolites by LC-ESI-MS/MS, after interaction with human phase II enzymes. Metabolites identified included quercetin-monoglucuronides, methyl quercetin monoglucuronides, a quercetin monoglucuronide sulfate, quercetin diglucuronides, a methyl quercetin diglucuronide, quercetin glucoside sulfates, methyl quercetin, free quercetin, and kaempferol monoglucuronides. Apart from free and methylated quercetin, the other 19 metabolites were also identified by the proper fragment ions, typically formed with the loss of the glucuronyl moiety and sulfate groups (Hong & Mitchell, 2004). However, these studies did not evaluate the absorption level of onion flavonol, as they did not consider the total flavonol excreted in urine with respect to the orally consumed dose. Therefore, Mullen *et al.* (2006) carried out a feeding study with six volunteers, in which plasma and urinary metabolites were quantitatively analyzed in samples collected at time points up to 24 hours from onion ingestion. Samples were analyzed by HPLC with a diode array coupled to MS/MS detection. Eleven quercetin metabolites were identified in plasma, but only five of these attained a sufficient concentration to enable pharmacokinetic profiles. The quantified metabolites were Q3′S, Q3GA, isorhamnetin-3-glucuronide, a quercetin diglucuronide, and a quercetin glucuronide sulfate. Submicromolar peak plasma concentrations of Q3′S, Q3GA, isorhamnetin-3-glucuronide, and quercetin diglucuronide were observed 0.6–0.8 hours after ingestion. However, quercetin glucuronide sulfate reached maximum plasma peak at 2.5 hours. Among plasma metabolites, the metabolites Q3′S and Q3GA reached the highest concentrations, equal to 665 and 351 nM, respectively. However, these two quercetin metabolites were not the major contributors to the total urinary excretion, especially the 3′-sulfate, which was present in trace amounts. This marked difference among plasma and urine profiles also occurred for several other metabolites. Indeed, the urinary components quercetin-3′-glucuronide, two quercetin glucoside sulfates, and a methylquercetin diglucuronide were absent or were present in only trace amounts in the bloodstream. This can be explained by their rapid circulation removal through kidney-mediated excretion. The total urinary excretion of quercetin metabolites was close to 13 μmol, corresponding to 4.7% of total intake of quercetin (Mullen *et al.*, 2006). Based on these data, a large fraction of the ingested quercetin was not absorbed in the upper gastrointestinal tract and thus reached the colon, where the resident microbiota was able to hydrolyze the quercetin derivatives and then break down the flavonol skeleton, finally generating several low-molecular-weight metabolites (Selma *et al.*, 2009). These modifications were shown using an *in vitro* fecal fermentation model, in which rutin underwent deglycosylation, ring fission, and dehydroxylation (Aura *et al.*, 2002). The main microbial metabolites produced by human colon microflora were 3′,4′-dihydroxyphenylacetic acid and its dehydroxylated counterpart, 3′-hydroxyphenylacetic acid.

Another *in vitro* study investigated modifications of colon-derived microflora towards rutin, using the fecal samples of three human donors (Jaganath *et al.*, 2009). 3′,4′-dihydroxyphenylacetic acid appeared within 2 hours of incubation and was the most abundant ring-fission catabolite, accounting for 60–97% of the break-down products. 3′-hydroxyphenylacetic acid first appeared after 4 hours and accounted for 15–20% of the break-down products. However, the modifications led to formation of other minor catabolites, such as 3,4-dihydroxybenzoic acid (also known as protocatechuic acid) and 4-hydroxybenzoic acid. The latter catabolite appeared in fecal samples 2–4 hours after its related dihydroxy counterpart because of the further modification necessary for its formation. In contrast, 3-(3′-hydroxyphenyl)propionic acid appeared in the fecal slurries in varying amounts and at different times depending upon the fecal sample source (Jaganath *et al.*, 2009). 3′,4′-dihydroxyphenylacetic acid was also recently identified after *in vitro* colonic fermentation of onions, as was phloroglucinol (Dall'Asta *et al.*, 2012a). The phloroglucinol was proposed to be a quercetin metabolite, assessed by incubation of quercetin-3-glucoside with two bacteria that usually colonize the human large intestine, such as *Enterococcus casseliflavus* and *Eubacterium ramulus*. These fecal bacteria have shown different metabolism capabilities. The strains of *E. casseliflavus* were able to hydrolyze the sugar moiety without affecting the aglycone degradation. Conversely, *E. ramulus* showed the ability to break down the flavonoid rings, catalyzing the formation of 3′,4′-dihydroxyphenylacetic acid and phlorglucinol (Schneider *et al.*, 1999).

The key contribution of the large intestine in flavonol metabolism and absorption was also established in a study carried out with healthy and ileostomy subjects (Jaganath *et al.*, 2006). The aim of this work was to investigate rutin metabolism and bioavailability by feeding 300 mL of tomato juice containing 176 μmol of rutin (after fortification). A maximum of nine methylated and glucuronidated quercetin metabolites were detected in the urine of six healthy volunteers, even though some of these excreted only three or four metabolites. Surprisingly, no sulfate conjugates were detected in biological samples. Only two plasma metabolites have been quantified, Q3GA and isorhamnetin-3-glucuronide. The first reached a peak plasma concentration of 12 nM, almost 33 higher than the second. However, for the same circulating metabolites, the concentration was at least 26-fold lower than in the onion study carried out by Mullen *et al.* (2006). Indeed, the two plasma metabolites reached maximum plasma concentration 5 hours post-intake. In contrast to quercetin glucosides, the rutinose moiety of rutin is not cleaved by either LPH or CBG associated with the wall of the gastrointestinal tract (Németh *et al.*, 2003). As a consequence, most of the rutin passes intact through the small intestine without being absorbed and will travel further down the gastrointestinal tract, reaching the colon of healthy subjects, or emerge in the ileal fluid in the case of ileostomists. In fact, rutin was not recovered in urine and plasma, besides the presence of 86% of the ingested rutin in the ileal fluid of subjects without a functioning colon. Because of the large fraction of rutin that reaches the colon, the different urinary excretions of phenolic acids between healthy and ileostomy subjects was evaluated. In volunteers with an intact colon, but not ileostomists, rutin was mainly catabolized to form 3′-methoxy-4′-hydroxyphenylacetic acid (also known as homovanillic acid), 3′-hydroxyphenylacetic acid, and 3′,4′-dihydroxyphenylacetic acid, and their urinary excretion accounted for 22% of the flavonol intake (Jaganath *et al.*, 2006). Therefore, in recent years it has been possible to gain knowledge of the extensive metabolism of flavonols by human gut microflora, leading to updates on the overall bioavailability of this flavonoid group.

Fig. 8.3 shows the chemical structures of the main phase II metabolites of quercetin (Fig. 8.3a) and its colon-derived metabolites (Fig. 8.3b).

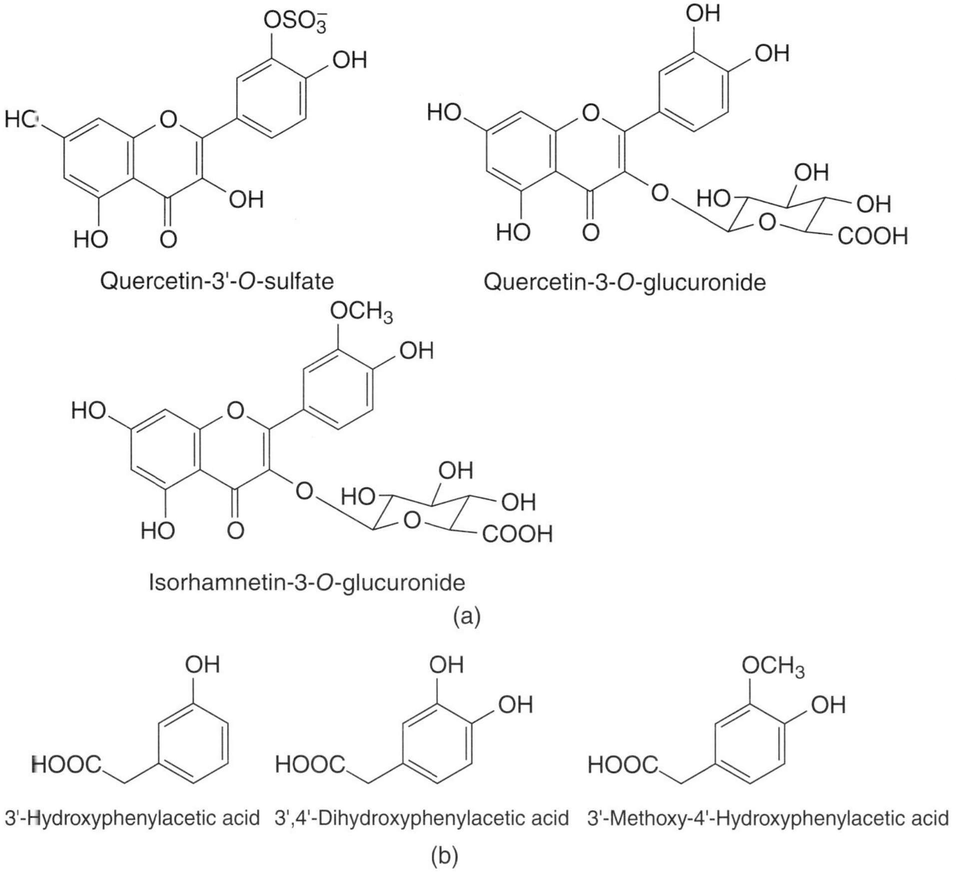


**Fig. 8.3** Chemical structures of the main phase II metabolites of quercetin (Fig. 8.3a) and its colon-derived metabolites (Fig. 8.3b).

## 8.4 *In vitro* studies

The dietary flavonoids are associated with beneficial health effects in humans, again based on *in vitro* studies. In the literature, the large number of *in vitro* investigations performed is focused on the evaluation of the biological effects of polyphenol-rich foods, their extracts, and the molecules present *in planta* (aglycones and their conjugated forms) (Del Rio *et al.*, 2010b). As a matter of fact, *in vivo* feeding studies conducted in recent years suggest that the molecules circulating in the organism are different from the molecules found *in planta* (Crozier *et al.*, 2010). The polyphenol metabolites circulating in humans are the sulfated, methyl-, and glucuronidated forms of the aglycones or the compounds (mainly phenolic acids) derived from microbiota metabolism and absorbed through the colon, as previously described (Crozier *et al.*, 2010). Therefore, these molecules have been tested in a μM–mM concentration range that largely exceeds the real values detected in plasma (even at Cmax) in *in vivo* studies (Del Rio *et al.*, 2010b). This scientific approach has generated confusing results regarding the real health-enhancing potential of these natural molecules (Williamson & Clifford, 2010). Therefore, the aim of this section is to review studies based on an evaluation of the biological activities of the flavan-3-ol and flavonol phase II and microbial metabolites, focusing on the bioactivities involved in the prevention of several chronic pathologies.

## 8.4.1 Flavan-3-ols

### 8.4.1.1 Phase II metabolites

The most studied bioactivity associated with the flavan-3-ols and their metabolites is antioxidant activity, due to the implication of oxidative stress in the initiation and progression of several chronic diseases. In fact, several studies have assessed the antioxidant capacity of the flavan-3-ols and flavan-3-ol phase II and microbial metabolites (Table 8.1).

Regarding the flavan-3-ol phase II metabolites, it has been reported that EGCG-3′-*O*-glucuronide and ECG-3-*O*-gallate-3-*O*-glucuronide have an antioxidant capacity similar to that of EGCG, using the DPPH antioxidant assay. EGCG-4″-*O*-glucuronide, EGC-3-*O*-gallate-7-*O*-glucuronide, and EGC-7-*O*-glucuronide are less active than their parent compounds, and EGC-3′-*O*-glucuronide is much less active than EGC (Lu *et al.*, 2003).

A study conducted by Natsume *et al.* (2004) showed that EC and its 7-*O*-glucuronide presented a similar delay of $Cu^{2+}$-induced LDL oxidation, whereas the activities of the 3′-O-glucuronide and the 4′-*O*-methyl-3'-*O*-glucuronide were significantly lower. Another study investigated the differences in antioxidant capacity of the phase II metabolites 3′-*O*- and 4′-*O*-methyl-(+)-catechin and EC compared with the quercetin phase II metabolites through the ferric reducing ability of plasma (FRAP) assay and two methods based on the ability to scavenge the $ABTS^{\bullet+}$ radical cation at different pH values. The results obtained described an antioxidant capacity of the metabolites with B-ring *O*-methylation at physiological pH (7.4), but lower than that of the parent compounds (Dueñas *et al.*, 2010). Recently, the antioxidant capacity of the parental compound EGCG and the *O*-methyl form, 4′,4″-di-*O*-methyl-EGCG were compared by means of FRAP expressed as μM Trolox equivalents. The study demonstrated that dimethylation of EGCG causes a reduction in the antioxidant activity of this metabolite (Lotito *et al.*, 2011).

Flavan-3-ols and their human metabolites were also studied *in vitro* for their bioactivities in vascular functionality and the prevention of cardiovascular disease (CVD). One interesting study conducted by Koga & Meydani (2001) investigated the effects of plasma metabolites of (+)-catechin and quercetin on the modulation of monocyte adhesion to human aortic endothelial cells (HAECs) and on the production of reactive oxygen species (ROS). In this report, the exposure of HAECs to the (+)-catechin metabolites inhibited U937 monocytic cell adhesion to IL-1$\beta$-stimulated cells, whereas incubation with intact (+)-catechin did not generate any effect. The production of ROS in hydrogen peroxide-stimulated HAECs was inhibited by (+)-catechin and its metabolites, whereas ROS generation in IL-1$\beta$-stimulated HAECs was inhibited only by (+)-catechin metabolites. The conclusion of this report was that the metabolic modification of the flavonoids also alters their biological activity, suggesting that the metabolites could be more active than the intact forms, especially in the prevention of CVD.

**Table 8.1** Biological activities of phase II and microbial metabolites derived from flavan-3-ols.

| Phase II metabolites | | | | |
|---|---|---|---|---|
| Metabolite | *In vitro* model | Expected effect | Actual effect | References |
| EGCG-7-*O*-glucuronide<br>EGCG-3′-*O*-glucuronide<br>EGCG-3″-*O*-glucuronide<br>EGCG-4″-*O*-glucuronide<br>EGC-3′-*O*-glucuronide<br>EGC-7-*O*-glucuronide | DPPH radical scavenging activity | **Antioxidant** | EGCG-3″-*O*-glucuronide, EGCG-3′-*O*-glucuronide, EGCG>ECG > EGC > EGCG-7-*O*-glucuronide, EGCG-4″-*O*-glucuronide> EGC-7-*O*-glucuronide, EC > EGC-3′-*O*-glucuronide | Lu *et al.* (2003) |
| 4′-*O*-methyl-EC-3′-*O*-glucuronide<br>EC-3′-*O*-glucuronide<br>EC-7-*O*-glucuronide | Inhibition of $Cu^{2+}$-induced LDL oxidation | **Antioxidant** | EC=EC-7-*O*-glucuronide> 4′-*O*-methyl-EC-3′-glucuronide, EC-3′-*O*-glucuronide | Natsume *et al.* (2004) |
| 4′-*O*-methyl-EC<br>3′-*O*-methyl-EC<br>4′-*O*-methyl-C<br>3′-*O*-methyl-C | ABTS/peroxide assay<br>ABTS/persulfate assay<br>FRAP assay | **Antioxidant** | C, EC> 4′-*O*-methyl-EC, 3′-*O*-methyl-EC, 4′-*O*-methyl-C, 3′-*O*-methyl-C | Dueñas *et al.* (2010) |
| EGCG<br>4′,4″-di-*O*-methyl-EGCG | FRAP assay | **Antioxidant** | EGCG>4′,4″-di-*O*-methyl-EGCG | Lotito *et al.* (2011) |
| EGCG-7-*O*-glucuronide<br>EGCG-3′-*O*-glucuronide<br>EGCG-3″-*O*-glucuronide<br>EGCG-4″-*O*-glucuronide<br>EGC-3′-*O*-glucuronide<br>EGC-7-*O*-glucuronide<br>EC-5-*O*-glucuronide<br>4′-*O*-methyl-EC-*O*-glucuronides (7, 5, and 3′)<br>EC-7-*O*-glucuronide | Inhibition of release of arachidonic acid from HT-29 from human colon cancer cells | **Anti-inflammatory** | EGC-3′-*O*-glucuronide < EGC-7-*O*-glucuronide, EGC<br>EGCG-7-*O*-glucuronide < EGCG, EGCG-3′-*O*-glucuronide, EGCG-3″-*O*-glucuronide, EGCG-4″-*O*-glucuronide | Lu *et al.* (2003) |

*(continued overleaf)*

**Table 8.1** (*continued*)

| Microbial metabolites | | | | |
|---|---|---|---|---|
| Metabolite | *In vitro* model | Expected effect | Actual effect | References |
| 5-(3′,4′-dihydroxyphenyl)-γ-valerolactone<br>5-(3′-methoxy-4′-hydroxyphenyl)-γ-valerolactone | FRAP assay<br>Inhibition of enzymatic activity on matrix metalloproteinases (MMPs) from human monocytes | **Antioxidant**<br>**Anti-inflammatory** | 5-(3′,4′-dihydroxyphenyl)-γ-valerolactone > 5-(3′-methoxy-4′-hydroxyphenyl)-γ-valerolactone<br>5-(3′,4′-dihydroxyphenyl)-γ-valerolactone, 5-(3′-methoxy-4′-hydroxyphenyl)-γ-valerolactone>C | Grimm *et al.* (2004) |
| 5-(3′,4′-dihydroxyphenyl)-γ-valerolactone<br>5-(3′,5′-dihydroxyphenyl)-γ-valerolactone<br>5-(3′,4′,5′-trihydroxyphenyl)-γ-valerolactone<br>5-(3′-hydroxyphenyl)-γ-valerolactone | ABTS assay | **Antioxidant** | 5-(3′,4′,5′-trihydroxyphenyl)-γ-valerolactone > 5-(3′,5′-dihydroxyphenyl)-γ-valerolactone > 5-(3′,4′-dihydroxyphenyl)-γ-valerolactone > 5-(3′-hydroxyphenyl)-γ-valerolactone | Takagaki *et al.* (2011) |
| 5-(3′,4′,5′-trihydroxyphenyl)-γ-valerolactone<br>5-(3′,4′-dihydroxyphenyl)-γ-valerolactone | Growth inhibition of human esophageal squamous-cell carcinoma cells (KYSE150), human colon adenocarcinoma cells (HT-29 and HCT-116), immortalized human intestinal epithelial cells (INT-407), immortalized rat intestinal epithelial cell line (IEC-6) | **Antiproliferative** | 5-(3′,4′,5′-trihydroxyphenyl)-γ-valerolactone >5-(3′,4′-dihydroxyphenyl)-γ-valerolactone | Lambert *et al.* (2005) |

EGCG, (−)-epigallocatechin-3-*O*-gallate; EGC, (−)-epigallocatechin; EC, (−)-epicatechin; C, (+)-catechin.

Once again, the study conducted by Lu *et al.* (2003) demonstrated the involvement in the inflammatory processes of the major metabolites derived from (−)-EGCG and (−)-EGC, the most important green-tea catechins. In the case of EGCG, the results showed the ability of the metabolite glucuronidated at C-7 to inhibit the production of NO or the arachidonic acid metabolism in HT29 cells compared to the aglycone, but it was not affected in the case of glucuronidation at C-3′, C-3″, or C-4′. In the case of EGCG, only the metabolite glucuronidation at C-3′ inhibits the radical scavenging activity and release of arachidonic acid from HT-29 human colon cancer cells, compared to the algycone form, while the C-7 glucuronided form has no effect on these capacities.

### 8.4.1.2 Microbe-derived metabolites

The biological activities of microbial metabolites derived from the catabolism of flavan-3-ols are still largely unknown, except for some studies published on the main microbial metabolites from flavan-3-ols: the valerolactones.

One interesting investigation focused on the evaluation of the effect of two metabolites derived from the metabolism of maritime pine bark extract, rich in proanthocyanidins. The aim was to elucidate the anti-inflammatory and antioxidant effects of the microbial-derived metabolites 5-(3′,4′-dihydroxyphenyl)-γ-valerolactone and 5-(3′-methoxy-4′-hydroxyphenyl)-γ-valerolactone. The anti-inflammatory effect was assessed by investigation of the inhibitory activity of the enzymatic activity on matrix metalloproteinases (MMPs). The results showed a greater inhibitory effect of the two metabolites towards the activity of MMP-1, MMP-2, and MMP-9 than that of the precursor (+)-catechin. In particular, both valerolactones (at 0.5 μM) inhibited about 50% of MMP-9 secretion. Moreover, in the radical-scavenging test, the 5-(3′,4′-dihydroxyphenyl)-γ-valerolactone metabolite demonstrated a higher capacity in superoxide scavenging than (+)-catechin, ascorbic acid, and Trolox, while no antioxidant effect was detected for the metabolite 5-(3′-methoxy-4′-hydroxyphenyl)-γ-valerolactone (Grimm *et al.*, 2004).

Therefore, another study conducted by Takagaki *et al.* (2011) investigated the antioxidant activity of EGCG microbial metabolites derived from the degradation of the rat microflora: 5-(3′,4′-dihydroxyphenyl)-γ-valerolactone, 5-(3′,5′-dihydroxyphenyl)-γ-valerolactone, 5-(3′,4′,5′-trihydroxyphenyl)-γ-valerolactone, and 5-(3′-hydroxyphenyl)-γ-valerolactone. The results showed an increasing antioxidant activity as follow: 5-(3′,4′,5′-trihydroxyphenyl)-γ-valerolactone > 5-(3′,5′-dihydroxyphenyl)-γ-valerolactone > 5-(3′,4′-dihydroxyphenyl)-γ-valerolactone > 5-(3′-hydroxyphenyl)-γ-valerolactone.

The antiproliferative effect of 5-(3′,4′,5′-trihydroxyphenyl)-γ-valerolactone and 5-(3′,4′-dihydroxyphenyl)-γ-valerolactone and their methoxy-derivatives on immortalized and malignant human cell lines was also investigated. 5-(3′,4′,5′-trihydroxyphenyl)-γ-valerolactone was the most active compound on human esophageal squamous-cell carcinoma cells (KYSE150), human colon adenocarcinoma cells (HT-29 and HCT-116), immortalized human intestinal epithelial cells (INT-407), and an immortalized rat intestinal epithelial cell line (IEC-6), supporting the hypothesis that this compound is active against some malignant human cells and is involved in the prevention of cancer (Lambert *et al.*, 2005).

### 8.4.2 Flavonols

#### 8.4.2.1 Phase II metabolites

As previously described, quercetin is the most representative molecule of the flavonol phenolic class, and a large number of studies are based on the evaluation of the bioactivity of this compound and its metabolites (Shirai *et al.*, 2002; Al-Shalmani *et al.*, 2011; Boesch-Saadatmandi *et al.*, 2011) (Table 8.2).

The antioxidant effects of quercetin and the quercetin phase II metabolite Q3GA were compared and measured by their inhibitory effect on the production of $H_2O_2$-induced intracellular ROS in mouse fibroblast 3T3 cells. Quercetin exerted an inhibitory effect on ROS production when incubated with $H_2O_2$, whereas Q3GA did not. When the cells were pretreated with Q3GA and quercetin, only Q3GA was active (Shirai *et al.*, 2002). Another study found that quercetin metabolites from plasma acted as effective antioxidants when subject to metal ion-induced lipid peroxidation. The oxidative power of plasma obtained from rats after intragastric administration of quercetin was studied to determine whether or not quercetin acts as an *in vivo* antioxidant after metabolic conversion. Quercetin was raised in the rat blood plasma, essentially as glucuronide and/or sulfate conjugates, the main phase II metabolite forms. The plasma obtained from rats after quercetin administration was more resistant against copper sulfate-induced lipid peroxidation than the control plasma on the basis of the accumulation of cholesteryl ester hydroperoxides and the consumption of $\alpha$-tocopherol (da Silva *et al.*, 1998).

When the antioxidant activity was tested by means of FRAP and the ABTS assays, methylated quercetin metabolites (namely 3′-*O*- and 4′-*O*-methyl-quercetin) showed some antioxidant activity, but less than the parent compound (Dueñas *et al.*, 2010); in another study, 3′-*O*-methyl-quercetin and Q3′S were shown to have lower reducing activity than quercetin by FRAP assay (Lotito *et al.*, 2011).

The antioxidant activity of Q3GA was also tested on lipid peroxidation using phosphatidylcholine large unilamellar vesicles (PC LUVs) as a biomembrane model. Iron ion, an aqueous peroxyl radical generator, a peroxynitrite generator, or lipoxygenase was used as the inducer of lipid peroxidation. In all cases, Q3GA inhibited lipid peroxidation significantly, although its inhibitory effect was lower than that of quercetin aglycone (Shirai *et al.*, 2001).

Flavonols and their metabolites were also studied for their anti-inflammatory capacities, supporting the implication of these compounds in the prevention of different inflammatory-derived pathologies, such as atherosclerosis and diabetes (Dall'Asta *et al.*, 2012b).

Boesch-Saadatmandi *et al.* (2011) evaluated the effect of quercetin, isorhamnetin, and Q3GA on the inflammatory gene expression in murine RAW264.7 macrophages stimulated with LPS. Quercetin and isorhamnetin were able to significantly decrease the mRNA and protein levels of TNF-$\alpha$, IL-1$\beta$, IL-6, macrophage inflammatory protein-1$\alpha$, and iNOS. In contrast, Q3GA was not able to influence the cells' gene expression. Furthermore, pro-inflammatory miRNA-155 was down-regulated after exposure to quercetin and isorhamnetin, but not to Q3GA.

**Table 8.2** Biological activities of phase II and microbial metabolites of flavonols.

| | | Phase II metabolites | | |
|---|---|---|---|---|
| **Metabolite** | *In vitro* model | Expected effect | Actual effect | References |
| Q3GA | Inhibition of the production of hydrogen peroxide ($H_2O_2$)-induced intracellular reactive oxygen species in mouse fibroblast3T3 cultured cells | **Antioxidant** | Q3GA > QUE when pretreated with the metabolites and exposed to $H_2O_2$<br>QUE > Q3GA when contemporaneously exposure of metabolites and $H_2O_2$ | Shirai *et al.* (2002) |
| Isorhamnetin<br>Q3GA | Inhibition of TNF$\alpha$, IL-1$\beta$, IL-6, protein-1$\alpha$ iNOS, microRNA-155 gene expressions in murine RAW264.7 macrophages stimulated with LPS | **Anti-inflammatory** | QUE, Isorhamnetin > Q3GA (no effect) | Boesch-Saadatmandi *et al.* (2011) |
| 3′-*O*- methyl-QUE<br>4′-*O*-methyl-QUE | ABTS/peroxide assay<br>ABTS/persulfate assay<br>FRAP assay | **Antioxidant** | 3′-*O*-methyl-QUE, 4′-*O*-methyl-QUE < QUE | Dueñas *et al.* (2010) |
| 3′-*O*-methyl-QUE<br>Q3′S | FRAP assay | **Antioxidant** | 3′-*O*-methyl-QUE, Q3′S < QUE | Lotito *et al.* (2011) |
| Q3GA | Inhibition of lipid peroxidation using phosphatidylcholine large unilamellar vesicles (PC LUV) as a biomembrane model | **Antioxidant** | Q3GA < QUE | Shirai *et al.* (2001) |
| Q3GA<br>Q3′S<br>Isorhamnetin-3-glucuronide | Inhibition of the inflammation-induced up-regulated expression of VCAM-1, ICAM-1, and the chemokine, monocyte chemoattractant protein-1 (MCP-1) in human umbilical vein endothelial cells (HUVECs) | **Anti-inflammatory** | Q3GA , Q3′S, isorhamnetin 3-glucuronide < QUE | Tribolo *et al.* (2008) |
| Q3GA<br>Q3′S<br>Myricetin | Changes in the isometric tension of segments and NO production in porcine isolated coronary artery | **Anti-inflammatory** | QUE, Q3GA, Q3′S > myricetin | Al-Shalmani *et al.* (2011) |

(*continued overleaf*)

**Table 8.2** (*continued*)

| **Metabolite** | *In vitro* model | Expected effect | Actual effect | References |
|---|---|---|---|---|
| | | Phase II metabolites | | |
| Q3GA<br>Q3′S | Inhibition of angiogenesis induced by vascular endothelial growth factor (VEGF) | **Antiproliferative** | QUE, Q3GA > Q3′S | Donnini *et al.* (2006) |
| Q3′S<br>Q3GA<br>Isorhamnetin-3-glucuronide | Inhibition of IL-1$\beta$-stimulated iNOS gene expression in RINm5F $\beta$-cells | **Anti-inflammatory** | QUE > Q3GA, Q3′S, isorhamnetin-3-glucuronide | Cho *et al.* (2012) |
| | | **Microbial metabolites** | | |
| 4′-hydroxyphenyl-acetic acid<br>3′-hydroxyphenyl-acetic acid<br>homovanillic acid<br>3,4-dihydroxytoluene<br><br>3′,4′-dihydroxyphenyl-acetic acid<br>3,4-dihydroxytoluene | DPPH assay | **Antioxidant** | 3′,4′-dihydroxyphenylacetic acid > 3,4-dihydroxytoluene > homovanillic acid > 4′-hydroxyphenylacetic acid, 3′-hydroxyphenylacetic acid | Glässer *et al.* (2002) |
| 3′,4′-dihydroxyphenyl-acetic acid<br>3,4-dihydroxybenzoic acid<br>3′-hydroxyphenylacetic acid<br>4-hydroxybenzoic acid<br>3-(3′-hydroxyphenyl)-propionic acid | FRAP assay | **Antioxidant** | QUE, 3′,4′-dihydroxyphenylacetic acid > 3,4-dihydroxybenzoic acid > 3′-hydroxyphenylacetic acid, 4-hydroxybenzoic acid, 3-(3′-hydroxyphenyl)propionic acid | Jaganath *et al.* (2009) |
| 3′,4′-dihydroxyphenyl-acetic acid<br>3,4-dihydroxytoluene | Inhibition of lipoperoxidation in differentiated PC12 cells as a model system of neuronal cells | **Antilipoperoxidant** | 3,4-dihydroxytoluene> 3′,4′-dihydroxyphenylacetic acid | Pavlica & Gebhardt (2010) |

| Microbial metabolites | | | | |
|---|---|---|---|---|
| **Metabolite** | ***In vitro* model** | **Expected effect** | **Resulted effect** | **References** |
| 3′,4′-dihydroxyphenyl-acetic acid | Human neuroblastoma SK-N-MC clonal cell line | **Neuroprotection** | No effect | Verzelloni *et al.* (2011) |
| 3′-hydroxyphenyl-acetic acid | Prevention of serum albumin glycation | **Antiglycative** | | |
| 3-(3′,4′-dihydroxyphenyl) propionic acid | | | | |
| homovanillic acid | | | | |
| 3′,4′-dihydroxyphenyl-acetic acid | Inhibition of AGEs production | **Antiglycative** | 3′,4′-dihydroxyphenylacetic acid, 3,4-dihydroxytoluene > homovanillic acid, 3′-hydroxyphenylacetic acid | Pashikanti *et al.* (2010) |
| 3,4-dihydroxytoluene | | | | |
| 3′-hydroxyphenylacetic acid | | | | |
| homovanillic acid | | | | |
| 3′,4′-dihydroxyphenyl-acetic acid | Up-regulation GSTT2 expression and decreasing COX-2 in human adenoma cells LT97 | **Anti-inflammatory** | 3′,4′-dihydroxyphenylacetic acid=3-(3′,4′-dihydroxyphenyl) propionic acid | Miene *et al.* (2011) |
| 3-(3′,4′-dihydroxyphenyl) propionic acid | | | | |

QUE, quercetin; Q3GA, quercetin-3-*O*-glucuronide; Q3′S, quercetin-3′-*O*-sulfate.

Epidemiological studies suggest that the consumption of flavonoid-rich diets decreases the risk of CVDs. A study conducted by Tribolo *et al.* (2008) investigated the effect of quercetin and its human metabolites on the adhesion of circulating monocytes to vascular endothelial cells, a critical step in both inflammation and atherosclerosis. The aim of the study was to evaluate the activity of these molecules at physiological concentrations (2 and 10 μM), and to attenuate the inflammation-induced up-regulated expression of VCAM-1, ICAM-1, and the chemokine, monocyte chemoattractant protein-1 (MCP-1), in human umbilical vein endothelial cells (HUVECs) at the protein and transcript levels. Quercetin treatment reduced the inflammation-induced over-expression of VCAM-1 and ICAM-1 (protein and transcript) in HUVECs. Quercetin also inhibited MCP-1 gene expression. However, Q3′S, Q3GA, and isorhamnetin-3-glucuronide generally exhibited either a reduced ability to inhibit the expression of these molecules compared with the parent aglycone or had no effect. All three metabolites did inhibit VCAM-1 cell-surface expression at 2 μM (Tribolo *et al.*, 2008). By inhibiting cytokine and inducible nitric oxide synthase expression through inhibition of the NF-kappaB pathway, quercetin obtained after glycoside's cleavage of quercitrin by fecal starters exerted a stronger down-regulation of the inflammatory responses than its glycosylated forms present in vegetables (Comalada *et al.*, 2005).

A study examined the effect of quercetin and its principal human metabolites on inflammatory changes in the porcine isolated coronary artery. The results were that 10 μM Q3′S and 10 μM Q3GA prevented LPS-induced changes in vascular responses in porcine coronary artery segments, possibly through a mechanism involving inhibition of NFkB, while myricetin (10 μM) was inactive. In contrast, quercetin (0.1 – 10.0 μM) opposed LPS-induced changes in vascular responses, nitrite production, and expression of inducible nitric oxide synthase (Al-Shalmani *et al.*, 2011).

In order to evaluate the potential effect of phase II metabolites on endothelial function, Lodi *et al.* (2009) conducted a study using thoracic aortic rings isolated from Wistar rats. They analyzed the vasorelaxant effects and the role in NO bioavailability and endothelial function of quercetin and its conjugated metabolites (Q3GA, isorhamnetin-3-glucuronide, and Q3′S) in rat aorta. Quercetin and the conjugated metabolites had no direct vasorelaxant effect and did not modify endothelial function or the biological activity of NO. However, all metabolites (at 10 μM) prevented, at least partially, the impairment of endothelial-derived NO response under conditions of high oxidative stress. Quercetin and Q3 S ($\geq$10 μM) or Q3GA (100 μM) inhibited NADPH oxidase-derived $O_2^-$ release, and quercetin and Q3GA (1 μM) prevented the endothelial dysfunction induced by incubation with endothelin-1.

Quercetin sulfate and glucuronide were also tested for their ability to prevent high glucose-induced apoptosis of HUVEC at 100 nM, 300 nM, and 1 μM concentration. The results showed that they inhibited it in a dose-dependent manner. The effects of quercetin sulfate and glucuronide on $H_2O_2$ quenching, inhibition of JNK, and caspase-3 activity at nanomolar concentration (300 nM) wre similar to those of ascorbic acid at micromolar concentrations (100 μM) (Chao *et al.*, 2009).

The phase II metabolites Q3GA and Q3′S showed a divergent effect during *in vivo* angiogenesis induced by vascular endothelial growth factor (VEGF). While quercetin and Q3GA inhibited VEGF-induced endothelial cell functions and angiogenesis, Q3′S promoted endothelial cell proliferation and angiogenesis. The inhibitory effect elicited by Q3GA was linked to inhibition of extracellular signal-regulated kinases 1/2 (ERK1/2) phosphorylation elicited by VEGF. The activation of endothelial cells by Q3′S was associated with stimulation of VEGF receptor-2 and with down-stream signaling activation (phosphatidylinositol-3 kinase/Akt and nitric oxide synthase pathways), which was ultimately responsible for ERK1/2 phosphorylation. These data indicate that the effects of circulating quercetin conjugates on angiogenesis are different depending on the nature of the conjugate (Donnini *et al.*, 2006).

Another study evaluated the inflammatory effect of quercetin and its metabolites on pancreatic islets, the cells implicated in the pathogenesis of diabetes mellitus. The aim of the study was to investigate the effects of quercetin and its metabolites Q3′S, Q3GA, and isorhamnetin 3-glucuronide on IL-1$\beta$-stimulated iNOS gene expression in RINm5F $\beta$-cells. The nitrite level, iNOS protein, and its mRNA expression levels and iNOS promoter activity were measured. In addition, I$\kappa$B$\alpha$ protein phosphorylation, nuclear translocation of NF-$\kappa$B, and NF-$\kappa$B DNA binding activity were determined and adenosine 5′-triphosphate disodium salt-induced insulin release was measured. Quercetin significantly reduced IL-1$\beta$-induced nitrite production, iNOS protein, and its mRNA expression levels, and it also inhibited IL-1$\beta$-induced I$\kappa$B$\alpha$ phosphorylation, NF-$\kappa$B activation, and iNOS promoter activity. Additionally, quercetin significantly restored the inhibition of insulin secretion by IL-1$\beta$. Meanwhile, quercetin metabolites did not show any effect on IL-1$\beta$-induced iNOS gene expression or on insulin secretion (Cho *et al.*, 2012).

Quercetin glucuronides were tested in an *in vitro* model of human lung cancer (cell line NCI-H209) and the results showed that quercetin glucuronides inhibited proliferation through G2/M arrest of the cell cycle and induced apoptosis via a caspase-3 cascade (Yang *et al.*, 2006).

Quercetin and its phase II metabolites were recently tested in human A549 lung cancer cells. Quercetin metabolite-enriched plasma (QMP) obtained from Mongolian gerbils 2 hours after quercetin feeding (100 mg/kg b.w./week) were incubated with the cells. The results confirmed that QMP, but not control plasma (CP), reduced cell growth in A549 cells. The study showed a QMP-mediated arrest of the cell cycle. Furthermore, QMP – but not CP or quercetin itself – significantly increased PPAR-$\gamma$ expression, suggesting that the activation of PPAR-$\gamma$ plays an important role in the antiproliferative effects of quercetin metabolites (Yeh *et al.*, 2011).

### 8.4.2.2 Microbe-derived metabolites

The main microbial-derived metabolites after eating quercetin-rich foods are phenolic acids. The most important phenolic acid derived from microbial metabolism is 3′,4′-dihydroxyphenylacetic acid, which has been studied for its antioxidant activity. A report by Glässer *et al.* (2002) compared quercetin and several of its possible metabolites for their antioxidant activity. The phenolic acid 3′,4′-dihydroxyphenylacetic acid and 3,4-dihydroxytoluene were the most effective microbial metabolites showing radical scavenging activity in the DPPH assay. 3′,4′-dihydroxyphenylacetic acid, 3,4-dihydroxybenzoic acid, and taxifolin were also tested for their antioxidant capacity by means of the FRAP assay. 3′,4′-dihydroxyphenylacetic acid was found to be the most active compound and was described as one of the catabolites that might play a key role in the overall antioxidant capacity of the colonic lumen after the ingestion of quercetin-rich foods (Jaganath *et al.*, 2009).

The metabolites 3′,4′-dihydroxyphenylacetic acid and 3,4-dihydroxytoluene were also tested in a model of differentiated PC12 cells as a model system of neuronal cells as protective compounds against the oxidative stress. In prevention of cell death and production of ROS assays, all tested compounds showed excellent protection. Lipid peroxidation levels induced by all stressors were reduced more potently during coincubation than during preincubation with polyphenols. While the lipophilic metabolite 3,4-dihydroxytoluene exerted excellent antilipoperoxidant activity, the hydrophilic metabolite 3′,4′-dihydroxyphenylacetic acid was less effective (Pavlica & Gebhardt, 2010). 3′,4′-dihydroxyphenylacetic acid was also tested in a different model of neuroblastoma cells and resulted in no activity against oxidative stress or in prevention of serum albumin glycation (Verzelloni *et al.*, 2011).

In contrast, when 3′,4′-dihydroxyphenylacetic acid and 3,4-dihydroxytoluene were tested as inhibitors of histone H1 glycation by the powerful glycating agent ADP-ribose, it was found that the two microbial metabolites effectively inhibited glycation (Pashikanti *et al.*, 2010).

The microbial metabolite from quercetin, 3′,4′-dihydroxyphenylacetic acid, and 3-(3′,4′-dihydroxyphenyl)propionic acid were also tested in a model of human adenoma cells LT97, suggesting their role in chemoprevention in the gut by modulation of the expression of cyclooxygenase 2 (COX-2) and glutathione S-transferase (GSTT2) (Miene *et al.*, 2011). Again, 3′,4′-dihydroxyphenylacetic acid, along with other colonic polyphenol metabolites like 3′,4′-dihydroxyphenylpropionic acid and trans-4-hydroxy-3-methoxycinnamic acid, is able to inhibit inflammation and in particular dextran sodium sulfate-induced colitis lipid peroxidation and DNA damage in colon mucosa. To a different degree, and with 3′,4′-dihydroxyphenylpropionic acid as the best performer, these substances also down-regulate fundamental cytokines involved in the inflammatory process (TNF-$\alpha$, IL-1b, and IL-8), reduce hyperalgesia, and inhibit $PGE_2$ production (Larrosa *et al.*, 2009).

## 8.5 *In vivo* studies

Despite the abundance of polyphenol *in vitro* investigations and the relative availability of animal studies performed on rats and pigs, research regarding the benefits of polyphenols in human health has only recently begun. Phenolic compounds, including flavonoids, have received increasing attention because of significant epidemiological evidence of associations between the consumption of polyphenol-rich foods and the prevention of chronic diseases such as cancer, CVDs, and neurodegenerative diseases. However the cause–effect relationship between polyphenol consumption and prevention of several pathologies is not completely understood. In this section, we describe the impact of flavonols and flavan-3-ols on human health and prevention of inflammation, CVD, neuropathologies, and cancer.

## 8.5.1 Cardiovascular and endothelial protection

### 8.5.1.1 Flavan-3-ols

Epidemiological and medical investigations suggest that flavanol-rich foods exert cardiovascular health benefits, supporting an association between the intake of flavanol-containing foods and a decreased risk of diseases, in particular CVDs (Bayard *et al.*, 2007; Desch *et al.*, 2010; Heiss *et al.*, 2010).

Cocoa is one food that is particularly rich in flavan-3-ols. It has been studied due to its implication in CVD prevention (Murphy *et al.*, 2003; Fraga *et al.*, 2005; Grassi *et al.*, 2005a,b, 2008; Taubert *et al.*, 2007a,b; Crews *et al.*, 2008; Hooper *et al.*, 2008; Muniyappa *et al.*, 2008; Desch *et al.*, 2010). Dark chocolate showed similar effects to cocoa after 2 weeks of consumption, but without changes in blood pressure or oxidative and lipid profiles (Engler *et al.*, 2004; Shiina *et al.*, 2009). Several meta-analyses were conducted in order to evaluate the effects of dark chocolate on modification of some physiological parameters linked to cardiovascular protection. A meta-analysis conducted by Desch *et al.* (2010) selected 10 randomized controlled trials comprising healthy, normotensive, pre- and hypertensive individuals. The results reported a decrease of 4.5 mm Hg in systolic and 2.5 mm Hg in diastolic blood pressure after intake of flavanol-rich chocolate or flavanol-rich cocoa for 2–18 weeks. In contrast, a more recent meta-analysis of 20 randomized controlled studies showed that systolic blood pressure decreased by 1.63 mm Hg but diastolic blood pressure and heart rate were unchanged after the consumption of flavanol-rich cocoa (Shrime *et al.*, 2011). One of the parameters considered an important marker for vascular health is endothelial function. The ingestion of flavanol-rich cocoa was associated with acute elevations in levels of circulating NO species, an enhanced flow-mediated vasodilation (FMD) response of conduit arteries, and an augmented microcirculation. In addition, the concentrations and chemical profiles of circulating flavanol metabolites were determined, and multivariate regression analyses identified EC and its metabolite, EC-7-*O*-glucuronide, as independent predictors of vascular effect after flavanol-rich cocoa ingestion (Schroeter *et al.*, 2006). Similarly, Vlachopoulos *et al.* (2005) described an improvement of endothelial function, measured by FMD of the brachial artery, aortic augmentation index (AIx), and carotid-femoral pulse wave velocity (PWV), up to 3 hours after consumption of dark chocolate in healthy adults. Another recent meta-analysis conducted by Hooper *et al.* (2012) included 42 acute or short-term chronic (≥18 weeks) randomized controlled trials comprising 1297 participants. They found consistent acute and chronic benefits of chocolate or cocoa on FMD and previously unreported promising effects on other important CVD risk factors, such as insulin and homeostasis model assessment of insulin resistance (HOMA-IR). Cocoa intake was also investigated for its contribution to platelet function, and in particular for its capacity to inhibit platelet activation (Rein *et al.*, 2000; Pearson *et al.*, 2002; Murphy *et al.*, 2003; Holt *et al.*, 2006). For example, Rein *et al.* (2000) described cocoa consumption suppressing

ADP- or epinephrine-stimulated platelet activation and platelet microparticle formation in 30 healthy subjects before and 2 and 6 hours after ingestion of a cocoa beverage.

Green tea is another flavanol-rich product that has been extensively studied for its possible implication in CVD prevention, also as a consequence of its widespread and regular consumption in specific populations (Kuriyama *et al.*, 2006a). A population-based prospective cohort study (the Ohsaki Study) was performed to examine the association between green-tea consumption and mortality from CVD in 40 530 persons in Japan. Green-tea consumption was found to be inversely associated with mortality due to all causes and due to CVD (Kuriyama *et al.*, 2006a). The FMD was improved in humans 30 minutes after consumption of 450 mL of green tea (Alexopoulos *et al.*, 2008). In healthy smokers, the forearm blood flow was improved upon consumption of 400 mL of green tea containing almost 250 mg EGCG (Nagaya *et al.*, 2004). Black tea with high amounts of EGCG could also ameliorate endothelial function in humans, to an equivalent degree to green tea (Jochmann *et al.*, 2008). Besides polyphenol-rich foods, the putative effect of phenolics on vascular function was also assessed by consumption of pure chemicals. Oral intake of pure EC in humans closely emulated the acute vascular effects of flavanol-rich cocoa (Schroeter *et al.*, 2006). Furthermore, treatment with 300 mg EGCG in subjects with coronary heart disease improved their endothelial function (Widlansky *et al.*, 2007). However, evidence coming from the administration of pure polyphenols should be interpreted with caution as its translation to dietary habits may be extremely difficult or even misleading. In fact, as recently demonstrated, the role of other phytochemicals and fibers present in whole-consumed fruits, vegetables, and plant-derived foods should not be overlooked. These co-ingested substances may play a relevant role in absorption, distribution, and availability through the digestive tract, and ultimately in the bioavailability of bioactive polyphenols and their metabolites. In particular, the binding of polyphenolic substances to polysaccharides of the plant cell wall may be beneficial for their bioactivities and alter their metabolization as, if compared to pure polyphenols, these bound compounds can be transported with higher efficacy through the stomach and small intestine to the colon, where a higher portion can undergo biotransformation from the resident microflora (Padayachee *et al.*, 2012).

The data on the potential health benefits of dietary flavan-3-ols, especially in the context of cardiovascular health, are considerable and continue to accumulate. However, a comprehensive understanding of the properties of flavan-3-ols is still lacking and further *in vivo* studies should be carried out, keeping in consideration the contribution of their metabolites, produced and circulating after human ingestion.

#### *8.5.1.2 Flavonols*

Quercetin has also been reported to be associated with protection against CVD (Hooper *et al.*, 2008; Egert *et al.*, 2009). The study conducted by Egert *et al.* (2009) was undertaken to examine the effects of quercetin supplementation on blood pressure, lipid metabolism, markers of oxidative stress, inflammation, and body composition in an at-risk population of 93 overweight or obese subjects aged 25–65 years with metabolic syndrome traits. Subjects were randomized to receive 150 mg quercetin per day in a double-blinded, placebo-controlled crossover trial with 6-week treatment periods separated by a 5-week washout period. Quercetin decreased systolic blood pressure by 2.6 mm Hg in the entire study group, by 2.9 mm Hg in the subgroup of hypertensive subjects, and by 3.7 mm Hg in the subgroup of younger adults aged 25–50 years. Quercetin also decreased serum HDL-cholesterol concentrations, while total cholesterol, TAG, and the LDL : HDL-cholesterol and TAG : HDL-cholesterol ratios were unaltered. Quercetin significantly decreased plasma concentrations of atherogenic oxidized LDL, but did not affect TNF-$\alpha$ and C-reactive protein when compared with placebo (Egert *et al.* 2009).

A study conducted by Loke *et al.* (2008) assessed a randomized, placebo-controlled, crossover trial in 12 healthy men to compare the acute effects of the oral administration of 200 mg quercetin on nitric oxide, endothelin-1, and oxidative stress after nitric oxide. Quercetin was shown to increase nitric oxide status and reduce endothelin-1 concentration, and may therefore improve endothelial function.

A meta-analysis of prospective cohort studies assessed quantitatively the strength of the association between flavonol intake and stroke incidence. Data from six cohorts involving 111 067 persons with at least 2155 nonfatal and fatal cases were pooled. A high intake of flavonols was inversely associated with non-fatal and fatal stroke (Hollman *et al.*, 2010).

## 8.5.2 *Neuroprotection*

### 8.5.2.1 *Flavan-3-ols*

Flavan-3-ols are involved in the prevention of neurodegenerative pathologies (Weinreb *et al.*, 2012). Several reports describe green-tea catechins reaching the brain and exerting a neuroprotective role (Levites *et al.*, 2001; Rezai-Zadeh *et al.*, 2008), but human clinical trials evaluating the impact of green tea on neurodegeneration are scarce.

A study conducted by Kuriyama *et al.* (2006b) analyzed cross-sectional data from a community-based comprehensive geriatric assessment (CGA) conducted in 2002 in order to evaluate the association between green-tea consumption and cognitive function in humans. The study (1003 Japanese subjects) demonstrated that higher consumption of green tea was associated with a lower prevalence of cognitive impairment. Moreover, another, more recent cross-sectional study examined the relationship between green-tea intake and cognitive impairment and decline among community-living Chinese adults. The results confirm that green-tea consumption is inversely associated with cognitive impairment but not with cognitive decline (Ng *et al.*, 2008). One of the few clinical dietary trials carried out was conducted by the Michael J. Fox Foundation. A multicenter, double-blind, randomized, placebo-controlled study was assessed to investigate the safety, tolerability, and potential neuroprotective effects of green-tea polyphenols in patients with early Parkinson's disease. The data indicate that the antioxidant/metal-chelating attributes of the catechin polyphenols are unlikely to serve as the sole explanation for their neuroprotective and neurorescue capacity (Chan *et al.*, 2009).

## 8.5.3 *Cancer prevention*

### 8.5.3.1 *Flavan-3-ols*

Recently, flavonoids have been studied for their potential activities against several cancers. Epidemiological studies investigated the association between cancer incidence and intake of fruit, vegetables, and tea (Favero *et al.*, 1998; Feskanich *et al.*, 2000; Zhang *et al.*, 2000; González *et al.*, 2006; Benetou *et al.*, 2008), though some described no association (Yuan *et al.*, 2011). The regular consumption of fruit, vegetable, or green tea may guarantee a high intake of flavan-3-ols because of their large distribution in plants. In an Italian case–control study mainly focused on oligomeric flavan-3-ols, an inverse association with colorectal cancer was estimated by a high intake of proanthocyanidins, which emerged for all DP (also for more than decamers) except monomeric flavan-3-ols (Rossi *et al.*, 2010). Two trials have investigated the effects of green-tea extracts on oral cancer. One was a double-blind intervention trial conducted in patients with oral mucosa leukoplakia using a mixed-tea product: 59 oral-mucosa leukoplakia patients were randomly divided into a treated group and treated with tea extract for 6 months. The results provide some direct evidence of the protective effects of tea on oral cancer. In contrast, the second study observed no reductions in oral premalignant lesions or histology after a 12-week green tea-extract supplementation, nor any changes in cancer incidence 27 months after trial completion. In another double-blind, placebo-controlled study, 60 volunteers with prostate intraepithelial neoplasia (HG-PIN) were treated with daily administration of three GTCs capsules, 200 mg each (total 600 mg/day). After 1 year, the tumor was diagnosed in only one subject of the treated group, whereas nine cancers were diagnosed in the 30-strong control group. Overall, the data available on the chemopreventive properties of green tea neither confirm nor refute a definitive cancer-preventive role, in particular when experimental data are translated into epidemiological evidence (Yuan *et al.*, 2011). The role of bioavailability in this regard has not yet been determined.

#### *8.5.3.2 Flavonols*

Quercetin has been described in *in vitro* studies as a chemopreventing compound. Its cancer-preventive effects have been attributed to various mechanisms, including the induction of cell-cycle arrest and/or apoptosis and its antioxidant functions in tumor cell lines, but data on *in vivo* studies are still too poor to provide conclusive results.

A recent study investigated the association between quercetin and risk of anatomic (cardia/noncardia) and histologic (intestinal and diffuse) subtypes of gastric cancer, using data from a large Swedish population-based case–control study of gastric cancer (505 cases and 1116 controls). High dietary quercetin intake was inversely related to the risk of noncardia gastric adenocarcinoma, and the protection appears to be particularly strong for women exposed to oxidative stress, such as tobacco smoking (Ekström *et al.*, 2011). Another large case–control study conducted in metropolitan Detroit, MI ($n = 2664$) aimed to evaluate dietary intake of quercetin and risk of colorectal cancer. Increased quercetin intake had no protective effects when tea intake was high. Interestingly, increased intake of quercetin was associated with increased risk of distal colon cancer when total fruit intake was low (OR for the highest vs. the lowest quartile = 1.99). These results suggest that quercetin can have disparate effects on colon-cancer risk depending on whether dietary intakes of fruit or tea is high, and that quercetin has protective effects only on proximal, not distal, colon cancer (Djuric *et al.*, 2012).

## 8.6 Conclusions

A number of human epidemiological and cohort studies have outlined an inverse correlation between the incidence of several human diseases, including cardiovascular, metabolic, and neurodegenerative disorders, and the adherence to specific diets, consumption of fruits, vegetables, wholemeal foods, and teas rich in polyphenols. Human intervention studies have supported this association, but an unambiguous and exhaustive pattern has not yet emerged concerning the mechanisms underlying the diet-related health properties ascribed to polyphenols. It is now increasingly evident that the direct action of secondary metabolites available *in planta* is insufficient (if not misleading) to fully explain the advantages of dietary regimens rich in polyphenolic compounds. To fully elucidate this scenario, a better comprehension of the interplay between human and microbial metabolism and dietary polyphenols might be required. With very few exceptions, polyphenols undergo rearrangements and modifications once they enter the digestive tract and the properties and physiological concentrations of these metabolites almost certainly differ from those of their parent compounds.

Flavonols and flavan-3-ols do not represent an exception in this regard, as their monomers (irrespective of whether or not they are glycosylated), oligomers, and polymers are extensively metabolized by the intestinal microflora and by human enzymatic pools. These substances and their human metabolites have been shown to possess some interesting biological properties, but knowledge of their behavior and disposition is still far from complete. In many cases, the contribution and bioavailability of human and intestinal metabolites and catabolites should not be overlooked, particularly when the goal is the assessment of the overall contribution of polyphenols to human health. In fact, despite the growing amount of knowledge on the bioavailability of flavonols and flavan-3-ols and on the health-promoting properties of their metabolites, issues still remain to be resolved. For example, in many human intervention studies an extensive variability in the response of both intestinal microflora and phase II human enzymes has been reported, both at individual and at population level (Lampe, 2009). An improved characterization of genetic polymorphisms and microbial factors contributing to inter-individual differences in flavonoid disposition may clarify the role of dietary flavonols and flavan-3-ols in the reduction of disease risk. This better understanding is critical to developing dietary strategies designed to properly enhance the nutritional benefits of flavonoids. In this regard, the recent definition of different enterotypes, and in particular of three clearly distinguishable population types in the gut microbiome, may offer a realistic starting point and pave the way for interesting correlations and hypotheses concerning the colonic metabolism of polyphenols (Arumugam *et al.*, 2011).

Furthermore, the described behavior of flavonols and flavan-3-ols in the human body is not yet fully comprehensive, as equal dietary intake of polyphenols does not guarantee equal exposure of the human internal compartments to the same metabolites. Central to this problem is the need to better define the proper markers of polyphenol intake and to track their actual bioavailability in plasma, urine, and tissues. Referring to biomarkers like enteric valerolactones, benzoic acids, phenylacetic acids, phenylpropionic acids, and their glucuronides and/or sulfates may overcome some of the distinctive shortcomings in the measurement of dietary intake, particularly with respect to self-reporting in epidemiological and clinical studies.

Some of these metabolites may reach relevant concentrations in plasma and thus may exert significant biological effects, so that at least some of the purported benefits ascribed to flavonoids might be a consequence of their action. It is therefore essential that future human studies aimed at the evaluation of the health-promoting properties of flavonoids keep track of these metabolites. Finally, colonic metabolism and the function of catabolites generated therein has not yet been fully elucidated: some pathways in the microbial catabolism of flavonoids seem to be convergent, as some final products are common to many of the major substrates. This implies that the overall fingerprint of flavonoid catabolites produced and available at the systemic level may reveal a lower complexity and a limited number of chemical structures and active moieties compared to the spectrum of parent plant-derived polyphenols ingested through a varied diet.

## References

Alexopoulos, N., Vlachopoulos, C., Aznaouridis, K., *et al.* (2008) The acute effect of green tea consumption on endothelial function in healthy individuals. *European Journal of Cardiovascular Prevention and Rehabilitation*, 15, 300–305.

Al-Shalmani, S., Suri, S., Hughes, D.A., *et al.* (2011) Quercetin and its principal metabolites, but not myricetin, oppose lipopolysaccharide-induced hyporesponsiveness of the porcine isolated coronary artery. *British Journal of Pharmacology*, 162, 1485–1497.

Appeldoorn, M.M., Vincken, J.P., Aura, A.M., *et al.* (2009) Procyanidin dimers are metabolized by human microbiota with 2-(3,4-dihydroxyphenyl)acetic acid and 5-(3,4-dihydroxyphenyl)-gamma-valerolactone as the major metabolites. *Journal of Agricultural and Food Chemistry*, 57, 1084–1092.

Aron, P.M. & Kennedy, J.A. (2008) Flavan-3-ols: nature, occurrence and biological activity. *Molecular Nutrition and Food Research*, 52, 79–104.

Arts, I.C. & Hollman, P.C. (2005) Polyphenols and disease risk in epidemiologic studies. *American Journal of Clinical Nutrition*, 81, 317S–325S.

Arumugam, M., Raes, J., Pelletier, E. *et al.* (2011) Enterotypes of the human gut microbiome. *Nature*, 473, 174–180.

Auger, C., Mullen, W., Hara, Y. & Crozier, A. (2008) Bioavailability of polyphenon E flavan-3-ols in humans with an ileostomy. *Journal of Nutrition*, 138, 1535–1542.

Aura, A.M., O'Leary, K.A., Williamson, G., *et al.* (2002) Quercetin derivatives are deconjugated and converted to hydroxyphenylacetic acids but not methylated by human fecal flora in vitro. *Journal of Agricultural and Food Chemistry*, 50, 1725–1730.

Bayard, V., Chamorro, F., Motta, J. & Hollenberg, N.K. (2007) Does flavanol intake influence mortality from nitric oxide-dependent processes? Ischemic heart disease, stroke, diabetes mellitus, and cancer in Panama. *International Journal of Medical Sciences*, 4, 53–58.

Bazzocco, S., Mattila, I., Guyot, S., *et al.* (2008) Factors affecting the conversion of apple polyphenols to phenolic acids and fruit matrix to short-chain fatty acids by human faecal microbiota in vitro. *European Journal of Nutrition*, 47, 442–452.

Benetou, V., Orfanos, P., Lagiou, P., *et al.* (2008) Vegetables and fruits in relation to cancer risk: evidence from the Greek EPIC cohort study. *Cancer Epidemiology, Biomarkers and Prevention*, 17, 387–392.

Bettuzzi, S., Brausi, M., Rizzi, F., *et al.* (2006) Chemoprevention of human prostate cancer by oral administration of green tea catechins in volunteers with high-grade prostate intraepithelial neoplasia: a preliminary report from a one-year proof-of-principle study. *Cancer Research*, 66, 1234–1240.

Boesch-Saadatmandi, C., Loboda, A., Wagner, A.E., *et al.* (2011) Effect of quercetin and its metabolites isorhamnetin and quercetin-3–glucuronide on inflammatory gene expression: role of miR-155. *Journal of Nutritional Biochemistry*, 22, 293–239.

Bonaccorsi, P., Caristi, C., Gargiulli, C. & Leuzzi, U. (2005) Flavonol glucoside profile of southern Italian red onion (Allium cepa L.) *Journal of Agricultural and Food Chemistry*, 53, 2733–2740.

Calani, L., Del Rio, D., Callegari, M.L., *et al.* (2012) Updated bioavailability and 48 h excretion profile of flavan-3-ols from green tea in humans. *International Journal of Food Science and Nutrition*, 63, 513–521.

Chan, P., Qin, Z., Zheng, Z., *et al.* (2009) A randomized, double-blind, placebo controlled, delayed start study to assess safety, tolerability and efficacy of green tea polyphenols in Parkinson's disease. *Proceedings of the XVIII WFN World Congress on Parkinson's Disease and Related Disorders*, Miami Beach, FL, USA, vol. 15, 145.

Chao, C.L., Hou, Y.C., Chao, P.D., *et al.* (2009) The antioxidant effects of quercetin metabolites on the prevention of high glucose-induced apoptosis of human umbilical vein endothelial cells. *British Journal of Nutrition*, 101, 1165–1170.

Cho, J.M., Chang, S.Y., Kim, D.B., *et al.* (2012) Effects of physiological quercetin metabolites on interleukin-1$\beta$-induced inducible NOS expression. *Journal of Nutritional Biochemistry*, 23, 1394–1402.

Chun, O.K., Chung, S.J. & Song W.O. (2007) Estimated dietary flavonoid intake and major food sourced of US adults. *Journal of Nutrition*, 137, 1244–1252.

Comalada, M., Camuesco, D., Sierra, S., *et al.* (2005) In vivo quercitrin anti-inflammatory effect involves release of quercetin, which inhibits inflammation through down-regulation of the NF-kappaB pathway. *European Journal of Immunology*, 35, 584–592.

Crews, W.D., Harrison, D.W. & Wright, J.W. (2008) A double-blind, placebo-controlled, randomized trial of the effects of dark chocolate and cocoa on variables associated with neuropsychological functioning and cardiovascular health: clinical findings from a sample of healthy, cognitively intact older adults. *American Journal of Clinical Nutrition*, 87, 872–880.

Crozier, A., Jaganath, I.B. & Clifford, M.N. (2009) Dietary phenolics: chemistry, bioavailability and effects on health. *Natural Product Reports*, 26, 1001–1043.

Crozier, A., Del Rio, D. & Clifford, M.N. (2010) Bioavailability of dietary flavonoids and phenolic compounds. *Molecular Aspects of Medicine*, 31, 446–467.

Dall'Asta, M., Calani, L., Tedeschi, M., *et al.* (2012a) Identification of microbial metabolites derived from in vitro fecal fermentation of different polyphenolic food sources. *Nutrition*, 28, 197–203.

Dall'Asta, M., Derlindati, E., Ardigò, D., *et al.* (2012b) Macrophage polarization: the answer to the diet/inflammation conundrum? *Nutrition, Metabolism and Cardiovascular Diseases*, 22, 387–392.

Day, A.J., Cañada. F.J., Díaz. J.C., *et al.* (2000) Dietary flavonoid and isoflavone glycosides are hydrolysed by the lactase site of lactase phlorizin hydrolase. *FEBS Letters*, 468, 166–170.

Day, A.J., Mellon, F., Barron, D., *et al.* (2001) Human metabolism of dietary flavonoids: identification of plasma metabolites of quercetin. *Free Radical Research*, 35, 941–952.

Del Rio, D., Stewart, A.J., Mullen, W., *et al.* (2004) HPLC-MS$^n$ analysis of phenolic compounds and purine alkaloids in green and black tea. *Journal of Agricultural and Food Chemistry*, 52, 2807–2815.

Del Rio, D., Calani, L., Cordero, C., *et al.* (2010a) Bioavailability and catabolism of green tea flavan-3-ols in humans. *Nutrition*, 26, 1110–1116.

Del Rio, D., Costa, L.G., Lean, M.E. & Crozier, A. (2010b) Polyphenols and health: what compounds are involved? *Nutrition, Metabolism and Cardiovascular Diseases*, 20, 1–6.

Del Rio, D., Calani, L., Dall'Asta, M. & Brighenti, F. (2011) Polyphenolic composition of hazelnut skin. *Journal of Agricultural and Food Chemistry*, 59, 9935–9941.

Déprez, S., Brezillon, C., Rabot, S., *et al.* (2000) Polymeric proanthocyanidins are catabolized by human colonic microflora into low-molecular-weight phenolic acids. *Journal of Nutrition*, 130, 2733–2738.

Desch, S., Schmidt, J., Kobler, D., *et al.* (2010) Effect of cocoa products on blood pressure: systematic review and meta-analysis. *American Journal of Hypertension*, 23, 97–103.

Djuric, Z., Severson, R.K. & Kato, I. (2012) Association of dietary quercetin with reduced risk of proximal colon cancer. *Nutrition and Cancer*, 64, 351–360.

Donnini, S., Finetti, F., Lusini, L., *et al.* (2006) Divergent effects of quercetin conjugates on angiogenesis. *British Journal of Nutrition*, 95, 1016–1023.

Dueñas, M., González-Manzano, S., González-Paramás, A. & Santos-Buelga, C. (2010) Antioxidant evaluation of O-methylated metabolites of catechin, epicatechin and quercetin. *Journal of Pharmaceutical and Biomedical Analysis*, 51, 443–449.

Egert, S., Bosy-Westphal, A., Seiberl, J., *et al.* (2009) Quercetin reduces systolic blood pressure and plasma oxidised low-density lipoprotein concentrations in overweight subjects with a high-cardiovascular disease risk phenotype: a double-blinded, placebo-controlled cross-over study. *British Journal of Nutrition*, 102, 1065–1074.

Ekström, A.M., Serafini, M., Nyrén, O., *et al.* (2011) Dietary quercetin intake and risk of gastric cancer: results from a population-based study in Sweden. *Annals of Oncology*, 22, 438–443.

Engler, M.B., Engler, M.M., Chen, C.Y., *et al.* (2004) Flavonoid-rich dark chocolate improves endothelial function and increases plasma epicatechin concentrations in healthy adults. *Journal of the American College of Nutrition*, 23, 197–204.

Erlund, I. (2004) Review of the flavonoids quercetin, hesperetin, and naringenin. Dietary sources, bioactivities, bioavailability, and epidemiology. *Nutrition Research*, 24, 851–874.

Favero, A., Parpinel, M. & Franceschi, S. (1998) Diet and risk of breast cancer: major findings from an Italian case-control study. *Biomedicine and Pharmacotherapy*, 52, 109–115.

Feskanich, D., Ziegler, R.G., Michaud, D.S., *et al.* (2000) Prospective study of fruit and vegetable consumption and risk of lung cancer among men and women. *Journal of the National Cancer Institute*, 92, 1812–1823.

Fraga, C.G., Actis-Goretta, L., Ottaviani, J.I., *et al.* (2005) Regular consumption of a flavanol-rich chocolate can improve oxidant stress in young soccer players. *Clinical and Developmental Immunology*, 12, 11–17.

Glässer, G., Graefe, E.U., Struck, F., *et al.* (2002) Comparison of antioxidative capacities and inhibitory effects on cholesterol biosynthesis of quercetin and potential metabolites. *Phytomedicine*, 9, 33–40.

González, C.A., Pera, G., Agudo, A., *et al.* (2006) Fruit and vegetable intake and the risk of stomach and oesophagus adenocarcinoma in the European Prospective Investigation into Cancer and Nutrition (EPIC-EURGAST). *International Journal of Cancer*, 118, 2559–2566.

Grassi, D., Lippi, C., Necozione, S., *et al.* (2005a) Short-term administration of dark chocolate is followed by a significant increase in insulin sensitivity and a decrease in blood pressure in healthy persons. *American Journal of Clinical Nutrition*, 81, 611–614.

Grassi, D., Necozione, S., Lippi, C., *et al.* (2005b) Cocoa reduces blood pressure and insulin resistance and improves endothelium-dependent vasodilation in hypertensives. *Hypertension*, 46, 398–405.

Grassi. D., Desideri, G., Necozione , *et al.* (2008) Blood pressure is reduced and insulin sensitivity increased in glucose-intolerant, hypertensive subjects after 15 days of consuming high-polyphenol dark chocolate. *Journal of Nutrition*, 138, 1671–1676.

Grimm, T., Schäfer, A. & Högger, P. (2004) Antioxidant activity and inhibition of matrix metalloproteinases by metabolites of maritime pine bark extract (pycnogenol). *Free Radical Biology and Medicine*, 36, 811–822.

Gu, L., Kelm, M.A., Hammerstone, J.F., *et al.* (2004) Concentrations of proanthocyanidins in common foods and estimations of normal consumption. *Journal of Nutrition*, 134, 613–617.

Heiss, C., Keen, C.L. & Kelm, M. (2010) Flavanols and cardiovascular disease prevention. *European Heart Journal*, 31, 2583–2592.

Hellström, J.K., Törrönen, A.R. & Mattila, P.H. (2009) Proanthocyanidins in common food products of plant origin. *Journal of Agricultural and Food Chemistry*, 57, 7899–7906.

Hertog, M.G., Feskens, E.J., Hollman, P.C., *et al.* (1993a) Dietary antioxidant flavonoids and risk of coronary heart disease: the Zutphen Elderly Study. *Lancet*, 342, 1007–1011.

Hertog, M.G.L., Hollman, P.C.H., Katan, M.B. & Kromhout, D. (1993b) Intake of potentially anticarcinogenic flavonoids and their determinants in adults in The Netherlands. *Nutrition and Cancer*, 20, 21–29.

Hertog, M.G., Kromhout, D., Aravanis, C., *et al.* (1995) Flavonoid intake and long-term risk of coronary heart disease and cancer in the seven countries study. *Archives of Internal Medicine*, 155, 381–386.

Hirota, S., Nishioka, T., Shimoda, T., *et al.* (2001) Quercetin glucosides are hydrolyzed to quercetin in human oral cavity to participate in peroxidase-dependent scavenging of hydrogen peroxide. *Food Science and Technology Research*, 7, 239–245.

Hollman, P.C., Geelen, A. & Kromhout, D. (2010) Dietary flavonol intake may lower stroke risk in men and women. *Journal of Nutrition*, 140, 600–604.

Holt, R.R., Lazarus, S.A., Sullards, M.C., *et al.* (2002) Procyanidin dimer B2 [epicatechin-(4beta-8)-epicatechin] in human plasma after the consumption of a flavanol-rich cocoa. *American Journal of Clinical Nutrition*, 76, 798–804.

Holt, R.R., Actis-Goretta, L., Momma, T.Y. & Keen, C.L. (2006) Dietary flavanols and platelet reactivity. *Journal of Cardiovascular Pharmacology*, 47, S187–196, disc. S206–209.

Hong, Y.J. & Mitchell, A.E. (2004) Metabolic profiling of flavonol metabolites in human urine by liquid chromatography and tandem mass spectrometry. *Journal of Agricultural and Food Chemistry*, 52, 6794–6801.

Hooper, L., Kroon, P.A., Rimm, E.B., *et al.* (2008) Flavonoids, flavonoid-rich foods, and cardiovascular risk: a meta-analysis of randomized controlled trials. *American Journal of Clinical Nutrition*, 88, 38–50.

Hooper, L., Kay, C., Abdelhamid, A., *et al.* (2012) Effects of chocolate, cocoa, and flavan-3-ols on cardiovascular health: a systematic review and meta-analysis of randomized trials. *American Journal of Clinical Nutrition*, 95, 740–751.

Inocencio, C., Rivera, D., Alcaraz, F. & Tomas-Barberan, F.A. (2000) Flavonoid content of commercial capers (*Capparis spinosa*, *C. sicula* and *C. orientalis*) produced in Mediterranean countries. *European Food Research and Technology*, 212, 70–74.

Jaganath, I.B., Mullen, W., Edwards, C.A. & Crozier, A. (2006) The relative contribution of the small and large intestine to the absorption and metabolism of rutin in man. *Free Radical Research*, 40, 1035–1046.

Jaganath, I.B., Mullen, W., Lean, M.E., *et al.* (2009) In vitro catabolism of rutin by human fecal bacteria and the antioxidant capacity of its catabolites. *Free Radical Biology and Medicine*, 47, 1180–1189.

Jochmann, N., Lorenz, M., Krosigk, A., *et al.* (2008) The efficacy of black tea in ameliorating endothelial function is equivalent to that of green tea. *British Journal of Nutrition*, 99, 863–868.

Kahle, K., Huemmer, W., Kempf, M., *et al.* (2007) Polyphenols are intensively metabolized in the human gastrointestinal tract after apple juice consumption. *Journal of Agricultural and Food Chemistry*, 55, 10605–10614.

Kalili, K.M. & de Villiers, A. (2010) Off-line comprehensive two-dimensional hydrophilic interaction x reversed phase liquid chromatographic analysis of green tea phenolics. *Journal of Separation Science*, 33, 853–863.

Katz, D.L., Doughty, K. & Ali, A. (2011) Cocoa and chocolate in human health and disease. *Antioxidants and Redox Signaling*, 15, 2779–2811.

Knaze, V., Zamora-Ros, R., Luján-Barroso, L., *et al.* (2011) Intake estimation of total and individual flavan-3–ols, proanthocyanidins and theaflavins, their food sources and determinants in the European Prospective Investigation into Cancer and Nutrition (EPIC) study. *British Journal of Nutrition*, 108, 1095–1108.

Koga, T. & Meydani, M. (2001) Effect of plasma metabolites of (+)-catechin and quercetin on monocyte adhesion to human aortic endothelial cells. *American Journal of Clinical Nutrition*, 73, 941–948.

Kuriyama, S., Hozawa, A., Ohmori, K., *et al.* (2006a) Green tea consumption and cognitive function: a cross-sectional study from the Tsurugaya Project 1. *American Journal of Clinical Nutrition*, 83, 355–361.

Kuriyama, S., Shimazu, T., Ohmori, K., *et al.* (2006b) Green tea consumption and mortality due to cardiovascular disease, cancer, and all causes in Japan: the Ohsaki study. *Journal of the American Medical Association*, 296, 1255–1265.

Kutschera, M., Engst, W., Blaut, M. & Braune, A. (2011) Isolation of catechin-converting human intestinal bacteria. *Journal of Applied Microbiology*, 111, 165–175.

Lakenbrink, C., Engelhardt, U.H. & Wray, V. (1999) Identification of two novel proanthocyanidins in green tea. *Journal of Agricultural and Food Chemistry*, 47, 4621–4624.

Lambert, J.D., Rice, J.E., Hong, J., *et al.* (2005) Synthesis and biological activity of the tea catechin metabolites, M4 and M6 and their methoxy-derivatives. *Bioorganic and Medicinal Chemistry Letters*, 15, 873–876.

Lampe, J. W. (2009) Interindividual differences in response to plant-based diets: implications for cancer risk. *American Journal of Clinical Nutrition*, 89, 1553S–1557S.

Larrosa, M., Luceri, C., Vivoli, E., *et al.* (2009) Polyphenol metabolites from colonic microbiota exert anti-inflammatory activity on different inflammation models. *Molecular Nutrition and Food Research*, 53, 1044–1054.

Leth, T. & Justesen, U. (1998) Analysis of flavonoids in fruits, vegetables and beverages by HPLC-UV and LC- MS and estimation of the total daily flavonoid intake in Denmark. In: *Polyphenols in Food* (eds R. Armadò, H. Andersson, S. Bardócz & F. Serra), pp. 39–40. Office for Official Publications of the European Communities, Luxembourg.

Levites, Y., Weinreb, O., Maor, G., *et al.* (2001) Green tea polyphenol (−)-epigallocatechin-3-gallate prevents N-methyl-4-phenyl-1,2,3,6-tetrahydropyridine-induced dopaminergic neurodegeneration. *Journal of Neurochemistry*, 78, 1073–1082.

Li, C., Lee, M.J., Sheng, S., *et al.* (2000) Structural identification of two metabolites of catechins and their kinetics in human urine and blood after tea ingestion. *Chemical Research in Toxicology*, 13, 177–184.

Llorach, R., Garrido, I., Monagas, M., *et al.* (2010) Metabolomics study of human urinary metabolome modifications after intake of almond (*Prunus dulcis* (Mill.) D.A. Webb) skin polyphenols. *Journal of Proteome Research*, 9, 5859–5867.

Lodi, F., Jimenez, R., Moreno, L., *et al.* (2009) Glucuronidated and sulfated metabolites of the flavonoid quercetin prevent endothelial dysfunction but lack direct vasorelaxant effects in rat aorta. *Atherosclerosis*, 204, 34–39.

Loke, W.M., Hodgson, J.M., Proudfoot, J.M., *et al.* (2008) Pure dietary flavonoids quercetin and (−)-epicatechin augment nitric oxide products and reduce endothelin-1 acutely in healthy men. *American Journal of Clinical Nutrition*, 88, 1018–1025.

Lotito, S.B., Zhang, W.J., Yang, C.S., *et al.* (2011) Metabolic conversion of dietary flavonoids alters their anti-inflammatory and antioxidant properties. *Free Radical Biology and Medicine*, 51, 454–463.

Lu, H., Meng, X., Li, C., *et al.* (2003) Glucuronides of tea catechins: enzymology of biosynthesis and biological activities. *Drug Metabolism and Disposition*, 31, 452–461.

Manach, C., Scalbert, A., Morand, C., *et al.* (2004) Polyphenols: food sources and bioavailability. *American Journal of Clinical Nutrition*, 79, 727–747.

Meng, X., Sang, S., Zhu, N., *et al.* (2002) Identification and characterization of methylated and ring-fission metabolites of tea catechins formed in humans, mice, and rats. *Chemical Research and Toxicology*, 15, 1042–1050.

Miene, C., Weise, A. & Glei, M. (2011) Impact of polyphenol metabolites produced by colonic microbiota on expression of COX-2 and GSTT2 in human colon cells (LT97) *Nutrition and Cancer*, 63, 653–662.

Monagas, M., Garrido, I., Lebrón-Aguilar, R., *et al.* (2009) Comparative flavan-3-ol profile and antioxidant capacity of roasted peanut, hazelnut, and almond skins. *Journal of Agricultural and Food Chemistry*, 57, 10590–10599.

Mullen, W., Boitier, A., Stewart, A.J. & Crozier, A. (2004) Flavonoid metabolites in human plasma and urine after the consumption of red onions: analysis by liquid chromatography with photodiode array and full scan tandem mass spectrometric detection. *Journal of Chromatography A*, 1058, 163–168.

Mullen, W., Edwards, C.A & Crozier, A. (2006) Absorption, excretion and metabolite profiling of methyl-, glucuronyl-, glucosyl- and sulpho-conjugates of quercetin in human plasma and urine after ingestion of onions. *British Journal of Nutrition*, 96, 107–116.

Mullen, W., Borges, G., Donovan, J.L., *et al.* (2009) Milk decreases urinary excretion but not plasma pharmacokinetics of cocoa flavan-3-ol metabolites in humans. *American Journal of Clinical Nutrition*, 89, 1784–1791.

Muniyappa, R., Hall, G., Kolodziej, T.L., *et al.* (2008) Cocoa consumption for 2 wk enhances insulin-mediated vasodilatation without improving blood pressure or insulin resistance in essential hypertension. *American Journal of Clinical Nutrition*, 88, 1685–1696.

Murphy, K.J., Chronopoulos, A.K., Singh, I., *et al.* (2003) Dietary flavanols and procyanidin oligomers from cocoa (Theobroma cacao) inhibit platelet function. *American Journal of Clinical Nutrition*, 77, 1466–1473.

Nagaya, N., Yamamoto, H., Uematsu, M., *et al.* (2004) Green tea reverses endothelial dysfunction in healthy smokers. *Heart*, 90, 1485–1486.

Natsume, M., Osakabe, N., Yasuda, A., *et al.* (2004) In vitro antioxidative activity of (−)-epicatechin glucuronide metabolites present in human and rat plasma. *Free Radical Research*, 38, 1341–1348.

Németh, K., Plumb, G.W., Berrin, J.G., *et al.* (2003) Deglycosylation by small intestinal epithelial cell beta-glucosidases is a critical step in the absorption and metabolism of dietary flavonoid glycosides in humans. *European Journal of Nutrition*, 42, 29–42.

Ng, T.P., Feng, L., Niti, M., *et al.* (2008) Tea consumption and cognitive impairment and decline in older Chinese adults. *American Journal of Clinical Nutrition*, 88, 224–231.

Padayachee, A., Netzel, G., Netzel, M., *et al.* (2012) Binding of polyphenols to plant cell wall analogues – Part 1: Anthocyanins. *Food Chemistry*, 134, 155–161.

Palafox-Carlos, H., Ayala-Zavala, J.F. & González-Aguilar, G.A. (2011) The role of dietary fiber in the bioaccessibility and bioavailability of fruit and vegetable antioxidants. *Journal of Food Science*, 76, 6–15.

Pashikanti, S., de Alba, D.R., Boissonneault, G.A. & Cervantes-Laurean, D. (2010) Rutin metabolites: novel inhibitors of nonoxidative advanced glycation end products. *Free Radical Biology and Medicine*, 48, 656–663.

Pavlica, S. & Gebhardt, R. (2010) Protective effects of flavonoids and two metabolites against oxidative stress in neuronal PC12 cells. *Life Sciences*, 86, 79–86.

Pearson, D.A., Paglieroni, T.G., Rein, D., *et al.* (2002) The effects of flavanol-rich cocoa and aspirin on ex vivo platelet function. *Thrombosis Research*, 106, 191–197.

Prasain, J.K. & Barnes, S. (2007) Metabolism and bioavailability of flavonoids in chemoprevention: current analytical strategies and future prospectus. *Molecular Pharmacology*, 4, 846–864.

Rein, D., Paglieroni, T.G., Wun, T., *et al.* (2000) Cocoa inhibits platelet activation and function. *American Journal of Clinical Nutrition*, 72, 30–35.

Rezai-Zadeh, K., Arendash, G.W., Hou, H., *et al.* (2008) Green tea epigallocatechin-3-gallate (EGCG) reduces beta-amyloid mediated cognitive impairment and modulates tau pathology in Alzheimer transgenic mice. *Brain Research*, 1214, 177–187.

Rios, L.Y., Gonthier, M.P., Rémésy, C., *et al.* (2003) Chocolate intake increases urinary excretion of polyphenol-derived phenolic acids in healthy human subjects. *American Journal of Clinical Nutrition*, 77, 912–918.

Roowi, S., Stalmach, A., Mullen, W., *et al.* (2010) Green tea flavan-3-ols: colonic degradation and urinary excretion of catabolites by humans. *Journal of Agricultural and Food Chemistry*, 58, 1296–1304.

Rossi, M., Negri, E., Parpinel, M., *et al.* (2010) Proanthocyanidins and the risk of colorectal cancer in Italy. *Cancer Causes Control*, 21, 243–250.

Sang, S., Lee, M.J., Yang, I., *et al.* (2008) Human urinary metabolite profile of tea polyphenols analyzed by liquid chromatography/electrospray ionization tandem mass spectrometry with data-dependent acquisition. *Rapid Communications in Mass Spectrometry*, 22, 1567–1578.

Scalbert, A., Johnson, I.T. & Saltmarsh, M. (2005) Polyphenols: antioxidants and beyond. *American Journal of Clinical Nutrition*, 81, 215S–217S.

Schneider, H., Schwiertz, A., Collins, M.D. & Blaut, M. (1999) Anaerobic transformation of quercetin-3-glucoside by bacteria from the human intestinal tract. *Archives of Microbiology*, 171, 81–91.

Schroeter, H., Heiss, C., Balzer, J., *et al.* (2006) (−)-epicatechin mediates beneficial effects of flavanol-rich cocoa on vascular function in humans. *Proceedings of the National Academy of Sciences*, 103, 1024–1029.

Selma, M.V., Espín, J.C. & Tomás-Barberán, F.A. (2009) Interaction between phenolics and gut microbiota: role in human health. *Journal of Agricultural and Food Chemistry*, 57, 6485–6501.

Shiina, Y., Funabashi, N., Lee, K., *et al.* (2009) Acute effect of oral flavonoid-rich dark chocolate intake on coronary circulation, as compared with non-flavonoid white chocolate, by transthoracic Doppler echocardiography in healthy adults. *International Journal of Cardiology*, 131, 424–429.

Shirai, M., Moon, J.H., Tsushida, T. & Terao, J. (2001) Inhibitory effect of a quercetin metabolite, quercetin 3-O-beta-D-glucuronide, on lipid peroxidation in liposomal membranes. *Journal of Agricultural and Food Chemistry*, 49, 5602–5608.

Shirai, M., Yamanishi, R., Moon, J.H., *et al.* (2002) Effect of quercetin and its conjugated metabolite on the hydrogen peroxide-induced intracellular production of reactive oxygen species in mouse fibroblasts. *Biosciences, Biotechnology and Biochemistry*, 66, 1015–1021.

Shrime, M.G., Bauer, S.R., McDonald, A.C., *et al.* (2011) Flavonoid-rich cocoa consumption affects multiple cardiovascular risk factors in a meta-analysis of short-term studies. *Journal of Nutrition*, 141, 1982–1988.

da Silva, E.L., Piskula, M.K., Yamamoto, N., *et al.* (1998) Quercetin metabolites inhibit copper ion-induced lipid peroxidation in rat plasma. *FEBS Letters*, 3, 405–408.

Slimestad, R., Fossen, T. & Vågen, I.M. (2007) Onions: a source of unique dietary flavonoids. *Journal of Agricultural and Food Chemistry*, 55, 10067–10080.

Stalmach, A., Troufflard, S., Serafini, M. & Crozier A. (2009) Absorption, metabolism and excretion of Choladi green tea flavan-3-ols by humans. *Molecular Nutrition and Food Research*, 53, S44–S53.

Stalmach, A., Mullen, W., Steiling, H., *et al.* (2010) Absorption, metabolism, and excretion of green tea flavan-3-ols in humans with an ileostomy. *Molecular Nutrition and Food Research*, 54, 323–334.

Stoupi, S., Williamson, G., Drynan, J.W., *et al.* (2010) A comparison of the in vitro biotransformation of (−)-epicatechin and procyanidin B2 by human faecal microbiota. *Molecular Nutrition and Food Research*, 54, 747–759.

Takagaki, A., Otani, S. & Nanjo, F. (2011) Antioxidative activity of microbial metabolites of (−)-epigallocatechin gallate produced in rat intestines. *Biosciences, Biotechnology and Biochemistry*, 75, 582–585.

Taubert, D., Roesen, R., Lehmann, C., *et al.* (2007a) Effects of low habitual cocoa intake on blood pressure and bioactive nitric oxide: a randomized controlled trial. *Journal of the American Medical Association*, 298, 49–60.

Taubert, D., Roesen, R. & Schömig, E. (2007b) Effect of cocoa and tea intake on blood pressure: a meta-analysis. *Archives of Internal Medicine*, 167, 626–634.

Tribolo, S., Lodi, F., Connor, C., *et al.* (2008) Comparative effects of quercetin and its predominant human metabolites on adhesion molecule expression in activated human vascular endothelial cells. *Atherosclerosis*, 197, 50–56.

Tsao, R., Yang, R., Young, J.C. & Zhu, H. (2003) Polyphenolic profiles in eight apple cultivars using high-performance liquid chromatography (HPLC) *Journal of Agricultural and Food Chemistry*, 51, 6347–6353.

Urpi-Sarda, M., Monagas, M., Khan, N., *et al.* (2009) Targeted metabolic profiling of phenolics in urine and plasma after regular consumption of cocoa by liquid chromatography-tandem mass spectrometry. *Journal of Chromatography A*, 1216, 7258–7267.

Urpi-Sarda, M., Llorach, R., Khan, N., *et al.* (2010) Effect of milk on the urinary excretion of microbial phenolic acids after cocoa powder consumption in humans. *Journal of Agricultural and Food Chemistry*, 58, 4706–4711.

Verzelloni, E., Pellacani, C., Tagliazucchi, D., *et al.* (2011) Antiglycative and neuroprotective activity of colon-derived polyphenol catabolites. *Molecular Nutrition and Food Research*, 55, S35–S43.

Vlachopoulos, C., Aznaouridis, K., Alexopoulos, N., *et al.* (2005) Effect of dark chocolate on arterial function in healthy individuals. *American Journal of Hypertension*, 18, 785–791.

Vrhovsek, U., Rigo, A., Tonon, D. & Mattivi, F. (2004) Quantitation of polyphenols in different apple varieties. *Journal of Agricultural and Food Chemistry*, 52, 6532–6538.

Walle, T., Browning, A.M., Steed, L.L., *et al.* (2005) Flavonoid glucosides are hydrolyzed and thus activated in the oral cavity in humans. *Journal of Nutrition*, 135, 48–52.

Weinreb, O., Amit, T., Youdim, M.B.H. & Mandel, S. (2012) Green tea flavan-3-ols and their role in protecting against Alzheimer's and Parkinson's disease pathophysiology. In: *Flavonoids and Related Compounds: Bioavailability and Function* (eds J.P.E. Spencer & A. Crozier), vol. 30, pp. 331–362. Oxidative Stress and Disease, CRC Press, Boca Raton, FL.

Widlansky, M.E., Hamburg, N.M., Anter, E., *et al.* (2007) Acute EGCG supplementation reverses endothelial dysfunction in patients with coronary artery disease. *Journal of the American College of Nutrition*, 26, 95–102.

Williamson, G. & Clifford, M.N. (2010) Colonic metabolites of berry polyphenols: the missing link to biological activity? *The British Journal of Nutrition*, 104, S48–S66.

Williamson, G. & Manach, C. (2005) Bioavailability and bioefficacy of polyphenols in humans. II. Review of 93 intervention studies. *American Journal of Clinical Nutrition*, 81, 243S–255S.

Wittig, J., Herderich, M., Graefe, E.U. & Veit, M. (2001) Identification of quercetin glucuronides in human plasma by high-performance liquid chromatography-tandem mass spectrometry. *Journal of Chromatography B – Biomedical Sciences Applications*, 753, 237–243.

Wojdylo, A., Oszmiański, J. & Laskowski, P. (2008) Polyphenolic compounds and antioxidant activity of new and old apple varieties. *Journal of Agricultural and Food Chemistry*, 56, 6520–6530.

Yang, J.H., Hsia, T.C., Kuo, H.M., *et al.* (2006) Inhibition of lung cancer cell growth by quercetin glucuronides via G2/M arrest and induction of apoptosis. *Drug Metabolism and Disposition*, 34, 296–304.

Yeh, S.L., Yeh, C.L., Chan, S.T. & Chuang, C.H. (2011) Plasma rich in quercetin metabolites induces G2/M arrest by upregulating PPAR-$\gamma$ expression in human A549 lung cancer cells. *Planta Medica*, 77, 992–998.

Yuan, J.M., Sun, C. & Butler, L.M. (2011) Tea and cancer prevention: epidemiological studies. *Pharmacological Research*, 64, 123–135.

Zamora-Ros, R., Knaze, V., Luján-Barroso, L., *et al.* (2011). Estimated dietary intakes of flavonols, flavanones and flavones in the European Prospective Investigation into Cancer and Nutrition (EPIC) 24 hour dietary recall cohort. *British Journal of Nutrition*, 106, 1915–1925.

Zhang, S.M., Hunter, D.J., Rosner, B.A., *et al.* (2000) Intakes of fruits, vegetables, and related nutrients and the risk of non-Hodgkin's lymphoma among women. *Cancer, Epidemiology Biomarkers and Prevention*, 9, 477–485.

Chapter 9

# Inhibition of VEGF Signaling by Polyphenols in Relation to Atherosclerosis and Cardiovascular Disease

Rebecca L. Edwards[1] and Paul A. Kroon[2]

[1] *Institute of Food Research, Norwich Research Park, Norwich, UK*

[2] *Polyphenols and Health Group, Institute of Food Research, Norwich Research Park, Norwich, UK*

**Abstract:** Angiogenesis is an important component of the process of atherosclerosis. It has recently been demonstrated that VEGF, a potent angiogenic growth factor, increases the rate of atherosclerosis in animal models. There is a significant body of literature concerned with the ability of polyphenols to alter VEGF signaling and angiogenesis. However, the potential for polyphenols to influence atherosclerosis as a result of inhibiting VEGF signaling and VEGF-dependent angiogenesis has not been considered. This chapter summarizes and reviews the existing literature concerning the effects of polyphenols on VEGF signaling, angiogenesis, and atherosclerosis. The data suggest that some polyphenols can inhibit VEGF signaling, as well as angiogenesis and atherosclerosis, via a variety of mechanisms and, in some cases, at physiologically relevant concentrations. However, there is a lack of published data from studies with polyphenol metabolites, and the majority of reports are from studies using aglycones at supraphysiological concentrations. Data from epidemiological studies show there is a positive correlation between polyphenol consumption and reduced risk of cardiovascular disease, but the role of specific compounds is difficult to determine from these. Future research should focus on establishing whether or not the reported antiatherosclerotic effects of polyphenol-rich foods and extracts are due to the polyphenols inhibiting VEGF signaling. Further research is required to elucidate the influence of bioavailability, metabolism, and combinations of polyphenols found in foods on VEGF signaling.

**Keywords:** cardiovascular disease, angiogenesis, dietary polyphenols, flavonoids, diet and health, human metabolism

---

*Recent Advances in Polyphenol Research, Volume 4*, First Edition.
Edited by Annalisa Romani, Vincenzo Lattanzio, and Stéphane Quideau.

## 9.1 Introduction

Cardiovascular diseases (CVDs) are a major cause of human mortality, and also cause substantial morbidity associated with significant reductions in quality of life. Indeed, CVD is the single biggest cause of death in Europe (47% of the total in 2012). Evidence from epidemiological studies investigating the relationships between habitual diets and disease generally shows an inverse relationship between the quantities of flavonoids and other polyphenols consumed and the incidence of CVD. In addition, there are now hundreds of reports from randomized controlled trials (RCTs) describing significant beneficial effects of consuming certain polyphenols and polyphenol-rich foods on CVD risk factors and biomarkers of vascular health. However, evidence to support plausible mechanisms underlying the observed health benefits in humans is largely lacking.

Atherosclerosis is the major cause of CVD. Atherosclerosis is the process by which arterial tissues become inflamed and damaged and take up lipid-laden macrophages, leading to the formation of atherosclerotic plaques. These can grow over long periods of time and may eventually become unstable, leading to rupture and a very high risk that a major vessel such as the aortic or carotid artery becomes blocked. There are a number of possible mechanisms by which polyphenols might interfere with atherosclerosis and hence reduce the risk of a subsequent cardiovacular event. A major difficulty has been establishing that mechanisms that can be demonstrated *in vitro* are relevant *in vivo*, and especially that the low concentrations of polyphenols found in blood post-consumption are sufficient to elicit the responses observed in the laboratory.

Recently, our search for physiologically relevant targets of polyphenols led us to discover that some polyphenols could inhibit vascular endothelial growth factor (VEGF) signaling at nanomolar concentrations (Paul Kroon, Christina Moyle, Ana Belen Cerezo-Lopez, Rebecca Edwards, Wendy Hollands, Mark Winterbone, unpublished data). In addition, there are numerous published reports describing the ability of polyphenols to inhibit VEGF signaling via a number of mechanisms, and these data are reviewed here. This is an exciting observation for several reasons, most notably that the data are consistent with inhibition of VEGF signaling *in vivo* following consumption of polyphenol-rich foods and diets. But further, it has previously been demonstrated that (i) angiogenesis has a central role in atherosclerosis, (ii) increased VEGF concentrations dramatically increase the rate of growth of established atherosclerotic plaques and their propensity to rupture, and (iii) circulating VEGF concentrations are increased in high CVD risk groups such as diabetics and those with a history of vascular disease, and exacerbated angiogenesis is observed in diabetic atherosclerosis. In this chapter, the evidence to support the notion that dietary polyphenols can inhibit VEGF signaling, and that this results in reduced atherosclerosis and CVD risk, will be explored.

## 9.2 VEGF and VEGF signaling

VEGF (also called VEGF-A) is a growth factor belonging to the cystine-knot superfamily. Members of this family of growth factors are characterized by the presence of eight

**Table 9.1** VEGF isoforms and their functions, giving the isoforms of the main VEGF molecule VEGF-A, their ability to initiate VEGFR signaling, and a structural background explaining these functions.

| Isoform | Exon 6 | Exon 7 | ECM binding | Receptor recognition | Functionality |
|---|---|---|---|---|---|
| $VEGF_{121}$ | × | × | No binding; found in circulation | VEGFR-2 | Poor functionality due to lack of heparin binding |
| $VEGF_{145}$ | ✓ | × | Moderate binding to ECM | VEGFR-2 | Able to activate VEGFR-2 but not any other receptors |
| $VEGF_{165}$ | × | ✓ | Moderate binding to ECM | VEGFR-2, VEGFR1, neuropilins | The major active isoform, with moderate heparin binding |
| $VEGF_{189}$ | ✓ | ✓ | Strong binding to ECM | N/A | Rapidly sequestered due to high affinity for heparin; not seen in circulation |
| $VEGF_{209}$ | ✓ | ✓ | Strong binding to ECM | N/A | Rapidly sequestered due to high affinity for heparin; not seen in circulation |
| $VEGF_{110}$ (formed by proteolysis) | × | × | No binding; found in circulation | VEGFR-2 | Acts like $VEGF_{121}$ with poor functionality due to lack of heparin binding |

ECM, extracellular matrix; N/A, not applicable.

conserved cysteine residues, which are involved in forming disulfide bonds or cystine linkages (R–S–S–R); the polypeptide sequence connecting two of the disulfide bonds forms a loop through which a third disulfide bond passes, which effectively forms a stabilized rotaxane substructure. There are five structurally related forms of VEGF found in humans: VEGF-A, VEGF-B, VEGF-C, VEGF-D, and placental growth factor (PlGF). Each form is involved in eliciting different downstream outcomes. VEGF-A is the predominant form in humans and is usually referred to simply as VEGF, as will be done from here on.

VEGF has been found as five different isoforms: $VEGF_{121}$, $VEGF_{145}$, $VEGF_{165}$, $VEGF_{189}$, and $VEGF_{206}$, with $VEGF_{165}$ being the major one (Ferrara *et al.*, 2003). The various isoforms are the result of differential exon splicing events, which give rise to different polypeptide sequences, each with variations in function; these are summarized in Table 9.1. A further isoform can be produced by plasmin cleavage, which gives rise to $VEGF_{110}$; this behaves in a similar way to $VEGF_{121}$, as exon 6 and exon 7 are excised during cleavage (Houck *et al.*, 1992). To initiate VEGF receptor (VEGFR) signaling, a moderate binding to the extracellular matrix (ECM) is desired, allowing the VEGF molecule to be attracted to the functional location without being totally sequestered. The VEGF protein is generally found as a homodimer, although a VEGF-A/PLGF heterodimer with a threefold decrease in mitogenic activity has also been observed (DiSalvo *et al.*, 1995). The two monomers are organized antiparallel and fixed by disulfide bonds, with a receptor binding site present at both poles of the molecule (Muller *et al.*, 1997a,b).

VEGF is a mitogenic ligand that exerts its effects by binding to VEGFRs. Although there is a small family of VEGFRs (VEGFR-1, VEGFR-2, and VEGFR-3), by far the most

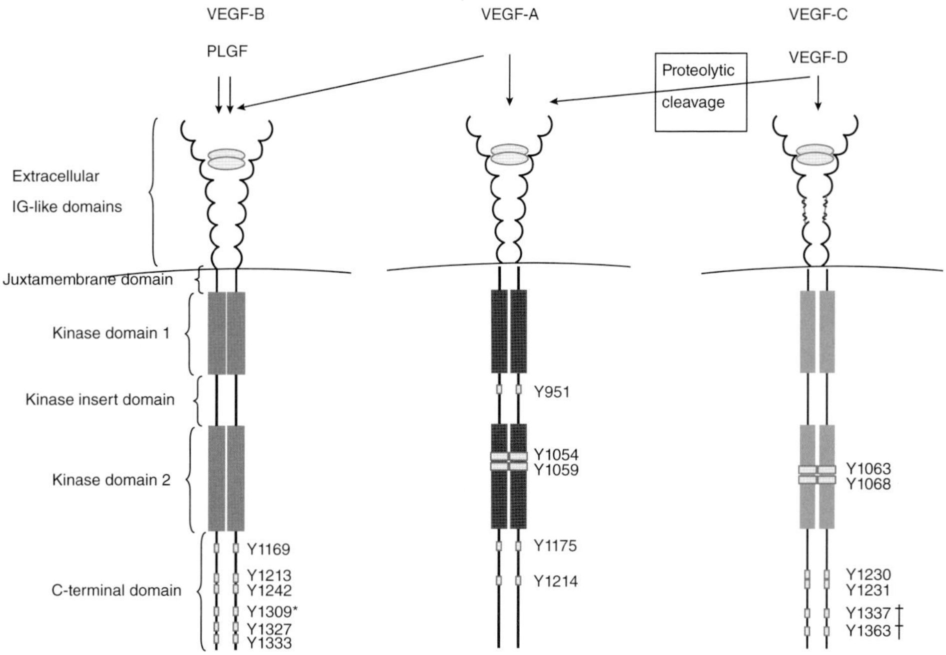


**Fig. 9.1** VEGFR structure and main autophosphorylation sites. VEGFR-1 is shown on the left with its main ligand, VEGF-B. VEGFR-2 is in the center with its main ligand, VEGF-A. VEGFR-3 is on the right with its main ligands, VEGF-C and VEGF-D. –S–S– indicates a disulfide bond. Arrows show other receptors each ligand can bind to and the ovals represent the binding location of the antiparallel ligand. The known/partially known functional phosphorylation sites on each receptor are indicated with the residue number to the right. *Phosphorylation and signaling from this residue occur only when a PlGF homo- or heterodimer ligand is bound. †Phosphorylation at this residue occurs only when the receptor is a homodimer. IG, immunoglobulin; PlGF, placental growth factor; VEGF, vascular endothelial growth factor; VEGFR, VEGF receptor; S, sulfur; Y, tyrosine.

important receptor involved in binding VEGF-A and transmitting strong downstream signaling activity is VEGFR-2. Hence, VEGF-A–VEGFR-2 interactions are the single most important VEGF-mediated initiator of pro-angiogenic signals in multiple tissues, including vascular tissues and cancers. VEGFR protein structures are relatively conserved, with each containing an extracellular Ig-like domain-containing region, a short transmembrane domain, a juxtamembrane kinase domain, and a receptor tyrosine kinase domain separated into two parts by a kinase insert region (Fig. 9.1). Receptors may be homodimeric or heterodimeric (Huang *et al.*, 2001) and it has been suggested that each combination can produce varying signal pathways, due to differences in the blocking of phosphorylation sites (Dixelius *et al.*, 2003). Soluble forms of VEGFR-1 and VEGFR-2 are formed through alternative splicing or proteolytic cleavage and regulate signaling by "mopping up" secreted VEGF before the ligand can reach the receptor (Barleon *et al.*, 2001; Ebos *et al.*, 2004).

VEGF dimer binding to the VEGFR causes receptor coupling, bringing the phosphorylation regions on each receptor monomer into close proximity with one another. This causes autophosphorylation at Y1054/1059 sites leading to the regulation of phosphorylation at the other tyrosine sites, stimulating the primary signalling molecules to begin

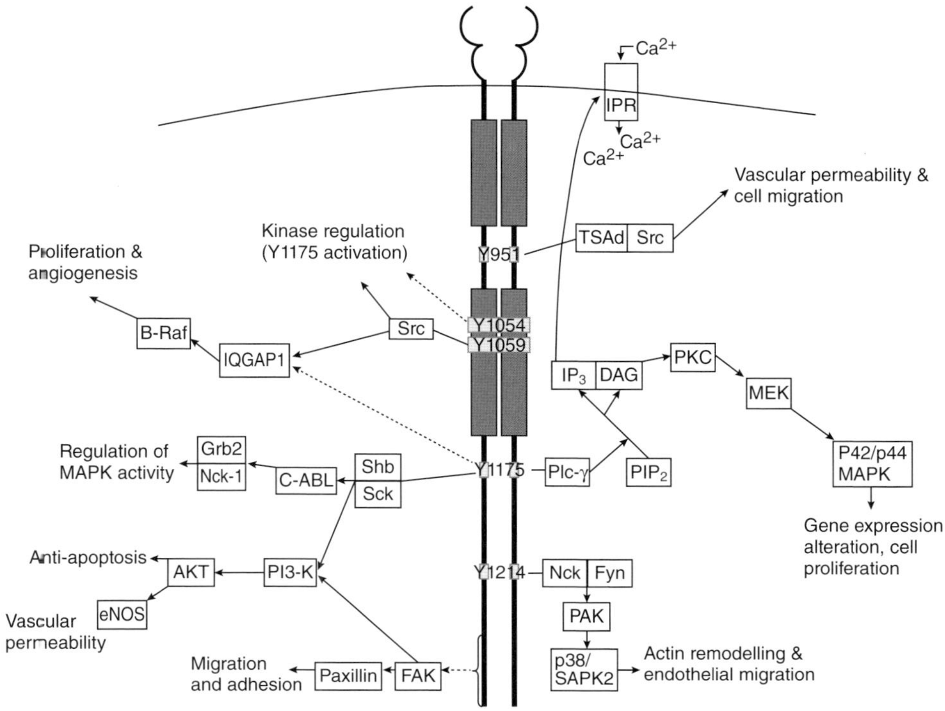


**Fig. 9.2** VEGF-A/VEGFR-2 signaling pathways in endothelial cells. The receptor is depicted as in Fig. 9.1. A solid line indicates a primary signaling molecule binding to the phosphorylation site. A solid arrow indicates known binding/activation. A dashed arrow indicates an unknown pathway leading to a known molecule/outcome. Written annotations shows the physiological outcomes of cell signaling. AKT, protein kinase B; B-Raf, serine/threonine protein kinase B; $Ca^{2+}$, calcium; C-ABL, Abelson Murine leukemia viral oncogene homolog (mammalian); DAG, diacyl glycerol; eNOS, endothelial nitric oxide synthase; ERK, extracellular signal-related kinase; FAK, focal adhesion kinase; Fyn, proto-oncogene tyrosine-protein kinase Fyn; Grb2, growth factor receptor-bound protein 2; IQGAP1, IQ motif containing GTPase activating protein 1; $IP_3$, inositol triphosphate; IPR, IP receptor; MAPK, mitogen activated protein kinase; MEK, MAPK and ERK 1/2 kinase; Nck, Nck adaptor protein; p38/SAPK2, MAPK 11; p42/p44, p42/p44 MAPK; PAK, P21 protein (Cdc42/Rac)-activated kinase 1; PI3-K, phosphatidylinositol 3-kinase; $PIP_2$, phosphatidylinositol biphosphate; PKC, protein kinase C; Plc-$\gamma$, phospholipase C gamma; Sck, SHC-like protein; Shb, SH2 domain containing adaptor protein B; Src, v-src avian sarcoma (Schmidt-Ruppin A-2) viral oncogene homolog; TSAd, T-cell-specific adapter molecule; Y, tyrosine.

downstream signaling and leading to physiological outcomes such as nitric oxide (NO) production and increased vascular permeability, cell survival, and angiogenesis (Fig. 9.2). Binding of VEGF-family proteins is aided by interactions with neuropilins, coreceptors to the VEGFRs, which assist in presenting the VEGF molecule to the VEGFR with consequent signaling.

Many of the downstream signaling pathways of VEGFR-1 remain to be elucidated, but VEGFR-1 signaling strongly stimulates recruitment of hematopoietic progenitor cells (HPCs), vascular permeability, and hematopoiesis, as well as maintaining normal vascular development and regulating VEGFR-2 signaling. VEGFR-3 is mainly involved in lymphangiogenesis and is typically only expressed in such tissues. The main angiogenic signaling occurring in the vasculature is VEGFR-2-dependent triggering of cell proliferation, cell migration, antiapoptotic pathways, and release of vasoregulatory molecules. Heterodimerization of the receptors produces altered signaling patterns, which has been suggested to be a regulatory process, although experimental data to support this notion are still limited. A summary of VEGFRs, ligands, and signaling is provided in Table 9.2.

Further regulation of VEGF signaling occurs through a variety of systems. Under hypoxic conditions, stabilized hypoxia-inducible factors (HIFs) can bind readily to a region on the VEGF-A promoter up-regulating VEGF-A and VEGFR-1 (Pugh & Ratcliffe, 2003). Phosphotyrosine kinases and normal receptor internalization/proteasome degradation are also thought to regulate VEGF signaling, as well as structural orientation; for example, phosphorylation of VEGFR-1 can be restricted by the juxtamembrane region or point mutations, resulting in lack of function through altered protein conformation.

## 9.3 VEGF signaling and angiogenesis

The importance of VEGF signaling in vascular function and angiogenesis has been demonstrated through knock-out studies. *Vegfr1*$^{-/-}$ mice display embryonic lethality between days E8.5 and E9. *Vegfr1*(TK)$^{-/-}$ mutants, which can still bind VEGF without causing a signal cascade, have reduced macrophage migration capabilities but develop normal vasculature (Hiratsuka *et al.*, 1998), supporting the idea that VEGFR-1 regulates VEGFR-2 angiogenesis during development (Roberts *et al.*, 2004) and modulates inflammation. *Vegfb*$^{-/-}$ mice displayed reduced vascular function but not pre- or postnatal mortality, suggesting VEGF-B/VEGFR-1 coupling is not essential for vascular development. *Vegfr-2*$^{-/-}$ displayed strong embryonic lethality due to defective vascular organization and formation, as well as limited hematopoiesis (Shalaby *et al.*, 1995).

Vegfa$^{+/-}$ *and Vegfa*$^{-/-}$ were shown to be genetically lethal between prenatal days E11–12 and E9.5–10.5, respectively, due to lack of vascular development. Isoform-specific VEGF transgenic mice only displayed normal phenotype in $VEGF_{164/164}$ (equivalent of human $VEGF_{165}$) mutants. VEGF-C and VEGF-D knock-outs affect only lymphatic development. Vascular defects were observed in coreceptor NRP1/2 knock-out or over-expression models, suggesting that alteration of the tight regulation of VEGF signaling can affect angiogenesis and vascular formation (Kitsukawa *et al.*, 1995).

## 9.4 Angiogenesis and atherosclerosis

The importance of angiogenesis and neovascularization within CVD has been known for a long time (see reviews by Moreno *et al.*, 2006; Mulligan-Kehoe, 2010). Normal intimal thickness requires no extra supply of blood vessels as the necessary nutrients are able to diffuse across the tissue, but this capability is limited to a distance of around 100 μm (Khurana *et al.*, 2005). During atherosclerotic development the intima expands beyond this diffusible distance and cannot receive the necessary supply of nutrients to continue growth, which explains the requirement for vascularization in advancing atherosclerotic plaques. A correlation between neovascularization and presence of foam cells and necrosis

**Table 9.2** Summary of VEGFRs, VEGF ligands, VEGF signaling, and functional outcomes, collating all known information on VEGF/VEGFR binding, phosphorylation sites, and primary signaling molecules and linking this with signaling pathways and physiological outcomes.

| Receptor dimer composition | Ligand | Activated tyrosine residues | Primary signaling molecule | Functionality |
|---|---|---|---|---|
| **VEGFR-1/ VEGFR-1** | VEGF, VEGF-B, PIGF, VEGF/PIGF | 1169 | PLC$\gamma$ | Mild activation of MAPK and cell proliferation |
| | | 1213/1333 | PI3K | Cell proliferation and survival. Negative regulation of VEGFR-2 |
| | | 1242/1327 | Unknown | Pathways remain to be elucidated |
| | | Unknown | Unknown | VEGFR-1-dependent hematopoiesis and HPC recruitment |
| | PIGF, VEGF/PIGF | 1309 | Unknown | Specific stimulation by PIGF ligands, downstream signals unknown |
| **VEGFR-2/ VEGFR-2** | VEGF, VEGF/PIGF | 951 | TSad | Vascular permeability and cell proliferation |
| | | 1054/1059 | Unknown | Kinase regulation |
| | | 1175 | Shb/Sck | AKT activation and antiapoptosis |
| | | | PLC$\gamma$ | PKC activation of MAPK and cell proliferation. eNOS stimulation and NO production |
| | | 1214 | Nck | P38/SAPK2 signalling and cell proliferation |
| | Altered VEGF-C, Altered VEGF-D | Unknown | PI3K | Reduced but sustained VEGFR-2 signaling. AKT and ERK1/2 activation of cell proliferation, survival, and increased NO levels |
| **VEGFR-3/ VEGFR-3** | VEGF-C, VEGF-D | 1063/1068 | Unknown | Kinase regulation |
| | | 1230/1231/ 1265 | Unknown | Unidentified pathways leading to lymphangiogenesis |
| | | 1337 | Shc | Only activated in VEGFR-3 homodimer. Grb2 recruitment and Ras-dependent MAPK signaling and proliferation |
| | | 1363 | unknown | Only activated in VEGFR-3 homodimer. Unknown function |
| **VEGFR-1/ VEGFR-2** | VEGF, PIGF | Unknown | Unknown | Vessel homeostasis. Limited proliferation but increased p-eNOS |
| | VEGF | Unknown | Unknown | P42/44 MAPK stimulation of COX-1/2 dependent prostacylin release |
| **VEGFR-2/ VEGFR-3** | VEGF-C/D, VEGF | 1063/1068 (R3), 0154/ 1059 (R2) | Unknown | Kinase regulation sites |
| | | 1230/1231/ 1265 | Unknown | Unidentified pathways leading to lymphangiogenesis |

AKT, protein kinase B; $Ca^{2+}$, calcium; eNOS, endothelial nitric oxide synthase; ERK, extracellular signal-related kinase; MAPK, mitogen activated protein kinase; MEK, MAPK and ERK 1/2 kinase; Nck, Nck adaptor protein; p38/SAPK2, MAPK 11; p42/p44 = p42/p44 MAPK; p-eNOS, phosphorylated eNOS; PI3-K, phosphatidylinositol 3-kinase; PKC, protein kinase C; Plc-$\gamma$, phospholipase C gamma; Sck, SHC-like protein; Shb, SH2 domain containing adaptor protein B; Src, v-src avian sarcoma (Schmidt-Ruppin A-2) viral oncogene homolog; TSAd, T-cell-specific adapter molecule; HPC, hematopoietic progenitor cells.

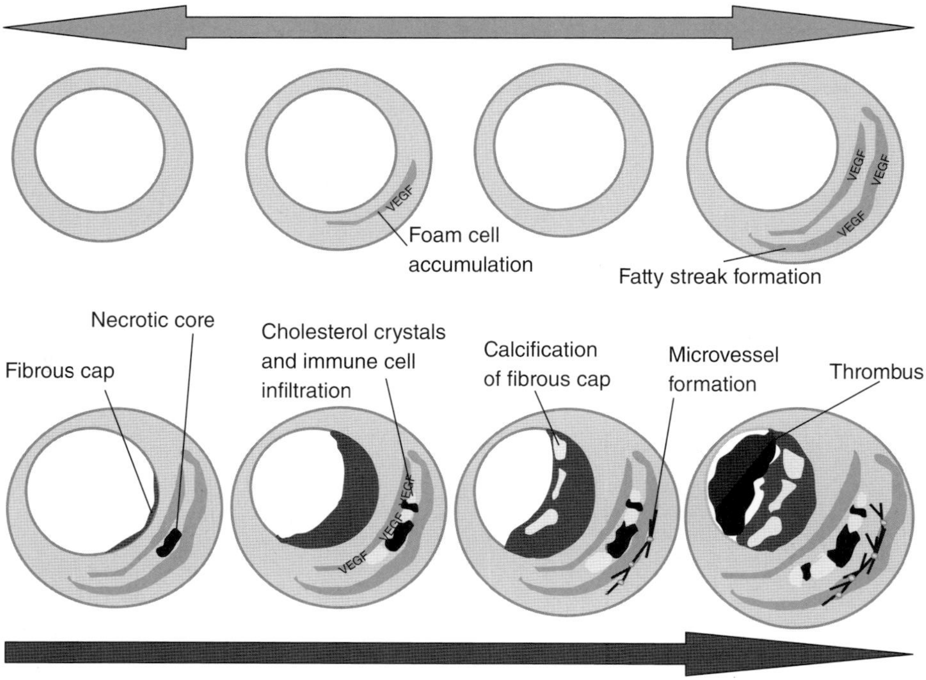


**Fig. 9.3** Initiation, regression, and continuation to progressive atherosclerosis, showing the known progression of atherosclerosis from a regressive state typical of young persons (double-headed arrow) to the non-regressive state typically seen in ages 50+ (single-headed arrow). The key factors in the progression of the disease are labeled on first appearance with a descriptive text. VEGF, vascular endothelial growth factor.

in plaque tissue obtained during endarterectomy procedures has been observed, suggesting a limitation to plaque development without microvessel formation (Hiyama *et al.*, 2010). A higher number of microvessels has been shown to correlate with intraplaque hemorrhage, leading to thrombus and fibrous-layer formation (Hiyama *et al.*, 2010) and calcification (Demer & Tintut, 2008), again suggesting the need for a vascular network and blood supply in the later stages of atherosclerotic disease. Further, aortic samples taken from atheroma and non-atheroma sites in $ApoE^{-/-}$ mice demonstrated significantly different angiogenic potential in an aortic-ring assay, with maximal sprouting of microvessels seen in the rings cut from atherosclerotic regions without growth-factor stimulation (Moulton *et al.*, 2003). A schematic of the process of atherosclerosis is shown in Fig. 9.3.

Mounting evidence supports the notion that growth factors and particularly VEGF contribute to atherosclerosis and can drive increases in rates of atherogenesis. For example, in a study in which plasma VEGF levels were compared with patient angiograms, it was observed that plasma VEGF levels were significantly higher in the critical legion group but not the non-critical legion group when compared to the control. This supports the theory that angiogenesis is important in the later stages of atherosclerosis progression (Kucukardali *et al.*, 2008). VEGFR-2 has also been shown to be present within human atherectomy samples, where the luminal cells from atheromatous sections display significantly increased staining compared with cells from normal tissues. Inohue *et al.* (1998) reported significantly higher expression of VEGF and VEGFR-1/2 in atherosclerotic regions of human carotid arteries, correlating to neovascularization and greater macrophage density.

Arguably the most important studies investigating the role of VEGF in atherosclerosis have been those from Michael Dake's laboratory (Stanford, CA). In particular, the data from a series of three studies provide clear evidence that VEGF can drive substantial increases in the growth and degree of plaque formation, and also demonstrate that anti-VEGF therapies blocked the pro-atherosclerotic effects induced by the hyper-VEGF conditions used in the experimental models. First, it was reported that the size (total cross-sectional area, cross-sectional plaque circumference, maximal plaque thickness) of atherosclerotic plaques in the aortic arches of high cholesterol-fed ApoB$^{-/-}$ ApoE$^{-/-}$ mice, a model of atherosclerosis, were substantially and significantly increased by a single low-dose (2 μg/kg b.w.) intraperitoneal injection of hrVEGF$_{165}$ (Celletti *et al.*, 2001a). Increased vascularity and greater infiltration of macrophages into plaques was also observed 3 weeks after the VEGF injection compared to the sham-injected control animals. The ability of a single rhVEGF$_{165}$ injection to induce hyperatherosclerosis was confirmed in another diet-induced atherosclerosis model (New Zealand white rabbits fed a high-cholesterol diet) (Celletti *et al.*, 2001b). Soon after these seminal findings were published, the Dake group reported that paclitaxel (mitotic growth inhibitor) and angiostatin (specific angiogenesis inhibitor) treatments were both able to prevent the VEGF-induced (2 μg/kg b.w. single injection) increases in intima–media ratios, macrophage infiltrate, and endothelial cell density in the diet-induced rabbit model of atherosclerosis (Celletti *et al.*, 2002). Together, these reports demonstrate that rhVEGF$_{165}$ drives hyperatherosclerosis by increasing the angiogenic potency of vascular tissues, and that antiangiogenic therapies can block this pro-atherosclerotic action of VEGF.

VEGF-A stimulation has also been shown to induce plaque progression by increasing membrane-bound immune adherence molecules such as vascular cellular adhesion molecule 1 (VCAM-1), which promote the infiltration of immune cells. Vaccination with VEGFR-2 DNA encoded in live attenuated bacteria resulted in a 77% reduction in the formation of new lesions and limited progression of atherosclerosis by 66% in hypercholesteremic models (Hauer *et al.*, 2007).

## 9.5 Polyphenols in foods and diets, and their absorption and metabolism

There are numerous polyphenols that are important components of plants, including those consumed as foods or used to make beverages. Here, as has become rather widespread, the term "polyphenols" will be used as a "catch-all" for all plant phenolics. The major types of dietary polyphenol are the flavonoids and the phenolic acids, which contribute approximately one-third and two-thirds of average total polyphenol intake, respectively. There are six classes of flavonoid: flavonols, such as quercetin and kaempferol; flavan-3-ols, such as catechin, epicatechin, epigallocatechin gallate, and proanthocyanidins; flavanones, such as hesperitin and naringenin; anthocyanidins, such as cyanidin, delphinidin, and pellargonidin; flavones, such as apigenin; and isoflavones, such as genistein and daidzein. Phenolic acids include various derivatives of benzoic acid (e.g. gallic, syringic, vanillic, and protocatechuic acids) and cinnamic acids (e.g. ferulic, *p*-coumaric, and sinapic acids). Other types of polyphenol include stilbenes (e.g. resveratrol), ellagitannins, and ellagic acid. Example structures are depicted in Fig. 9.4. Knowledge of what types of polyphenol are present in plant foods and their concentrations has improved markedly over the past decade, and there are several databases available containing extensive polyphenol composition data for foods (Gry *et al.*, 2007; Neveu *et al.*, 2010; Scalbert *et al.*, 2011; US Department of Agriculture, 2013). Using these resources, it has been estimated that the average polyphenol intake in France is around 1.2 g per day, with hydroxycinnamates, esters, and proanthocyanidins consumed in the largest amounts (Pérez-Jiménez *et al.*, 2011).

Only a fraction of any ingested polyphenol is absorbed from the gut and can be accounted for in plasma or urine, and there are large differences in bioavailability between different polyphenols. For example, the isoflavones daidzein and genistein are highly bioavailable and the flavanol epicatechin is moderately bioavailable, whereas quercetin is poorly bioavailable and oligomeric proanthocyanidins are hardly absorbed at all (Manach *et al.*, 2005). Except for the flavanols (catechins and procyanidins) that are synthesized by plants without additional sugars, all polyphenols are found in plants and foods as glycosides (typically O-glycosides, occasionally as C-glycosides) or as esters (e.g. caffeoyl-quinates=chlorogenic acids). The sugar residues on flavonoids are usually removed during absorption by human or microbial *β*-glucosidases (Németh *et al.*, 2003). Esterases are also present in the human intestine, where they split ester bonds to release cinnamic and phenolic acids from small soluble esters such as chlorogenic acids (Andreasen *et al.*, 2001), but also from the esters linking them to large insoluble fiber components such as arbinozylans (Kroon *et al.*, 1997). Other important considerations are human metabolism (especially phase 2 conjugation) and microbial catabolism, which can take place when polyphenols interact with the gut flora. Phase 2 conjugation appears to be highly efficient for both flavonoids and phenolic acids, because, almost without exception, the vast majority of polyphenols found in blood are in the form of sulfated and glucuronidated conjugates of the aglycones or their methylated derivatives (Kroon *et al.*, 2004). The few exceptions include minute quantities of anthocyanidin-glycosides (=anthocyanins), which are absorbed as such without cleavage of the sugar moiety, and galloylated catechins such as those found in green tea, which are usually absorbed without phase 2 conjugation. The reader is alerted to the fact that the majority of the *in vitro* studies (e.g. with cultured cells) reported in the scientific literature are performed with aglycones and not with the phase 2 conjugates that are the physiologically relevant forms (Kroon *et al.*, 2004).

## 9.6 Effects of polyphenols on VEGF signaling, angiogenesis, and atherosclerosis

There are a very large number of published reports of the effects of polyphenols on VEGF signaling, angiogenesis, and atherosclerosis. The research focuses on inhibition of VEGF-A signaling due to its function as the primary VEGF signaling factor in angiogenesis. Tables 9.3, 9.4, and 9.5 give the results of a survey of the literature for relevant reports.

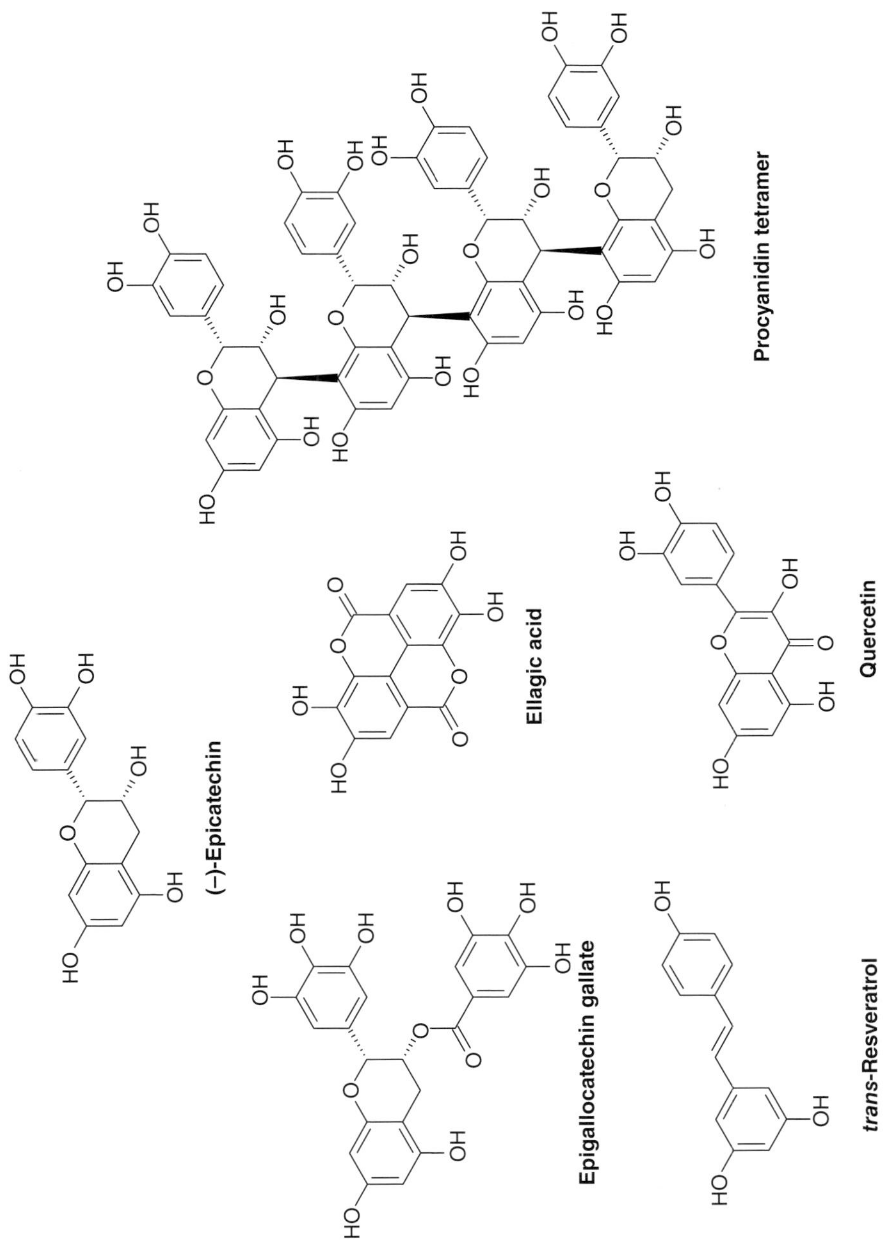


**Fig. 9.4** Structures of several of the polyphenols covered in this chapter.

**Table 9.3** Reported effects of polyphenols on VEGF signaling.

| PP-induced VEGF signaling-associated response | Assay | Compound | Concentration (% inhibition/ activation)[a] | References |
|---|---|---|---|---|
| ***Reduced VEGF gene expression*** | VEGF gene expression in resting HUVECs stimulated with PP for 24 hours; measurement of gene expression by RT-PCR and gel electrophoresis | Rutin | 25 μg/mL (almost complete) | Guruvayoorappan & Kuttan (2007) |
| | VEGF gene expression in HUAECs stimulated with hypoxia (low $O_2$) and PP for 6 hours; quantification of gene expression by RT-PCR and western blot | Apigenin | 20 μM (100%) | Osada *et al.* (2004) |
| | VEGF gene expression in hepatic tissue from animals fed NDEA hepatocellular carcinogen alongside pre- and post-feeding of PP; quantified by qRT-PCR | Naringenin | Pre-treat: 200 mg/kg (81.25%)<br>Post-treat: 200 mg/kg (87.5%) | Subramanian & Arul (2012) |
| | VEGF gene expression in OVCAR-3 cells stimulated with PP for 16 hours; quantification by qRT-PCR | Kaempferol | 40 μM (43%), 20 μM (27%) | Luo *et al.* (2009) |
| | VEGF gene expression in A2780/CP70 cells stimulated with PP for 16 hours; quantification by qRT-PCR | Kaempferol | 40 μM (28%), 20 μM (19%) | Luo *et al.* (2009) |
| | VEGF gene expression in MCF-7 cells stimulated with PP for 18 hours; quantification of gene expression by reporter gene luciferase activity | Quercetin<br>Silibinin<br>Apigenin<br>Morin<br>Naringenin | 30 μM (~74%)<br>30 μM (~65%)<br>30 μM (~47%)<br>30 μM (~43%)<br>30 μM (~28%) | Oh *et al.* (2010) |
| | VEGF gene expression in 24-hour SW837 cells stimulated with PP for 3 hours; quantification by qRT-PCR | EGCg | 25 μM (~66%) | Shimizu *et al.* (2010) |
| | VEGF gene expression in SW837 xenograft from BALB/c nude mice fed PP in drinking water for 35 days; quantified by qRT-PCR | EGCg | 0.1% in water (~53%) | Shimizu *et al.* (2010) |
| | VEGF gene expression from HUVECs stimulated with $H_2O_2$ with PP for 24 hours; quantification by qRT-PCR | Quercetin<br>Resveratrol | 1 μM (8%)<br>1 μM (12%) | Nicholson *et al.* (2010) |

| | | | | |
|---|---|---|---|---|
| ***Increased VEGF gene expression*** | VEGF gene expression from HUVECs stimulated with PP for 24 hours; quantification by qRT-PCR | Resveratrol | 1 µM (54%) | Nicholson *et al.* (2010) |
| | VEGF gene expression in HUVECs stimulated with PP for 4 hours; quantified by RT-PCR | Ferulic acid | 10 µM (~119%) | Lin *et al.* (2010) |
| | VEGF gene expression from HUVECs stimualted with $H_2O_2$ with PP for 24 hours; quantification by qRT-PCR | Quercetin | 1 µM (8%) | Nicholson *et al.* (2010) |
| | VEGF gene expression from HUVECs stimulated with PP for 24 hours; quantification by qRT-PCR | Quercetin | 1 µM (50%) | Nicholson *et al.* (2010) |
| ***Reduced VEGF protein expression*** | VEGF protein expression in macrophages stimulated with PP for 24 hours; VEGF protein quantified by ELISA | Rutin | 25 µg/mL (26.81%) | Guruvayoorappan & Kuttan (2007) |
| | VEGF protein expression in CLL B cells stimulated with PP for 24 hours; VEGF protein quantified by ELISA | EGCg | 2.5 µg/mL (~27%) | Lee *et al.* (2004) |
| | VEGF protein expression in TAMR-MCF-7 cells stimulated with PP for 24 hours; VEGF protein quantified by ELISA | Quercetin | 30 µM (~82%) | Oh *et al.* (2010) |
| | VEGF protein expression in OVCAR-3 cells stimulated with PP for 24 hours; VEGF protein quantified by ELISA | Kaempferol | 10 µM (20%) | Luo *et al.* (2009) |
| | VEGF protein expression in A2780/CP70 cells stimulated with PP for 24 hours; VEGF protein quantified by ELISA | Kaempferol | 10 µM (18%) | Luo *et al.* (2009) |
| | VEGF protein expression in resting and $CoCl_2$ hypoxia-induced RL952 cells stimulated with PP for 24 hours; VEGF protein quantified by ELISA | Resveratrol | Resting: 100 µM (65.6%)<br>Hypoxic: 100 µM (62.7%) | Dann *et al.* (2009) |
| | VEGF protein expression in resting and $CoCl_2$ hypoxia-induced RL952 cells stimulated with PP for 24 hours; VEGF protein quantified by ELISA | EGCg | Resting: 100 µM (81.1%)<br>Hypoxic: 100 µM (56.1%) | Dann *et al.* (2009) |
| | VEGF protein expression in PC-3 cells stimulated with PP for 24 hours; VEGF protein quantified by ELISA | Quercetin | 40 µM (50%) | Pratheeshkumar *et al.* (2012) |
| | VEGF protein expression in VSMCs stimulated with PP for 30 minutes, then stimulated with PDGF for 24 hours; VEGF protein quantified by ELISA | Delphinidin<br>Cyanidin | 10 µM (~45%), 30 µM (65–75%)<br>10 µM (~36%), 30 µM (~76%), 100 µM (~92%) | Oak *et al.* (2006) |

(*continued overleaf*)

**Table 9.3** *(continued)*

| PP-induced VEGF signaling-associated response | Assay | Compound | Concentration (% inhibition/ activation)[a] | References |
|---|---|---|---|---|
| ***Increased VEGF protein expression*** | VEGF protein expression in HUVECs treated with PP for 6 hours; VEGF protein quantified by immunoblot | Ferulic acid | 10 μM (∼140%) | Hseu *et al.* (2011) |
| | VEGF protein expression in VSMCs stimulated with PP for 30 minutes; VEGF protein quantified by ELISA | Gallic acid | 30 μM (175 ± 27%) | Oak *et al.* (2006) |
| | VEGF protein expression in VSMCs stimulated with PP for 30 minutes, then PDGF for 24 hours; VEGF protein quantified by ELISA | Quercetin | 100 μM (301 ±16%) | Oak *et al.* (2006) |
| ***Inhibition of VEGFR-1 phosphorylation*** | p-VEGFR-1 in immunoprecipitate from leukeamic B cells stimulated with PP for 24 hours before 30 minutes' VEGF stimulation; p-VEGFR-1 quantified by immunoblot | EGCg | 25 μg/mL (almost complete) | Lee *et al.* (2004) |
| | p-VEGFR-1 in immunoprecipitate from BAECs stimulated with PP for 24 hours, then stimulated with VEGF for 1 minute; p-VEGFR-1 protein quantified by immunoblot | EGCg | 25 μM (∼100%) | Lamy *et al.* (2002) |
| | P-VEGFR-1 in immunoprecipitate from HUAECs stimulated with PP for 4 hours, then stimulated with VEGF; p-VEGFR-1 protein quantified by immunoblot | EGCg | 50 μM (∼100%) | Neuhaus *et al.* (2004) |
| ***Inhibition of VEGFR-2 phosphorylation*** | p-VEGFR-2 in immunoprecipitate from leukaemic B cells stimulated with PP for 24 hours, then stimulated with VEGF for 30 minutes; p-VEGFR-2 protein presence visualized by immunoblot | EGCg | 25 μg/mL (nearly complete) | Lee *et al.* (2004) |
| | p-VEGFR-2 in immunoprecipitate from BAECs stimulated with PP for 18 hours, then stimulated with VEGF for 1 minute; p-VEGFR-2 protein presence observed by immunoblot | Ellagic acid | 5 μM (48%), 10 μM (83%) | Labrecque *et al.* (2005) |
| | p-VEGFR-2 in immunoprecipitate from HUVECs stimulated with PP for 18 hours, then stimulated for 1 minute with VEGF; p-VEGFR-2 protein presence observed by western blot | Delphinidin | 2 μM (50%) | Lamy *et al.* (2006) |
| | p-VEGFR-2 in immunoprecipitate from PAE cells stimulated with PP for up to 60 minutes, then stimulated with VEGF for 8 minutes; p-VEGFR-2 protein presence observed by western blot | Delphinidin | 50 μM (50%) | Teller *et al.* (2009) |

| | | | | |
|---|---|---|---|---|
| | p-VEGFR-2 activity in solution after stimulation by *in vitro* tyrosine kinase assay for 1 hour; quantified by peroxidase immunostain and spectroscopy | Delphinidin | 2.8 ± 1.3 μM (~50%) | Teller *et al.* (2009) |
| | p-VEGFR-2 in lysate from HUVECs stimulated with PP for 1 hour, then stimulated for 10 minutes with VEGF; p-VEGFR-2 was quantified by western blot | Quercetin | 40 μM (almost complete) | Pratheeshkumar *et al.* (2012) |
| | p-VEGFR-2 in immunoprecipitate from BAECs stimulated with PP from 24 hours, then stimulated with VEGF for 1 minute; p-VEGFR-2 protein quantified by immunoblot | EGCg<br>CG<br>ECg | 10 μM (94.4%)<br>10 μM (93.7%)<br>10 μM (74.8%) | Lamy *et al.* (2002) |
| | p-VEGFR-2 activity of immunoprecipitate after stimulation of BAECs with PP for 24 hours and VEGF for 1 minute; p-VEGFR-2 activity measured by *in vitro* tyrosine kinase assay and electrophoeresis | EGCg<br>CG<br>ECG | 10 μM (not complete)<br>10 μM (not complete)<br>50 μM (not shown) | Lamy *et al.* (2002) |
| | P-VEGFR-2 in immunoprecipitate in HUAECs stimulated with PP for 4 hours, then stimulated with VEGF; p-VEGFR-2 protein quantified by immunoblot | EGCg | 50 μM (~100%) | Neuhaus *et al.* (2004) |
| | p-VEGFR-2 in lysate from HUVECs stimulated with PP for 24 hours, then stimulated with VEGF for 5 minutes; p-VEGFR-2 quantified by western blot | Scopoletin | 100 μM (partially reduced) | Pan *et al.* (2011) |
| | P-VEGF-2 SW837 cells stimulated with PP for 6 hours; quantified by western blot | EGCg | 25 μg/mL (complete) | Shimizu *et al.* (2010) |
| | p-VEGFR-2 gene expression in SW837 xenograft from BALB/c nude mice fed PP in drinking water for 35 days; quantified by western blot | EGCg | 0.1% in water (~77%) | Shimizu *et al.* (2010) |
| ***Inhibition of VEGFR-2 protein expression*** | VEGFR-2 in lysates from SW837 cells stimulated with PP for 24 hours; VEGFR-2 presence observed after western blot | EGCg | 25 μg/mL (almost complete) | Shimizu *et al.* (2010) |
| ***Reduced VEGF/receptor coupling*** | The proportion of bound tagged VEGF (I-VEGF) to HUVECs after 90 minutes' VEGF incubation following 30 minutes with PP stimulation; quantified by gamma-spectrophotometry | EGCg | 25 μM (~40%) | Kondo *et al.* (2002) |

(*continued overleaf*)

**Table 9.3** (*continued*)

| PP-induced VEGF signaling-associated response | Assay | Compound | Concentration (% inhibition/activation)[a] | References |
|---|---|---|---|---|
| ***Inhibition of downstream signaling protein activation*** | p-ERK quantity in immunoprecipitate from BAECs stimulated with PP for 18 hours, then stimulated with VEGF for 1 minute; p-ERK quantified by western blot | Ellagic acid | 10 μM (66%) | Labrecque *et al.* (2005) |
| | p-PLC-γ quantity in immunoprecipitate from BAECs stimulated with PP for 18 hours, then stimulated with VEGF for 1 minute; p-PLC-γ observed by western blot | Ellagic acid | 10 μM (almost complete) | Labrecque *et al.* (2005) |
| | p-ERK in the lysate from HUVECs stimulated with PP for 18 hours, then stimulated with VEGF for 10 minutes; p-ERK was quantified by western blot | Delphinidin | 11.8 μM (50%) | Lamy *et al.* (2006) |
| | p-JNK in VSMCs stimulated with PP for 30 minutes, then stimulated with PDGF for 30 minutes; quantified by western blot | Delphinidin<br>Cyanidin | 30 μM (~66%)<br>30 μM (~73%) | Oak *et al.* (2006) |
| | P-ERK in immunoprecipitate in HUAECs stimulated with PP for 4 hours, then stimulated with VEGF; p-ERK protein quantified by immunoblot | EGCg | 50 μM (complete) | Neuhaus *et al.* (2004) |
| | p-AKT, p-mTOR, p-S6K1 in HUVECs stimulated with PP for 1 hour, then stimulated with VEGF for 10 minute; p-AKT, p-mTOR, and p-S6K1 protein observed by western blot | Quercetin | 40 μM (almost complete) | Pratheeshkumar *et al.* (2012) |
| | p-AKT, p-mTOR in PC-3 cells stimulated with PP for 6 hours; p-AKT, p-mTOR, and p-S6K1 protein observed by western blot | Quercetin | 40 μM (almost complete) | Pratheeshkumar *et al.* (2012) |
| | p-S6K1 in PC-3 cells stimulated with PP for 6 hours; p-AKT, p-mTOR, and p-S6K1 protein observed by western blot | Quercetin | 40 μM (complete) | Pratheeshkumar *et al.* (2012) |
| | NF-κB in HuH7 cells stimulated with PP for 24 hours, then transfected with luciferase NF-κB reporter gene and stimulated with IL-1β for 18 hours; quantified by luciferase reporter assay | Quercetin | 10 μM (56.4 ± 5.2%), 30 μM (12.1 ± 2.7%) | Kleemann *et al.* (2006) |
| | p-AKT in TAMR-MCF-7 cells stimulated with PP for 1 hour; protein presence observed by western blot | Quercetin | 30 μM (almost complete) | Oh *et al.* (2010) |

| | | | | |
|---|---|---|---|---|
| | P-ERK in immunoprecipitate in HUVECs stimulated with PP and VEGF for 2 hours; p-ERK protein quantified by western blot | Resveratrol | 10 µM (almost complete) | Lin *et al.* (2003) |
| | p-AKT quantity in lysate from OVCAR-3 cells stimulated with PP for 24 hours; p-AKT quantified by ELISA | Kaempferol | 20 µM (48.6%) | Luo *et al.* (2009) |
| | p-ERK quantity in lysate from OVCAR-3 and A2780/CP70 cells stimulated with PP for 2h; p-ERK quantified by ELISA | Kaempferol | 10 µM (~50–67%) | Lou *et al.* (2012) |
| | p-AKT level in HUVECs after stimulated with PP for 24 hours then stimulated by VEGF for 30 minutes; presence detected by western blot | EGCg | 2 µM (42%), 5 µM (54%), 10 µM (61%) | Rodriguez *et al.* (2006) |
| | p-AKT in immunoprecipitate from HMVECs stimulated with PP for 24 hours, then with VEGF for 90 minutes; p-AKT quantified by western blot | EGCg | 20 µM (complete) | Tang *et al.* (2003) |
| | p-Akt gene expression in SW837 xenograft from BALB/c nude mice fed PP in drinking water for 35 days; quantified by western blot | EGCg | 0.1% in water (~60%) | Shimizu *et al.* (2010) |
| | p-ERK gene expression in SW837 xenograft from BALB/c nude mice fed PP in drinking water for 35 days; quantified by western blot | EGCg | 0.1% in water (~71%) | Shimizu *et al.* (2010) |
| | p-ERK in lysate from HUVECs stimulated with PP for 18 hours, then IL-6 for 10 minutes; quantified by western blot | Delphinidin<br>Ellagic acid<br>EGCg | 20 µM (~51%)<br>20 µM (~44%)<br>20 µM (~32%) | Lamy *et al.* (2012) |
| | Oligonucleotide-tagged functional NF-κB in nuclear extract from HUVECs stimulated with PP for 24 hours then VEGF for 30 minutes; measured by gel electrophoresis | EGCg | 2 µM (39%), 5 µM (46%), 10 µM (69%) | Rodriguez *et al.* (2006) |
| ***Induction of downstream signaling protein activation*** | p-ERK quantity in lysate from OVCAR-3 cells stimulated with PP for 24 hours; p-ERK quantified by ELISA | Kaempferol | 20 µM (42.6%) | Luo *et al.* (2009) |
| | p-ERK in lysate from HUVECs stimulated with PP for 18 hours, then IL-6 for 10 minutes; quantified by western blot | Apigenin<br>Luteolin | 20 µM (210%)<br>20 µM (140%) | Lamy *et al.* (2012) |
| | P-Akt in BAECs stimulated with PP for 15 minutes; observed by western blot | EGCg | 50 µM (strong) | Kim *et al.* (2007) |

(*continued overleaf*)

**Table 9.3** (*continued*)

| PP-induced VEGF signaling-associated response | Assay | Compound | Concentration (% inhibition/ activation)[a] | References |
|---|---|---|---|---|
| ***Reduced expression of downstream signaling proteins*** | Rats fed PP 1 week before (and throughout expt. 16 weeks total) and 5 weeks after consumption of NDEA hepatocellular carcinogen (16 weeks); quantification of NF-$\kappa$B mRNA in liver tissue by qRT-PCR | Naringenin | 200 mg/kg pre-treat (70.59%) 200 mg/kg post-treat (76.47%) | Subramanian & Arul (2012) |
| | NF-$\kappa$B quantity in lysate from OVCAR-3 cells stimulated with PP for 24 hours; p-ERK quantified by western blot | Kaempferol | 20 µM (~24%), 40 µM (~47%), 80 µM (~73%) | Lou *et al.* (2012) |
| | NF-$\kappa$B quantity in lysate from A2780/CP70 cells stimulated with PP for 24 hours; p-ERK quantified by western blot | Kaempferol | 40 µM (~78%), 80 µM (~87%) | Lou *et al.* (2012) |
| | Gene down-regulation of VEGF signaling proteins in aortas from WT mice fed HF-HC diet ± PP for 18 weeks; identification by microarray, GO, and database | Naringin | 0.02% diet (4 genes altered) | Chanet *et al.* (2012) |

[a]Values in parentheses indicate the percentage change in the parameter indicated in response to polyphenol treatment, or in cases where numerical data were not available, the current author's interpretation of the image information (e.g. western blots) as textual information.

ELISA, enzyme-linked immunosorbent assay; GO, gene ontology; HF-HC, high fat–high cholesterol; PP, polyphenol; HUVEC, human umbilical endothelial cell; HUAEC, human umbilical aortic endothelial cell; BAEC, bovine aortic endothelial cell; HMVEC, human microvascular endothelial cell; PDGF, platelet-derived growth factor; PAE, porcine aortic endothelial (cell); VEGF, vascular endothelial growth factor; VEGFR(1/2), vascular endothelial growth factor receptor (1/2); C, catechin; EC, epicatechin; EGC, epigallocatechin; ECG, epicatechin gallate; EGCg, epigallocatechin gallate; VSMC, vascular smooth-muscle cells; CAM, chorioallantoic membrane assay; WT, wild type.

**Table 9.4** Reported effects of polyphenols on angiogenesis and related functions.

| Angiogenesis or associated response | Assay details | Compound | Concentration (% inhibition/ activation to control) | References |
|---|---|---|---|---|
| ***Halted cell cycle*** | The proportion of EA.hy 927 cells stimulated with PP for 1 hour, then stimulated with VEGF for 24 hour in G0/G1 cell-cycle arrest; quantified by flow cytometry | Gallic acid | 5 µg/mL | Hseu *et al.* (2011) |
| | The increase in proportion of HUVECs stimulated, with PP for 18 hours in G0/G1 cell-cycle arrest; quantified by flow cytometry | Delphinidin | 10 µg/mL (17 ± 0.4%) | Favot *et al.* (2003) |
| | The increase in proportion of HUVECs stimulated, with VEGF and PP for 18 hours in G0/G1 cell-cycle arrest; quantified by flow cytometry | Delphinidin | 10 µg/mL (25 ± 1.1%) | Favot *et al.* (2003) |
| ***Reduced cell proliferation*** | Thyamidine incorporation into HUVECs stimulated with PP for 24 hours then incubated with [$^3$H] Thymidine for 18 hours; thyamidine incorporation measured by a counter | Rutin | 25 µg/mL (41.02%) | Guruvayoorappan & Kuttan (2007) |
| | Thyamidine incorporation into SMCs stimulated with PP for 24 hours then incubated with [$^3$H] Thymidine for 18 hours; thyamidine incorporation measured by a counter | Avenanthramide | 120 µM (50%) | Nie *et al.* (2006) |
| | Number of SMCs after stimulation with PP for 96 hours; quantified by hemocytometer | Avenanthramide | 40 µM (41%), 80 µM (80%), 120 µM (73%) | Nie *et al.* (2006) |
| | Number of HUVECs after stimulation with PP for 24 hours; quantified using a hemocytometer | EGCg | 20 µM (~27%) | Liu *et al.* (2008) |
| | Number of HUVECs after stimulation with PP for 3 days; quantified by colorimetric assay | Delphinidin | 26 µM (22 ± 3.9%) | Favot *et al.* (2003) |
| | Alteration to the expression profile of genes associated with negative regulation of cell proliferation in HUVECs stimated with PP for 24 hours; identified by gene ontology | EGCg | 20 µM, 2/78 (2.6%) | Liu *et al.* (2008) |
| | Cell number in ASMCs, isolated from aortas of mice fed HF-HC diet ± PP for 18 weeks, after stimulation with PP for 3 days; quantified by cell counter | Naringenin | 1 µM (25%) | Chanet *et al.* (2012) |

(*continued overleaf*)

**Table 9.4** *(continued)*

| Angiogenesis or associated response | Assay details | Compound | Concentration (% inhibition/ activation to control) | References |
|---|---|---|---|---|
| | Cell number in A2780/CP70 cells stimulated with PP for 16 hours; quantified by hemocytometer | Kaempferol | 20 μM (~36%) | Lou *et al.* (2012) |
| | HUVEC cell number after stimulation with oxLDL ± PP for 48 hours; quantified my MTT assay | Quercetin<br>Resveratrol | 10 μM (~33%)<br>10 μM (~42%) | Kostyuk *et al.* (2011) |
| ***Reduced cell migration*** | Measurement of areas infiltrated by HUVECs stimulated with PP for 48 hours after score-injury; quantified by crystal violet stain and microscope count | Rutin | 25 μg/mL (87.5%), 10 μg/mL (42.18%), 5 μg/mL (31.8%) | Guruvayoorappan & Kuttan (2007) |
| ***Reduced tube formation*** | Number of tubes formed by HUVECs seeded between two collagen layers stimulated with PP for 24 hours; tube formation quantified using image analysis | Ellagic acid | 5 μM (38%), 10 μM (46%) | Labrecque *et al.* (2005) |
| | Number of tubes formed by HUVECs pre-stimulated with PP for 6 hours, seeded on to matrigel wells and incubated for 18 hours; tube formation quantified by image analysis | Delphinidin | 15 μM (76%) | Lamy *et al.* (2006) |
| | Number of tubes formed by HUVECs in a matrigel-coated well stimulated with PP for 48 hours; observation of tube formation by image analysis | Rutin | 25 μg/mL (maximum) | Guruvayoorappan & Kuttan (2007) |
| | Number of tubes formed by HMEC-1 cells seeded in matrigel-coated wells stimulated with PP for 18 hours; tube formation quantified by image analysis | EGCg<br>Cg | 50 μM (91.6%)<br>50 μM (82.5%) | Lamy *et al.* (2002) |
| | Number of tubes formed by HUVECs seeded between two collagen layers stimulated with PP for 48 hours; tube formation quantified using image analysis | EGCg<br>ECg<br>EGC<br>EC | 25 μM (~70%)<br>25 μM (~50%)<br>100 μM (~40%)<br>100 μM (~40%) | Kondo *et al.* (2002) |
| | Tube formation by HUVECs pre-stimulated with PP for 24 hours, seeded on a matrigel layer and incubated for 48 hours; tube length/branch points quantified from microscopy images | EGCg | Tube length: 20 μM (58%)<br>Branch points: 20 μM (42%) | Liu *et al.* (2008) |

| | | | | |
|---|---|---|---|---|
| ***Increased tube formation*** | Number of tubes formed by HUVECs seeded in matrigel-coated wells stimulated with PP for 18 hours; tube formation quantified by 3D image analysis | Ferulic acid | 10 μM (~290%) | Lin *et al.* (2010) |
| ***Reduced VEGF-stimulated cell proliferation*** | Number of HUVECs after stimulation with VEGF for 24 hours, then stimulation with PP for 72 hours; cell number quantified by counting | EGCg<br>ECg<br>EGC<br>EC | 100 μM (55%)<br>100 μM (50%)<br>100 μM (20–25%)<br>100 μM (20–25%) | Kondo *et al.* (2002) |
| | Number of HUVECs after stimulation with PP for 24 hours, then stimulation with VEGF for 24 hours; cell number quantified by hemocytometer | EGCg | 10 μM (~35%), 20 μM (~53%) | Liu *et al.* (2008) |
| | Proliferation ratio of HUVECs stimulated with VEGF with PP for 24, 48, and 72 hours; quantified by optical density after MMT assay | Scopeletin | 24 hours: 100 μM (~9%)<br>48 hours: 100 μM (~18%)<br>72 hours: 30 μM (~10%), 100 μM (~19%) | Pan *et al.* (2011) |
| | | Quercetin | 24 hours: 10 μM (11%), 20 μM (32%), 30 μM (44%), 40 μM (64%) | Pratheeshkumar *et al.* (2012) |
| | Number of HUVECs after stimulation with VEGF and PP for 3 days; cell number quantified by colorimetric assay | Delphinidin | 26 μM (21 ± 5.4%) | Favot *et al.* (2003) |
| | HUVECs stimulated with PP for 4 hours, then stimulated with VEGF and incubated with [$^3$H] thymidine for 18 hours; thymidine incorporation measured using scintillation counter | EGCg | 20 μM (~61%), 50 μM (~75%) | Neuhaus *et al.* (2004) |
| | Number of HUVECs after stimulation with PP for 4 hours, then stimulation with VEGF; cell number quantified by cell counter | EGCg | 20 μM (~27%), 50 μM (~37%) | Neuhaus *et al.* (2004) |
| | HUVECs stimulated with VEGF and PP for 24 hours, then incubated with [$^3$H] thymidine for 6 hours; thyamidine incorporation measured using scintillation counter | Resveratrol | 10 μM (35–40%) | Lin *et al.* (2003) |

(*continued overleaf*)

**Table 9.4** Reported effects of polyphenols on angiogenesis and related functions.

| Angiogenesis or associated response | Assay details | Compound | Concentration (% inhibition/ activation to control) | References |
|---|---|---|---|---|
| ***Reduced VEGF-stimulated cell migration*** | Measurement of areas not infiltrated by HUVECs stimulated with PP for 48 hours after score-injury; quantified by the amount of colloidal gold-coated coverslip remaining | EGCg<br>ECg<br>EGC<br>EC | 100 μM (~80%)<br>100 μM (~70%)<br>100 μM (~60%)<br>100 μM (~50%) | Kondo *et al.* (2002) |
| | Number of HUVECs stimulated with PP in cell suspension and VEGF with PP in well media for 6 hours, migrated to the lower layer of transwell plate; cell number quantified manually by staining and microscopy | Scopeletin | 30 μM (~12%), 100 μM (~24%) | Pan *et al.* (2011) |
| | Number of HUVECs stimulated with PP for 2 hours then stimulated by addition of VEGF to the lower chamber for 4 hours and migrated to the lower layer of the transwell plate; cell number quantified manually by staining and microscopy | Ellagic acid | 10 μM (~81%) | Labrecque *et al.* (2005) |
| | Number of BAECs stimulated with PP for 2 hours then stimulated by addition of VEGF to the lower chamber for 4 hours and migrated to the lower layer of the transwell plate; cell number quantified manually by staining and microscopy | Ellagic acid | 10 μM (~67%) | Labrecque *et al.* (2005) |
| | Number of HUVECs stimulated with PP for 2 hours then stimulated by addition of VEGF to the lower chamber for 3 hours and migrated to the lower layer of the transwell plate; cell number quantified by staining and microscopy | Delphinidin | 25 μM (~75%) | Lamy *et al.* (2006) |
| | Migration of HUVECs stimulated with PP for 24 hours after score-injury; quantified from microscopy images | Delphinidin | 13 μM (17 ± 2.2%), 26 μM (30 ± 2.9%), 52 μM (50 ± 2.4%) | Favot *et al.* (2003) |
| | | Resveratrol<br>Quercetin | 1 μM (50–60%)<br>20 μM (~29%), 30 μM (~42%), 40 μM (~64%) | Lin *et al.* (2003), Pratheeshkumar *et al.* (2012) |

| | | | | |
|---|---|---|---|---|
| | Number of HUVECs stimulated with PP in cell suspension and VEGF with PP in well media for 24 hours, then migrated to the lower layer of the transwell plate; cell number quantified by staining and microscopy | Quercetin | 20 μM (~35%), 30 μM (~54%), 40 μM (~72%) | Pratheeshkumar *et al.* (2012) |
| ***Reduced VEGF-stimulated tube formation*** | Tube formation by HUVECs seeded on matrigel-coated well stimulated with VEGF and PP in 10 mg/mL for 18 hours; tube length quantified from microscopy images | Scopoletin | 30 μM (~29%), 100 μM (52%) | Pan *et al.* (2011) |
| | Tube formation by HUVECs pre-stimulated with PP for 18 hours then grown on matrigel-coated wells for 6 hours; quantification thread length from microscopy images | Delphinidin | 25 μM (complete) | Lamy *et al.* (2006) |
| | Tube formation by HUVECs seeded between fibrin layers stimulated with PP for 1 hour then stimulated with VEGF for 6 hours; microscopy used to observe capillary tube formation | Delphinidin | 25 μM (appears complete) | Lamy *et al.* (2006) |
| | Tube formation by HUVECs seeded on a matrigel layer stimulated with PP and VEGF for 12 hours; tube length measured from microscopy images | Resveratrol | 1 μM (50–60%) | Lin *et al.* (2003) |
| | Tube formation by HUVECs pre-stimulated with PP for 24 hours and seeded on a matrigel layer stimulated with VEGF for 48 hours; tube length/branch points quantified from microscopy images | EGCg | Tube length: 20 μM (44%)<br>Branch points: 20 μM (30%) | Liu *et al.* (2008) |
| | 3D tube formation by HMVECs seeded between two collagen layers stimulated with PP for 24 hours, then with VEGF for 72 hours; tube formation quantified by microscopy | C<br>EC<br>EGC<br>ECg<br>EGCg | 20 μM (54%)<br>20 μM (47%)<br>5 μM (52%), 20 μM (67%)<br>2 μM (75%), 5 μM (88%)<br>2 μM (39%), 5 μM (74%), 20 μM (83%) | Tang *et al.* (2003) |
| | Tube formation by HUVECs on basement membrane stimulated with conditioned media from 24-hour PP-treated A2780/CP70 cells for 6 hours; tube formation observed through stained microscopy images | Kaempferol | 40 and 80 μM (appear complete) | Lou *et al.* (2012) |

(*continued overleaf*)

**Table 9.4** Reported effects of polyphenols on angiogenesis and related functions.

| Angiogenesis or associated response | Assay details | Compound | Concentration (% inhibition/ activation to control) | References |
|---|---|---|---|---|
| | Tube formation by HUVECs seeded on a matrigel layer stimulated with VEGF and PP for 24 hours; microscopy used to observe capillary tube formation | Quercetin | 40 μM (almost complete) | Pratheeshkumar *et al.* (2012) |
| ***Lower VE-cadherin activation*** | p-VE-cadherin in immunoprecipitate from HUVECs stimulated with VEGF and PP for 1 hour; quantified by western blot | Resveratrol | 2.5 μM (almost complete) | Lin *et al.* (2003) |
| | p-VE-cadherin in immunoprecipitate from HMVECs stimulated with PP for 24 hours then stimulated with VEGF for 30 minutes; quantified by western blot | EGCg | 20 μM (60%) | Tang *et al.* (2003) |
| | VEGFR-2/VE-cadherin complex formation in VE-cadherin immunoblot from HUVECs stimulated with PP for 24 hours then VEGF for 30 minutes; quantified by western blot | EGCG | 2 μM (43%), 5 μM (45%), 10 μM (67%) | Rodriguez *et al.* (2006) |
| ***Lower MMP expression*** | MMP-2 mRNA extracted from hepatic tissue from animals fed NDEA hepatocellular carcinogen alongside pre- and post-feeding of PP; quantified by qRT-PCR | Naringenin | 200 mg/kg pre-treat (82.61%)<br>200 mg/kg post-treat (86.52%) | Subramanian & Arul (2012) |
| | MMP-9 mRNA extracted from hepatic tissue from animals fed NDEA hepatocellular carcinogen alongside pre- and post-feeding of PP; quantified by qRT-PCR | Naringenin | 200 mg/kg pre-treat (72.35%)<br>200 mg/kg post-treat (78.24%) | Subramanian & Arul (2012) |
| ***Reduced angiogenesis in matrigel-plug assay*** | bFGF stimulated angiogenesis in matrigel plug injected into mouse flank with PP for 7 days; hemoglobin content of plug quantified by spectroscopy | Delphinidin | 300 μM (~57%) | Lamy *et al.* (2006) |
| | Matrigel containing ovarian cancer cells ± polyphenol treatment injected into the CAM and grown for 5 days, followed by quantification of weight and blood vessel number | Kaempferol | Weight: 20 μM (45.7%)<br>Blood vessels: 20 μM (38.5%) | Luo *et al.* (2009) |

| | | | | |
|---|---|---|---|---|
| | Angiogenesis in matrigel plug, with VEGF and PP, embedded in chorioallantoic membrane for 72 hours; hemoglobin content of plug quantified by spectroscopy | Quercetin | 20 µg (27%) | Pratheeshkumar *et al.* (2012) |
| ***Reduced new-vessel development in aortic-ring assay*** | Collagen-embedded rings cut from rat aorta cultured in VEGF-supplemented media with or without polyphenol and number of new microvessels counted | Quercetin | 20 µM (~29%), 30 µM (~43%), 40 µM (~73%) | Pratheeshkumar *et al.* (2012) |
| | Microvessel sproutings from rat aorta embedded in collagen stimulated with conditioned media with PP; microvessel number quantified by phase-contrast microscopy | Rutin | 25 µg/mL (52.8%) | Guruvayoorappan & Kuttan (2007) |
| ***Reduced neovascualrisation in CAM*** | Angiogenesis measured in CAM stimulated with PP-soaked filter paper for 48 hours; quantification of number of vessel branches by image analysis | Rutin<br>Quercetin | 10, 30, 100 nM/egg<br>20 µM/L (26.32%), 40 µM/L (47.37%) | Guruvayoorappan & Kuttan (2007), Pratheeshkumar *et al.* (2012) |
| | Angiogenesis in 1 cm$^2$ region of CAM stimulated with PP for 2 days; observation of effect by image analysis | Delphinidin | 50 µg (hemorrhagic vessels only) | Favot *et al.* (2003) |
| | Angiogenesis measured in CAM stimulated with PP-soaked filter paper for 48 hours; quantification of number of vessel branches by image analysis | Gallic acid | 5 µg/mL (~63%) | Hseu *et al.* (2011) |
| | Angiogenesis in tumor after TAMR-MCF-7 cells implanted into CAM ± VEGF and polyphenol for 72 hours; number of blood vessels quantified from microscopy image | Quercetin | 10 µM/egg (~39%), 30 µM/egg (~46%) | Oh *et al.* (2010) |
| ***Increased neovascularisation in CAM*** | 0.8 cm-diameter straws placed on to the CAM and 150 µL treatment solution ± PP placed inside; after 48 hours, new blood vessels quantified by image analysis | Ferulic acid | 10 µM (50%) | Lin *et al.* (2010) |
| ***Alterations in expression of angiogenic genes*** | HUVECs in full media treated with PP for 4 hours; mRNA then extracted from cells and converted to cDNA; gene-expression profiles determined using Affymetrix U133 Plus 2.0 chips; compared with vehicle-only control | EC<br>Apple procyanidin (av. dp = 3.9) | No significant changes; 1318 gene significantly altered; top functions were angiogenic | Garcia-Conesa *et al.* (2009) |

ELISA, enzyme-linked immunosorbent assay; GO, gene ontology; HF-HC, high fat–high cholesterol; PP, polyphenol; HUVEC, human umbilical endothelial cell; HUAEC, human umbilical aortic endothelial cell; BAEC, bovine aortic endothelial cell; HMVEC, human microvascular endothelial cell; PDGF, platelet-derived growth factor; PAE, porcine aortic endothelial (cell); VEGF, vascular endothelial growth factor; VEGFR(1/2), vascular endothelial growth factor receptor (1/2); C, catechin; EC, epicatechin; EGC, epigallocatechin; ECG, epicatechin gallate; EGCg, epigallocatechin gallate; VSMC, vascular smooth-muscle cells; CAM, chorioallantoic membrane assay; WT, wild type.

**Table 9.5** Reported effects of polyphenols on atherosclerosis.

| Atherosclerosis-related response | Assay | Compound | Concentration (% inhibition/activation) | References |
|---|---|---|---|---|
| ***Reduced plaque size*** | Atherosclerotic lesion size in ApoE3L mice fed HC diet ± PP for 15 weeks; quantified by total atherosclerotic area in cross-sectioned aortic valve | Quercetin | 0.1% in diet (40%) | Kleemann *et al.* (2006) |
| | Developing atherosclerotic lesion size and number in ApoE −/− mice on HC diet, stimulation of carotid injury ± daily PP IP injections; legions at site of injury quantified by intima size | EGCg | Day 21: 10 mg/kg daily (55%)<br>Day 42: 10 mg/kg daily (73%) | Chyu *et al.* (2004) |
| | Fatty streak containing lesions, stained by oil red O, identified in the descending thoracic aortas of NZW rabbits fed a HC diet ± PP for 8 weeks; quantified by percentage of intima stained | Naringin<br>Naringenin | 0.1% in diet (~68%)<br>0.05% in diet (~70%) | Lee *et al.* (2001) |
| | Fatty streak containing lesions in the aorta of Syrian golden hamsters fed HC diet ± PP for 12 weeks; quantified by oil red O stain and micsroscopy | Catechin<br>Quercetin<br>Resveratrol | 2.856 mg/kg/daily (84%)<br>0.1428 mg/kg/daily (80%)<br>0.1428 mg/kg/daily (76%) | Auger *et al.* (2005) |
| | Aortic plaque size in aortic root and descending aorta of WT mice fed HF-HC diet ± PP for 18 weeks; quantification of lesion area after cryosectioning, staining, and microscopy | Naringin | 0.02% in diet (41%) | Chanet *et al.* (2012) |
| | Total plaque coverage of abdominal aorta in sedated LDLR −/− mice fed PP in diet for 8 weeks; quantified by ultrasound bio-microscopy | Quercetin | 12.5–25.0 mg/dL (52–65%) | Leckey *et al.* (2010) |

| | | | | |
|---|---|---|---|---|
| | Percentage plaque coverage of abdominal aorta, opened longitudinally and stained with Sudan IV, in LDLR −/− mice fed PP in diet for 4 and 8 weeks; quantified by image-analysis software | Quercetin | 4 weeks: 12.5 mg/dL (18%), 18.75 mg/dL (32%), 25 mg/dL (35%)<br>8 weeks: 12.5 mg/dL (52%), 18.75 mg/dL (57%), 25 mg/dL (61%) | Leckey *et al.* (2010) |
| | Lesion area in aortic sinus from ApoE −/− mice fed PP for 26 weeks; quantified by sectioning, staining, and microscopy | Quercetin<br>Theaflavin | 1.3 mg/day (∼80%)<br>1.3 mg/day (∼60%) | Loke *et al.* (2010) |
| | Lesion area in thoracic aorta from ApoE −/− mice fed PP for 26 weeks; quantified by sectioning, staining, and microscopy | Quercetin<br>Theaflavin | 1.3 mg/day (∼45%)<br>1.3 mg/day (∼55%) | Loke *et al.* (2010) |
| | Plaque area on intima in aorta of rabbit fed HC diet ± PP for 12 weeks; area quantified by oil red O and sudan III observed by microscopy | Quercetin | 0.05 mg/kg daily (0.7 ± 1.28%) | Juzwiak *et al.* (2005) |
| ***Reduced intima thickness*** | Intima thickening aorta between first and second intercostal arteries of NZW rabbits fed an HC diet ± PP for 8 weeks; semi-quantified using a 1–3 scale of foam cell/macrophage inclusion, n ≥ 20 | Naringin<br>Naringenin | 0.1% in diet (∼56%)<br>0.05% in diet (∼57%) | Lee *et al.* (2001) |
| ***Reduced expression of adhesion molecules*** | VCAM-1 protein in HUVECs stimulated with PP for 18 hours, then stimulated with TNF-$\alpha$ for 6 hours; protein quantification by ELISA | Quercetin<br>Genistein | 30 μM (64.0 ± 2.0%)<br>92.5 μM (56.3 ± 7.5%) | Kleemann *et al.* (2006) |

(*continued overleaf*)

**Table 9.5** (*continued*)

| Atherosclerosis-related response | Assay | Compound | Concentration (% inhibition/activation) | References |
|---|---|---|---|---|
| | VCAM-1 RNA in portion of thoracic aorta from NZW rabbits fed a HC diet ± PP for 8 weeks; quantified by RT-PCR and gel electrophoresis | Naringin<br>Naringenin | 0.1% in diet (~56%)<br>0.05% in diet (~48%) | Lee *et al.* (2001) |
| | VCAM-1 expression in HUVECs stimulated with PP for 45 minutes, then LPS+TNF-$\alpha$ for 24 hours; VCAM-1=expressing cells immunotagged and quantified by flow cytometry | Quercetin<br>Quercetin-3-glucuronide<br>Quercetin-3-sulfate<br>Isorhamnetin-3-glucuronide | 10 μM (~80%)<br>2 μM (~30%)<br>2 μM (~30%)<br>2 μM (~30%) | Tribolo *et al.* (2008) |
| | VCAM-1 expression in HUVECs stimulated with PP for 45 minutes, then LPS+TNF-$\alpha$ for 6 hours; quantified by RT-PCR | Quercetin<br>Quercetin-3-glucuronide | 10 μM (~60%)<br>10 μM (~20%) | Tribolo *et al.* (2008) |
| | VCAM1 gene expression in HUVECs stimulated with oxLDL ± PP for 6 hours; quantified by RT-PCR | Verbscoside<br>Resveratrol | 10 μM (~46%)<br>10 μM (~42%) | Kostyuk *et al.* (2011) |
| | E-selectin protein in HUVECs stimulated with PP for 18 hours, then with TNF-$\alpha$ for 6 hours; protein quantification by ELISA | Quercetin<br>Genistein | 30 μM (35.0 ± 12.0%)<br>92.5 μM (62.0 ± 19.0%) | Kleemann *et al.* (2006) |
| | ICAM-1 expression in HUVECs stimulated with PP for 45 minutes, then LPS+TNF-$\alpha$ for 24 hours; VCAM-1-expressing cells immunotagged and quantified by flow cytometry | Quercetin<br>Quercetin-3-glucuronide | 10 μM (~50%)<br>2 μM (~20%) | Tribolo *et al.* (2008) |
| | ICAM-1 expression in HUVECs stimulated with PP for 45 minutes, then LPS+TNF-$\alpha$ for 6 hours; quantified by RT-PCR | Quercetin | 10 μM (~60%) | Tribolo *et al.* (2008) |
| | MCP-1 RNA in portion of thoracic aorta from NZW rabbits fed a HC diet ± PP for 8 weeks; quantified by RT-PCR and gel electrophoresis | Naringin<br>Naringenin | 0.1% in diet (~51%)<br>0.05% in diet (~53%) | Lee *et al.* (2001) |

| | | | |
|---|---|---|---|
| MCP-1 expression in HUVECs stimulated with PP for 45 minutes, then LPS+TNF-$\alpha$ for 6 hours; quantified by RT-PCR | Quercetin<br>Isorhamnetin-3-glucuronide | 10 μM (~40%)<br>10 μM (~25%) | Tribolo *et al.* (2008) |
| MCP-1 gene expression in HUVECs stimulated with oxLDL ± PP for 6 hours; quantified by RT-PCR | Resveratrol | 10 μM (~29%) | Kostyuk *et al.* (2011) |
| MCP-1 protein in HUVECs stimulated with oxLDL ± PP for 18 hours; quantified by ELISA | Resveratrol | 10 μM (~10%) | Kostyuk *et al.* (2011) |
| CCR2 gene expression in PBM from ApoE −/− mice stimulated with PP for 24 hours; quantified by RT-PCR | Cyanidin-3-O-glycoside<br>Protocatechuic acid | 0.5 μM (~51%)<br>0.5 μM (~63%) | Wang *et al.* (2011) |
| CCR2 protein levels in PBM from ApoE −/− mice stimulated with PP for 24 hours; quantified by western blot | Cyanidin-3-O-glycoside<br>Protocatechuic acid | 0.5 μM (~61%)<br>0.5 μM (~71%) | Wang *et al.* (2011) |
| CCR2 gene expression in PBM from WT −/− mice stimulated with PP for 24 hours; quantified by RT-PCR | Cyanidin-3-O-glycoside<br>Protocatechuic acid | 0.5 μM (~54%)<br>0.5 μM (~80%) | Wang *et al.* (2011) |
| CCR2 protein levels in PBM from WT −/−mice stimulated with PP for 24 hours; quantified by western blot | Cyanidin-3-O-glycoside<br>Protocatechuic acid | 0.5 μM (~60%)<br>0.5 μM (~88%) | Wang *et al.* (2011) |
| CCR2 gene expression in PBM from ApoE −/− mice fed PP through oral gavage once; quantified by RT-PCR | Cyanidin-3-O-glycoside<br>Protocatechuic acid | 0.5 μM (~55%)<br>0.5 μM (~46%) | Wang *et al.* (2011) |
| CCR2 protein levels in PBM from ApoE −/− mice fed PP through oral gavage once; quantified by western blot | Cyanidin-3-O-glycoside<br>Protocatechuic acid | 0.5 μM (~76%)<br>0.5 μM (~51%) | Wang *et al.* (2011) |

(*continued overleaf*)

**Table 9.5** *(continued)*

| Atherosclerosis-related response | Assay | Compound | Concentration (% inhibition/activation) | References |
|---|---|---|---|---|
| | p-selectin in plasma from ApoE −/− mice fed ± PP for 26 weeks; quantified by ELISA | Quercetin<br>Epicatechin<br>Theaflavin | 1.3 mg/daily (∼29%)<br>1.3 mg/daily (∼31%)<br>1.3 mg/daily (∼25%) | Loke *et al.* (2010) |
| ***Induced expression of adhesion molecules*** | VCAM1 gene expression in HUVECs stimulated with ± PP for 6 hours; quantified by RT-PCR | Quercetin | 10 μM (2.1-fold) | Kostyuk *et al.* (2011) |
| | MCP-1 gene expression in HUVECs stimulated with oxLDL ± PP for 6 hours; quantified by RT-PCR | Quercetin | 10 μM (∼28%) | Kostyuk *et al.* (2011) |
| | MCP-1 protein in HUVECs stimulated with oxLDL ± PP for 18 hours; quantified by ELISA | Quercetin | 10 μM (∼11%) | Kostyuk *et al.* (2011) |
| ***Reduced immune cell adhesion*** | Fluorescently labelled U937 cell adhesion to HAEC stimulated with PP for 20 hours, then with IL-1$\beta$L for 6 hours; quantified by spectroscopy | Catechin | 10 μM (∼41%) | Koga & Meydani (2001) |
| | U937 cell adhesion to HUVECs stimulated with PP for 24 hours, then with TNF-$\alpha$ for 4 hours; quantified by staining and microscopy counts | Naringenin | 1 μM (16%) | Chanet *et al.* (2012) |
| | Leukocyte adhesion molecule 1 expression in aortas from WT mice fed HF-HC diet ± PP for 18 weeks; measured by gene ontology | Naringin | 0.02% in diet (80%) | Chanet *et al.* (2012) |
| | vWF expression in aortas from WT mice fed HF-HC diet ± PP for 18 weeks; measured by gene ontology | Naringin | 0.02% in diet (76%) | Chanet *et al.* (2012) |
| ***Reduced immune cell migration*** | Movement of 24-hour PP-stimulated murine PBM from top layer to bottom layer of boyden chamber after incubation with CCR2 in bottom chamber for 4 hours; quantified by microscopy count | Cynidin-3-*O*-glucoside<br>Protocate-chuic acid | 0.5 μM (∼42%)<br>0.5 μM (∼62%) | Wang *et al.* (2011) |

| | | | | |
|---|---|---|---|---|
| | Movement of ApoE −/− PBM from mice fed PP through oral gavage once, from top layer to bottom layer of boyden chamber after incubation with CCR2 in bottom chamber for 4 hours; quantified by microscopy count | Cynidin-3-*O*-glucoside<br>Protocatechuic acid | 0.5 μM (~69%)<br>0.5 μM (~49%) | Wang *et al.* (2011) |
| ***Reduction in vaso-regulatory-associated protein expression*** | Endothelin-1 in HUVECs stimulated with PP for 24 hours; quantified by qRT-PCR | Resveratrol<br>Quercetin | 1 μM (35%)<br>1 μM (16%) | Nicholson *et al.* (2010) |
| | Endothelin-1 in HUVECs stimulated with PP and $H_2O_2$ for 24 hours; quantified by qRT-PCR | Resveratrol<br>Quercetin | 1 μM (51%)<br>1 μM (40%) | Nicholson *et al.* (2010) |
| | Endothelin-1 in HUVECs stimulated with PP for 4 hours; quantified by immunometric assay | Procyanidin D.p.3.9 | 25 μM (12.6%) | Garcia-Conesa *et al.* (2009) |
| | Endothelin-1 in urine from ApoE −/− mice fed ± PP for 26 weeks; quantified by immunoassay | Quercetin<br>Epicatechin | 1.3 mg/daily (~52%)<br>1.3 mg/daily (~51%) | Loke *et al.* (2010) |
| | NO levels in HUVECs stimulated with PP for 24 hours, then with VEGF for 1 hour; quantified by colorimetric assay | Scopoletin | 30 μM (~16%),<br>100 μM (~34%) | Pan *et al.* (2011) |
| | NO levels in HUVECs stimulated with PP for 6 hours; quantified by staining and spectroscopy | Verbscoside<br>resveratrol | 10 μM (~97%)<br>10 μM (~49%) | Kostyuk *et al.* (2011) |
| | NO levels in HUVECs stimulated with oxLDL ± PP for 1 hour; quantified by staining and spectroscopy | Quercetin<br>Verbscoside | 10 μM (~96%)<br>10 μM (~84%) | Kostyuk *et al.* (2011) |
| ***Increase in vaso-regulatory protein expression*** | eNOS in HUVECs stimulated with PP for 24 hours; quantified by qRT-PCR | Resveratrol<br>Quercetin | 1 μM (91%)<br>1 μM (46%) | Nicholson *et al.* (2010) |
| | eNOS in HUVECs stimulated with PP and $H_2O_2$ for 24 hours; quantified by qRT-PCR | Resveratrol<br>Quercetin | 1 μM (22%)<br>1 μM (9%) | Nicholson *et al.* (2010) |
| ***Increase in vaso-regulatory protein activation*** | eNOS in aortic tissue from ApoE−/− mice fed ± PP for 26 weeks; quantified by tagged arginine-conversion eNOS activity assay | Quercetin<br>Theaflavin | 1.3 mg/daily (~1500%)<br>1.3 mg/daily (~1050%) | Loke *et al.* (2010) |
| | P-eNOS in BAECs stimulated with PP for 15 minutes; observed by western blot | EGCg | 50 μM (strong) | Kim *et al.* (2007) |

(*continued overleaf*)

**Table 9.5** (*continued*)

| Atherosclerosis-related response | Assay | Compound | Concentration (% inhibition/activation) | References |
|---|---|---|---|---|
| | PI3-K/eNOS-dependent NO release from BAECs stimulated with EGCG for 5 minutes; observed by staining and microscopy | EGCg | 50 µM (strong) | Kim *et al.* (2007) |
| | eNOS mRNA in SMCs stimulated with PP for 24 hours; quantified by qPCR | Avenanthramide | 80 µM (~220%) | Nie *et al.* (2006) |
| | eNOS mRNA in HAECs stimulated with PP and $H_2O_2$ for 24 hours; quantified by qPCR | Avenanthramide | 80 µM (~240%) | Nie *et al.* (2006) |
| | NO concentration in SMCs stimulated with PP for 24 hours, then stimulated of NO synthesis; quantified by spectroscopy | Avenanthramide | 120 µM (~300%) | Nie *et al.* (2006) |
| | NO concentration in HAECs stimulated with PP for 24 hours, then stimulation of NO synthesis; quantified by spectroscopy | Avenanthramide | 80 µM (~1100%) | Nie *et al.* (2006) |
| | NO levels in HUVECs stimulated with PP for 1 hour; quantified by staining and spectroscopy | Resveratrol | 10 µM (48%) | Kostyuk *et al.* (2011) |
| | NO levels in HUVECs stimulated with oxLDL ± PP for 6 hours; quantified by staining and spectroscopy | Resveratrol<br>Quercetin | 10 µM (~591%)<br>10 µM (~700%) | Kostyuk *et al.* (2011) |
| ***Increase in vasodilation*** | Perfusion pressure in mesenteric vascular beds stimulated with increasing concentration of EGCG; quantified by reduction in pressure from continuous reading | EGCG | 100 µM (~32%) | Kim *et al.* (2007) |
| ***Reduced LDLC concentration/oxidation*** | Plasma levels of TC and VLDLC, and TC/HDLC ratio from WT mice fed HF-HC diet ± PP for 18 weeks; HDLC quantified by enzymatic analysis and VLDLC by FPLC | Naringin | TC: 0.02% in diet (13%)<br>TC/HDLC: 0.02% in diet (20%)<br>VLDLC: 0.02% in diet (~44%) | Chanet *et al.* (2012) |
| | TC, LDLC, and VLDLC in serum from rabbits fed HC diet ± PP for 12 weeks | Quercetin | 0.05 mg/kg daily (TC: 679%; LDLC: 1070%; VLDLC: 444%) | Juzwiak *et al.* (2005) |

| | | | | |
|---|---|---|---|---|
| | PON1 gene expression in hepatic tissue from LDLR −/− mice fed PP for 4 and 8 weeks; quantified by RT-PCR | Quercetin | 4 weeks: 12.5 mg/dL (39%), 25 mg/dL (51%)<br>8 weeks: 12.5 mg/dL (25%), 25 mg/dL (31%) | Leckey *et al.* (2010) |
| | PON1 activity in serum from LDLR −/− mice fed PP for 4 and 8 weeks; quantified by paraoxon conversion activity assay | Quercetin | 4 weeks: 12.5 mg/dL (90%), 25 mg/dL (80%)<br>8 weeks: 12.5 mg/dL (80%), 25 mg/dL (70%) | Leckey *et al.* (2010) |
| | LDL oxidation after incubation of LDL with MPO, nitrites, and H2O2 ± PP; quantified by peroxiDetect kit | Quercetin<br>Verbscoside | 5 μM (~97%), 10 μM (~98%)<br>10 μM (~23%), 40 μM (~74%) | Kostyuk *et al.* (2011) |
| ***Reduced oxidative stress*** | $F_2$ isoprostane in aorta from ApoE −/− mice fed ± PP for 26 weeks; quantified by GC-MS | Quercetin<br>Epicatechin | 1.3 mg/daily (~61%)<br>1.3 mg/daily (~76%) | Loke *et al.* (2010) |
| | Superoxide concentrations in aorta from ApoE −/− mice fed ± PP for 26 weeks; quantified by chemiluminescence | Quercetin<br>Epicatechin | 1.3 mg/daily (~39%)<br>1.3 mg/daily (~42%) | Loke *et al.* (2010) |

ELISA, enzyme-linked immunosorbent assay; GO, gene ontology; HF-HC, high fat–high cholesterol; PP, polyphenol; HUVEC, human umbilical endothelial cell; HUAEC, human umbilical aortic endothelial cell; BAEC, bovine aortic endothelial cell; HMVEC, human microvascular endothelial cell; PDGF, platelet-derived growth factor; PAE, porcine aortic endothelial (cell); VEGF, vascular endothelial growth factor; VEGFR(1/2), vascular endothelial growth factor receptor (1/2); C, catechin; EC, epicatechin; EGC, epigallocatechin; ECG, epicatechin gallate; EGCg, epigallocatechin gallate; VSMC, vascular smooth-muscle cells; CAM, chorioallantoic membrane assay; WT, wild type.

Table 9.3 contains data from studies of the effects of polyphenols on VEGF signaling, Table 9.4 contains data from studies of the effects of polyphenols on angiogenesis, and Table 9.5 contains data from studies of the effects of polyphenols on atherosclerosis.

### 9.6.1 VEGF signaling

The reported effects of polyphenol treatments on VEGF signaling have predominantly been investigated using cultured cell models, particularly human umbilical vein endothelial cells (HUVECs), which are a widely used primary human endothelial cell line. The data demonstrate that polyphenols can affect VEGF signaling in a number of different ways. There are numerous reports demonstrating inhibition of VEGF-induced phosphorylation (activation) of VEGFR-2, which occurs within minutes of treatment with VEGF. The most potent effects reported were for the anthocyanidin delphinidin, which was able to inhibit VEGFR-2 activation in HUVECs by about 50% at 2 μM (Lamy *et al.*, 2006), ellagic acid (48 and 83% at 5 and 10 μM, respectively; Labrecque *et al.*, 2005), and galloylated green-tea catechins at 10 μM (94, 94 and 75% by EGCg, CG, and ECg, respectively; Lamy *et al.*, 2002). Although these concentrations of polyphenols are low compared to those in many other reports, they are higher than have been observed in human peripheral blood following their consumption. The reader should note that there are also numerous reports of the effects of mixed polyphenol extracts on VEGF-induced pVEGFR-2 activation, but we have not included these in Table 9.3.

There are also numerous reports of polyphenol-mediated inhibition of VEGF and VEGFR-2 gene and protein expression. Effects on gene and protein expression will take longer to manifest than those on VEGFR-2 activation. It is noticeable that the inhibition of gene and protein expression appears less potent than the inhibition of pVEGFR-2 activation, with smaller effects at higher concentrations (typically ~30 μM reported). The exception is for the data of Nicholson *et al.* (2010), who report 54% inhibition of *VEGF* gene expression in HUVECs by resveratrol at 1 μM. It is also of interest that the same authors report increased expression of *VEGF* induced by low concentrations of quercetin (50% by 1 μM), and that increases in VEGF gene expression in HUVECs have also been observed in response to ferulic acid (Lin *et al.*, 2010). The major green-tea catechin EGCg has also been reported to inhibit VEGFR-2 protein expression and binding of VEGF to HUVECs, although at supraphysiological concentrations (54 and 25 μM, respectively). A number of polyphenols have also been reported to inhibit signaling events downstream of the receptor tyrosine kinase VEGFR-2. Inhibition of the activation of PLC$\gamma$, ERK, Akt, NF$\kappa$B, and other downstream targets has been reported, with EGCg and ellagic acid being particularly potent (see Table 9.3).

These data also make it clear that some polyphenols can interact to affect VEGF signaling and angiogenic functions in more than way. It is also apparent that mixtures of polyphenols could work effectively to modulate VEGF signaling and angiogenesis by interacting with different molecular targets involved in VEGF signaling and angiogenic function. There is also a lack of data concerned with the potential for microbial metabolites of polyphenols to inhibit VEGF signaling and downstream angiogenic functions. With the importance of microbial metabolites becoming increasingly clear, future research should include studies with these compounds.

### 9.6.2 Angiogenesis

Table 9.4 summarizes the effects of polyphenols in a number of models of angiogenesis and on markers of angiogenesis. It is noticeable that many of the reported effects occurred in response to supraphysiological concentrations of polyphenols; for example, a significant proportion of the reports concern responses to 100 μM polyphenol. However, significant *in vitro* antiangiogenic effects have been reported for 1–2 μM concentrations of naringenin on aortic smooth-muscle cells (ASMC) (Chanet *et al.*, 2012), of resveratrol on HUVECs (Lin *et al.*, 2003), and of the galloylated green-tea catechins EGCg and ECg on HUVECs (Tang *et al.*, 2003). It is interesting to note that the non-galloylated green-tea catechins, such as (−)-epicatechin (EC), (+)-catechin (C), and epigallocatechin (EGC), were either ineffective or only effective at much higher (nonphysiological) concentrations (see Table 9.4).

### 9.6.3 Atherosclerosis

Table 9.5 lists the reported effects of isolated polyphenols on atherosclerosis and atherosclerosis-related biomarkers. There are a number of animal models of atherosclerosis, with the most widely used being ApoE knock-out mice ($ApoE^{-/-}$) and high-cholesterol diet-fed rodents. These animal models exhibit accelerated atherosclerosis, and typically measurements are made of the plaque that develops in the aortic arch, where the changes in blood velocity provide a pro-atherosclerotic environment, leading to ready development of plaque. Several polyphenols (quercetin, naringin/naringenin, resveratrol, EGCg, catechin, and theaflavin) have been reported to significantly reduce the size of atherosclerotic lesions when incorporated into rodent diets (Table 9.5). The effects of these polyphenols on atherosclerotic lesion development are substantial (e.g. 45–80% reductions for quercetin, six studies; 41–70% reductions for naringenin, two studies; 55–73% reductions for EGCg, one study; 76% reduction for resveratrol, one study). It can also be observed that polyphenols that have been reported to be ineffective at inhibiting VEGF signaling have not been reported to inhibit atherosclerosis in rodent models. Although it is not evident whether this merely reflects the lack of such studies to date, it is possible that completed studies with impotent polyphenolic VEGF-signaling inhibitors such as epicatechin have not been published because of the challenges of publishing negative study results.

In addition to the data obtained in rodent models of atherosclerosis, there is also a large body of data showing that polyphenols inhibit various biomarkers related with atherosclerosis. These markers include expression of cellular adhesion molecules (e.g. ICAM-1, VCAM-1) involved in recruitment of macrophages to the subendothelial space. Of particular note is the study reported by Tribolo *et al.* (2008), which demonstrates that the major human phase 2 conjugates of quercetin (quercetin-3′-sulfate, quercetin-3-glucuronide, 3′-methylquercetin-3-glucuronide) at physiological concentrations (2 μM) significantly reduced VCAM-1 expression on the surface of HUVECs pre-stimulated with a mixture of lipopolysaccharide and TNF$\alpha$, with a similar effect of quercetin-3-glucuronide on ICAM-1 expression. The importance of utilizing human physiological metabolites at appropriate concentrations when investigating potential mechanisms of action of polyphenols *in vitro* has been highlighted elsewhere (Kroon *et al.*, 2004).

Another interesting feature of the data presented in Table 9.5 is the abundance of reports of polyphenol-induced increases in NO or NO synthetic capability of endothelial and smooth-muscle cells. NO is a potent inducer of vasodilation; mechanistically, this is achieved through its strong activating effect on guanylyl cyclase, which synthesizes cyclic guanosine monophosphate (cGMP), which diffuses into and relaxes smooth-muscle cells. The enzyme that makes NO in vascular endothelial and smooth-muscle cells is endothelial nitric oxide synthase (eNOS). The widely reported observation that polyphenol treatments increase eNOS activation and NO bioavailability is interesting because eNOS is downstream of VEGFR-2, and activation of this receptor tyrosine kinase leads to PI3K/Akt-dependent phosphorylation of eNOS, which has substantially increased catalytic capacity compared to nonphosphorylated eNOS. Hence, these observations are apparently not in keeping with the widely reported potent inhibitory effects of polyphenols on VEGF signaling, particularly the inhibition of VEGF-induced VEGFR-2 activation (see Table 9.5, top). It is apparent, therefore, that although certain polyphenols do strongly inhibit VEGF–VEGFR-2 signaling, they are at the same time activating eNOS through a distinct mechanism. That is, some polyphenols can inhibit VEGF signaling and accordingly cause reductions in angiogenesis-related functions such as neovascularization, but they can also activate eNOS and increase the cellular bioavailability of NO, and thus avoid the potentially deleterious effect of inhibiting eNOS activity/reducing NO bioavailability. This observation is consistent with polyphenols interacting with one or more other signaling pathways in addition to the VEGFR-2 pathway.

## 9.7 Relationships between polyphenol consumption and CVD risk

The previous sections presented and reviewed the large quantity of data showing that polyphenols can inhibit VEGF signaling, inhibit VEGF-mediated angiogenesis, and reduce the rate of growth of atherosclerotic plaques in animal models. VEGF is the most important angiogenic factor and is largely responsible for driving angiogenesis in many tissues, including vascular tissues (Celetti *et al.*, 2001; Belgore *et al.*, 2004). These observations suggest that there is potential for dietary polyphenols to inhibit VEGF-dependent angiogenesis and consequently angiogenesis-dependent growth of atherosclerotic plaques. Since development of atherosclerosis and rupture of atherosclerotic plaques are crucial processes underlying CVD, inhibition of VEGF activity by polyphenols is a possible mechanism by which long-term consumption of polyphenols could reduce CVD risk.

### *9.7.1 Epidemiological studies*

The very first reports of possible associations between consumption of polyphenols and subsequent risk of CVD involved the relatively small Zutphen Elderly cohort ($n = 805$) of elderly men in The Netherlands in the 1990s. The reports from this study only included three flavonols (quercetin, myricetin, and kaempferol), two flavones (apigenin and luteolin), and the catechins [(+)-catechin, (−)-epicatechin, epigallocatechin, epicatechin gallate, and epigallocatechin gallate] in the dietary assessment, reflecting the lack of extensive food-composition data for flavonoids and other polyphenols in foods at the time. The first report was published in 1993, after a 5-year follow-up period (Hertog *et al.*, 1993), and the data from 10 years of follow-up were reported in 1997 (Hertog *et al.*, 1997). Ilja Arts and Peter Hollman reviewed the data from the Zutphen Elderly cohort and several other prospective studies investigating similar compounds in 2005 (Arts & Hollman, 2005); at this time there were 12 published cohort studies on flavonoid intake and the risk of coronary artery disease (CAD) and 5 cohort studies on the risk of stroke. The authors noted that 7 of the 12 studies found a significant association between consumption of these flavonoids and CAD, while significant associations with stroke incidence were reported for 2 of the 5 studies. In the Zutphen Elderly study, a protective effect was observed for flavonols and flavones (Keli *et al.*, 1996) but not for catechins (Arts *et al.*, 2001). The major flavonol in the diet is quercetin, with much smaller quantities of kaempferol consumed. The major dietary sources of quercetin in The Netherlands, the United Kingdom, and the United States are onions, apples, tea, and broccoli. The major sources of catechins in the diet are green and black tea, chocolate, cocoa, and red wine.

More recent epidemiological studies have made use of the more extensive flavonoid composition data that have been made available online. The first to be made available were the USDA datasets, which now cover all the flavonoid subclasses (http://www.ars.usda.gov/Services/docs.htm?docid=6231; last accessed March 27, 2014), although data for the oligomeric proanthocyanidins subset of the flavan-3-ols are still lacking, largely due to the complexities of analyzing these compounds. Other excellent online food polyphenol composition resources that have been made available include PhenolExplorer (http://www.phenol-explorer.eu/; last accessed March 27, 2014), and EuroFIR-BASIS (http://ebasis.eurofir.org/Default.asp; last accessed March 27, 2014) (Gry *et al.*, 2007; Neveu *et al.*, 2010). The first reported use of the USDA data in an epidemiological study was in 2007, when they were used to assess the dietary intakes of 34 489 women from the Iowa Women's Health Study cohort (Mink *et al.*, 2007). After 16 years of follow-up, the authors reported 7091 total deaths, with 2316 due to CVD, 1329 due to coronary heart disease (CHD), and 469 due to stroke. Following multivariate adjustments, the following statistically significant associations were observed for flavonoid classes: lower CHD, CVD, and total mortality for any anthocyanidin intake versus no intake; highest versus lowest quintile of flavanone intake and CHD, flavones, and total mortality. For flavonoid-rich foods, the significant associations were for apples, pears, or both and red wine (with CHD and CVD), grapefruit (with CHD), strawberries (CVD), and chocolate (CVD). More recently, it was reported that higher versus lower intakes of five of the seven classes of flavonoid assessed (anthocyanidins, flavan-3-ols, flavones, flavonols, procanthocyanidins) were individually associated with reduced risk of fatal CVD in a study of the Cancer Prevention Study II (CPS-II) Nutrition Cohort (38 180 men and 60 289 women) (McCullough *et al.*, 2012). Noting that the data demonstrated significant associations even at intermediate intakes, the authors concluded that even relatively modest intakes may provide benefits.

The most powerful epidemiological studies are those with very large cohorts, such as the CPS-II cohort, and meta-analyses of data from multiple studies. A number of meta-analyses have been reported for flavonoids and CVD risk. Huxley & Neil (2003) reported a meta-analysis of seven cohort studies investigating associations between flavonol intakes and CHD mortality. The studies included 2087 fatal CHD events, and the statistical analysis of individuals in the highest third of intakes versus those in the lowest third of intakes indicated a 20% reduction in relative risk for the higher intakes (95% confidence interval 0.69–0.93). It has also been reported that higher flavonol intakes are associated with lower risk of stroke (both fatal and nonfatal), with a meta-analysis of available data indicating a 20% lower risk for high versus low intakes (Hollman *et al.*, 2010). A meta-analysis of prospective studies investigating chocolate intakes and cardiovascular mortality also provided support for the notion that higher intakes of specific flavonoid-rich foods are protective against CVDs, with a relative risk of 0.81 reported for a comparison of the highest versus lowest quartiles (Ding *et al.*, 2006).

It is clear from the various cohort studies investigating the relationships between flavonoid intakes and risk of cardiovascular events such as CHD, CVD, and stroke that higher flavonoid intakes are more often than not associated with significantly lower CVD events and deaths compared to lower flavonoid intakes. There is also evidence that higher levels of consumption of particular foods, such as chocolate and cocoa, tea, red wine, apples, and strawberries, are associated with reduced risk of fatal cardiovascular events.

### *9.7.2 Intervention studies*

There are now several hundred reports on the effects of acute, short-term, and to a lesser extent long-term consumption of flavonoids and flavonoid-rich foods on risk factors and biomarkers associated with cardiovascular function in human subjects. However, only a relatively small proportion of these studies are designed such that the reported effects of the interventions can be ascribed to the inherent flavonoids or other polyphenols. Nevertheless, a well-designed and high-quality meta-analysis was undertaken in order to identify those studies that were designed in a way that was considered sufficient to allow a cause-and-effect relationship between the observed effects on CVD risk factors (lipoproteins, blood pressure (BP), endothelial function measured using flow-mediated dilatation (FMD)) and the flavonoid component(s) of the intervention to be established (Hooper *et al.*, 2008). This generally required that a suitable placebo was used as a comparator for the flavonoid-rich food in a double-blinded RCT, because the number of studies with isolated flavonoids was extremely limited. From the almost 600 reports identified in online searches of various databases of published papers, the authors identified only 133 studies that met the strict inclusion criteria. Following a meta-analysis, it was shown that chocolate/cocoa consumption improved FMD after acute and chronic intake and reduced systolic and diastolic BP, and that green tea, cocoa, and dark chocolate reduced plasma LDL cholesterol. The lack of significant effects for a wider range of flavonoids/foods was largely due to there being insufficient numbers of high-quality studies investigating the same foods/endpoints. A subsequent follow-up meta-analysis of studies concerned with chocolate consumption and cardiovascular risk factors, which included data from studies investigating effects on insulin resistance, confirmed the previous findings and also indicated significant effects on insulin resistance, largely because of the effects on plasma insulin levels (Hooper *et al.*, 2012).

However, in trying to establish whether flavonoid consumption can reduce CVD risk via a mechanism involving inhibition of VEGF-mediated angiogenic promotion of atherosclerotic plaque growth, the appropriateness of the biomarkers needs to be considered. Arguably, the best biomarkers of atherosclerosis are those involving direct measurement of plaque size or volume, which can be done using imaging techniques, such as noninvasive ultrasound probes. However, there is a lack of published data from such studies for polyphenols. Indeed, at the time of publication the authors were only aware of one intervention study, in which intima media thickness of the common carotid artery (CCA-IMT) was assessed by ultrasonography before and after intervention with polyphenols (Curtis *et al.*, 2013). This study was completed by 93 patients who were postmenopausal women with type II diabetes, all of whom were on statin therapy and a subset of whom were on additional medication (e.g. to treat hypertension). No significant effects of the 12-month intervention with chocolate bars (35 g/day) supplemented with soy protein isolate (containing 100 mg isoflavones) and a flavanol-rich cocoa extract (containing 850 mg total flavanols, including 90 mg of epicatechin) on CCA-IMT were observed. However, the Curtis *et al.* (2013) study did not report data for changes in plaque size (e.g. plaque area or volume), which would be a direct marker of the effects of polyphenol consumption on plaque growth; this is the process that may be sensitive to polyphenol treatments, because of their ability to inhibit VEGF signaling and VEGF-dependent angiogenesis. Further, CCA-IMT thicknesses are small, and the expected changes in response to diet are very small, so the 12-month intervention period, although long relative to most other interventions with flavonoids, was rather short in terms of observing changes in CCA-IMT.

## 9.8 Conclusions

There is a substantial body of reported evidence demonstrating that some dietary polyphenols can potently inhibit VEGF signaling (Table 9.3), which is the major driver of angiogenesis. Angiogenesis, driven by VEGF through its interaction with VEGFR-2 in vascular tissues, underpins neovascularization and therefore plays a key role in diseases that require neovascularization, including cancers and atherosclerosis. There are also a substantial number of reports of significant antiangiogenic activity by polyphenols, and in the main these are consistent with reports of polyphenol anti-VEGF signaling activity in terms of which polyphenols are effective and most potent (Table 9.4). It is now known that VEGF-dependent angiogenesis can drive atherosclerotic plaque development, and consistent with this paradigm, many of the polyphenols that have been shown to inhibit VEGF signaling and angiogenesis have also been reported to inhibit atherosclerosis in rodents fed polyphenols (Table 9.5). Overall, the available evidence is consistent with the notion that polyphenol-mediated inhibition of atherosclerosis is mediated through inhibition of VEGF signaling. The available data suggest that the crucial interaction occurs at the level of VEGFR-2 (that is, at the earliest stage of VEGF signaling), since VEGF-induced activation of VEGFR-2 is potently inhibited by polyphenols that have also been reported to inhibit atherosclerosis.

## Acknowledgments

The authors received financial support from the Biotechnology and Biological Sciences Research Council (UK) through an Institute Strategic Programme Grant ("Food & Health," Grant Number BB/J004545/1) to the Institute of Food Research, and from the European Union Seventh Framework Programme (FP7/2007-2013) under grant agreement no. 312090 (BACCHUS).

## References

Andreasen, M.F., Kroon, P.A., Williamson, G. & Garcia-Conesa, M.T. (2001) Esterase activity able to hydrolyse dietary antioxidant hydroxycinnamates is distributed along the intestine of mammals. *Journal of Agricultural and Food Chemistry*, 49, 5679–5684.

Arts, I.C. & Hollman, P.C. (2005) Polyphenols and disease risk in epidemiologic studies. *American Journal of Clinical Nutrition*, 81, 317S–325S.

Arts, I.C., Hollman, P.C., Feskens, E.J., *et al.* (2001) Catechin intake might explain the inverse relation between tea consumption and ischemic heart disease: the Zutphen Elderly Study. *American Journal of Clinical Nutrition*, 74, 227–232.

Auger, C., Teissedre, P.-L., Gerain, P., *et al.* (2005) Dietary wine phenolics catechin, quercetin, and resveratrol efficiently protect hypercholesterolemic hamsters against aortic fatty streak accumulation. *Journal of Agricultural and Food Chemistry*, 53, 2015–2021.

Barleon, B., Reusch, P., Totzke, F., *et al.* (2001) Soluble VEGFR-1 secreted by endothelial cells and monocytes is present in human serum and plasma from healthy donors. *Angiogenesis*, 4, 143–154.

Belgore, F., Blann, A., Neil, D., *et al.* (2004) Localisation of members of the vascular endothelial growth factor (VEGF) family and their receptors in human atherosclerotic arteries. *Journal of Clinical Pathology*, 57, 266–272.

Celletti, F.L., Waugh, J.M., Amabile, P.G., *et al.* (2001a) Vascular endothelial growth factor enhances atherosclerotic plaque progression. *Nature Medicine*, 7, 425–429.

Celletti, F.L., Hilfiker, P.R., Ghafouri, P. & Dake, M.D. (2001b) Effect of human recombinant vascular endothelial growth factor 165 on progression of atherosclerotic plaque. *Journal of the American College of Cardiology*, 37, 2126–2130.

Celletti, F.L., Waugh, J.M., Amabile, P.G., *et al.* (2002) Inhibition of vascular endothelial growth factor-mediated neointima progression with angiostatin or paclitaxel. *Journal of Vascular and Interventional Radiology*, 13, 703–707.

Chanet, A., Milenkovic, D., Deval, C., *et al.* (2012) Naringin, the major grapefruit flavonoid, specifically affects atherosclerosis development in diet-induced hypercholesterolemia in mice. *The Journal of Nutritional Biochemistry*, 23, 469–477.

Chyu, K. Y., Babbidge, S.M., Zhao X., *et al.* (2004) Differential effects of green tea-derived catechin on developing versus established atherosclerosis in apolipoprotein E-null mice. *Circulation*, 109, 2448–2453.

Curtis, P.J., Potter, J., Kroon, P.A., *et al.* (2013) Vascular function and atherosclerosis progression after 1 y of flavonoid intake in statin-treated postmenopausal women with type 2 diabetes: a double-blind randomized controlled trial. *American Journal of Clinical Nutrition*, 97, 936–92.

Dann, J.M., Sykes, P.H., Mason, D.R. & Evans, J.J. (2009) Regulation of vascular endothelial growth factor in endometrial tumour cells by resveratrol and EGCG. *Gynecologic Oncology*, 113, 374–378.

Demer, L.L. & Tintut, Y. (2008) Vascular calcification: pathobiology of a multifaceted disease. *Circulation*, 117, 2938–2948.

Ding, E.L., Hutfless, S.M., Ding, X. & Girotra, S. (2006) Chocolate and prevention of cardiovascular disease: a systematic review. *Nutrition & Metabolism (Lond)*, 3, 2, doi:10.1186/1743-7075-3-2.

DiSalvo, J., Bayne, M.L., Conn, G., *et al.* (1995) Purification and characterization of a naturally occurring vascular endothelial growth factor. Placenta growth factor heterodimer. *The Journal of Biological Chemistry*, 270, 7717–7723.

Dixelius, J., Makinen, T., Wirzenius, M., *et al.* (2003) Ligand-induced vascular endothelial growth factor receptor-3 (VEGFR-3) heterodimerization with VEGFR-2 in primary lymphatic endothelial cells regulates tyrosine phosphorylation sites. *The Journal of Biological Chemistry*, 278, 40973–40979.

Ebos, J.M., Bocci, G., Man, S., *et al.* (2004) A naturally occurring soluble form of vascular endothelial growth factor receptor 2 detected in mouse and human plasma. *Molecular Cancer Research*, 2, 315–326.

Favot, L., Martin, S., Keravis, T., *et al.* (2003) Involvement of cyclin-dependent pathway in the inhibitory effect of delphinidin on angiogenesis. *Cardiovascular Research*, 59, 479–487.

Ferrara, N., Gerber, H.P. & LeCouter, J. (2003) The biology of VEGF and its receptors. *Nature Medicine* 9, 669–676.

Garcia-Conesa, M.T., Tribolo, S., Guyot, S., *et al.* (2009) Oligomeric procyanidins inhibit cell migration and modulate the expression of migration and proliferation associated genes in human umbilical vascular endothelial cells. *Molecular Nutrition & Food Research*, 53, 266–276.

Gry, J., Black, L., Eriksen, F.D., *et al.* (2007) EuroFIR-BASIS – a combined composition and biological activity database for bioactive compounds in plant based foods. *Trends in Food Science Technology*, 18, 434–445.

Guruvayoorappan, C. & Kuttan, G. (2007) Antiangiogenic effect of rutin and its regulatory effect on the production of VEGF, IL-lß and TNF-$\alpha$ in tumor associated macrophages. *Journal of Biological Sciences*, 7, 1511–1519.

Hauer, A.D., van Puijvelde, G.H., Peterse, N., *et al.* (2007) Vaccination against VEGFR2 attenuates initiation and progression of atherosclerosis. *Arteriosclerosis, Thrombosis, and Vascular Biology*, 27, 2050–2057.

Hertog, M.G., Feskens, E.J., Hollman, P.C., *et al.* (1993) Dietary antioxidant flavonoids and risk of coronary heart disease –the Zutphen elderly study. *Lancet*, 342, 1007–1011.

Hertog, M.G., Feskens, E.J. & Kromhout, D. (1997) Antioxidant flavonols and coronary heart disease risk. *Lancet*, 349, 699.

Hiratsuka, S., Minowa, O., Kuno, J., *et al.* (1998) Flt-1 lacking the tyrosine kinase domain is sufficient for normal development and angiogenesis in mice. *Proceedings of the National Academy of Sciences USA*, 95, 9349–9354.

Hiyama, T., Tanaka, T., Endo, S., *et al.* (2010) Angiogenesis in atherosclerotic plaque obtained from carotid endarterectomy: association between symptomatology and plaque morphology. *Neurologia Medico-Chirurgica (Tokyo)*, 50, 1056–1061.

Hollman, P.C., Geelen, A. & Kromhout, D. (2010) Dietary flavonol intake may lower stroke risk in men and women. *Journal of Nutrition*, 140, 600–604.

Hooper, L., Kroon, P.A., Rimm, E.B., *et al.* (2008) Flavonoids, flavonoid-rich foods and cardiovascular risk: a meta-analysis of randomized controlled trials. *American Journal of Clinical Nutrition*, 88, 38–50.

Hooper, L., Kay, C., Abdelhamid, A., *et al.* (2012) Effects of chocolate, cocoa, and flavan-3-ols on cardiovascular health: a systematic review and meta-analysis of randomized trials. *American Journal of Clinical Nutrition*, 35, 226–232.

Houck, K.A., Leung, D.W., Rowland, A.M., *et al.* (1992) Dual regulation of vascular endothelial growth factor bioavailability by genetic and proteolytic mechanisms. *Journal of Biological Chemistry*, 267, 26031–26037.

Hseu, Y.-C., Chen, S.-C., Lin, W.-H., *et al.* (2011) Toona sinensis (leaf extracts) inhibit vascular endothelial growth factor (VEGF)-induced angiogenesis in vascular endothelial cells. *Journal of Ethnopharmacology*, 134, 111–121.

Huang, K., Andersson, C., Roomans, G.M., *et al.* (2001) Signaling properties of VEGF receptor-1 and -2 homo- and heterodimers. *International Journal of Biochemistry & Cell Biology*, 33, 315–324.

Huxley, R.R. & Neil, H.A. (2003) The relation between dietary flavonol intake and coronary heart disease mortality: a meta-analysis of prospective cohort studies. *European Journal of Clinical Nutrition*, 57, 904–908.

Juzwiak, S., Wojcicki, J., Mokrzycki, K., *et al.* (2005) Effect of quercetin on experimental hyperlipidemia and atherosclerosis in rabbits. *Pharmacology Reports*, 57, 604–609.

Keli, S.O., Hertog, M.G., Feskens, E.J. & Kromhout, D. (1996) Dietary flavonoids, antioxidant vitamins, and incidence of stroke: the Zutphen study. *Archives of International Medicine*, 156, 637–642.

Khurana, R., Simons, M., Martin, J.F. & Zachary, I.C. (2005) Role of angiogenesis in cardiovascular disease: a critical appraisal. *Circulation*, 112(12), 1813–1824.

Kim, J.A., Formoso, G., Li, Y., *et al.* (2007) Epigallocatechin gallate, a green tea polyphenol, mediates NO-dependent vasodilation using signaling pathways in vascular endothelium requiring reactive oxygen species and Fyn. *The Journal of Biological Chemistry*, 282, 13736–13745.

Kitsukawa, T., Shimono, A., Kawakami, A., *et al.* (1995) Overexpression of a membrane protein, neuropilin, in chimeric mice causes anomalies in the cardiovascular system, nervous system and limbs. *Development*, 121, 4309–4318.

Kleemann, R., Verschuren, L., Morrison, M., *et al.* (2011) Anti-inflammatory, anti-proliferative and anti-atherosclerotic effects of quercetin in human in vitro and in vivo models. *Atherosclerosis*, 218, 44–52.

Koga, T. & Meydani, M. (2001) Effect of plasma metabolites of (+)-catechin and quercetin on monocyte adhesion to human aortic endothelial cells. *The American Journal of Clinical Nutrition*, 73, 941–948.

Kondo, T., Ohta, T., Igura, K., *et al.* (2002) Tea catechins inhibit angiogenesis in vitro, measured by human endothelial cell growth, migration and tube formation, through inhibition of VEGF receptor binding. *Cancer Letters*, 180, 139–144.

Kostyuk, V.A., Potapovich, A.I., Suhan, T.O., *et al.* (2011) Antioxidant and signal modulation properties of plant polyphenols in controlling vascular inflammation. *European Journal of Pharmacology*, 658, 248–256.

Kroon, P.A., Faulds, C.B., Ryden, P., *et al.* (1997) Release of bound ferulic acid from fiber in the human colon. *Journal of Agricultural and Food Chemistry*, 45, 661–667.

Kroon, P.A., Clifford, M.N., Crozier, A., *et al.* (2004) How should we assess the effects of exposure to dietary polyphenols in vitro? *American Journal of Clinical Nutrition*, 80, 15–21.

Kucukardali, Y., Aydogdu, S., Ozmen, N., *et al.* (2008) The relationship between severity of coronary artery disease and plasma level of vascular endothelial growth factor. *Cardiovascular Revascularization Medicine*, 9, 66–70.

Labrecque, L., Lamy, S., Chapus, A., *et al.* (2005) Combined inhibition of PDGF and VEGF receptors by ellagic acid, a dietary-derived phenolic compound. *Carcinogenesis*, 26, 821–826.

Lamy, S., Gingras, D. & Beliveau, R. (2002) Green tea catechins inhibit vascular endothelial growth factor receptor phosphorylation. *Cancer Research*, 62, 381–385.

Lamy, S., Blanchette, M., Michaud-Levesque, J., *et al.* (2006) Delphinidin, a dietary anthocyanidin, inhibits vascular endothelial growth factor receptor-2 phosphorylation. *Carcinogenesis*, 27, 989–996.

Lamy, S., Akla, N., Ouanouki, A., *et al.* (2012) Diet-derived polyphenols inhibit angiogenesis by modulating the interleukin-6/STAT3 pathway. *Experimental Cell Research*, 318, 1586–1596.

Leckey, L.C., Garige, M., Varatharajalu, R., *et al.* (2010) Quercetin and ethanol attenuate the progression of atherosclerotic plaques with concomitant up regulation of paraoxonase1 (PON1) gene expression and PON1 activity in LDLR−/− mice. *Alcoholism: Clinical and Experimental Research*, 34, 1535–1542.
Lee, C.H., Jeong, T.S., Choi Y.K., *et al.* (2001) Anti-atherogenic effect of citrus flavonoids, naringin and naringenin, associated with hepatic ACAT and aortic VCAM-1 and MCP-1 in high cholesterol-fed rabbits. *Biochemical and Biophysical Research Communications*, 284, 681–688.
Lee, Y.K., Bone, N. D., Strege, A.K., *et al.* (2004) VEGF receptor phosphorylation status and apoptosis is modulated by a green tea component, epigallocatechin-3-gallate (EGCG), in B-cell chronic lymphocytic leukemia. *Blood*, 104, 788–794.
Lin, M.T., Yen, M.L., Lin, C.Y. & Kuo, M.L. (2003) Inhibition of vascular endothelial growth factor-induced angiogenesis by resveratrol through interruption of Src-dependent vascular endothelial cadherin tyrosine phosphorylation. *Molecular Pharmacology*, 64, 1029–1036.
Lin, C.-M., Chiu, J.-H., Wu, I.H., *et al.* (2010) Ferulic acid augments angiogenesis via VEGF, PDGF and HIF-1$\alpha$. *The Journal of Nutritional Biochemistry*, 21, 627–633.
Liu, L., Lai, C.Q., Nie, L., *et al.* (2008) The modulation of endothelial cell gene expression by green tea polyphenol-EGCG. *Molecular Nutrition & Food Research*, 52, 1182–1192.
Loke, W.M., Proudfoot, J.M., Hodgson, J. M., *et al.* (2010) Specific dietary polyphenols attenuate atherosclerosis in apolipoprotein E-knockout mice by alleviating inflammation and endothelial dysfunction. *Arteriosclerosis, Thrombosis, and Vascular Biology*, 30, 749–757.
Luo, H., Rankin, G.O., Liu, L., *et al.* (2009) Kaempferol inhibits angiogenesis and VEGF expression through both HIF dependent and independent pathways in human ovarian cancer cells. *Nutrition and Cancer*, 61, 554–563.
Luo, H., Rankin, G.O., Juliano, N., *et al.* (2012) Kaempferol inhibits VEGF expression and in vitro angiogenesis through a novel ERK-NF$\kappa$B-cMyc-p21 pathway. *Food Chemistry*, 130, 321–328.
Manach, C., Williamson, G., Morand, C., *et al.* (2005) Bioavailability and bioefficacy of polyphenols in humans. I. Review of 97 bioavailability studies. *American Journal of Clinical Nutrition*, 81, 230S–242S.
McCullough, M.L., Peterson, J.J., Patel, R., *et al.* (2012) Flavonoid intake and cardiovascular disease mortality in a prospective cohort of US adults. *American Journal of Clinical Nutrition*, 95, 454–464.
Mink, P.J., Scrafford, C.G., Barraj, L.M., *et al.* (2007) Flavonoid intake and cardiovascular disease mortality: a prospective study in postmenopausal women. *American Journal of Clinical Nutrition*, 85, 895–909.
Moreno, P.R., Purushothaman, K.R., Sirol, M., *et al.* (2006) Neovascularization in human atherosclerosis. *Circulation*, 113, 2245–2252.
Moulton, K.S., Vakili, K., Zurakowski, D., *et al.* (2003) Inhibition of plaque neovascularization reduces macrophage accumulation and progression of advanced atherosclerosis. *Proceedings of the National Academy of Sciences USA*, 100, 4736–4741.
Muller, Y.A., Li, B., Christinger, H.W., *et al.* (1997a) Vascular endothelial growth factor: crystal structure and functional mapping of the kinase domain receptor binding site. *Proceedings of the National Academy of Sciences USA*, 94, 7192–7197.
Muller, Y.A., Christinger, H.W., Keyt, B.A. & de Vos, A.M. (1997b) The crystal structure of vascular endothelial growth factor (VEGF) refined to 1.93 Å resolution: multiple copy flexibility and receptor binding. *Structure*, 5, 1325–1338.
Mulligan-Kehoe, M.J. (2010) The vasa vasorum in diseased and nondiseased arteries. *American Journal of Physiology – Heart and Circulatory Physiology*, 298, H295–H305.
Németh, K., Plumb, G.W., Berrin, J-G., *et al.* (2003) Deglycosylation by small intestinal epithelial cell glucosidases is a critical step in the absorption and metabolism of dietary flavonoid glycosides in humans. *European Journal of Nutrition*, 42, 29–42.
Neuhaus, T., Pabst, S., Stier, S., *et al.* (2004) Inhibition of the vascular-endothelial growth factor-induced intracellular signaling and mitogenesis of human endothelial cells by epigallocatechin-3 gallate. *European Journal of Pharmacology*, 483, 223–227.

Neveu, V., Perez-Jiménez, J., Vos, F., *et al.* (2010) Phenol-explorer: an online comprehensive database on polyphenol contents in foods. *Database* (Oxford), bap024, doi: 10.1093/database/bap024.

Nicholson, S.K., Tucker, G.A. & Brameld, J.M. (2010) Physiological concentrations of dietary polyphenols regulate vascular endothelial cell expression of genes important in cardiovascular health. *British Journal of Nutrition*, 103, 1398–1403.

Nie, L., Wise, M.L., Peterson, D.M. & Meydani, M. (2006) Avenanthramide, a polyphenol from oats, inhibits vascular smooth muscle cell proliferation and enhances nitric oxide production. *Atherosclerosis*, 186, 260–266.

Oak, M.H., Bedoui, J.E., Madeira, S.V., *et al.* (2006) Delphinidin and cyanidin inhibit PDGF(AB)-induced VEGF release in vascular smooth muscle cells by preventing activation of p38 MAPK and JNK. *British Journal of Pharmacology*, 149, 283–290.

Oh, S.J., Kim, O., Lee, J.S., *et al.* (2010) Inhibition of angiogenesis by quercetin in tamoxifen-resistant breast cancer cells. *Food and Chemical Toxicology*, 48, 3227–3234.

Osada, M., Imaoka, S. & Funae, Y. (2004) Apigenin suppresses the expression of VEGF, an important factor for angiogenesis, in endothelial cells via degradation of HIF-1$\alpha$ protein. *FEBS Letters*, 575, 59–63.

Pan, R., Dai, Y., Gao, X.-H., *et al.* (2011) Inhibition of vascular endothelial growth factor-induced angiogenesis by scopoletin through interrupting the autophosphorylation of VEGF receptor 2 and its downstream signaling pathways. *Vascular Pharmacology*, 54, 18–28.

Pérez-Jiménez, J., Fezeu, L., Touvier, M., *et al.* (2011) Dietary intake of 337 polyphenols in French adults. *American Journal of Clinical Nutrition*, 93, 1220–1228.

Pratheeshkumar, P., Budhraja, A., Son, Y.-O., *et al.* (2012) Quercetin inhibits angiogenesis mediated human prostate tumor growth by targeting VEGFR-2 regulated AKT/mTOR/P70S6K signaling pathways. *PLoS ONE*, 7(10), e47516.

Pugh, C.W. & Ratcliffe, P. J. (2003) Regulation of angiogenesis by hypoxia: role of the HIF system. *Nature Medicine*, 9, 677–684.

Roberts, D.M., Kearney, J. B., Johnson, J.H., *et al.* (2004) The vascular endothelial growth factor (VEGF) receptor Flt-1 (VEGFR-1) modulates Flk-1 (VEGFR-2) signaling during blood vessel formation. *American Journal of Pathology*, 164, 1531–1535.

Rodriguez, S.K., Guo, W., Liu, L., *et al.* (2006) Green tea catechin, epigallocatechin-3-gallate, inhibits vascular endothelial growth factor angiogenic signaling by disrupting the formation of a receptor complex. *International Journal of Cancer*, 118, 1635–1644.

Scalbert, A., Andres-Lacueva, C., Arita, M., *et al.* (2011) Databases on food phytochemicals and their health promoting effects. *Journal of Agricultural and Food Chemistry*, 59, 4331–4348.

Shalaby, F., Rossant, J., Yamaguchi, T.P., *et al.* (1995) Failure of blood-island formation and vasculogenesis in Flk-1-deficient mice. *Nature*, 376, 62–66.

Shimizu, M., Shirakami, Y., Sakai, H., *et al.* (2010) (−)-epigallocatechin gallate inhibits growth and activation of the VEGF/VEGFR axis in human colorectal cancer cells. *Chemico-Biological Interactions*, 185, 247–252.

Subramanian, P. & Arul, D. (2012) Attenuation of NDEA-induced hepatocarcinogenesis by naringenin in rats. *Cell Biochemistry and Function*, 31, 511–517.

Tang, F.Y., Nguyen, N. & Meydani, M. (2003) Green tea catechins inhibit VEGF-induced angiogenesis in vitro through suppression of VE-cadherin phosphorylation and inactivation of Akt molecule. *International Journal of Cancer*, 106, 871–878.

Teller, N., Thiele, W., Boettler, U., *et al.* (2009) Delphinidin inhibits a broad spectrum of receptor tyrosine kinases of the ErbB and VEGFR family. *Molecular Nutrition & Food Research*, 53, 1075–1083.

Tribolo, S., Lodi, F., Connor, C., *et al.* (2008) Comparative effects of quercetin and its predominant human metabolites on adhesion molecule expression in activated human vascular endothelial cells. *Atherosclerosis*, 197, 50–56.

US Department of Agriculture, Agricultural Research Service (2013) *USDA Database for the Flavonoid Content of Selected Foods*, Release 3.1. Nutrient Data Laboratory Home Page: http://www.ars.usda.gov/main/site_main.htm?modecode=12-35-45-00 (last accessed April 2, 2014).

Wang, D., Zou, T., Yang, Y., *et al.* (2011) Cyanidin-3-O-$\beta$-glucoside with the aid of its metabolite protocatechuic acid, reduces monocyte infiltration in apolipoprotein E-deficient mice. *Biochemical Pharmacology*, 82, 713–719.

Chapter 10

# Phenolic Compounds from a Sex-Gender Perspective

Ilaria Campesi[1], Annalisa Romani[2], Maria Marino[3], and Flavia Franconi[4]
[1]*Laboratory of Sex-Gender Medicine, National Institute of Biostructures and Biosystems, Osilo, Italy*
[2]*Department of Pharmaceutical Sciences, University of Florence, Florence, Italy*
[3]*Department of Biology, University Roma Tre, Rome, Italy*
[4]*Department of Biomedical Science, Centre of Excellence for Biotechnology Development and Biodiversity, University of Sassari, Sassari, Italy*

**Abstract:** Food and beverage intake is the only behavior that animals, including human beings, must continually repeat in order to provide energy and chemical materials to the body. Recent epidemiological findings are poised to usher the science of nutrition into in a new era in which the consumption of specific whole foods, now called functional foods, is strongly associated with reduced risk of chronic diseases, including cancer and cardiovascular diseases. These findings have led to the hypothesis that specific food components, in particular phenolic compounds, interact with the human organism at the organ, cellular, and molecular levels to trigger a preventive action against chronic diseases. Numerous bioactive compounds have been identified and isolated from fruits, vegetables, and whole grains and are now present in all drugstores as dietary supplements to the usual intake of amino acids, vitamin, botanical compounds, and so on, although much basic information about their action in humans remains limited. For example, no stratification by sex has been performed in studies of their effects. Although human beings are mainly made up of men and women, in recent years it has become accepted that both biological sex and gender are prominent players in health and medicine. Until the last decade of the 20th century, research on women was neglected and results obtained in men were directly translated to women, both in medicine and in nutrition. Different strategies have been developed in order to maintain male and female body homeostasis, implying that there are important differences in the bioavailability, metabolism, distribution, and elimination of bioactive compounds in males and in females. This article will review

*Recent Advances in Polyphenol Research, Volume 4*, First Edition.
Edited by Annalisa Romani, Vincenzo Lattanzio, and Stéphane Quideau.

some of these differences underlying the possibility that plant-derived phenolic compounds can differently influence the health of male and female individuals.

**Keywords:** phenolic compound, sex-gender, health

## 10.1 Introduction

The relatively consistent epidemiological finding that consumption of specific whole foods, now called functional foods, such as fruits, vegetables, and whole grains, is strongly associated with reduced risk of chronic diseases, including cancers, cardiovascular diseases, and others, has led to the hypothesis that specific phytochemicals, in particular phenolic compounds, may be responsible for the observed preventive action (Manson, 2003; Ramos, 2008; Perez-Vizcaino & Duarte, 2010; Siow & Mann, 2010). As a result, it has been suggested that the adoption of "recommended diets together with maintenance of physical activity and appropriate body mass, could in time reduce the pathologies incidence by 30–40%" (World Health Organization, 2004). Hence, the dietary advice to individuals is to "choose predominantly plant-based diets rich in a variety of vegetables and fruit, pulses (legumes) and minimally processed starchy staple foods" (Manson, 2003). According to the European Commission's Concerted Action on Functional Food Science in Europe (FuFoSE), coordinated by the International Life Science Institute (ILSI) Europe, "a food can be regarded as 'functional' if it is satisfactorily demonstrated to affect beneficially one or more target functions in the body, beyond adequate nutritional effects, in a way that is relevant to either an improved state of health and well-being and/or reduction of risk of disease. Functional foods must remain foods and they must demonstrate their effects in amounts that can normally be expected to be consumed in the diet: they are not pills or capsules, but part of a normal food pattern" (Diplock, 1999).

Furthermore, numerous bioactive compounds have been identified and isolated from plants and are now present in all drugstores as dietary supplements to the usual intake of amino acids, vitamins, botanical compounds, and so on (Office of Dietary Supplements, 1994; Doyon & Labrecque, 2008). Very basic information about their action in humans remains limited.

Humans are mainly made up of men and women, but it is important to remember that other categories, such as transgender, also exist, as has been well testified by the Institute of Medicine of the National Institutes of Health (NIH) (Institute of Medicine, 2011). Only in recent years has been well accepted that biological sex and gender are prominent players in health and medicine (Wizemann & Pardue, 2001). In the context of biomedical sciences, the term "sex" indicates the biological maleness or femaleness of a given subject, while the term "gender" refers to social or cultural expectations associated with femininity or masculinity (Wizemann & Pardue, 2001; Marino *et al.*, 2011). Considering that most of the differences between men and women are a function of the interaction between biology and environment, many authors (Franconi *et al.*, 2011a; Marino *et al.*, 2011; Regitz-Zagrosek, 2012) have proposed joining the two terms; thus the term "sex-gender."

Body responses to xenobiotics (i.e. foods, supplements, drugs, etc.) rely on many characteristics, including genomic imprinting, body and organ size, body fat and water, absorption, metabolism, elimination, and many others. Thus, it is obvious that the age-dependent physiological differences between men and women can influence xenobiotic metabolism (xenokinetics) and its mechanisms of action (xenodynamics) (Gandhi *et al.*, 2004; Franconi *et al.*, 2007; Bulzomi & Marino, 2011). In addition, the study of sex-gender differences should also consider the changes in ovarian function that occur during a woman's life (including pregnancy), the use of oral contraceptives, and the hormone replacement treatment of postmenopausal women, all of which can affect xenobiotic responses (Gandhi *et al.*, 2004; Franconi *et al.*, 2007).

Up until recently, only a few preclinical and clinical studies took into account the influence of sex-gender on biomedical studies (Wizemann & Pardue, 2001; Franconi *et al.*, 2007, 2011b,c; Bulzomi & Marino, 2011; Marino *et al.*, 2011). As a consequence, knowledge of the effect of xenobiotics in males and in females is not very clear.

## 10.2 Phenolic compound classification and molecular mechanisms

Plant phenolic and polyphenolic compounds are secondary metabolites of plants possessing one or more aromatic rings that bear hydroxyl substituent(s). These compounds (i.e. several aromatic rings) are the most abundant "minor components" in the diet, being found in many foods and beverages of plant origin, such as fruits, vegetables, coffee, tea, beer, wine, and chocolate (Beecher, 2003; Manach *et al.*, 2004; Marino & Bulzomi, 2009; Tsao, 2010; Xiao *et al.*, 2011). They may be classified into different groups as a function of the number of phenol rings that they contain and of the structural elements that bind these rings to one another (Manach *et al.*, 2004). Certainly the most studied class of plant phenols is flavonoids, since some epidemiological studies suggest that they may play a role in the prevention of degenerative diseases. Flavonoids may themselves be divided into six subclasses as a function of the type of heterocycle involved: flavonols, flavones, isoflavones, flavanones, anthocyanidins, and flavanols (catechins and proanthocyanidins) (Beecher, 2003; Galluzzo & Marino, 2006; Virgili & Marino, 2008; Dixon & Pasinetti, 2010). On the basis of the 24-hour dietary recall in the 1999–2002 National Health and Nutrition Examination Survey (NHANES), and from the database of the US Department of Agriculture, the average total flavonoid intake of subjects older than 19 years is about 190 mg/day (Chun *et al.*, 2007), but other authors have calculated higher intakes (up to 1000 mg/day) of total phenolic compounds, in the presence of a dietary profile rich in fruit and vegetables (Scalbert & Williamson, 2000). Unfortunately, no stratification by sex-gender was performed in these studies.

It has been proposed that phenolic compounds might protect against cancer and/or heart disease by behaving as antioxidants (Fraga *et al.*, 2010 and references therein). However, antioxidant capacity as a chemical property of a given molecule or of a whole food item is not necessarily associated with an equivalent biological function. Accordingly, more recently it has been found that the molecular basis of the activity of phenolic compounds is much larger than was originally believed (Halliwell, 2007). Examples include the capacity to bind a member of the nuclear-receptor superfamily (i.e. sex steroid hormone receptors, retinoid X receptor (RXR), peroxisome proliferator-activated receptors (PPAR), pregnane X-receptor (PXR), and constitutive androstane receptor (CAR)) (Kretschmer & Baldwin, 2005; Galluzzo & Marino, 2006; Virgili & Marino, 2008; di Masi *et al.*, 2009), modulation of redox-sensitive transcription factors (including Nrf2, NF-$\kappa$B, and AP-1) (Andreadi *et al.*, 2006; Brunelleschi *et al.*, 2007), inhibition of signal-transduction proteins (i.e. EGFR, PKC, PI3K, ERK) (Marino & Bulzomi, 2009), modulation of enzyme activity by either competitive or allosteric interaction (Jakobs *et al.*, 2006), glutathione biosynthesis (Moskaug *et al.*, 2005), and gene expression in general (Wung *et al.*, 2005).

Many of the previous studies do not consider sex-gender aspects, and therefore we do not know whether the molecular mechanism is the same in males and in females.

## 10.3 Sex-gender and the xenokinetics of phenolic compounds

Physiological differences between men and women include body dimension and composition, gastric pH, gastrointestinal motility, hepatic metabolism, and renal functions, and all of these parameters impact on pharmacokinetic parameters (absorption, distribution, metabolism, and elimination) (Franconi *et al.*, 2007).

Regarding the importance of sex-gender on the pharmacokinetics of phenolic compounds, there are not univocal data. Numerous phenolic compounds such as flavonoids are ingested as glycosides, which are hydrolyzed to form the sugar group or **glycone** and the non-sugar group or **aglycone** by bacterial $\beta$-glucosidases and by mammalian lactase phlorizin hydrolase in the gastro-enteric tract (Rowland *et al.*, 2003). They can then be absorbed by the small and large intestines (Manach *et al.*, 2005; Kahle *et al.*, 2010) or further metabolized by intestinal microflora. For example, in a certain percentage of humans (about 30–35% among the Caucasian population), and in many animals, daidzein can be enzymatically converted to a more active compound (equol) (Juniewicz *et al.*, 1988; Lundh, 1995; Xu *et al.*, 1995; Rowland *et al.*, 2000; Ohta *et al.*, 2002; Setchell *et al.*, 2002). Interestingly, after 15 days of soy intake some women develop the capacity to metabolize daidzein into equol (Lu *et al.*, 1996), but this does not seem to occur in men (Lu *et al.*, 1995). Intuitively, this equol-producing status might affect the response after the use of daidzein or soy. Indeed, a 2-year study in postmenopausal women showed a major hypocholesterolemic effect in equol producers compared with nonproducers (Setchell *et al.*, 2002). Interestingly, *O*-desmethylangolensin can also be produced, and its plasmatic level seems to be higher in men than in women (Wiseman *et al.*, 2004).

The aglycones can be adsorbed by intestinal mucosa and in the liver, where they can be biotransformed by conjugation. Three main types of conjugation have been described: glucuronidation, sulfation, and methylation.

(i) Aglycones are mainly conjugated with glucuronic acid by UDP-glucuronosyltransferases (UGTs) (Bock, 2012), which are made up of different families and are regulated in a tissue-specific way and by ligand-activated transcription factors (LATFs). UGT isoforms conjugate different but partially overlapping endogenous compounds (bilirubin, estrogens, testosterone, serotonin) and xenobiotics, including food components (polyphenols, such as flavonoids) and drugs (Bock, 2012). Some UGTs, namely UGT2B17, are more expressed in men than in women (Gallagher *et al.*, 2010; Wu *et al.*, 2011). This might potentially result in different biotransformations of phenolic compound in men versus women.

(ii) A smaller amount are conjugated with sulfate in the liver by sulfotransferases. This reaction can be controlled by estrogens. Thus, it is not surprising that, in rodents, oral bioavailability of genistein appears to be dependent on estrous phase. Notably, estradiol administration in ovariectomized female rats significantly reduces the oral bioavailability of total genistein, mostly via increased sulfation (Kulkarni *et al.*, 2012).

(iii) Catechol-*O*-methyltransferase (COMT) is present in a wide range of tissues. It metabolizes neurotransmitters and hormones such as dopamine, norepinephrine, and epinephrine, as well as catechol drugs. It has been demonstrated that women have about 25% lower levels of enzyme activity than men (Boudikova *et al.*, 1990). Phenolic compounds are methylated predominantly in the 3′ position, but a minor proportion of 4′-*O*-methylated product is also formed. Note that a substantial amount of 4′-methylepigallocatechin is detected in human plasma after ingestion of tea, whereas methylation of the catechol group occurs only to a limited extent (Bao & Fenwick, 2004; Manach *et al.*, 2004). Methylation is also implicated in the metabolism of quercetin in humans, accounting for about 20–30% (Manach *et al.*, 1998; Day *et al.*, 2000). Despite this, there are no specific data regarding the existence of sex-gender differences in plant phenolic compound methylation, but considering the different levels of COMT expression in men and women, sex-gender differences could be plausible.

Furthermore, some phenolic compounds, such as genistein and daidzein, can also be metabolized by the hepatic cytochrome p450 enzymes (CYP), which are responsible for the oxidative metabolism of various and numerous endogenous and exogenous molecules. Sex-gender differences in drug metabolism are well established for some human CYP isoforms. CYP2B6, CYP2A6, and CYP3A have greater levels of activity in women than in men, while CYP2D6, CYP2E1, and CYP1A2 have slightly higher activity in men (Franconi *et al.*, 2007, 2011b and references therein).

The excretion of conjugated phenolic compounds occurs by renal (7–30 %) and biliary (10%) routes (Kelly *et al.*, 1993; Xu *et al.*, 1995); their biological half-life ranges from 11 to 28 hours (Manach *et al.*, 2005). Considering that less renal excretion occurs in women than in men (Franconi *et al.*, 2007), this could be of some importance in their elimination. Indeed, during 1 month of daily soy ingestion (80–210 mg of each isoflavone), women initially excrete more isoflavone conjugates in urine than do men, but they also experience a progressive decrease in urinary excretion of genistein and daidzein, unlike men (Lu & Anderson, 1998). Thus, isoflavone metabolism and disposition are affected by the duration of soy ingestion in women but not in men. In addition, it has been found that men excrete more enterolactone and less enterodiol than women, while no differences have been found in the excretion of isoflavonoids (Kirkman *et al.*, 1995).

With regard to plasma levels, there are no conclusive age or sex-gender differences in the bioavailability of genistein and daidzein (Cassidy *et al.*, 2006). After administration of [$^{14}$C]genistein, the maximal plasma concentrations of radioactivity are significantly lower in female rats than in males, and the half-lives are 8.5 and 12.4 hours, respectively, while the concentration of radioactivity is greater in the liver of female rats than in males of the species (Coldham & Sauer, 2000). However, not all authors find sex-gender differences in plasma levels of genistein and daidzein and their metabolites (Kelly *et al.*, 1993; Lampe *et al.*, 1998).

Unfortunately, pharmacokinetic data are scarce and do not include all phenolic compounds, and furthermore they often do not consider sex-gender, making plausible conclusions impossible. Physiological differences between men and women could well impact the pharmacokinetics of phenolic compounds, so pharmacokinetic studies that include both men and women are needed.

Notably, polyphenols can modify the metabolism of endogenous and exogenous compounds, and this may be sex-gender dependent. Itraconazole intestinal metabolism appears to be sex-specific when associated with grapefruit juice, a selective enteric CYP3A4 inhibitor, with the influence of grapefruit juice being greater in women than in men (Gubbins *et al.*, 2008). A characteristic of the CYP superfamily is its relatively low activity in the absence of substrate, which steeply increases in the presence of substrates or other inducer molecules. In particular, members of CYP1A, CYP2B, CYP2C/H, CYP3A, and CYP4A are highly inducible in tissue- and age-specific ways (Handschin & Meyer, 2003). It is important to remember that phase 2 enzymes are also inducible.

The induction of phase 1 and phase 2 enzymes mainly occurs through xenosensing of nuclear receptors or nuclear transcription factors; as already mentioned, phenols can bind many of them. In particular, PXR, which is implicated in the activation of CYP3A, CYP2B6, CYP2C9, sulfotransferase, UGT1A1, glutathione S-transferases, and multidrug resistance protein 2 (Meijerman *et al.*, 2006 and di Masi *et al.*, 2009, and references therein), is activated by hyperforin, a component of extracts from the St John's wort (Meijerman *et al.*, 2006 and references therein), and by isoflavones (genistein, daidzein, equol) and polymethoxylated flavone (tangeretin) (Satsu *et al.*, 2008; Li *et al.*, 2009), while it is inhibited by coumestrol (Wang *et al.*, 2008). At the same time, CAR, which has as its main target the CYP2B gene, followed by CYP2C9 (Meijerman *et al.*, 2006; di Masi *et al.*, 2009), can be activated by flavonoids (quercetin, kaempferol, and tamarixetin). Importantly, flavonoid binding to CAR and PXR is not specific, in that the ligand-binding capacity of these receptors can overlap (Li *et al.*, 2009); as an example, flavonoids can activate both CAR and PRX (Meijerman *et al.*, 2006; Moon *et al.*, 2006; Yao *et al.*, 2011). Finally, Nrf2 is necessary to the induction of NADPH:quinone oxidoreductase-1 glutathione S-transferase isoforms, and UGT and some phenolic compounds can bind this transcription factor (Andreadi *et al.*, 2006). The binding of phenolic compounds with the previous targets may be clinically relevant because it may be the basis of many food and drug interactions. Considering the role of sexual hormones on the previous targets, the interactions may occur in a sex-gender-specific way.

## 10.4 Sex-gender differences in xenodynamics

Of the numerous activities of polyphenols, we shall discuss only those which have been studied with a view to sex-gender aspects. Phenolic compounds in red wine, for example, relax aortic rings in the endothelium in a sex-gender-dependent way (Kane *et al.*, 2009). In fact, relaxation is more pronounced in female than in male rats. Furthermore, genistein and daidzein inhibit formation of neointimal cells more actively in male than in female rabbit aortic rings (Finking *et al.*, 1999). Nagata *et al.* (2003) found an inverse correlation between soy intake and diastolic blood pressure in Japanese men, and a marginally positive association in peri- and postmenopausal women. In vascular smooth-muscle cells obtained from males and females, the polyphenols of extra-virgin olive oil increased glutathione reductase, peroxidase, and GSH levels, activating Nrf2, in a sex-gender-specific manner (Ambrosini *et al.*, 2012). In addition, the polyphenols of extra-virgin olive oil lower apoptosis in male cells only, and potentially bolster autophagic cytoprotection mainly in female cells (Romani *et al.*, 2012). Moreover, it has also been shown (Alonso & Martınez-Gonzalez, 2004) that the antihypertensive effect of extra-virgin olive oil is present only among men. Collectively, these findings indicate that sex-gender is important for the effects of phenols in vessels.

Regarding the heart, sex-gender differences in the acute cardiac actions of genistein have been detected, and they can be attributed to genistein's effect on the intracellular $Ca^{2+}$ cycle (Liew *et al.*, 2004). These results suggest that genistein may afford greater cardioprotection in females than in males. In contrast, soy phytoestrogen supplementation augments cardiac growth in male mice but not in females (Stauffer *et al.*, 2006), and a soy-based diet worsens hypertrophic cardiomyopathy more in male mice than in females (Haines *et al.*, 2012a). These results strongly suggest that caution should be used before recommending estrogenic compound supplementation, at least in patients with cardiac disease (Haines *et al.*, 2012b).

The beneficial health effects of phytoestrogens remain equivocal. A few years ago the American Heart Association reversed its endorsement of soy because it did not find a significant beneficial effect of soy proteins or plant estrogens on lipids, blood pressure, or menopausal changes (Sacks *et al.*, 2006). By contrast, some human studies indicate a potential beneficial effect on cardiovascular risk factors, including body mass index (Maskarinec *et al.*, 2008), lipid profiles (Anderson *et al.*, 1995; Welty *et al.*, 2007), glucose metabolism (Nagata *et al.*, 1998; Cheng *et al.*, 2004; Azadbakht *et al.*, 2007), and hypertension (Welty *et al.*, 2007), but the average reduction of low-density lipoproteins (LDLs) is very low. Indeed, a sex-gender effect of soy products on lipid metabolism has been described in the Chinese population, as the higher soy intake associates with a better plasma lipid profile in men than in women (Ho *et al.*, 2000). The association between bifurcation intima-media thickness and soy intake is different in middle-aged Chinese men versus women (Zhang *et al.*, 2008). Sex-specific effects of soy have also been observed in male and female animals (Blair *et al.*, 2002; Jenkins *et al.*, 2002). A recent study indicates that habitual intake of soy causes a moderately reduced risk of metabolic syndrome in middle-aged and elderly Chinese women, but elevates the risk in men. Soy significantly increased the risk for hyperglycemia but reduced the risk for hypertension in these men (Pan *et al.*, 2008). These results indicate that more controlled clinical trials that consider sex-gender aspects are required.

Phytoestrogens may also exert hormonal and metabolic effects in a sex-gender manner. Nutritional amounts of genistein induced adipose tissue deposition in 4-week-old male mice, but not in female rats (Penza *et al.*, 2006). Genistein, dose-dependently increased the glucose sensitivity of pancreatic islets and caused goiter in humans and animals through non-estrogenic effects (Doerge & Sheehan, 2002). Furthermore, in premenopausal women isoflavone-rich soy products decreased follicle-stimulating hormones and lutenizing hormones, and they could increase estradiol in postmenopausal women (Hooper *et al.*, 2009). Daidzein and genistein increased the body mineral density in young male mice, although daidzein inhibited bone growth by decreasing the bone formation rate in female mice, indicating that the effects of isoflavones depend on sex (Fujioka *et al.*, 2007). Genistein counteracted the induction of a marker of colonic premalignancy mediated by low nutritional calcium in both males and females. However, soy itself enhanced COX-2 and reduced Bak, but only in females (Bises *et al.*, 2007).

Sex-gender-dependent effects are also seen with other molecules. In fact, the associated administration of vitamin C + rutin, a flavonoid present in onions, apples, tea, and red wine, prevents the nephrotoxicity induced by a high-cholesterol diet to a greater degree in male than in female animals (Al-Rejaie *et al.*, 2012).

## 10.5 Conclusions

The reported evidence emphasizes that sex-gender could influence the biological activity of complex food-matrix components "as they are in food" and of isolated food compounds. In evaluating the relevance of sex-gender determinants, it is necessary to consider biotransformation, and possibly synergism and cross-activity between different components. In particular, sex-gender differences appear in the xenokinetics and xenodynamics of phenolic compounds; however, there is a lack of adequate sex-related studies. Considering that in Western countries the consumption of natural substances is widespread and that women are the greatest consumers (Al-Windi *et al.*, 2000; Durante *et al.*, 2001; Fennell, 2004; Kennedy, 2005), the inclusion of women in studies is clearly a need (Eisenberg *et al.*, 1998; Kaufman *et al.*, 2002; Singh & Levine, 2006). The experimental paradigms should also consider environmental factors, in view of the importance of socialcultural aspects in health.

## References

Al-Rejaie, S.S., Abuohashish, H.M., Alkhamees, O.A., *et al.* (2012) Gender difference following high cholesterol diet induced renal injury and the protective role of rutin and ascorbic acid combination in Wistar albino rats. *Lipids in Health and Disease*, 11, 41.

Al-Windi, A., Elmfeldt, D. & Svardsudd, K. (2000) The relationship between age, gender, well-being and symptoms, and the use of pharmaceuticals, herbal medicines and self-care products in a Swedish municipality. *European Journal of Clinical Pharmacology*, 56, 311–317.

Alonso, A. & Martinez-Gonzalez, M.A. (2004) Olive oil consumption and reduced incidence of hypertension: the SUN study. *Lipids*, 39, 1233–1238.

Ambrosini, G., Romani, A., Scazzocchio, B., *et al.* (2012) Sex-gender influences the expression of antioxidant/detoxifying enzyme induced by extra rirgin olive oil extracts in rat vascular smooth muscle cells. In: *XXVIth International Conference on Polyphenols, July, 23–26, 2012, Florence, Italy* (eds V. Lattanzio, N. Mulinacci, P. Pinelli & A. Romani), Vol. 2, pp. 437–438, Abstract T4.03. University of Florence, Florence.

Anderson, J.W., Johnstone, B.M. & Cook-Newell, M.E. (1995) Meta-analysis of the effects of soy protein intake on serum lipids. *The New England Journal of Medicine*, 333, 276–282.

Andreadi, C.K., Howells, L.M., Atherfold, P.A. & Manson, M.M. (2006) Involvement of Nrf2, p38, B-Raf, and nuclear factor-kappaB, but not phosphatidylinositol 3-kinase, in induction of hemeoxygenase-1 by dietary polyphenols. *Molecular Pharmacology*, 69, 1033–1040.

Azadbakht, L., Kimiagar, M., Mehrabi, Y., *et al.* (2007) Soy inclusion in the diet improves features of the metabolic syndrome: a randomized crossover study in postmenopausal women. *The American Journal of Clinical Nutrition*, 85, 735–741.

Bao, Y. & Fenwick, R. (2004) *Phytochemicals in Health and Disease*. Marcel Dekker, New York.

Beecher, G.R. (2003) Overview of dietary flavonoids: nomenclature, occurrence and intake. *Journal of Nutrition*, 133, 3248S–3254S.

Bises, G., Bajna, E., Manhardt, T., *et al.* (2007) Gender-specific modulation of markers for premalignancy by nutritional soy and calcium in the mouse colon. *Journal of Nutrition*, 137, 211S–215S.

Blair, R.M., Appt, S.E., Bennetau-Pelissero, C., *et al.* (2002) Dietary soy and soy isoflavones have gender-specific effects on plasma lipids and isoflavones in golden Syrian f(1)b hybrid hamsters. *Journal of Nutrition*, 132, 3585–3591.

Bock, K.W. (2012) Human UDP-glucuronosyltransferases: feedback loops between substrates and ligands of their transcription factors. *Biochemical Pharmacology*, 84, 1000–1006.

Boudikova, B., Szumlanski, C., Maidak, B. & Weinshilboum, R. (1990) Human liver catechol-O-methyltransferase pharmacogenetics. *Clinical Pharmacology & Therapeutics*, 48, 381–389.

Brunelleschi, S., Bardelli, C., Amoruso, A., *et al.* (2007) Minor polar compounds extra-virgin olive oil extract (MPC-OOE) inhibits NF-kappa B translocation in human monocyte/macrophages. *Pharmacological Research*, 56, 542–549.

Bulzomi, P. & Marino, M. (2011) Environmental endocrine disruptors: does a sex-related susceptibility exist? *Frontiers in Bioscience*, 16, 2478–2498.

Cassidy, A., Brown, J.E., Hawdon, A., *et al.* (2006) Factors affecting the bioavailability of soy isoflavones in humans after ingestion of physiologically relevant levels from different soy foods. *Journal of Nutrition*, 136, 45–51.

Cheng, S.Y., Shaw, N.S., Tsai, K.S. & Chen, C.Y. (2004) The hypoglycemic effects of soy isoflavones on postmenopausal women. *Journal of Women's Health (Larchmt)*, 13, 1080–1086.

Chun, O.K., Chung, S.J. & Song, W.O. (2007) Estimated dietary flavonoid intake and major food sources of U.S. adults. *Journal of Nutrition*, 137, 1244–1252.

Coldham, N.G. & Sauer, M.J. (2000) Pharmacokinetics of [(14)C]Genistein in the rat: gender-related differences, potential mechanisms of biological action, and implications for human health. *Toxicology and Applied Pharmacology*, 164, 206–215.

Day, A.J., Bao, Y., Morgan, M.R. & Williamson, G. (2000) Conjugation position of quercetin glucuronides and effect on biological activity. *Free Radical Biology & Medicine*, 29, 1234–1243.

di Masi, A., De Marinis, E., Ascenzi, P. & Marino, M. (2009) Nuclear receptors CAR and PXR: molecular, functional, and biomedical aspects. *Molecular Aspects of Medicine*, 30, 297–343.

Diplock, A.T. (1999) Scientific concepts of functional foods in Europe. *Consensus document. British Journal of Nutrition*, 81, Suppl. 1, S1–27.

Dixon, R.A. & Pasinetti, G.M. (2010) Flavonoids and isoflavonoids: from plant biology to agriculture and neuroscience. *Plant Physiology*, 154, 453–457.

Doerge, D.R. & Sheehan, D.M. (2002) Goitrogenic and estrogenic activity of soy isoflavones. *Environmental Health Perspectives*, 110, Suppl. 3, 349–353.

Doyon, M. & Labrecque, J. (2008) Functional foods: a conceptual definition. *British Food Journal*, 110, 1133–1149.

Durante, K.M., Whitmore, B., Jones, C.A. & Campbell, N.R. (2001) Use of vitamins, minerals and herbs: a survey of patients attending family practice clinics. *Clinical & Investigative Medicine*, 24, 242–249.

Eisenberg, D.M., Davis, R.B., Ettner, S.L., *et al.* (1998) Trends in alternative medicine use in the United States, 1990–1997: results of a follow-up national survey. *Journal of the American Medical Association*, 280, 1569–1575.

Fennell, D. (2004) Determinants of supplement usage. *Preventive Medicine*, 39, 932–939.

Finking, G., Wohlfrom, M., Lenz, C., *et al.* (1999) The phytoestrogens Genistein and Daidzein, and 17 beta-estradiol inhibit development of neointima in aortas from male and female rabbits in vitro after injury. *Coronary Artery Disease*, 10, 607–615.

Fraga, C.G., Galleano, M., Verstraeten, S.V. & Oteiza, P.I. (2010) Basic biochemical mechanisms behind the health benefits of polyphenols. *Molecular Aspects of Medicine*, 31, 435–445.

Franconi, F., Brunelleschi, S., Steardo, L. & Cuomo, V. (2007) Gender differences in drug responses. *Pharmacological Research*, 55, 81–95.

Franconi, F., Carru, C., Malorni, W., *et al.* (2011a) The effect of sex/gender on cardiovascular pharmacology. *Current Pharmaceutical Design*, 17, 1095–1107.

Franconi, F., Carru, C., Spoletini, I., *et al.* (2011b) A GENS-based approach to cardiovascular pharmacology: impact on metabolism, pharmacokinetics and pharmacodynamics. *Therapeutic Delivery*, 2, 1437–1453.

Franconi, F., Sanna, M., Straface, E., *et al.* (2011c) Pharmacokinetics and pharmacodynamics: the role of sex and gender. In: *Sex and Gender Aspects in Clinical Medicine* (eds S. Oertelt-Prigione & V. Regitz-Zagrosek), pp. 183–194. Springer-Verlag, London.

Fujioka, M., Sudo, Y., Okumura, M., *et al.* (2007) Differential effects of isoflavones on bone formation in growing male and female mice. *Metabolism*, 56, 1142–1148.

Gallagher, C.J., Balliet, R.M., Sun, D., *et al.* (2010) Sex differences in UDP-glucuronosyltransferase 2B17 expression and activity. *Drug Metabolism and Disposition*, 38, 2204–2209.

Galluzzo, P. & Marino, M. (2006) Nutritional flavonoids impact on nuclear and extranuclear estrogen receptor activities. *Genes & Nutrition*, 1, 161–176.

Gandhi, M., Aweeka, F., Greenblatt, R.M. & Blaschke, T.F. (2004) Sex differences in pharmacokinetics and pharmacodynamics. *Annual Review of Pharmacology and Toxicology*, 44, 499–523.

Gubbins, P.O., Gurley, B.J., Williams, D.K., *et al.* (2008) Examining sex-related differences in enteric itraconazole metabolism in healthy adults using grapefruit juice. *European Journal of Clinical Pharmacology*, 64, 293–301.

Haines, C.D., Harvey, P.A. & Leinwand, L.A. (2012a) Estrogens mediate cardiac hypertrophy in a stimulus-dependent manner. *Endocrinology*, 153, 4480–4490.

Haines, C.D., Harvey, P.A., Luczak, E.D., *et al.* (2012b) Estrogenic compounds are not always cardioprotective and can be lethal in males with genetic heart disease. *Endocrinology*, 153, 4470–4479.

Halliwell, B. (2007) Dietary polyphenols: good, bad, or indifferent for your health? *Cardiovascular Research*, 73, 341–347.

Handschin, C. & Meyer, U.A. (2003) Induction of drug metabolism: the role of nuclear receptors. *Pharmacological Reviews*, 55, 649–673.

Ho, S.C., Woo, J.L., Leung, S.S., *et al.* (2000) Intake of soy products is associated with better plasma lipid profiles in the Hong Kong Chinese population. *Journal of Nutrition*, 130, 2590–2593.

Hooper, L., Ryder, J.J., Kurzer, M.S., *et al.* (2009) Effects of soy protein and isoflavones on circulating hormone concentrations in pre- and post-menopausal women: a systematic review and meta-analysis. *Human Reproduction Update*, 15, 423–440.

Institute of Medicine (2011) The Health of Lesbian, Gay, Bisexual, and Transgender People: Building a Foundation for Better Understanding. National Academies Press, Washington, DC.

Jakobs, S., Fridrich, D., Hofem, S., *et al.* (2006) Natural flavonoids are potent inhibitors of glycogen phosphorylase. *Molecular Nutrition & Food Research*, 50, 52–57.

Jenkins, D.J., Kendall, C.W., Jackson, C.J., *et al.* (2002) Effects of high- and low-isoflavone soyfoods on blood lipids, oxidized LDL, homocysteine, and blood pressure in hyperlipidemic men and women. *The American Journal of Clinical Nutrition*, 76, 365–372.

Juniewicz, P.E., Pallante Morell, S., Moser, A. & Ewing, L.L. (1988) Identification of phytoestrogens in the urine of male dogs. *Journal of Steroid Biochemistry*, 31, 987–994.

Kahle, K., Kempf, M., Schreier, P., *et al.* (2010) Intestinal transit and systemic metabolism of apple polyphenols. *European Journal of Nutrition*, 50, 507–522.

Kane, M.O., Anselm, E., Rattmann, Y.D., *et al.* (2009) Role of gender and estrogen receptors in the rat aorta endothelium-dependent relaxation to red wine polyphenols. *Vascular Pharmacology*, 51, 140–146.

Kaufman, D.W., Kelly, J.P., Rosenberg, L., *et al.* (2002) Recent patterns of medication use in the ambulatory adult population of the United States: the Slone survey. *Journal of the American Medical Association*, 287, 337–344.

Kelly, G.E., Nelson, C., Waring, M.A., *et al.* (1993) Metabolites of dietary (soya) isoflavones in human urine. *Clinica Chimica Acta*, 223, 9–22.

Kennedy, J. (2005) Herb and supplement use in the US adult population. *Clinical Therapeutics*, 27, 1847–1858.

Kirkman, L.M., Lampe, J.W., Campbell, D.R., *et al.* (1995) Urinary lignan and isoflavonoid excretion in men and women consuming vegetable and soy diets. *Nutrition and Cancer*, 24, 1–12.

Kretschmer, X.C. & Baldwin, W.S. (2005) CAR and PXR: xenosensors of endocrine disrupters? *Chemico-Biological Interactions*, 155, 111–128.

Kulkarni, K.H., Yang, Z., Niu, T. & Hu, M. (2012) Effects of estrogen and estrus cycle on pharmacokinetics, absorption, and disposition of genistein in female sprague-dawley rats. *Journal of Agricultural and Food Chemistry*, 60, 7949–7956.

Lampe, J.W., Karr, S.C., Hutchins, A.M. & Slavin, J.L. (1998) Urinary equol excretion with a soy challenge: influence of habitual diet. *Proceedings of the Society for Experimental Biology and Medicine*, 217, 335–339.

Li, Y., Ross-Viola, J.S., Shay, N.F., *et al.* (2009) Human CYP3A4 and murine Cyp3A11 are regulated by equol and genistein via the pregnane X receptor in a species-specific manner. *Journal of Nutrition*, 139, 898–904.

Liew, R., Stagg, M.A., Chan, J., *et al.* (2004) Gender determines the acute actions of genistein on intracellular calcium regulation in the guinea-pig heart. *Cardiovascular Research*, 61, 66–76.

Lu, L.J. & Anderson, K.E. (1998) Sex and long-term soy diets affect the metabolism and excretion of soy isoflavones in humans. *The American Journal of Clinical Nutrition*, 68, 1500S–1504S.

Lu, L.J., Grady, J.J., Marshall, M.V., *et al.* (1995) Altered time course of urinary daidzein and genistein excretion during chronic soya diet in healthy male subjects. *Nutrition and Cancer*, 24, 311–323.

Lu, L.J., Lin, S.N., Grady, J.J., *et al.* (1996) Altered kinetics and extent of urinary daidzein and genistein excretion in women during chronic soya exposure. *Nutrition and Cancer*, 26, 289–302.

Lundh, T. (1995) Metabolism of estrogenic isoflavones in domestic animals. *Proceedings of the Society for Experimental Biology and Medicine*, 208, 33–39.

Manach, C., Morand, C., Crespy, V., *et al.* (1998) Quercetin is recovered in human plasma as conjugated derivatives which retain antioxidant properties. *FEBS Letters*, 426, 331–336.

Manach, C., Scalbert, A., Morand, C., *et al.* (2004) Polyphenols: food sources and bioavailability. *The American Journal of Clinical Nutrition*, 79, 727–747.

Manach, C., Williamson, G., Morand, C., *et al.* (2005) Bioavailability and bioefficacy of polyphenols in humans. I. Review of 97 bioavailability studies. *The American Journal of Clinical Nutrition*, 81, 230S–242S.

Manson, M.M. (2003) Cancer prevention – the potential for diet to modulate molecular signalling. *Trends in Molecular Medicine*, 9, 11–18.

Marino, M. & Bulzomi, P. (2009) Mechanisms at the root of flavonoid action in cancer: a step toward solving the Rubik's cube. In: *Flavonoids: Biosynthesis, Biological Effects and Dietary Sources* (ed. R.B. Keller), pp. 231–248. Nova Science, New York.

Marino, M., Masella, R., Bulzomi, P., *et al.* (2011) Nutrition and human health from a sex-gender perspective. *Molecular Aspects of Medicine*, 32, 1–70.

Maskarinec, G., Aylward, A.G., Erber, E., *et al.* (2008) Soy intake is related to a lower body mass index in adult women. *European Journal of Nutrition*, 47, 138–144.
Meijerman, I., Beijnen, J.H. & Schellens, J.H. (2006) Herb-drug interactions in oncology: focus on mechanisms of induction. *Oncologist*, 11, 742–752.
Moon, Y.J., Wang, X. & Morris, M.E. (2006) Dietary flavonoids: effects on xenobiotic and carcinogen metabolism. *Toxicology in Vitro*, 20, 187–210.
Moskaug, J.O., Carlsen, H., Myhrstad, M.C. & Blomhoff, R. (2005) Polyphenols and glutathione synthesis regulation. *The American Journal of Clinical Nutrition*, 81, 277S–283S.
Nagata, C., Takatsuka, N., Kurisu, Y. & Shimizu, H. (1998) Decreased serum total cholesterol concentration is associated with high intake of soy products in Japanese men and women. *Journal of Nutrition*, 128, 209–213.
Nagata, C., Shimizu, H., Takami, R., *et al.* (2003) Association of blood pressure with intake of soy products and other food groups in Japanese men and women. *Preventive Medicine*, 36, 692–697.
The Office of Dietary Supplements (1994) Dietary Supplement Health and Education Act of 1994. National Institutes of Health, Bethesda, MD.
Ohta, A., Uehara, M., Sakai, K., *et al.* (2002) A combination of dietary fructooligosaccharides and isoflavone conjugates increases femoral bone mineral density and equol production in ovariectomized mice. *Journal of Nutrition*, 132, 2048–2054.
Pan, A., Franco, O.H., Ye, J., *et al.* (2008) Soy protein intake has sex-specific effects on the risk of metabolic syndrome in middle-aged and elderly Chinese. *Journal of Nutrition*, 138, 2413–2421.
Penza, M., Montani, C., Romani, A., *et al.* (2006) Genistein affects adipose tissue deposition in a dose-dependent and gender-specific manner. *Endocrinology*, 147, 5740–5751.
Perez-Vizcaino, F. & Duarte, J. (2010) Flavonols and cardiovascular disease. *Molecular Aspects of Medicine*, 31, 478–494.
Ramos, S. (2008) Cancer chemoprevention and chemotherapy: dietary polyphenols and signalling pathways. *Molecular Nutrition & Food Research*, 52, 507–526.
Regitz-Zagrosek, V. (2012). *Sex and Gender Differences in Pharmacology*. Handbook of Experimental Pharmacology, Vol. 214. Springer-Verlag, Berlin, Heidelberg.
Romani, A., Vona, R., Masella, R., *et al.* (2012). Identification and characterization of Tuscan extra-virgin olive oil extracts and their biological activity in rat VSMC. In: *XXVIth International Conference on Polyphenols, July, 23–26, 2012, Florence, Italy* (eds V. Lattanzio, N. Mulinacci, P. Pinelli & A. Romani), Vol. 2, pp. 563–564, Abstract T4.66. University of Florence, Florence.
Rowland, I.R., Wiseman, H., Sanders, T.A., *et al.* (2000) Interindividual variation in metabolism of soy isoflavones and lignans: influence of habitual diet on equol production by the gut microflora. *Nutrition and Cancer*, 36, 27–32.
Rowland, I., Faughnan, M., Hoey, L., *et al.* (2003) Bioavailability of phyto-oestrogens. *British Journal of Nutrition*, 89, Suppl. 1, S45–S58.
Sacks, F.M., Lichtenstein, A., Van Horn, L., *et al.* (2006) Soy protein, isoflavones, and cardiovascular health: a summary of a statement for professionals from the American Heart Association Nutrition Committee. *Arteriosclerosis, Thrombosis, and Vascular Biology*, 26, 1689–1692.
Satsu, H., Hiura, Y., Mochizuki, K., *et al.* (2008) Activation of pregnane X receptor and induction of MDR1 by dietary phytochemicals. *Journal of Agricultural and Food Chemistry*, 56, 5366–5373.
Scalbert, A. & Williamson, G. (2000) Dietary intake and bioavailability of polyphenols. *Journal of Nutrition*, 130, 2073S–2085S.
Setchell, K.D., Brown, N.M. & Lydeking-Olsen, E. (2002) The clinical importance of the metabolite equol – a clue to the effectiveness of soy and its isoflavones. *Journal of Nutrition*, 132, 3577–3584.
Singh, S.R. & Levine, M.A. (2006) Natural health product use in Canada: analysis of the National Population Health Survey. *The Canadian Journal of Clinical Pharmacology*, 13, e240–250.
Siow, R.C. & Mann, G.E. (2010) Dietary isoflavones and vascular protection: activation of cellular antioxidant defenses by SERMs or hormesis? *Molecular Aspects of Medicine*, 31, 468–477.
Stauffer, B.L., Konhilas, J.P., Luczak, E.D. & Leinwand, L.A. (2006) Soy diet worsens heart disease in mice. *The Journal of Clinical Investigation*, 116, 209–216.

Tsao, R. (2010) Chemistry and biochemistry of dietary polyphenols. *Nutrients*, 2, 1231–1246.

Virgili, F. & Marino, M. (2008) Regulation of cellular signals from nutritional molecules: a specific role for phytochemicals, beyond antioxidant activity. *Free Radical Biology & Medicine*, 45, 1205–1216.

Wang, H., Li, H., Moore, L.B., *et al.* (2008) The phytoestrogen coumestrol is a naturally occurring antagonist of the human pregnane X receptor. *Molecular Endocrinology*, 22, 838–857.

Welty, F.K., Lee, K.S., Lew, N.S. & Zhou, J.R. (2007) Effect of soy nuts on blood pressure and lipid levels in hypertensive, prehypertensive, and normotensive postmenopausal women. *Archives of Internal Medicine*, 167, 1060–1067.

World Health Organization (2004) *Food and Health in Europe: A New Basis for Action* (eds A. Robertson, C. Tirado, T. Lobstein, *et al.*), WHO Regional Publications, European series, No. 96. WHO Regional Office for Europe, Copenhagen.

Wiseman, H., Casey, K., Bowey, E.A., *et al.* (2004) Influence of 10 wk of soy consumption on plasma concentrations and excretion of isoflavonoids and on gut microflora metabolism in healthy adults. *The American Journal of Clinical Nutrition*, 80, 692–699.

Wizemann, T.M. & Pardue, M.L. (2001) *Exploring the Biological Contributions to Human Health: Does Sex Matter?* National Academy Press, Washington, DC.

Wu, B., Kulkarni, K., Basu, S., *et al.* (2011) First-pass metabolism via UDP-glucuronosyltransferase: a barrier to oral bioavailability of phenolics. *Journal of Pharmaceutical Sciences*, 100, 3655–3681.

Wung, B.S., Hsu, M.C., Wu, C.C. & Hsieh, C.W. (2005) Resveratrol suppresses IL-6–induced ICAM-1 gene expression in endothelial cells: effects on the inhibition of STAT3 phosphorylation. *Life Sciences*, 78, 389–397.

Xiao, Z.P., Peng, Z.Y., Peng, M.J., *et al.* (2011) Flavonoids health benefits and their molecular mechanism. *Mini-Reviews in Medicinal Chemistry*, 11, 169–177.

Xu, X., Harris, K.S., Wang, H.J., *et al.* (1995) Bioavailability of soybean isoflavones depends upon gut microflora in women. *Journal of Nutrition*, 125, 2307–2315.

Yao, R., Yasuoka, A., Kamei, A., *et al.* (2011) Polyphenols in alcoholic beverages activating constitutive androstane receptor CAR. *Bioscience, Biotechnology, and Biochemistry*, 75, 1635–1637.

Zhang, B., Chen, Y.M., Huang, L.L., *et al.* (2008) Greater habitual soyfood consumption is associated with decreased carotid intima-media thickness and better plasma lipids in Chinese middle-aged adults. *Atherosclerosis*, 198, 403–411.

Chapter 11

# Thermodynamic and Kinetic Processes of Anthocyanins and Related Compounds and their Bio-Inspired Applications

Fernando Pina
*REQUIMTE, Department of Chemistry, Faculty of Science and Technology, New University of Lisbon, Lisbon, Portugal*

**Abstract:** This chapter revisits the reversible pH-dependent chemistry of anthocyanins and related compounds, proposing a step-by-step procedure by which to calculate all the rate and equilibrium constants of their respective networks of chemical reactions. Self-aggregation dramatically changes the values of the rate and equilibrium constants of a network. The equilibrium constant ($K_a$) increases and the hydration ($K_h$) and global ($K'_a$) constants decrease with increasing anthocyanin concentration. The hydration rate constant ($k_h$) decreases as the dehydration rate constant ($k_{-h}$) increases. Evidence for the formation of flavylium cation dimers aggregating according to a left-handed helix has been obtained from circular dichroism, while indirect evidence for a stronger aggregation in quinoidal base than in flavylium cation has been found. The chapter discusses the photochromism of synthetic flavylium compounds lacking or possessing a high *cis–trans* isomerization barrier; in the latter case, the system illustrates the concept of an optical memory capable of write–lock–read–unlock–erase. Extended conjugation in flavylium-derived compounds such as styryl-1-benzopyrylium and naphtoflavylium permits access to photochromic systems that present new switching colors. Dye-sensitized solar cells based on anthocyanins and synthetic flavylium analogs show the potential of flavylium compounds in practical devices. Two final sections explain in detail the steps required to construct the energy-level diagram of the network and the mole-fraction distribution of the respective species in solution.

**Keywords:** anthocyanins, flavylium, self-aggregation, photochromism, memories, solar cells

*Recent Advances in Polyphenol Research, Volume 4*, First Edition.
Edited by Annalisa Romani, Vincenzo Lattanzio, and Stéphane Quideau.

# 11.1 Introduction

The scientific community that deals with polyphenols, and in particular anthocyanins, is made up of researchers from a variety of backgrounds and with a variety of areas of expertise and interest, including health, plant physiology, the food industry, and physical chemistry. This is a result of the ubiquity of polyphenols in our lives. However, this plurality of disciplines makes dialog between the different researchers and the indispensable crossing of knowledge difficult. A huge number of works describing new anthocyanins and their antioxidant properties has been reported, for example (Anderson & Markham, 2006; Kong *et al.*, 2003; Zafra-Stone *et al.*, 2007; Pascual-Teresa *et al.*, 2010), while works dealing with anthocyanin thermodynamics and kinetics are relatively scarce (Pina *et al.*, 2012a). There is a tendency to think that this is an issue for chemists and physicists, of little importance for other research areas and applications. This is not the case, however. When we consider the health benefits of the anthocyanins present in edible fruits, for example, the performance of a system cannot be directly correlated to the activity of the flavylium cation. In fact, many anthocyanins are not consumed in the form of flavylium cation; for example, some of those that give a blue color are involved in supramolecular structures that stabilize the quinoidal base (Yoshida *et al.*, 2009), but the pH is only acidic during gastric emptying (Charman *et al.*, 1997). Therefore, the other species of the network of chemical reactions, in particular the hemiketal, should play a role in the biological activity of anthocyanins. Knowledge of the pH-dependent equilibria involving the different species of the anthocyanin network is thus crucial information that should be taken into account whenever applications of these compounds are considered, whether in the field of health or in any other scientific area in which anthocyanins and related compounds find application.

# 11.2 Anthocyanins in aqueous solution

Anthocyanins are generally represented in literature by the red flavylium cation. However, the flavylium cation is stable only at very acidic pH values. At moderately acidic pH values a series of more or less reversible chemical reactions takes place. This section describes the kinetics and thermodynamics of these reactions. Malvidin-3-glucoside (oenin) will be used as example, except where otherwise stated, as all of the anthocyanins follow the same network of chemical reactions.

Aqueous solutions of the flavylium cation are stable around pH 1 or less. When the pH is raised, two competing reactions take place: (i) deprotonation to give the quinoidal base (11.1) and (ii) hydration in position 2 of the flavylium cation, leading to the hemiketal (11.2).

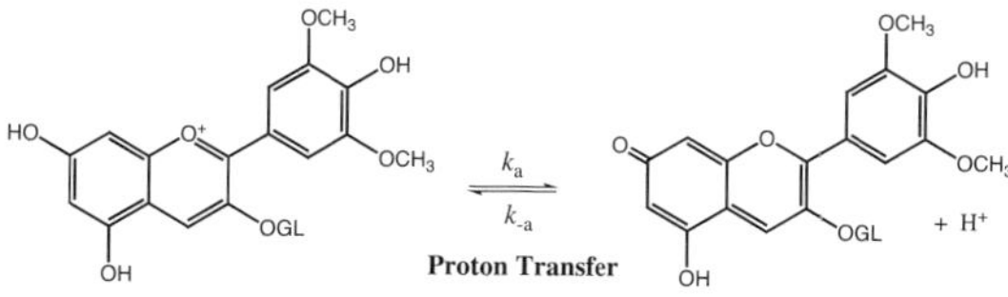


$$AH^+ + H_2O \rightleftharpoons A + H_3O^+ \qquad K_a \qquad (11.1)$$

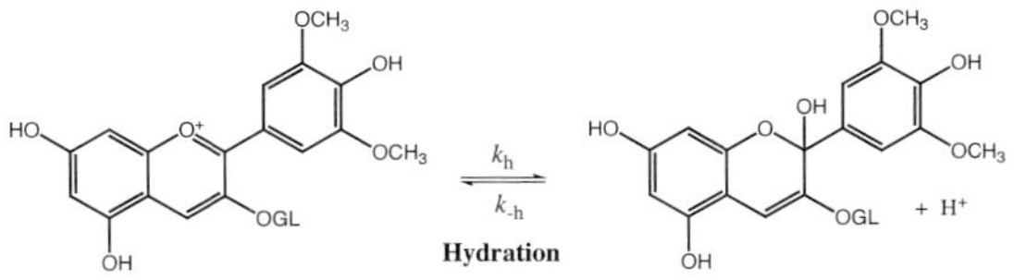


$$AH^+ + 2H_2O \rightleftharpoons B + H_3O^+ \qquad K_h \qquad (11.2)$$

In order to get more insight into the thermodynamics and kinetics of these reactions, it is convenient to introduce a thermodynamic energy-level diagram (Pina *et al.*, 1997b). This diagram is easily constructed from the relationship $\Delta G^0 = -RT \ln K$, where $\Delta G^0$ is the standard Gibbs energy, $R$ is the gas constant, $T$ is the temperature in Kelvin, and $K$ is the equilibrium constant for the process. Since energy-level diagrams are an extremely useful tool and will be used throughout this work, a detailed explanation of how they are constructed is given in Section 11.5. Scheme 11.1 shows the energy-level diagram of oenin, obtained by the procedure described in Section 11.5.

As can be seen in Scheme 11.1, the flavylium cation is stable only at very low pH values. When dissolved in less acidic solutions (pH > 2.5), the hemiketal B becomes more stable than the flavylium cation, while at higher pH values the same is true for quinoidal base A. A convenient way of studying the structural transformations of flavylium cation in moderately acidic pH values is to carry out pH jumps. In so-called "direct" pH jumps, strongly acidic solutions of flavylium cation (generally at pH 1) are rapidly brought to higher pH values, while in "reverse" pH jumps mildly acidic solutions are made more acidic. With a direct pH jump, the behavior of the system is very dependent on the final pH value, for thermodynamic and kinetic reasons. In the case of oenin, if the final pH of the jump is ~2.5, thermodynamic access is available to the hemiketal but not to the quinoidal base (see Scheme 11.1). This means that the kinetic process is essentially dependent on the hydration reaction. By contrast, if the final pH of the jump is ~3.8 or higher, the quinoidal base is thermodynamically available. However, proton transfer to water in order to form the base is much faster than the hydration reaction. In these conditions, the base is immediately formed upon the pH jump substitute and in spite of (thermodynamically) stable than the hemiketal. The quinoidal base is thus a kinetic product that in a second step will evolve to the more stable hemiketal. In other words, $AH^+$ and A are in fast equilibrium compared to the hydration reaction and behave as a single species that disappears to give B. One important detail of the mechanism (Brouillard & Dubois, 1977) is that the formation of B takes place exclusively through $AH^+$, meaning that A does not give B directly unless in basic medium, by means of the $OH^-$ attack on the quinoidal base. The kinetic process does not finish here. Unless the pH is very acidic (which is not the case for a direct pH jump), opening and closure of the C ring (tautomerization) is faster than hydration, and consequently, once B is formed, it leads to a small percentage of *cis*-chalcone (Cc):[1]

[1] In anthocyanins, the percentage of Cc is small, but this is not the case for many flavylium derivatives.

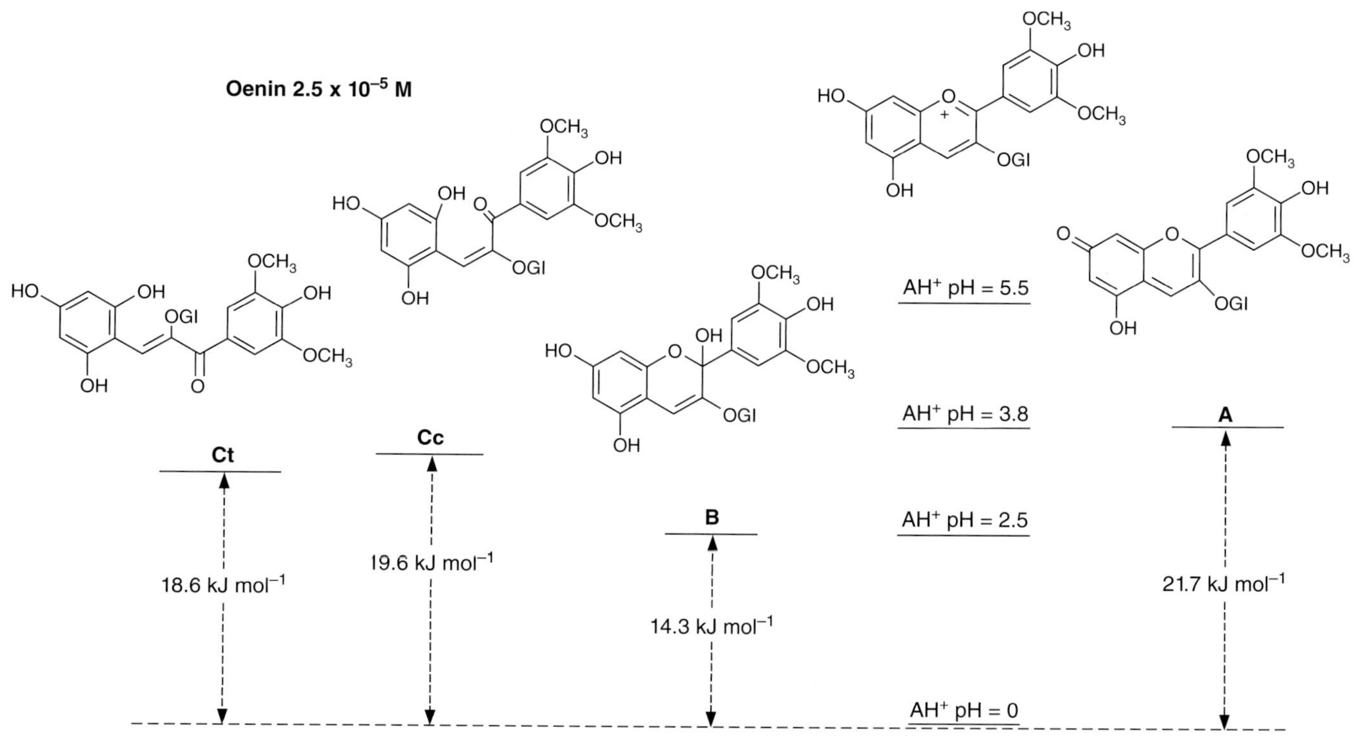


**Scheme 11.1** Energy-level diagram of oenin. At this concentration the self-aggregation can be neglected. (Leydet *et al.*, 2012. Reproduced with permission of Elsevier.) For color details, please see Plate 11.1.

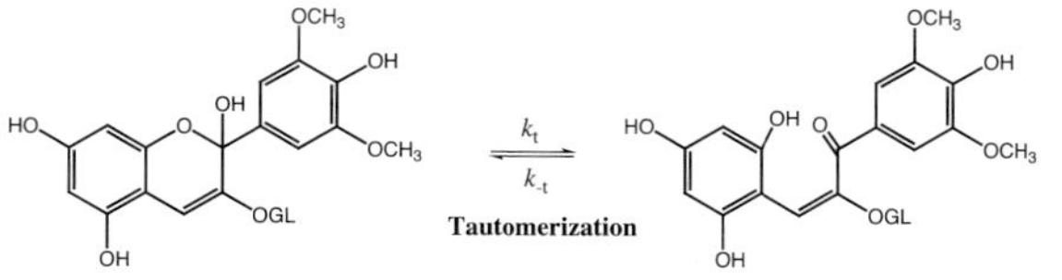


$$\mathrm{B} \rightleftharpoons \mathrm{Cc} \qquad K_\mathrm{t} \tag{11.3}$$

Finally, the system equilibrates in a few hours, due to the formation of a small mole fraction of *trans*-chalcone (Ct):

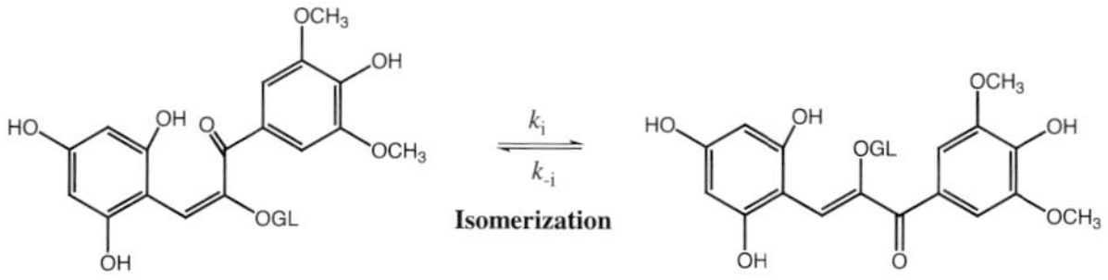


$$\mathrm{Cc} \rightleftharpoons \mathrm{Ct} \qquad K_\mathrm{i} \tag{11.4}$$

Analysis of the thermodynamic equilibria becomes easier if (11.5) is considered. This equation is equivalent to a single acid–base equilibrium between flavylium cation and a conjugate base (CB), defined as the sum of the concentrations of the other species in the network: [CB] = [a] + [B] + [Cc] + [Ct] (Brouillard & Delaporte, 1977; Brouillard *et al.*, 1978; Brouillard & Lang, 1990; Pina, 1998; Pina *et al.*, 2012a,b).

$$\mathrm{AH^+ + H_2O \rightleftharpoons CB + H_3O^+} \quad K'_a = K_a + K_\mathrm{h} + K_\mathrm{h}K_\mathrm{t} + K_\mathrm{h}K_\mathrm{t}K_\mathrm{i} \tag{11.5}$$

From the set of equations (11.1) to (11.5), it is easy to obtain the pH-dependent mole-fraction distribution of the network species:

$$\chi_{AH^+} = \frac{[H^+]}{[H^+]+K'_a}; \quad \chi_A = \frac{K_a}{[H^+]+K'_a}; \quad \chi_B = \frac{K_h}{[H^+]+K'_a};$$
$$\chi_{Cc} = \frac{K_hK_t}{[H^+]+K'_a}; \quad \chi_{Ct} = \frac{K_hK_tK_i}{[H^+]+K'_a} \tag{11.6}$$

In Section 11.6, the mathematical deduction of these expressions is reported. When a relatively high *cis–trans* isomerization barrier is present, as in anthocyanins, it is very useful to define a pseudo-equilibrium attained before formation of significant amounts of Ct:

$$\mathrm{AH^+ + H_2O \rightleftharpoons CB^{\hat{}} + H_3O^+} \quad K^{\hat{}}_a = K_a + K_h + K_hK_t \tag{11.7}$$
$$[CB^{\hat{}}] = [A] + [B] + [Cc]$$

The mole-fraction distribution of the species at the pseudo-equilibrim, $AH^+$, A, B, and Cc, can be obtained from the expressions of (11.6), substituting $K'_\mathrm{a}$ with $K^{\hat{}}_\mathrm{a}$.

Given the anthocyanin kinetic processes described later in the chapter, the first step (a direct pH jump corresponding to the formation of the base) takes place in microseconds, the second step (the (pseudo) equilibration of the species $AH^+$, A, B, and Cc) in seconds to minutes, and the third step (the formation of Ct) in a few hours.

The three kinetic steps towards the equilibration of anthoycanins can be expressed by simple mathematical expressions, due to their clear separation in time. It is possible to resolve the kinetic equations describing the system using more or less complex mathematical procedures (Brouillard & Lang, 1990; Petrov & Pina, 2010), but here we present an easy way of deducing them, based on the fact that in each step the reactions follow a first-order kinetics approaching the equilibrium, and consequently the observed rate constant is the sum of the forward and backward processes (Atkins, 1999).

Therefore, the first step is:

$$k_{obs1} = k_a + k_{-a}[H^+] \tag{11.8}$$

While the second is:

$$k_{obs2} = \frac{[H^+]}{K_a + [H^+]}k_h + \frac{1}{1+K_t}k_{-h}[H^+] \tag{11.9}$$

This equation presupposes that the proton transfer and the tautomerization are faster than the hydration, which is a valid assumption if the pH is not very acidic.[2] The first term results from the discovery of Brouillard & Dubois (1977): the hydration constant should be multiplied by the mole fraction of flavylium available. Regarding the backward reaction, it is multiplied by the mole fraction of B (i.e. the true species that gives back $AH^+$).

The system equilibrates as follows:

$$k_{obs3} = \frac{K_h K_t}{K_a + K_h + K_h K_t + [H^+]}k_i + k_{-i} \tag{11.10}$$

This equation assumes the existence of a pseudo-equilibrium involving all of the species except Ct, and the isomerization constant $k_i$ is multiplied by the respective mole fraction of Cc.

Fig. 11.1a shows the spectral variations following the pH jump of an equilibrated solution of oenin at pH 1.0–5.2. Immediately after the pH jump, the spectrum of the quinoidal base is observed.[3] The formation of the base is too fast to be followed even by stopped-flow UV-vis spectroscopy. The quinoidal base evolves to the pseudo-equilibrium in around 8 minutes at this pH value, but the process is faster at lower pHs. Representation of the respective rate constant as a function of pH is shown in Fig. 11.1b, together with the fitting carried out by means of (11.9).

More information can be achieved from the data in Fig.11.1a and Fig. 11.1c (top). When the direct pH jump is carried out to a pH sufficiently higher than $pK'_a$, ($K'_a >> [H^+]$), all of the initial flavylium cation is transformed into the quinoidal base, and the initial absorption band is that of the base in pure form. The ratio between the final and the initial absorbance values is the ratio $K_a : K'_a$:

$$\frac{Absorbance_{final}}{Absorbance_{initial}} = \frac{[A]}{C_0} = \frac{K_a}{[H^+] + K'_a} \approx \frac{K_a}{K'_a} \quad \text{for } K'_a >> [H^+] \tag{11.11}$$

One aspect of anthocyanin kinetics later reported by Brouillard & Lang (1990) that was not considered in the initial publications from the first author was the existence of the third

[2] The pH should be moderately acidic unless the species **A**, **B**, **Cc**, and **Ct** are not thermodynamically accessible.
[3] At pH 5.0 there is 94% quinodal base and 6% flavylium cation ($pK_a = 3.8$).

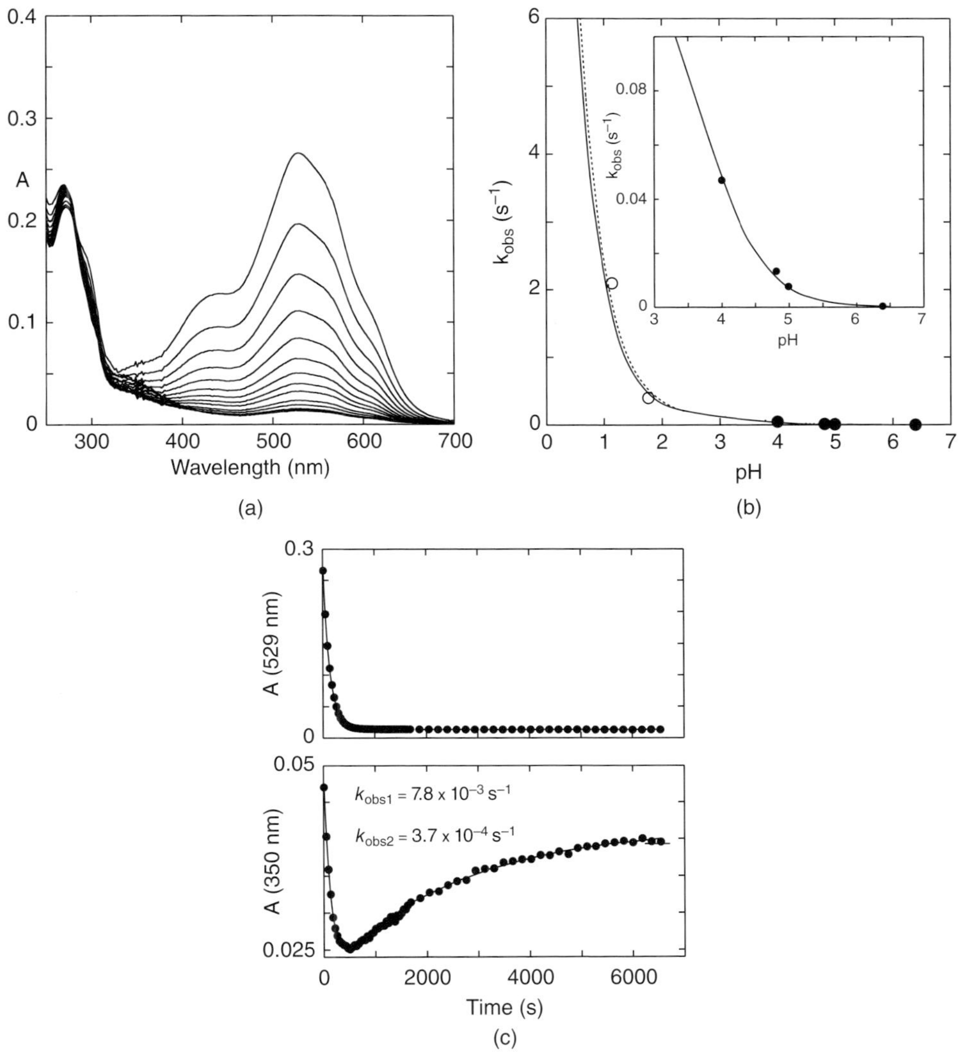


**Fig. 11.1** (a) Spectral variations of oenin ($2 \times 10^{-5}$ M) taken at around 1 minute after a pH jump from 1.0 to 5.0. (b) •: Representation of the rate constant of the second process as a function of pH after a direct pH jump. Fitting was achieved by means of (11.9). o: The same upon a reverse pH jump (11.12). (c) Traces of the kinetic processes measured at the absorption maximum of the quinoidal base (529 nm) and *trans*-chalcone (Ct) (350 nm). (Leydet *et al.*, 2012. Reproduced with permission of Elsevier.)

kinetic step that leads to Ct. This is due to the fact that in anthocyanins the mole fraction of Ct is small and Cc is elusive and hard to detect. However, the third process is clearly identified by the traces at 350 nm, as shown in Fig. 11.1c (down), and this behavior is general for all anthocyanins. Special attention should be paid to the papers from before 1990, regarding the physical chemistry of anthocyanins, because in the early papers the third process is associated with the tautomerization reaction (Pina *et al.*, 2012a). More information on the system can be obtained by carrying out reverse pH jumps, as shown in Fig. 11.2.

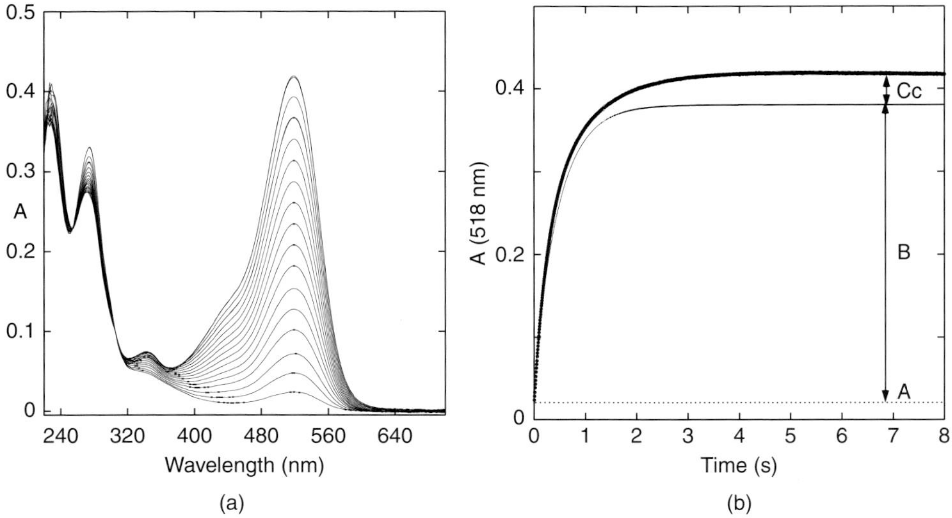


**Fig. 11.2** (a) Spectral variations upon a reverse pH jump from an equilibrated solution of oenin ($2 \times 10^{-5}$ M) at pH 5.2 to pH 1.1. (b) Kinetic trace at 518 nm, fitted with a bi-exponential law with $k_{obs1} = 2.1$ per second and $k_{obs2} = 0.5$ per second. (Leydet *et al.*, 2012. Reproduced with permission of Elsevier.)

Following a strategy reported by McClelland and coworkers (McClelland & Gedge, 1980; McClelland & McGall, 1982), pH jumps from equilibrated solutions of oenin at pH 5.2 to pH 1.1 were monitored by stopped-flow UV-vis spectroscopy, as shown in Fig. 11.2a. The raising of the flavylium absorption band takes place through a kinetic process exhibiting a bi-exponential dependence. After an initial increase in absorbance due to the very fast conversion of A into $AH^+$, the first kinetic process corresponds to flavylium cation formation from the hemiketal present in solution before the pH jump, as described by (11.12). This step is faster than tautomerization because the dehydration rate is proportional to $[H^+]$, and it can be made very fast at very high proton concentrations.[4,5] The second step is the formation of more flavylium cation from Cc through B, as given by (11.13), since there is no reversibility; that is, when B is formed from Cc, it disappears to give $AH^+$.

$$k_{obs2} = \frac{[H^+]}{K_a + [H^+]} k_h + k_{-h}[H^+] \tag{11.12}$$

$$k_{obs2} = k_{-t} \tag{11.13}$$

A last much slower process leads to flavylium cation from Ct (not shown in this figure). In the case of anthocyanins, the mole fraction of Cc is very small, and thus the experiment in Fig. 11.2b has some associated error and should be repeated several times in order to obtain a reliable value. From the amplitudes of the two exponentials, the amount of Cc and B in equilibrium can be calculated, and thus so too can the constant $K_t$ in (11.3).

[4] (11.11.12) is different from (11.11.9) in that in a reverse pH jump the equilibrium between B and Cc is not reached.

[5] Even an atypical small value of $k_{-h} = 40\,M/\text{second}$ leads to a rate of 4 per second at pH 1, which is fast compared with $k_{-t} \approx 0.5$–$0.8$ per second.

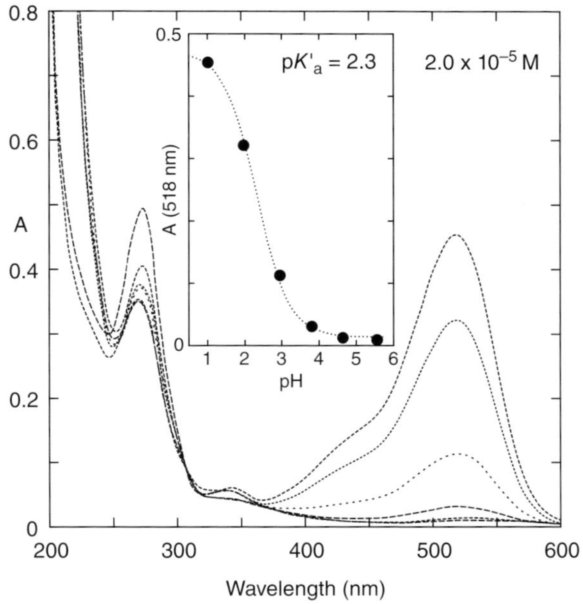


**Fig. 11.3** Absorption spectra of equilibrated solutions of oenin ($2 \times 10^{-5}$ M) as a function of pH. Inset: determination of p$K'_a$ using the absorbance at 518 nm. (Leydet *et al.*, 2012. Reproduced with permission of Elsevier.)

Finally, the absorption spectra of the mixture in thermodynamic equilibrium shown in Fig. 11.3 permit calculation of the global constant $K'_a$. Determination of all the rate and equilibrium constants can be carried out using a step-by-step procedure described in the next section.

## 11.2.1 *Step-by-step procedure for calculating rate and equilibrium constants*

### *11.2.1.1 Step 1: determination of the equilibrium constant* $K'_a$

This constant is obtained by recording the absorption spectra of the equilibrated solutions, as illustrated in Fig. 11.3, upon fitting the data to a single acid–base equilibrium, as defined by (11.5).

### *11.2.1.2 Step 2: determination of the equilibrium constant* $K_a$

This constant can be obtained from the absorption spectra of a set of direct pH jumps, monitored preferentially by means of a stopped-flow apparatus. Alternatively, (11.11) can be used. The rate constants $k_a$ and $k_{-a}$, and thus the equilibrium constant $K_a$, can be determined by temperature jumps (Brouillard & Dubois, 1977), or alternatively by laser-flash photolysis (Maçanita *et al.*, 2002).

#### 11.2.1.3 Step 3: determination of the equilibrium constant $K_t$ and the respective rate constants

This involves reverse pH jumps from equilibrated solutions or pseudo-equilibrated solutions back to pH < 1, as shown in Fig. 11.2. This step exhibits high error in the case of anthocyanins, due to the small fraction of Cc at the equilibrium, or even the pseudo-equilibrium. The rate of the slowest process of the reverse pH jump follows (11.13), allowing $k_{-t}$ and $k_t$ to be determined directly by substitution in the value of $K_t$.

#### 11.2.1.4 Step 4: determination of the hydration rate and equilibrium constants

This involves carrying out a series of direct and reverse pH jumps as a function of pH, followed by fitting of the respective rate constants by (11.9) (for hydration rates; see also Fig. 11.1b) and (11.12) (for equilibrium constants). If $K_t$ is very small, as in anthocyanins, the two curves resulting from the direct and reverse pH jumps are very similar. The two contributions of (11.9) are shown in Fig. 11.4a. From this step, the two rate constants are calculated and the equilibrium constant $K_h$ can be determined. Once $K_a$, $K_t$, and $K_h$ have been obtained, $K_i$ is determined from $K'_a$.

#### 11.2.1.5 Step 5: determination of the isomerization rate and equilibrium constants

Fitting of the slowest process leads to the determination of $k_i$ and $k_{-i}$. In some flavylium networks, as in some synthetic flavylium compounds, $k_{-i} << k_i$ and there is no accuracy in the determination of the former. In that case, the equilibrium constant $K_i$ should be used.

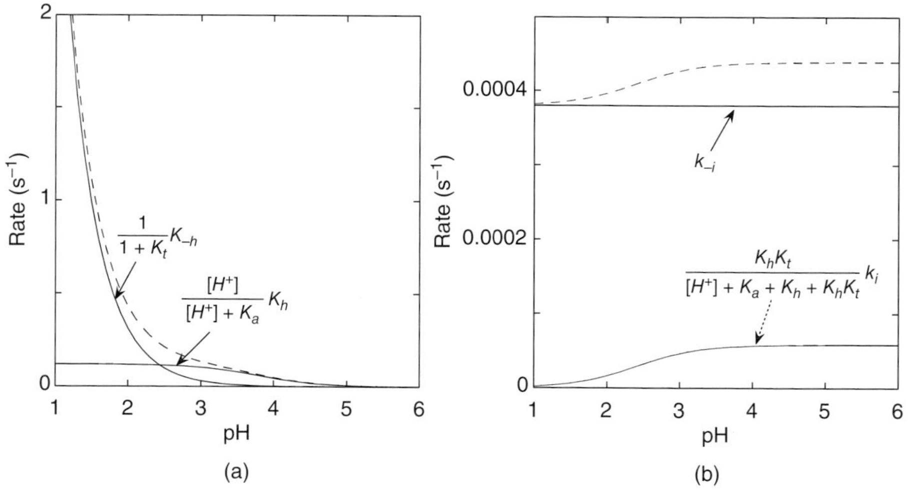


**Fig. 11.4** (a) Illustration of the contribution of the two terms of (11.9) for oenin upon amplification of Fig. 11.2b. Traced line represents the sum of the two contributions. (b) The same for (11.10).

### 11.2.1.6 Step 6: verification of the self-consistency of all the data

The ratio of the rate constants should be equal to the respective equilibrium constants, within experimental error.

## 11.3 Influence of anthocyanin self-aggregation on the determination of rate and equilibrium constants

When the thermodynamics and kinetics of the structural transformations of anthocyanins are investigated at different concentrations, the apparent values of the respective rate and equilibrium constants change as shown in Fig. 11.5 for the $pK_a$, $pK'_a$, and $pK_h$ of oenin (Leydet *et al.*, 2012).

According to Fig. 11.5, the acidity constant $K_a$ increases with increasing concentration of the anthocyanin, while the hydration constant $K_h$ decreases. Similar results were found for the other most common anthocyanidin 3-glucosides. This behavior is due to self-aggregation, a phenomenon previously reported for anthocyanidin 3,5-diglucosides by other authors and characterized by circular dichroism (Hoshino *et al.*, 1980, 1981a,b, 1982; Hoshino & Goto, 1990; Hoshino, 1991). There is no information on the intimate structure of the aggregates. However, hydrogen bonding, $\pi$-$\pi$ stacking, and the hydrophobic effect should be driving forces behind the association. In the case of the flavylium cation, the experimental results suggest the existence of dimers, while for the quinoidal base aggregates of higher order cannot be excluded. One possible explanation is the positive charge of the flavylium cation, which prevents further association (Gavara *et al.*, 2013).

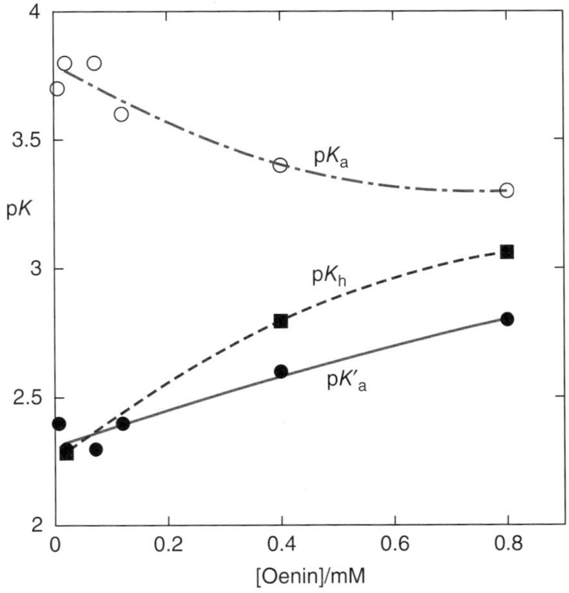


**Fig. 11.5** Variation of the equilibrium constants of oenin as a function of concentration. (Leydet *et al.*, 2012. Reproduced with permission of Elsevier.)

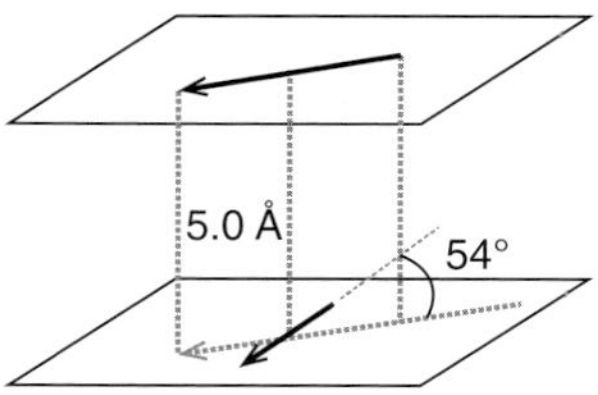


**Scheme 11.2** Representation of the relative position of the transition dipole of myrtyllin flavylium cations in a dimer. (Leydet *et al.*, 2012. Reproduced with permission of Elsevier.)

In order to detect the variations in the shapes of the absorption bands as a function of the anthocyanin concentration, the spectra of myrtillin presented in Fig. 11.6a were expressed in terms of molar absorption coefficient, $\epsilon\,(M/cm)$, by dividing the absorbance by the concentration and optical path (Fig. 11.6b). Fig. 11.6c shows that with increasing concentration the flavylium absorption band splits into two bands with maxima at ca. 440 and 600 nm. This type of split suggests the formation of aggregates exhibiting a shape between J and H[6] (McRae & Kasha, 1958). The fitting was achieved for a monomer–dimer equilibrium with a dimerization constant of 1260 per M.

The spectra of Fig. 11.6b were treated by multi-res program (Antonov & Petrov, 2002) and the following dimer parameters were calculated: $\alpha = 54°$ and $R = 5.0$ Å (Scheme 11.2). Additional information was obtained by $^1$H-NMR carried out in $H_2O$ (nondeuterated water), in order to avoid corrections due to the deuterium isotopic effect and to allow direct comparison of $^1$H-NMR and UV-vis data (Glasoe & Long, 1960; Shaka & Hwang, 1995; Popov *et al.*, 2006). The effect of the self-association of the six anthocyanidin 3-glucosides (see Scheme 11.3) in the flavylium form (pH 1.0) is well illustrated in Fig. 11.7.

In Fig. 11.7, the chemical shifts of protons H4, H6, and H8 are represented as a function of the concentration. It is worth noting that at lower concentrations the chemical shift does not depend significantly on the substitution pattern of the anthocyanins. This is an expected result if it is considered that the anthocyanin substituents are different only in ring B, having a small influence on protons H4, H6, and H8, if aggregation is negligible. The observed chemical shifts can thus be interpreted as a result of the shielding of these protons upon the aggregation processes, an effect previously reported by other authors (Hoshino *et al.*, 1980, 1981a,b, 1982; Hoshino & Goto, 1990; Hoshino, 1991).

The results were treated with a dimer–monomer model and good fittings were obtained using the same dimerization constants for each anthocyanin in each of the three curves (Table 11.1). Good fittings were also obtained by treating the data with the isodesmic model, which considers the formation of higher-order aggregates possessing the same association constant, and consequently the accuracy of the data permits neither discrimination between the monomer–dimer and isodesmic models nor conclusion of the formation of higher-order aggregates (Leydet *et al.*, 2012). Evidence for the aggregation of the quinoidal base was achieved through the experiments shown in Fig. 11.8.

---

[6] Upon aggregation, the absorption spectra of the monomer splits into two bands, one red-shifted (J) and the other blue-shifted (H).

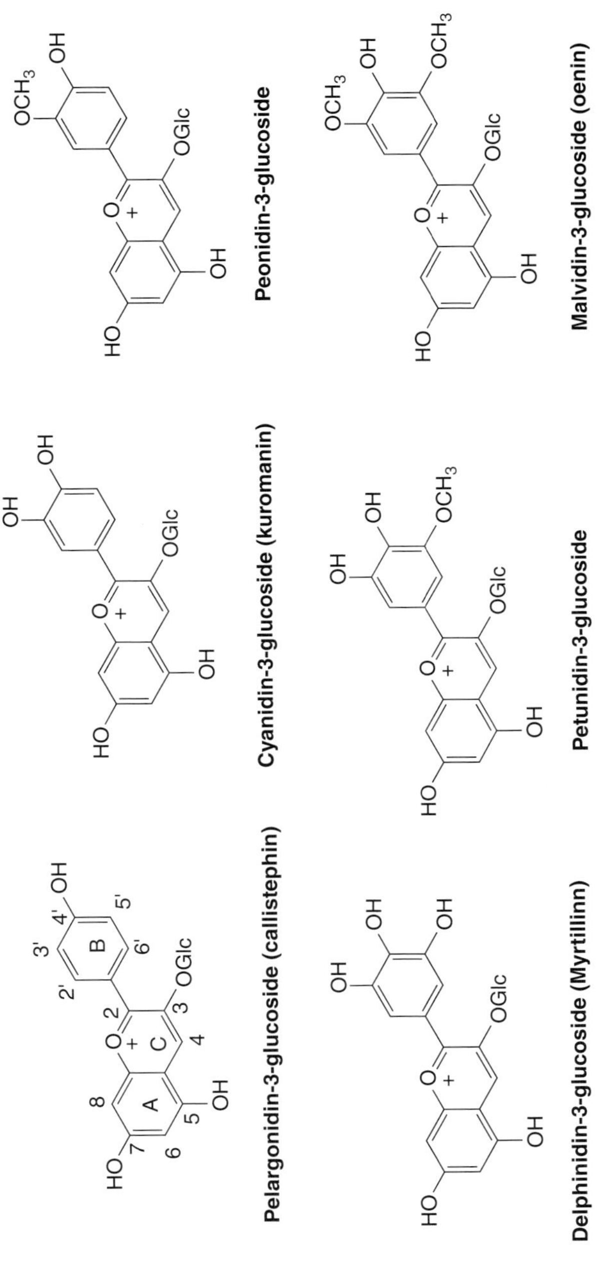


**Scheme 11.3** Structures of the six most common 3-glucoside anthocyanins.

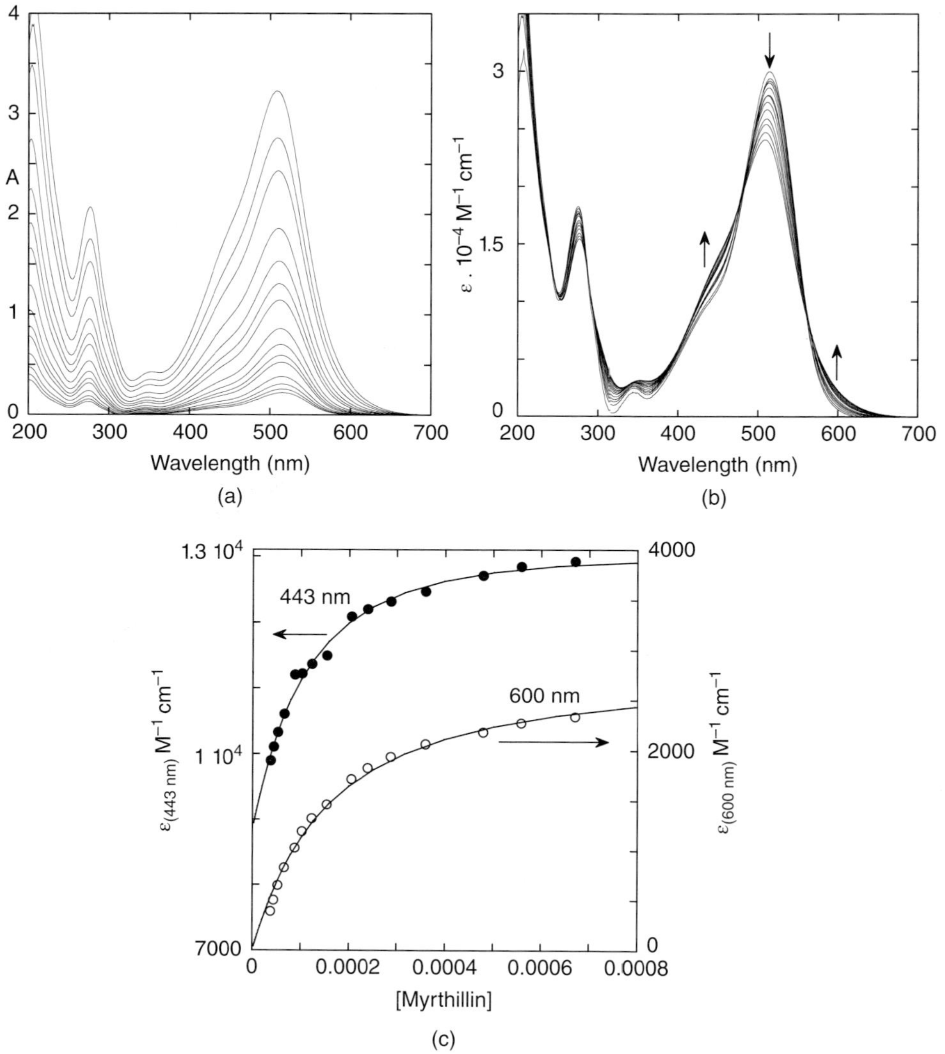


**Fig. 11.6** (a) Absorption spectra of Myrtillin as a function of concentration. Optical path 0.2 cm, in the concentration range $2.09 \times 10^{-5}$ M < [Myrtillin] < $6.2 \times 10^{-4}$ M, pH 1.0 (0.1 M HCl). (b) Normalization of the spectra to show characteristic spitting. (c) Split of the flavylium absorption band into two bands with maxima at ca. 440 and 600 nm, upon increasing concentration. (Leydet *et al.*, 2012. Reproduced with permission of Elsevier.)

In order to obtain these figures, identification of the $^1$H-NMR peaks for each component of the network was accomplished following some simple observations: (i) species A is in fast equilibrium with $AH^+$, and conversion of $AH^+$ into A through an increase in pH is reflected in the chemical shift of the respective peaks to higher field (Santos *et al.*, 1993); (ii) B is easily identified not only because it is the major species at higher pH values but also because of the existence of twin peaks related to the presence of epimers (due to the appearance of the uncontrolled stereocenter C2 after hydration); and (iii) the chalcones

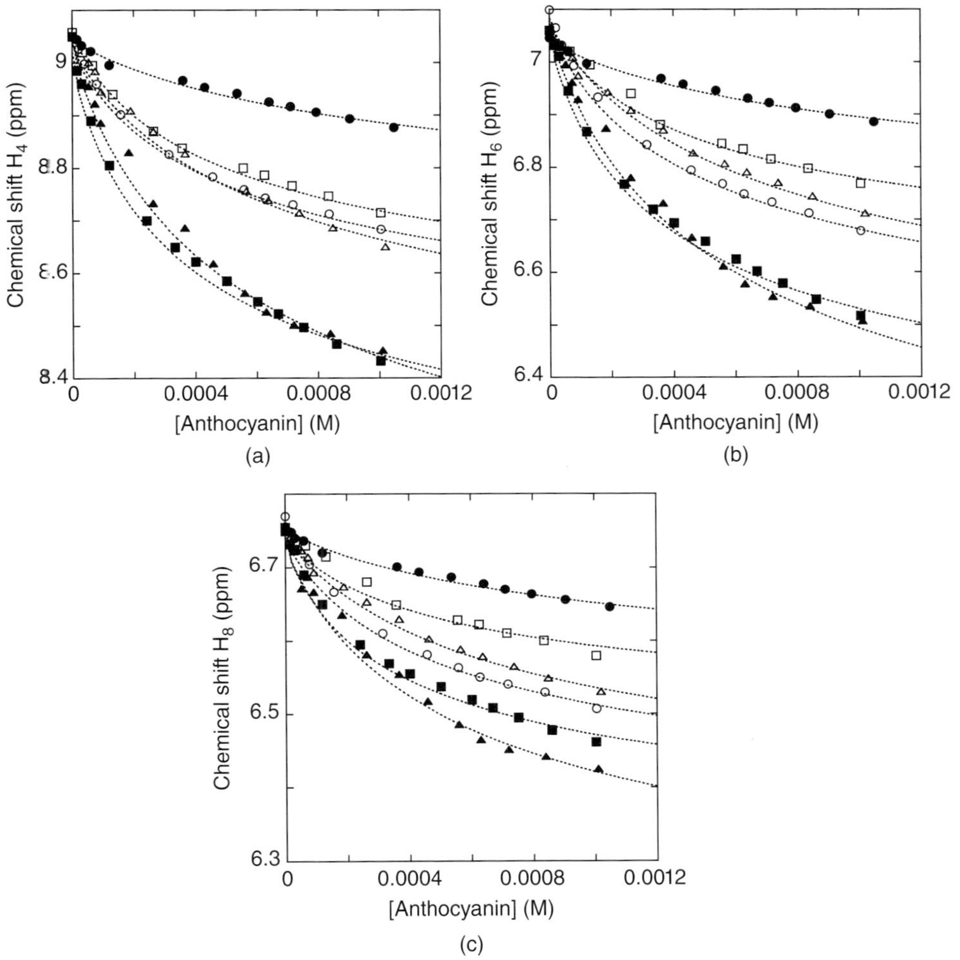


**Fig. 11.7** Representation of the chemical shifts of protons: (a) $H_4$, (b) $H_6$, and (c) $H_8$: callistephin (•), oenin (○), myrthillin (■), kuromanin (□), petunidin-3-glucoside ($\pi$), peonidin-3-glucoside (Δ). (Leydet *et al.*, 2012. Reproduced with permission of Elsevier.)

usually appear in low concentrations but Ct can easily be identified due to its formation from the isomerization of Cc, which is the slowest process.

Once the $^1$H-NMR peaks have been assigned, it is possible to calculate the mole-fraction distribution of each species of the network at the equilibrium, for different total concentrations of anthocyanin, as shown in Fig. 11.8. Inspection of Fig. 11.8 clearly shows that with increasing total anthocyanin concentration the mole fraction of A at the plateau increases, while that of B decreases. Moreover, the apparent acidity constant increases; that is, the flavylium ion tends to become a stronger acid at higher concentration. All these results can be interpreted as self-aggregation of the quinoidal base taking place more efficiently than that of the flavylium cation, which must be hindered by the repulsion of the positive charges. This presupposes that both aggregates involving the flavylium cation or quinoidal base are

**Table 11.1** Dimerization constants and chemical shifts obtained from fitting of the data reported in Fig. 11.7. (Leydet *et al.*, 2012. Reproduced with permission of Elsevier.)

| Anthocyanin | | H4 | | H6 | | H8 | |
|---|---|---|---|---|---|---|---|
| | $K_D$ (M) | $\delta_M$ (ppm) | $\delta_D$ (ppm) | $\delta_M$ (ppm) | $\delta_D$ (ppm) | $\delta_M$ (ppm) | $\delta_D$ (ppm) |
| Myrtillin | 1240 | 9.05 | 7.99 | 7.05 | 6.09 | 6.76 | 6.22 |
| Oenin | 976 | 9.05 | 8.32 | 7.09* | 6.27 | 6.76 | 6.27 |
| Petunidin-3-OGl | 900 | 9.05 | 7.78 | 7.05 | 5.85 | 6.76 | 6.07 |
| Kuromanin | 700 | 9.05 | 8.3 | 7.07* | 6.37 | 6.76 | 6.36 |
| Peonidin-3-OGl | 661 | 9.05 | 8.16 | 7.05 | 6.22 | 6.76 | 6.25 |
| Callistephin | 404 | 9.05 | 8.58 | 7.05 | 6.61 | 6.76 | 6.45 |

$\delta_M$, chemical shift of the monomer; $\delta_D$, chemical shift of the dimer.
[a]Fitting is slightly better with these values than 7.05.

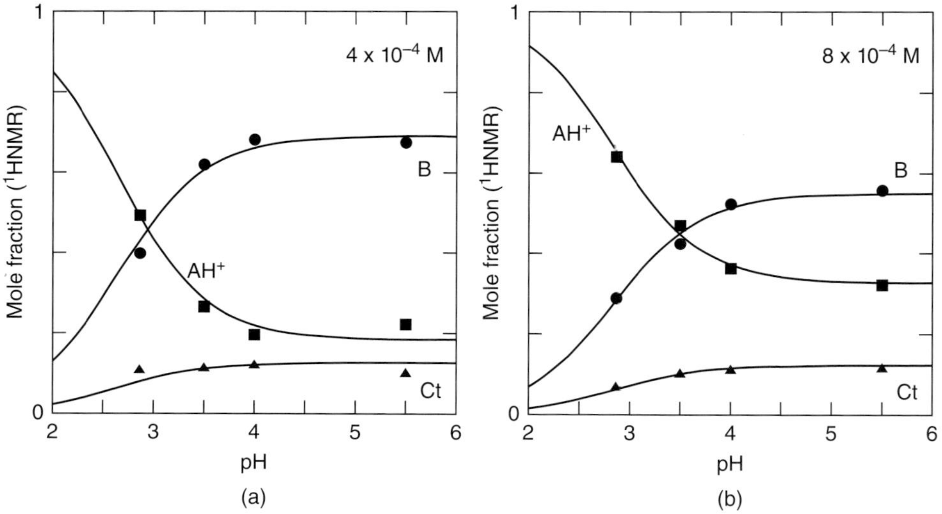


**Fig. 11.8** Mole-fraction distribution of $AH^+/A$ (■), B (•), and Ct (▲) as a function of pH for (a) [oenin] = 0.0004 M and (b) [oenin] = 0.0008 M. (Leydet *et al.*, 2012. Reproduced with permission of Elsevier.)

in fast exchange on the NMR timescale, since no additional peaks due to these aggregates are observed and only a continuous variation of the chemical shift is detected. In other words, flavylium cation, quinoidal base, and their aggregates are included in the same set of $^1$H-NMR peaks. A model assuming the existence of flavylium dimers and quinoidal base dimers was used to fit the data of Fig. 11.8. The constants obtained for the quinoidal base seem to have a small variation with concentration, which is an indication of the possibility of the formation of higher-order aggregates, at least for the quinoidal base. Finally, it should be stressed that the self-association is stronger with colored forms and thus protects color.

Further evidence for the formation of aggregates was achieved by circular dichroism. Fig. 11.9 shows the mole ellipticity as a function of oenin concentration. These results corroborate previous reports from Hoshino *et al.* (1982), who verified in the case of the quinoidal bases of anthocyanidin 3,5-diglucosides, the formation of a left-handed helix

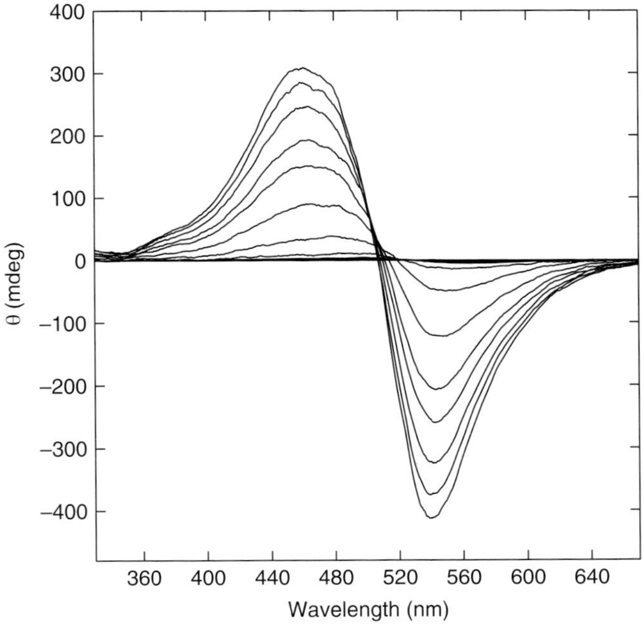


**Fig. 11.9** Ellipticity of petunidin 3-glucoside. (Gavara *et al.*, 2013. Reproduced with permission of Elsevier.)

for delphin, malvin, and peonin and of a right-handed one for cyanin and pelargonin. In contrast, all flavylium cations showed a left-handed screw self-association. Investigation of this subject is currently ongoing in our research group, and it can be anticipated that all six of the most common anthocyanidin 3-glucosides will present a CD spectrum that is similar to those of the 3,5-diglucoside parents, but with a generally higher absolute value of the molar ellipticity. This suggests the central role of the sugar in position 3 in the ellipticity value.

## 11.4 Photochromism: applications bio-inspired in anthocyanins

As mentioned earlier, the mole fraction of Ct in anthocyanins is very small and they are consequently not suitable to profit from a very interesting property of the flavylium networks: the possibility of exploring their photochromic properties (Figueiredo *et al.*, 1994). Conversely, photochromism has been observed in deoxyanthocyanins and in many synthetic flavylium compounds (Pina *et al.*, 2012a,b). A few examples to illustrate this phenomenon will be given in this section.

### *11.4.1 Systems lacking the* cis–trans *isomerization barrier*

The first example regards a flavylium derivative lacking the *cis–trans* isomerization barrier, the 3′-butoxy-7-hydroxyflavylium ion (Gago *et al.*, 2012) (Scheme 11.4). One interesting

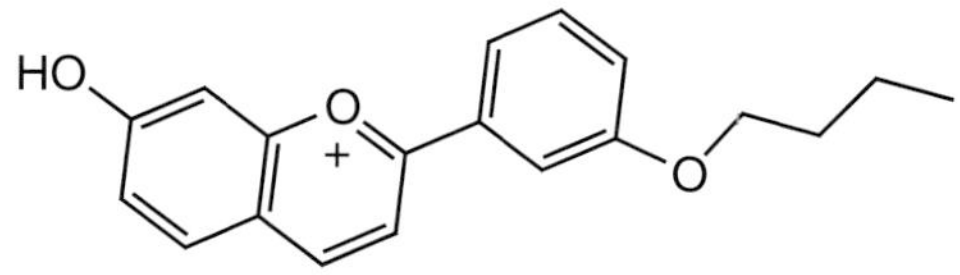


**Scheme 11.4** 3′-butoxy-7-hydroxyflavylium ion.

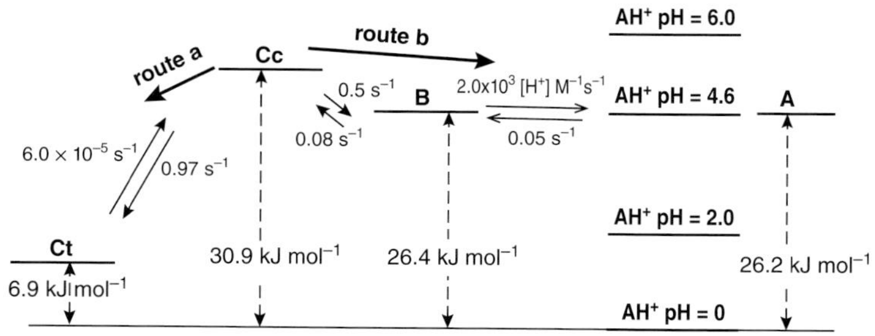


**Scheme 11.5** Energy-level diagram and rate constants for 3′-butoxy-7-hydroxyflavylium in ethanol : water 30 : 70, at 294 K. (Gago *et al.*, 2012. Reproduced with permission of Elsevier.)

characteristic of this compound is the extremely low value of its global constant, $pK'_a =$ 0.91. In other words, the flavylium cation is only stable in extremely acidic solutions. At pH 2.0, the major species at the equilibrium is already the Ct, as can be seen in the energy-level diagram given in Scheme 11.5.

Irradiation of equilibrated solutions of 3′-butoxy-7-hydroxyflvylium at pH 2.0 lead to the formation of flavylium cation, as shown in Fig. 11.10a. The absorbance due to the flavylium cation disappears in the dark, defining a photochromic system (Fig. 11.10b).

A very convenient way of studying the photochromism of flavylium systems is to use the flash-photolysis technique (Pina *et al.*, 1997a). This consists in monitoring the absorbance variations that follow the application of a light flash (Fig. 11.11). The results are easily interpreted by means of Scheme 11.5. At pH 4, the Ct species is thermodynamically stable, but upon the flash Ct is transformed into Cc, resulting in a very fast bleaching at 377 nm, as evidenced by Fig. 11.11, due to the lower molar absorption coefficient of Cc compared with Ct. Cc is not thermodynamically stable and has two alternative means of disappearing: backwards, to recover Ct, or forwards, leading to $AH^+/A$. The evolution of Cc depends on the relative kinetics of the two reactions. In sufficiently acidic solutions, the forward reaction competes with the backward reaction and the appearance of the colored species ($AH^+/A$) can thus be observed (Fig. 11.11a, top), as can a partial recovery of Ct (Fig. 11.11a, bottom) (faster kinetic steps). These two traces corresponding to first-order reactions have the same lifetime (within experimental error) expected for two parallel reactions. The second and slower kinetic step corresponds to the thermal reaction back to the equilibrium. However, when the photochemistry is carried out at higher pH values, as in Fig. 11.11b, the forward reaction is not able to compete with the backward reaction and practically all Cc goes

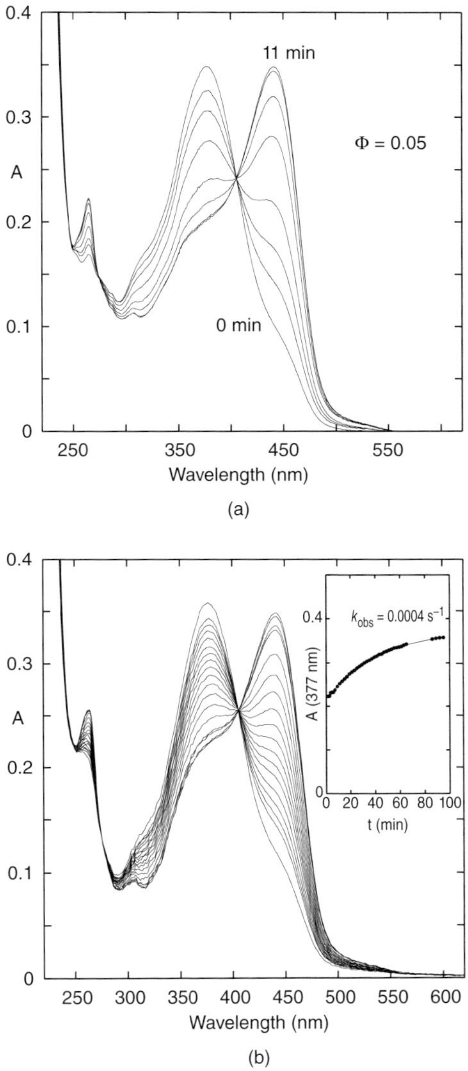


**Fig 11.10** (a) Spectral variations accompanying the continuous irradiation at 377 nm of 3′-butoxy-7-hydroxyflavylium at pH 2.0 ($1.91 \times 10^{-5}$ M, ethanol : water 30 : 70) for the following irradiation times: 0, 0.5, 1, 2.5, 5, 7, 9, 11 minutes ($I_0 = 2.19 \times 10^{-7}$ einsteins per minute). (b) Thermal recovery of the irradiated solution at 294 K. Inset: fitting of the absorbance at 438 nm as a monoexponential decay leads to $k_{obs} = 4.0\ (\pm 0.1) \times 10^{-4}$ per second. (Gago *et al.* 2012. Reproduced with permission of Elsevier.)

backwards to Ct. This behavior is a result of the proton dependence of the forward reaction through the term $k_{-h}[H^+]$, which for low proton concentrations practically vanishes. In conclusion, the forward reaction can be tuned from a situation where the color of the photochromic system is not observed (Fig. 11.11b) to one where the color appears (Fig. 11.11a). The colored species formed after the flash are not stable, and in a further kinetic step the system equilibrates to the initial state (Fig. 11.11a). As shown in Fig. 11.11b, the system is equilibrated after around 20 seconds.

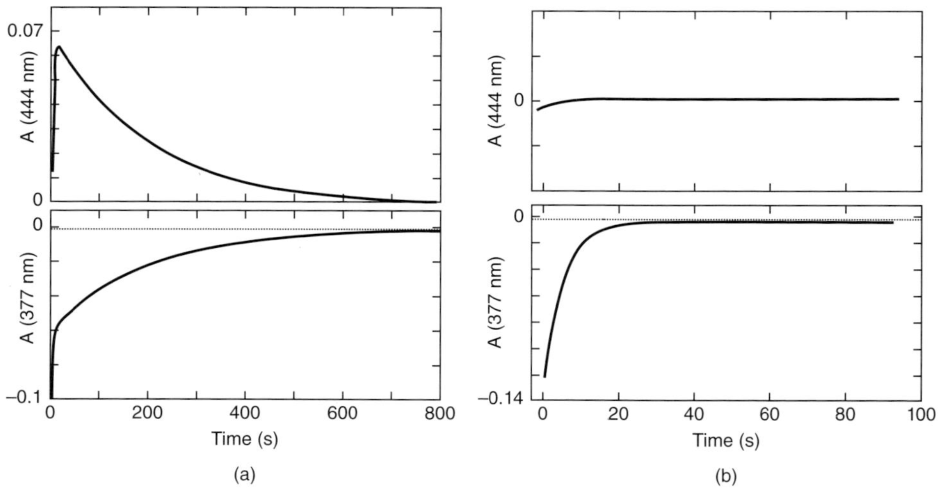


**Fig. 11.11** Flash photolysis of 3′-butoxy-7-hydroxyflavylium ($3.3 \times 10^{-5}$ M) in ethanol : water 30 : 70 at (a) pH 4.09 and (b) pH 6.2. (Gago *et al.*, 2012. Reproduced with permission of Elsevier.)

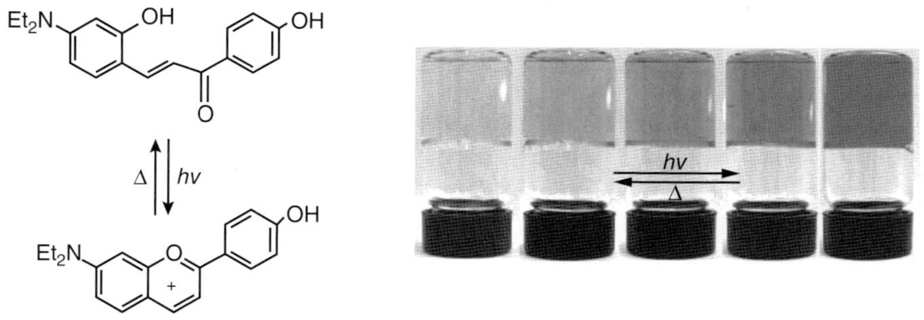


**Fig. 11.12** Highly transparent photochromic gels in the visible region can be prepared by the incorporation of flavylium compounds in Pluronic F-127 matrices. The chromophore 7-(N,N-diethylamino)-4-hydroxyflavylium switches from its yellow chalcone form to the red flavylium cation upon irradiation and reverts back to the chalcone form in the dark. (Pina and Hatton, 2008. Reproduced with permission of American Chemical Society.) For color details, please see Plate 11.12.

The quantum yield of the present system is relatively small, as a consequence of the high isomerization rate, which competes efficiently with the forward process even at low pH values. In other systems, quantum yields of formation of colored species up to 0.2 have been observed (Pina *et al.*, 2012a).

The flavylium-based photochromic systems can operate in matrices different from solution, such as micelles, polymers, zeolites, and gels (Pina *et al.*, 2012a). As an example, beautiful changes of color can be achieved when 7-diethylamino-4′-hydroxyflavylium is incorporated in Pluronic F-127 (Pina & Hatton, 2008) (Fig. 11.12).

### 11.4.2 *Systems exhibiting high* cis–trans *isomerization barriers*

In the case of flavylium-based photochromic systems exhibiting a high *cis–trans* isomerization barrier, as for example 4′-hydroxyflavylium and 4′-methoxyflavylium, the system is kinetically controlled by the forward reaction and two situations can take place: (i) if the pH is sufficiently high, Cc persists for a long time, depending on the rate of the *cis–trans* isomerization; or (ii) if the pH is sufficiently low, the colored species $AH^+$/a can be formed and persists for a long period of time.

#### *11.4.2.1 The concept of right–lock–read–unlock–erase optical memories*

The concept of optical memory was first proposed by Hirshberg (1956) and was recast by Lehn (1990) some years later. Light is used to form a new species, for example an isomer, which should have different physico-chemical properties in order to permit its differentiation from the original state and allow the next step: reading the information (e.g. measuring a different color). The cycle is closed when the information is erased, and the memory is repaired for a new one. In the examples described by Hirshberg and Lehn, the systems revert back to the initial state through irradiation of the photochemical product. This means that the light used to read the information causes its erasure. The flavylium-based system allows the modification of the concept (Pina *et al.*, 1997b) (Scheme 11.6). The pH of the system should be moderately acidic if it is to have Ct as the major species. The following sequence of steps defines the optical memory: (i) The light absorbed by Ct forms Cc, which in moderately acidic solutions is relatively stable due to the high isomerization barrier on one side and to the thermodynamically unfavorable formation of the flavylium cation on the other. In this step, the system should not be interrogated, or else Cc will revert to Ct, and as in the previous systems the information will be erased upon reading. (ii) The system is locked, which involves a decrease in the pH to allow for complete and spontaneous transformation of Cc into $AH^+$. (iii) It is now possible to read the information, because the flavylium cation is stable. (iv) The unlock step is carried out in order to erase the information; that is, the

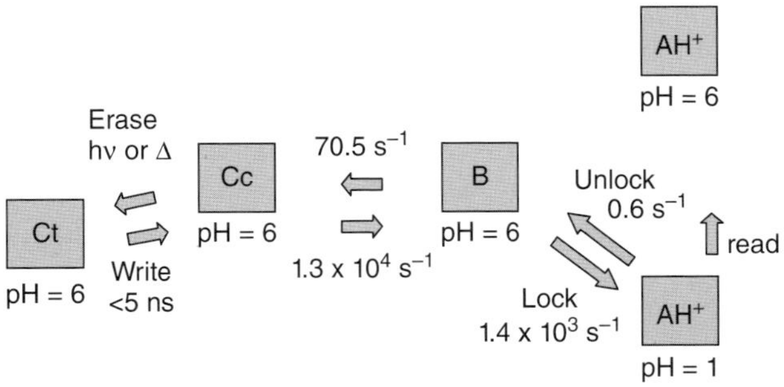


**Scheme 11.6** Write–lock–read–unlock–erase cycle of an optical memory based in the flavylium reaction network. (Adapted from Melo *et al.*, 2010.)

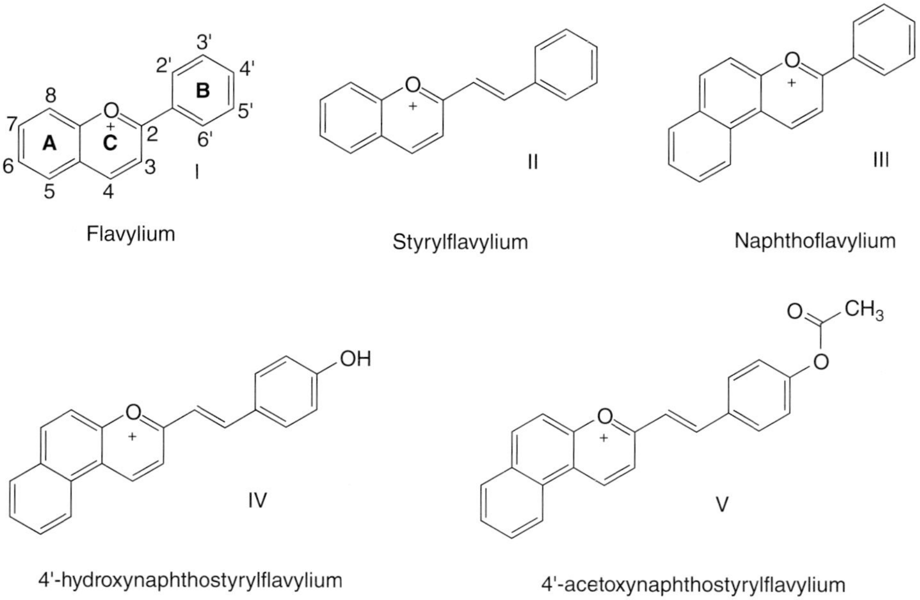


**Scheme 11.7** Chemical structures of 2-(4-hydroxystyryl)-1-naphthopyrylium, 2-(4-acetoxystyryl)-1-naphthopyrylium, and analogous compounds.

system reverts to the initial pH and Cc is formed again. (v) Finally, Cc is erased to give Ct, by light or by heating.

### *11.4.3 Styryl-1-benzopyrylium (styryl flavylium) and naphthoflavylium*

The network of flavylium derivatives can be extended to naphthoflavylium and styryl-1-benzopyrylium (styryl flavylium) (Gavara *et al.*, 2012; Petrov *et al.*, 2012) (Scheme 11.7). All of these compounds follow the same network of chemical reactions as common flavylium derivatives, opening the possibility for new applications in several fields, particularly photochromic systems, due to the access to different colors with high hues and red shifts in comparison with flavylium analogs (Pina *et al.*, 2012a,b). In particular, compound V encompasses the characteristics of styrylflavylium and naphthoflavylium ions (Gavara *et al.*, 2012).

A nice property of styrylflavylium derivatives is the possibility of obtaining different flavylium isomers from the same asymmetric Ct (Petrov *et al.*, 2012) (Scheme 11.8). While symmetric 2,2′-styrylchalcones lead independently to the same flavylium cation on the side where the formation of Cc, and therefore of the flavylium cation, takes place, the flavylium cations derived from asymmetric chalcones are different. It is possible to partially convert one flavylium into the other according to which stimulus is used, light or pH.

### *11.4.4 Dye-sensitized solar cells based on anthocyanins*

The use of anthocyanins as light collectors in dye-sensitized solar cells (DSSCs) was reported by Cherepy *et al.* (1997) (Scheme 11.9). Upon light absorption, the dye injects

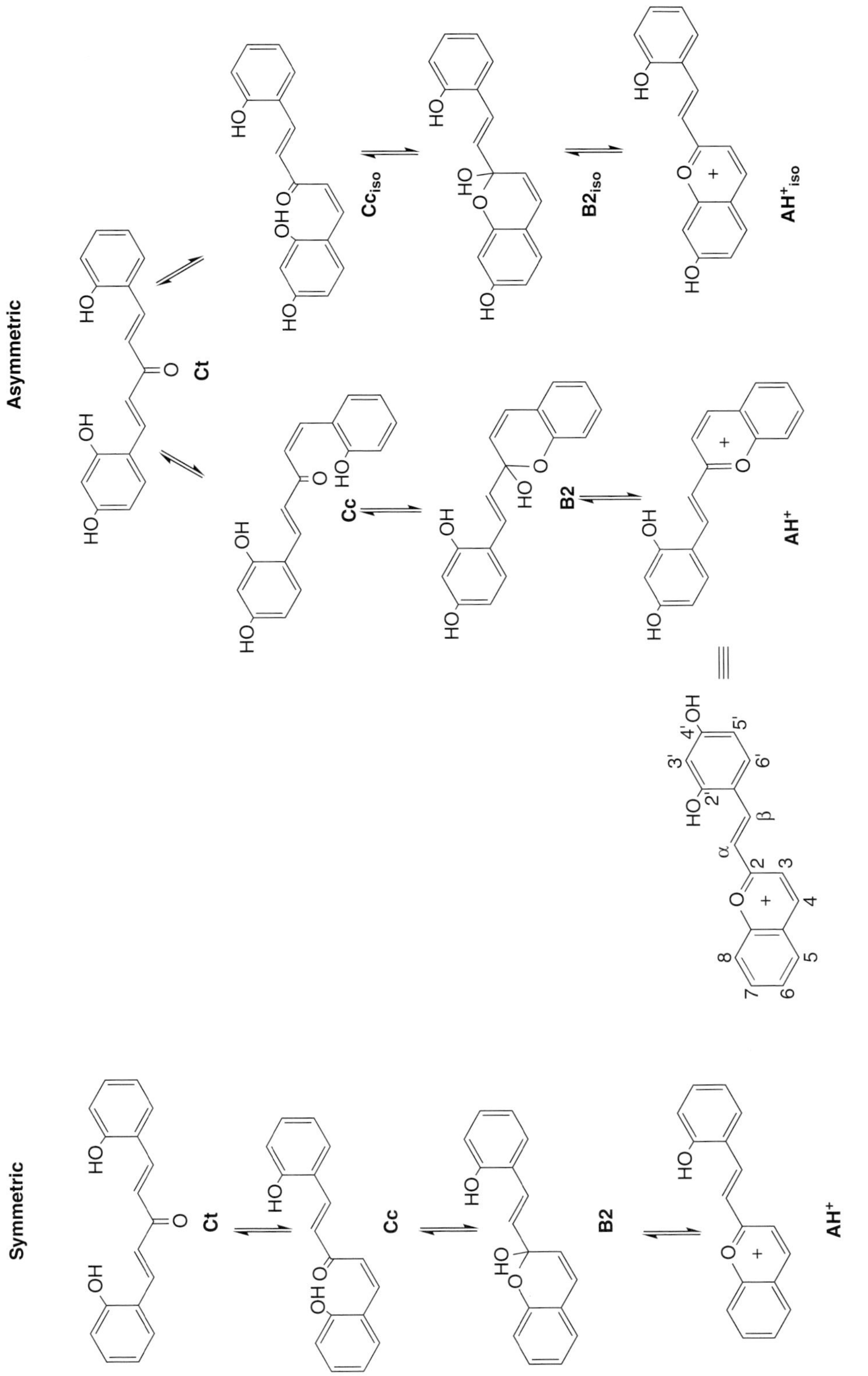


**Scheme 11.8** Symmetric and asymmetric *cis*-chalcone (Cc) formation. (Petrov *et al.*, 2012. Reproduced with permission of American Chemical Society.)

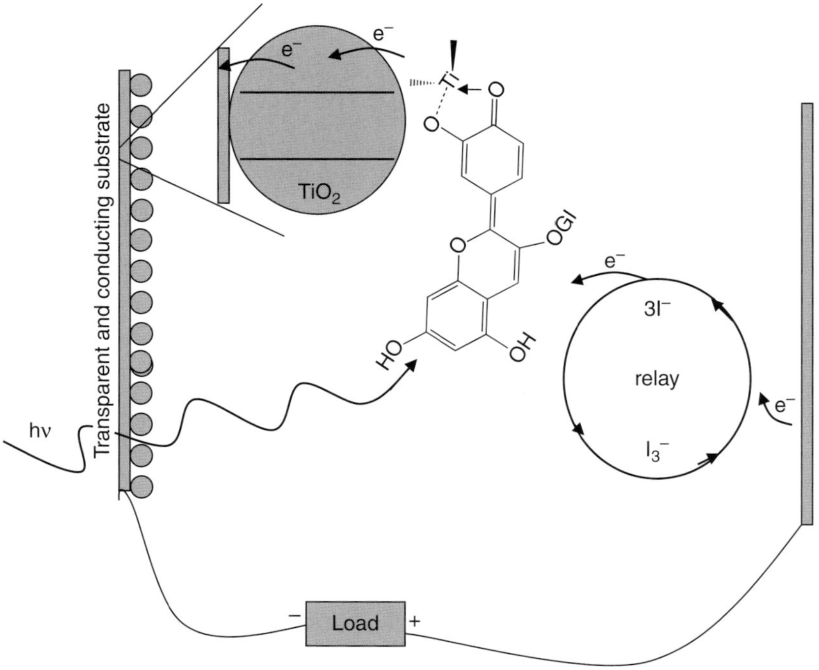


**Scheme 11.9** Schematic of a DSSC with cyanidine-3-glucoside (kuromanine) as the dye sensitizer. (Melo *et al.*, 2010. Reproduced with permission of Applied Science Innovations Private Ltd, India.)

an electron into $TiO_2$ (which remains oxidized). The electron is conduced through the transparent electrode to the other one, where it is trapped by the relay $I_3^-$ (leading to $I^-$). Finally, the $I^-$ transfers the electron back to the oxidized dye to complete the cycle. Work using synthetic flavylium analogs is under development (Calogero *et al.*, 2013). The best performance was obtained by a DSSC based on the novel compound 7-(N,N-Diethylamino)-3′, 4′-dihydroxyflavylium, which produced a 2.15% solar energy–electricity conversion efficiency under AM 1.5 irradiation (100 mW/cm), with a short-circuit current density (Jsc) of 12.0 mA/cm, a fill factor of 0.5, and an open-circuit voltage (Voc) of 0.355 V; its incident photocurrent efficiency of 51% at the peak of the visible absorption band of the dye is remarkable (Calogero *et al.*, 2013).

To summarize, besides their contribution to our understanding of the details of anthocyanin chemistry, flavylium compounds bio-inspired in anthocyanins offer many application possibilities, from photochromism to solar-energy conversion.

## 11.5 How to construct an energy-level diagram

Given an equilibrium between two species R and P (11.14), with equilibrium constant K, the standard Gibbs free energy $\Delta G^\circ$ is defined by (11.15), where $R$ is the gas constant and

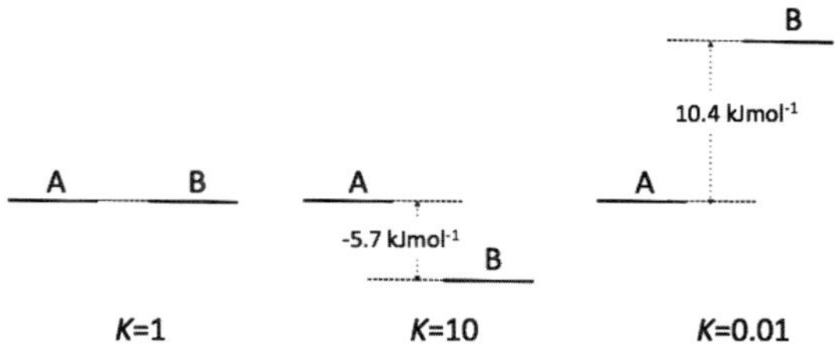


**Scheme 11.10** Definition of the standard Gibbs free energy ($\Delta G°$) for a chemical reaction as (11.14).

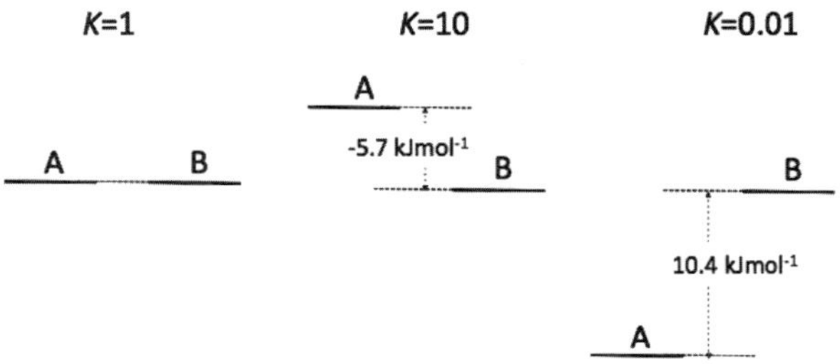


**Scheme 11.11** As Scheme 11.10, maintaining the energy level of P as reference.

$T$ the absolute temperature (K).

$$\mathrm{R} \overset{K}{\rightleftharpoons} \mathrm{P} \tag{11.14}$$

$$\Delta G° = -RT\ln K \tag{11.15}$$

If the equilibrium constant is $K = 1$, $\Delta G° = 0$ and the energy levels of both species in (11 14) are equal. If the equilibrium constant is $K = 10$, $\Delta G° = -5.7\,\mathrm{kJ/mol}$ at 25 °C, while if $K = 0.01$, $\Delta G° = +11.4\,\mathrm{kJ/mol}$, as illustrated in Scheme 11.10.

Due to the fact that the values of $\Delta G°$ are relative, it is more convenient for the following developments to maintain the energy level of $P$ as the reference (Scheme 11.11).

When the equilibrium is dependent on pH, as is the case for the proton-transfer and hydration reactions in the flavylium network of chemical reactions (11.16), a similar reasoning can be made assuming that for each pH value there is an apparent equilibrium constant (which is different for different pH values).

$$\mathrm{R} \overset{K}{\rightleftharpoons} \mathrm{P} + \mathrm{H}^+ \tag{11.16}$$

$$K_{ap} = \frac{K}{[H^+]} = \frac{[A]}{[AH^+]} \tag{11.17}$$

$$\Delta G^{\circ} = -RT\ln\frac{K}{[H^+]} = -RT\ln K - 2.303RT\mathrm{pH} \tag{11.18}$$

When pH = p$K_a$, $\Delta G^\circ = 0$. Consequently:

$$RT \ln K = -2.303RT\text{p}K_a \tag{11.19}$$

and (11.18) can be rearranged as:

$$\Delta G^\circ = -2.303RT \ln pK_a - 2.303RT\text{pH} \tag{11.20}$$

According to (11.20), it is possible to calculate $\Delta G^\circ$ for each pH value. In particular, if pH is one unit higher than p$K_a$, $\Delta G^\circ$ increases by 5.7 kJ/mol, while if it is one unit lower, $\Delta G^\circ$ decreases by the same amount (Scheme 11.12).

At this point, we have all the tools we need to apply these ideas to the flavylium network of chemical reactions. Taking pH 0 as reference, $\Delta G^\circ = -5.7\,\text{p}K_a$ kJ/mol. A scale in pH values can be constructed, and the base located at pH = p$K_a$. In the case of hemiketal (B) formation, the procedure is identical. Final Cc can be positioned from B, taking into account that $\Delta G^\circ = -RT \ln K_t$, and Ct forms Cc, assuming that $\Delta G^\circ = -RT \ln K_i$ (Scheme 11.13).

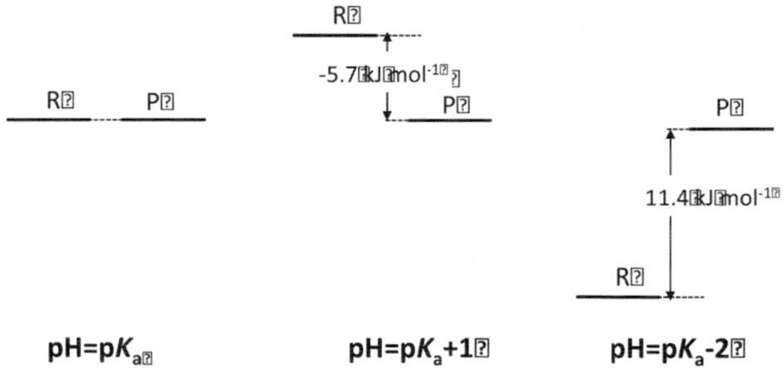


**Scheme 11.12** Definition of the standard Gibbs free energy ($\Delta G^\circ$) for an acid–base reaction.

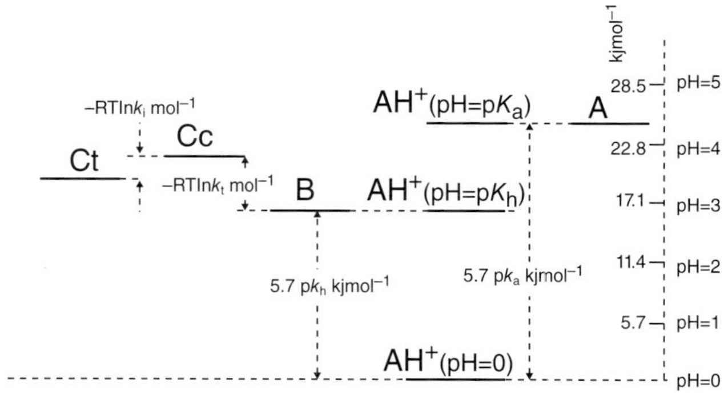


**Scheme 11.13** Definition of the energy-level diagram of the flavylium network of chemical reactions.

## 11.6 How to calculate the mole-fraction distribution of a network species

The following set of equations accounts for the different equilibria at which anthocyanins and related compounds are involved:

$$AH^+ + H_2O \rightleftharpoons A + H_3O^+ \quad K_a \tag{11.21}$$

$$AH^+ + 2H_2O \rightleftharpoons B + H_3O^+ \quad K_h \tag{11.22}$$

$$B \rightleftharpoons Cc \quad K_t \tag{11.23}$$

$$Cc \rightleftharpoons Ct \quad K_i \tag{11.24}$$

Considering now the equilibrium between AH+ and the remaining "basic species":

$$AH^+ + H_2O \rightleftharpoons CB + H_3O^+ \tag{11.25}$$

where

$$[CB] = [A] + [B] + [Cc] + [Ct] \tag{11.26}$$

$$K_a' = \frac{\{[A] + [B] + [Cc] + [Ct]\}[H^+]}{[AH^+]} \tag{11.27}$$

From (11.21) to (11.24):

$$K_a' = K_a + K_h + K_hK_t + K_hK_tK_i \tag{11.28}$$

The total concentration, $C_0$, can be written as:

$$C_0 = [AH^+] + [[A] + [B] + [Cc] + [Ct] = [AH^+]\left(1 + \frac{K_a}{[H^+]} + \frac{K_h}{[H^+]} + \frac{K_hK_t}{[H^+]} + \frac{K_hK_tK_i}{[H^+]}\right) \tag{11.29}$$

$$C_0 = [AH^+]\left(1 + \frac{K_a'}{[H^+]}\right) \tag{11.30}$$

From (11.30), the mole fraction of $AH^+$ is obtained:

$$\chi_{AH^+} = \frac{[AH^+]}{C_0} = \frac{[H^+]}{[H^+] + K_a'} \tag{11.31}$$

Using the rate constant of (11.21):

$$\chi_A = \frac{[A]}{C_0} = \frac{K_a[AH^+]}{[H^+]C_0} = \frac{K_a}{[H^+] + K_a'} \tag{11.32}$$

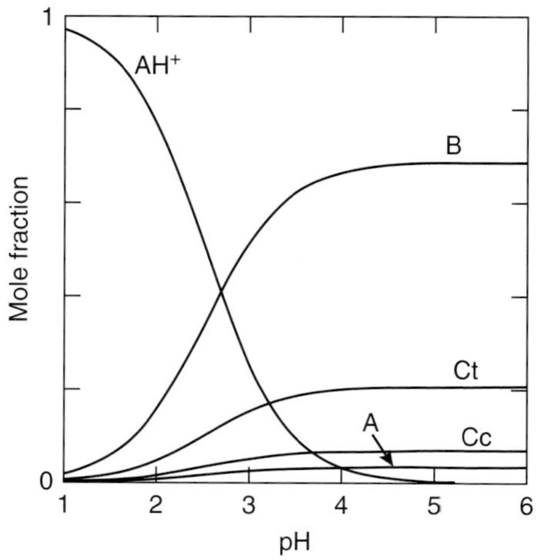


**Fig. 11.13** Mole-fraction distribution for the following set of equilibrium constants: $K_a = 10^{-4}$ M; $K_a = 2 \times 10^{-3}$ M; $K_t = 0.1$; $K_i = 3$ ($K_a' = 2.9 \times 10^{-3}$).

By an identical procedure, the mole fractions of the other species are also obtained:

$$\chi_B = \frac{K_h}{[H^+] + K_a'};\quad \chi_{Cc} = \frac{K_h K_t}{[H^+] + K_a'};\quad \chi_{Ct} = \frac{K_h K_t K_i}{[H^+] + K_a'} \tag{11.33}$$

The mole-fraction distribution of the several species can be calculated as a function of pH from (11.31) to (11.33) to give a simulation for $K_a = 10^{-4}$ M; $K_a = 2 \times 10^{-3}$ M; $K = 0.1$; and $K_i = 3$ ($K_a' = 2.9 \times 10^{-3}$) (Fig. 11.13).

# References


Anderson, Ø.M. & Markham, K.R. (2006) *Flavonoids – Chemistry, Biochemistry and Applications*. Taylor & Francis Group, CRC Press, New York.

Antonov, L. & Petrov, V. (2002) Quantitative analysis of undefined mixtures – "fishing net" alghoritm. *Analytical and Bioanalytical Chemistry*, 374, 1312–1317.

Atkins, P.W. (1999) *Physical Chemistry*. Oxford University Press, Oxford.

Brouillard, R. & Dubois, J.-E. (1977) Mechanism of structural transformations of anthocyanins in acidic media. *Journal of the American Chemical Society*, 99, 1359–1364.

Brouillard, R. & Delaporte, B. (1977) Chemistry of anthocyanins pigments. 2. Kinetic and thermodynamic study of proton transfer, hydration and tautomerization reactions of malvidin-3-glucoside. *Journal of the American Chemical Society*, 99, 8461–8468.

Brouillard, R. & Lang, J. (1990) The hemiacetal-cis-chalcone equilibrium of malvin, a natural anthocyanin. *Canadian Journal of Chemistry*, 68, 755–761.

Brouillard, R., Delaporte, B. & Dubois, J.-E. (1978) Chemistry of anthocyanin pigments. 3. Relaxation amplitudes in pH-jumps experiments. *Journal of the American Chemical Society*, 100, 6202–6205.

Calogero, G., Sinopoli, A., Citro, I., *et al.* (2013) Synthetic analogues of anthocyanins as sensitizers for dye-sensitized solar cells. *Photochemical & Photobiological Sciences*, 12, 883–94.

Charman, W.N., Porter, C.J.H., Mithani, S. & Dressman, J.B. (1997) Physicochemical and mechanism for the effects of food on drug absorption: the role of lipids and pH. *Journal of Pharmaceutical Sciences*, 86, 269–282.

Cherepy, N.J., Smestad, G.P., Gratzel, M. & Zhang, J.Z. (1997) Ultra fast elecron injection: implications for a photoelectrochemical cell utilizing dye-sensitized $TiO_2$ nanocrystaline electrode. *Journal of Physical Chemistry B*, 101, 9342–9351.

Figueiredo, P., Lima, J.C., Santos, H., *et al.* (1994) Photochromism of the synthetic 4′,7–dihydroxyflavylium chloride. *Journal of the American Chemical Society*, 116, 1249–1254.

Gago, S., Petrov, V., Parola, A.J. & Pina, F. (2012) Synthesis and characterization of 3′-butoxyflavylium derivatives. *Journal of Photochemistry and Photobiology, A: Chemistry*, 244, 54–64.

Gavara, R., Leydet, Y., Petrov, V. & Pina F. (2012) Photochemistry of 2-(4-hydroxystyryl)-1-naphthopyrylium. *Photochemical & Photobiological Sciences*, 11, 1691–1699.

Gavara, R., Petrov, V., Pina, F. & Quintas, A. (2013) Circular dicroism of anthocyanin 3-glucosides self-aggregates. *Phytochemistry*, 88, 92–98.

Glasoe, P.K. & Long, F.A. (1960) Use of glass electrodes to measure acidities in deuterium oxide. *Journal of Physical Chemistry*, 64, 188–190.

Hirshberg, Y. (1956) Reversible formation and eradication of colors by irradiation at low temperatures – a photochemical memory model. *Journal of the American Chemical Society*, 78, 2304–2312.

Hoshino, T. (1991) Anthocyanin self-aggregates. 6. An approximation estimate of self-association constants and the self stacking of malvin quinonoidal bases studied by H-1-NMR. *Phytochemistry*, 30, 2049–2055.

Hoshino, T. & Goto, T. (1990) Effects of pH and concentration on the self-association of malvidin quinonoidal base – electronic and circular dichroic studies. *Tetrahedron Letters*, 31, 1593–1596.

Hoshino, T., Matsumoto, U. & Goto, T. (1980) Evidence for self-association of anthocyanins. 1. Circular dichroism of cyanin anhydrobase. *Tetrahedron Letters*, 21, 1751–1754.

Hoshino, T., Matsumoto, U. & Goto, T. (1981a) Evidence for self-association of anthocyanins. 2. Self-association of some anthocyanins in neutral aqueous-solution. *Phytochemistry*, 120, 1971–1976.

Hoshino, T., Matsumoto, U., Harada, N. & Goto, T. (1981b) Chiral excitation couple stacking of anthocyanins. Interpretation of the origin of anomalous CD induced by anthocyanin association. *Tetrahedron Letters*, 22, 3621–3624.

Hoshino T., Matsumoto, U. & Goto, T. (1982) Evidence for self-association of anthocyanins. 4. PMR spectroscopic evidence for the vertical stacking of anthocyanin molecules. *Tetrahedron Letters*, 23, 433–436.

Kong, J.M, Chia, L.S., Goh, N.K., *et al.* (2003) Analysis and biological activities of anthocyanins. *Phytochemistry*, 64, 923–933.

Lehn, J.-M. (1990) Perspectives in supramolecular chemistry – from molecular recognition towards molecular information-processing and self-organization. *Angewandte Chemie International Edition*, 29, 1304–1319.

Leydet, L., Gavara, R., Petrov, V., *et al.* (2012) The effect of self-aggregation on the determination of the kinetic and thermodynamic constants of the network of chemical reactions in 3–glucoside anthocyanins. *Phytochemistry*, 83, 125–135.

Maçanita, A.L., Moreira, P., Lima, J.C., *et al.* (2002) Proton transfer in anthocyanins and related flavylium salts. Determination of ground-state rate constants with nanosecond-laser flash photolysis (light-pulse jump). *Journal of Physical Chemistry A*, 106, 1248–1255.

McClelland, R.A. & Gedge, S. (1980) Hydration of the flavylium ion. *Journal of the American Chemical Society*, 102, 5838–5848.

McClelland, R.A. & McGall, G.H. (1982) Hydration of the flavylium ion. 2. The 4′-hydroxyflavylium. *Journal of Organic Chemistry*, 47, 3730–3736.

McRae, E.G. & Kasha, M. (1958) Enhancement of phosphorescence ability upon aggregation of dye molecules. *Journal of Physical Chemistry*, 28, 721–722.

Melo, M.J., Parola, A.J., Lima, J.C. & Pina, F. (2010) Chromogenic materials based on 2-phenyl-1-benzopyrylium: from anthocyanins to synthetic flavylium compounds. In: *Chromic Materials, Phenomena and Their Technological Applications* (ed. P.R. Somani), Chapter 16, pp 537–576. Applied Science Innovations Private Limited, India.

Pascual-Teresa, S., Moreno, D.A. & Garcia-Viguera, C. (2010) Flavanols and anthocyanins in cardiovascular health: a review of current evidence. *International Journal of Molecular Sciences*, 11, 1679–1703.

Petrov, V. & Pina, F. (2010) Analytical resolution of the reaction rates of flavylium network by Laplace transform. *Journal of Mathematical Chemisrty*, 47, 1005–1026.

Petrov, V., Parola, A.J. & Pina, F. (2012) Isomerisation between 2-(2,4-dihydroxystyryl)-1-benzopyrylium and 7-hydroxy-2-(4-hydroxystyryl)-1-benzopyrylium. *Journal of Physical Chemistry*, 116, 8107–8118.

Pina, F. (1998) Photochromism of the synthetic 4′-7–dihydroxyflavylium chloride. *Journal of the Chemical Society, Faraday Transactions*, 94, 2109–2116.

Pina, F. & Hatton, T.A. (2008) Photochromic soft materials: flavylium compounds incorporated in pluronic F-127 hydrogel matrixes. *Langmuir*, 24, 2356–2364.

Pina, F., Melo, M.J., Ballardini, R., *et al.* (1997a) Flash photolysis of 4′,7–dihydroxyflavylium perchlorate. *New Journal of Chemistry*, 21, 969–976.

Pina, F. Melo, M.J., Maestri, M., *et al.* (1997b) Photochromism of 4′-methoxyflavylium perchorate. A “write-lock-read-unlock-erase” molecular switching device. *Journal of the American Chemical Society*, 119, 5556–5561.

Pina, F., Melo, M.J., Laia, C.A.T., *et al.* (2012a) Chemistry and applications of flavylium compounds: a handful of colours. *Chemical Society Reviews*, 41, 869–908.

Pina, F., Petrov, V. & Laia, C.A.T. (2012b) Photochromism of flavylium systems. An overview of a versatile multistate system. *Dyes and Pigments*, 92, 877–889.

Popov, K., Rönkkömäki, H. & Lajunen, L.H.J. (2006) Guidelines for NMR measurements for determination of high and low pK(a) values. *Pure and Applied Chemistry*, 78, 663–675.

Santos, H., Turner, D.L., Lima, J.C., *et al.* (1993) Elucidation of the multiple equilibria, of malvin in aqueous solution by one and two dimensional N.M.R. *Phytochemistry*, 33, 1227–1232.

Shaka, A.J. & Hwang, T.-L. (1995) Water suppression that works-excitation sculpting using arbitrary wave-forms and pulsed-field gradients. *Journal of Magnetic Resonance A*, 112, 275–279.

Yoshida, K., Mori, M. & Kondo, T. (2009) Blue flower color development by anthocyanins: from chemicalstructure to cell physiology. *Natural Product Reports*, 26, 884–915.

Zafra-Stone, S., Yasmin, T, Bagchi, M., *et al.* (2007) Berry anthocyanins as novel antioxidants in human health and disease prevention. *Molecular Nutrition and Food Research*, 51, 675–683.

Chapter 12

# Synthetic Strategies and Tactics for Catechin and Related Polyphenols

Ken Ohmori and Keisuke Suzuki
*Department of Chemistry, Tokyo Institute of Technology, Tokyo, Japan*

**Abstract:** Among flavone-derived polyphenols, flavan-3-ols, in particular catechin, epicatechin, epigallocatechin, and their gallate conjugates, have received considerable attention due to their significant bioactivities and their potential for human health. However, detailed studies of these compounds at molecular levels are hampered by the scarce availability of pure samples, as their natural sources generally comprise hard-to-separate mixtures of closely related compounds. Therefore, development of reliable synthetic methods is of keen necessity. This review summarizes stereoselective synthetic approaches to flavan-3-ols and their derivatives. The strategies can be classified into two major approaches, differing in the order of bond formation. The review also focuses on stereoselective methods for constructing a flavan skeleton with 2,3-*trans* and 2,3-*cis* stereochemistries.

**Keywords:** flavan-3-ol, catechin, EGCg, polyphenol, proanthocyanidin, total synthesis

## 12.1 Introduction

Flavonoids, which share a common structure depicted in Fig. 12.1, constitute one of the most abundant classes of natural products, occurring widely as various plant ingredients. Over the last few decades, these compounds have attracted considerable interest through their multiple biological properties, including antivirus and anticancer activities and health-promoting effects in the human diet. As an example, the French paradox – the observation of low coronary heart disease (CHD) death rates despite high intake of dietary cholesterol and saturated fat in France – has been attributed to the drinking of red wine, and in particular to the biological effects of the polyphenols contained therein.

*Recent Advances in Polyphenol Research, Volume 4*, First Edition.
Edited by Annalisa Romani, Vincenzo Lattanzio, and Stéphane Quideau.

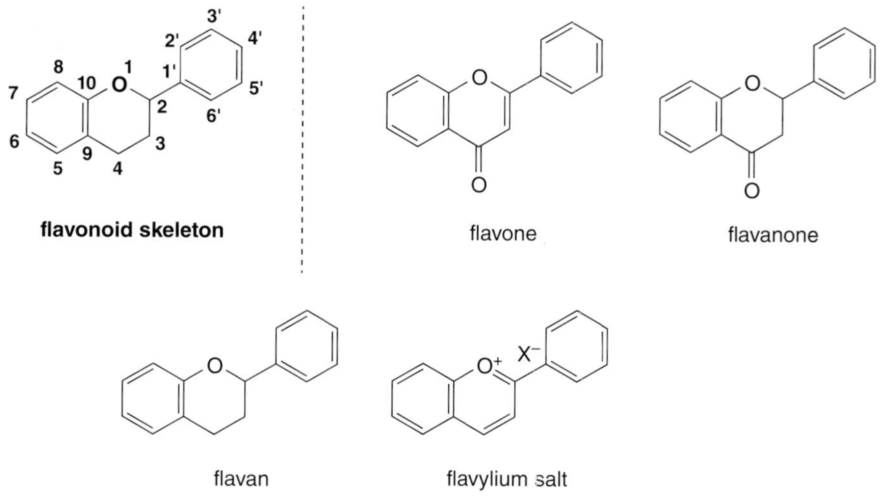


**Fig. 12.1** Common structure of flavonoids.

Among the natural flavonoids, increasing interest is currently focused on flavan-3-ols (Fig. 12.2), such as (+)-catechin (CA, **1**), (−)-epicatechin (EC, **2**), and (−)-epigallocatechin (EGC, **3**), as well as their 3-*O*-gallates, which are well known as antioxidant ingredients of green tea, wine, fruits, vegetables, and other plant-derived foodstuffs. These compounds constitute a "catechin library," differing in their substitution and oxidation pattern, as well as in their stereochemistry. In addition, they serve as the nucleophilic entities in the formation of oligomeric proanthocyanidins. Despite its potential as a rich source of compounds with biological or pharmaceutical relevance, this natural library has hardly been explored because of the difficulty in obtaining pure samples of individual compounds. Indeed, the natural samples are mostly obtained as complex mixtures of closely related compounds, which are difficult to separate even with the aid of modern techniques.

The situation calls for organic synthesis to supply enantiopure samples of natural flavan-3-ols and their derivatives for systematic biological testing with the aim of clarifying structure–activity relationships. In spite of such needs, the stereoselective synthesis of catechins and epicatechins remains a challenge, due to their chemical and stereochemical lability. The handling of free polyphenols is difficult, due to their high sensitivity to oxidants, acid, base, light, and metallic species. Fig. 12.3 exemplifies the chemical vulnerability of non-protected flavans. Under basic/acidic conditions, they undergo reversible opening of the pyran ring, leading to a transient quinone methide **A**, which results in stereochemical scrambling at C(2) (***cis*** ⇄ **A** ⇄ ***trans***). Even racemization (***cis*** ⇄ ***ent-cis***) can take place via the achiral enol **B**. In addition, rearrangement of the flavan skeleton into products such as catechinic acid is also worth mentioning.

This review summarizes research on the stereo-controlled synthesis, in particular total synthesis, of representative class of flavan-3-ols.

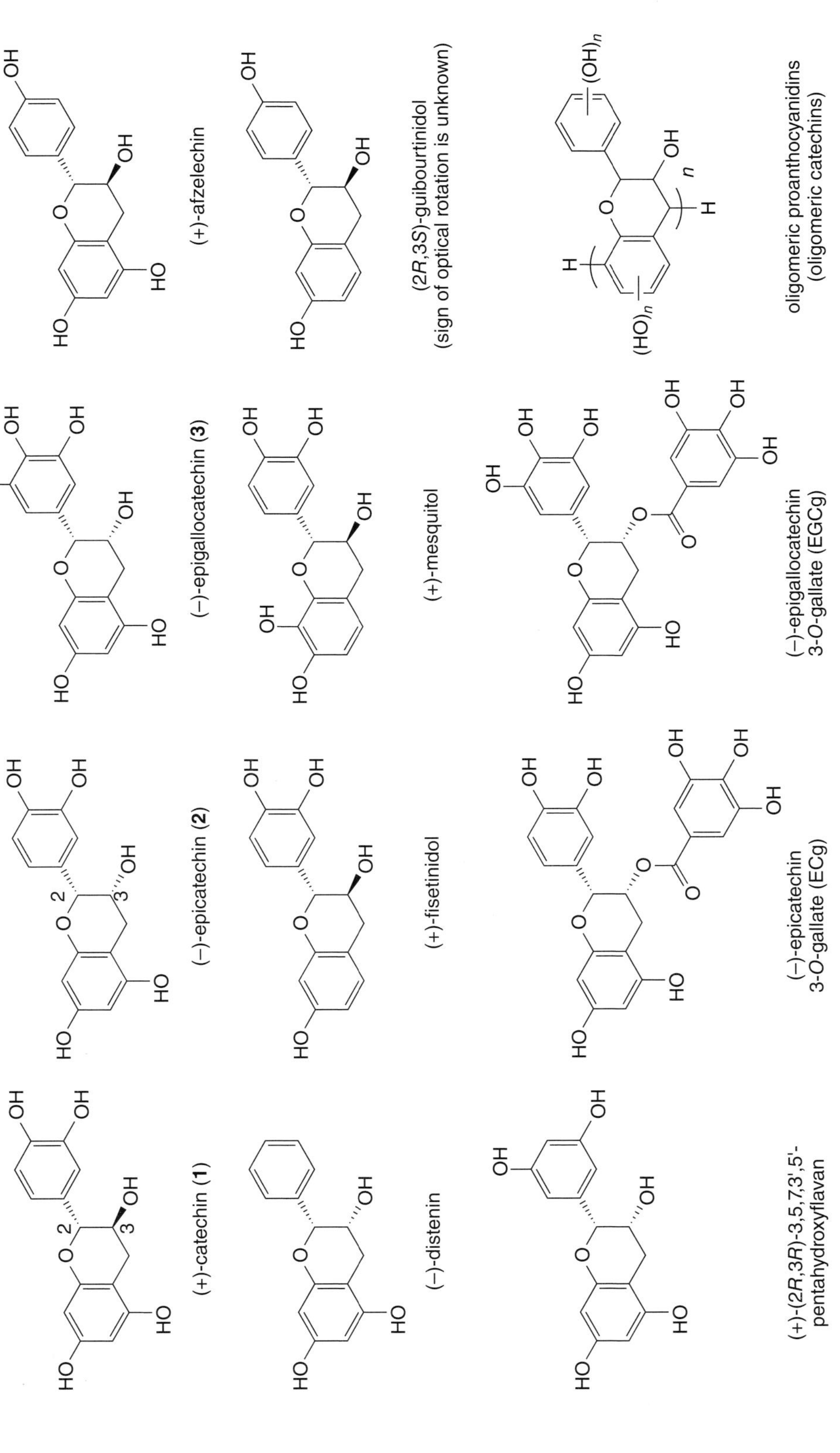


**Fig. 12.2** Naturally occurring flavan-3-ols and their 3-*O*-gallates.

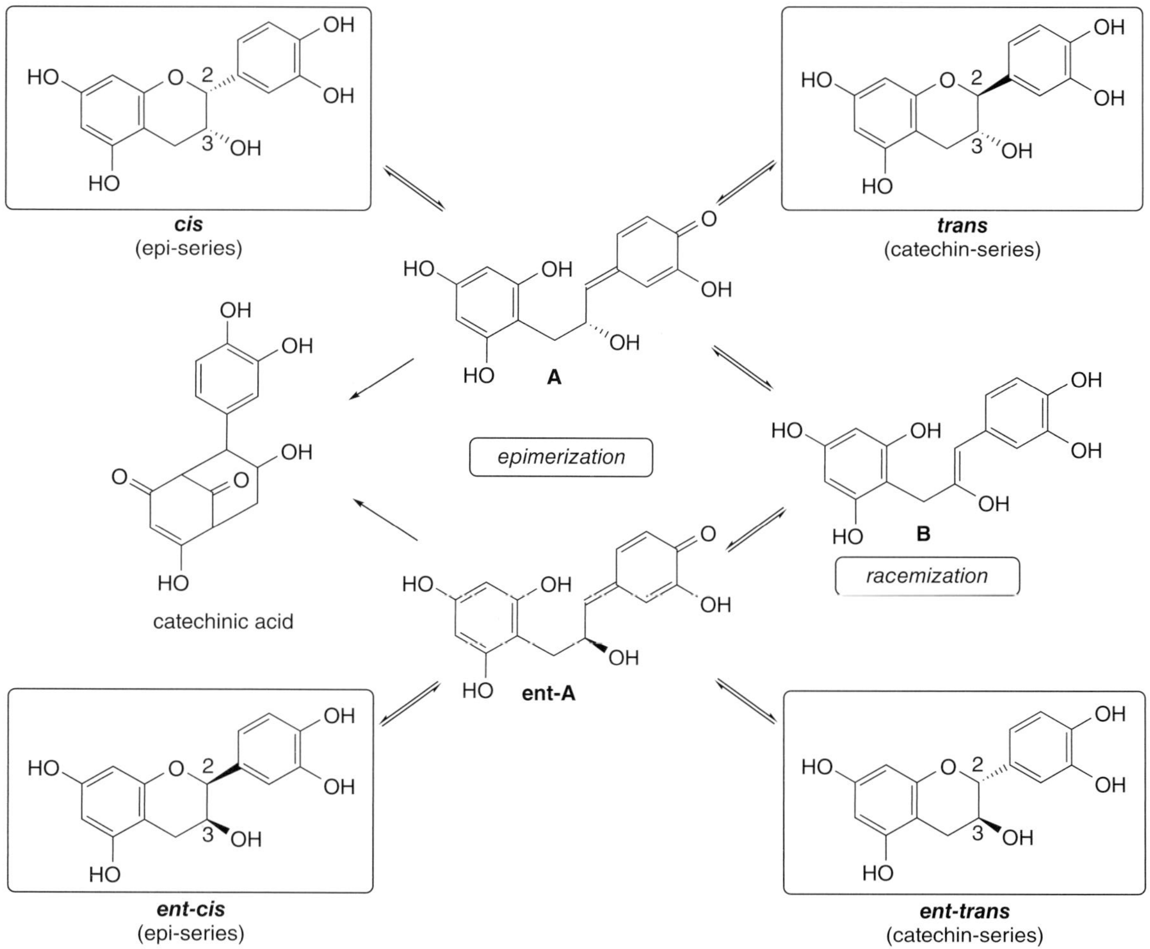


**Fig. 12.3** Skeletal and stereochemical lability of flavan-3-ols.

## 12.2 Early synthetic work

Although many synthetic protocols for naturally occurring flavones have been reported, the stereoselective synthesis of 3-hydroxy flavanones or flavan-3-ols, possessing two additional stereogenic centers at C(2) and C(3), has not been sufficiently explored. More surprisingly, the total synthesis of (+)-catechin with full stereochemical control was not reported until this century. In an early study aimed at obtaining optically active catechin (**1**), a semi-synthetic method was developed that encompassed the consecutive reduction of a protected 2,3-*trans*-dihydroquercetin derivative with $LiAlH_4$ and $H_2$/Pd-C (Fig. 12.4). The starting material was obtained from natural sources (Weinges, 1958; Freudenberg & Weinges, 1958). However, the scale at which this protocol can be executed is limited, because the starting material is not readily available.

As an alternative attempt to synthesize flavan-3-ols, Clark-Lewis & Skingle (1967) reported an efficient diastereoselective approach to racemic flavan-3,4-diols, in which the relative stereochemistry between C(2) and C(3) was rigorously controlled to be *trans*. By way of example, the authors synthesized the flavan-3,4-diol **8**, as shown in Fig. 12.5. Thus, upon treatment of the chalcone **6** with $NaBH_4$, 1,2-reduction of the enone moiety was followed by an intramolecular $S_N2'$ reaction to form the chromene **7**. Dihydroxylation of **7** with $OsO_4$ proceeded stereoselectively, giving the 2,3-*trans* product **8** in 34% overall yield from **6**. Unfortunately, this protocol was restricted to a few unnatural derivatives.

A synthesis of the flavan-3,4-diol derivative **13** with the oxygenation pattern of natural catechin (**1**), except at C(4), was realized by applying the Clark-Lewis protocol (Kawamoto *et al.*, 1989) (Fig. 12.6). Aldol condensation of acetophenone **9** with benzaldehyde **10** gave chalcone **11** in 95% yield. The subsequent two-step sequence of reductive cyclization and dihydroxylation afforded the flavan-3,4-diol **13** with 2,3-*trans* stereochemistry. This compound proved to be a good, albeit racemic, precursor for the oligomerization to form oligomeric proanthocyanidins such as procyanidins.

Fig. 12.7 shows a stereoselective approach to optically active 2,3-*trans*-3-hydroxyflavanone (i.e. (−)-ampelopsin (**18**)) through the exploitation of the asymmetric epoxidation of chalcone derivatives, which enabled the determination of the stereochemistry of the natural leucodelphinidins (Onda *et al.*, 1989): condensation of the acetophenone **14** with the benzaldehyde **15** under basic conditions (KOH, EtOH) followed by asymmetric epoxidation of the resulting chalcone **16** with *t*-BuOOH in the presence of the chiral

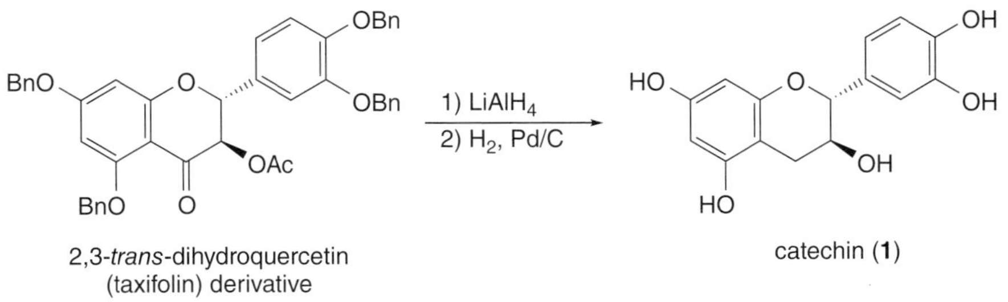


**Fig. 12.4** Semi-synthesis of catechin (**1**) from a natural flavonoid derivative.

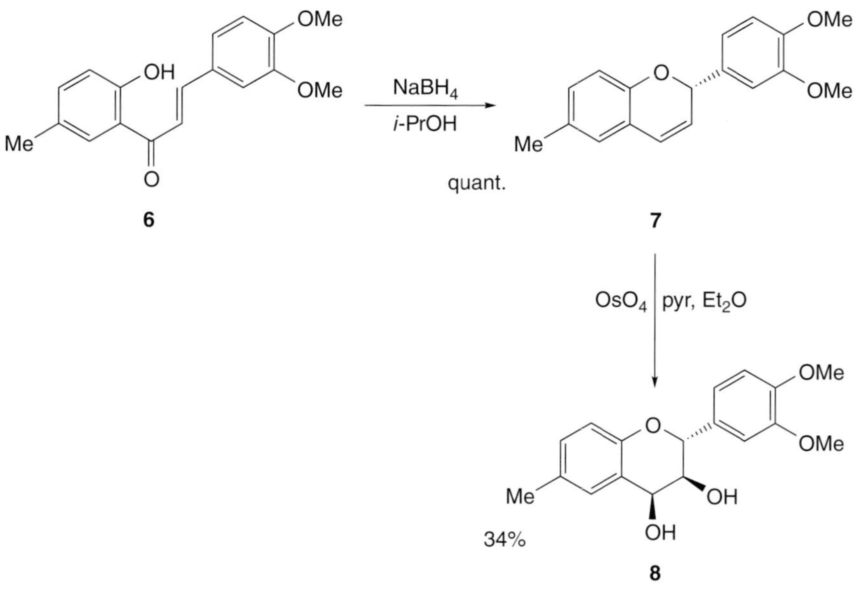


**Fig. 12.5** Clark-Lewis synthesis: $S_N2'$ route to 2,3-*trans* flavans.

phase-transfer catalyst 1-benzylquininium chloride (BQC) and NaOH gave the optically active epoxide **17** in 63% yield. The enantiomeric excess (*ee*) was assessed to be 57.4% by HPLC analysis on a chiral stationary phase. In contrast, the asymmetric epoxidation of **16** using 1-benzylquinidinium chloride (BQdC) instead of BQC under the same conditions yielded the antipode *ent*-**17** with 62.3% *ee* (60% yield). Furthermore, enantiopure **17** was obtained by optical resolution using preparative HPLC on the chiral stationary phase, Chiralpak OT(+). Treatment of **17** with HCl/MeOH resulted in complete removal of the methoxymethyl (MOM) protecting groups and simultaneous 6-*endo-trig* cyclization, obtaining (+)-ampelopsin (dihydromyricetin, **18**) stereospecifically in 70% yield.

A similar approach was employed for the synthesis of flavanone **21** (Augustyn *et al.*, 1990) (Fig. 12.8). The asymmetric epoxidation of a series of polyoxygenated chalcones was systematically examined using poly-L-alanine and its derivatives as chiral modifiers (for a review, see Marais *et al.*, 2005). The enantioselectivity showed a high dependence on the substituent pattern of the substrates. For example, epoxy ketone **20** was obtained with 84% *ee* upon treatment of chalcone **19** with $H_2O_2$ and NaOH in the presence of poly-L-alanine.

The acid-promoted cyclization of **20** gave the corresponding flavanone **21** in moderate yield. At this stage, however, considerable loss of the enantiomeric purity of the cyclized product was observed. This suggested the generation of a stabilized benzyl cation **D**, facilitating an $S_N1$ ionization to provide a *trans*/*cis* mixture of **21**. Subsequently, the latter diastereomer, ***cis*-21**, suffered epimerization at C(2), leading to the thermodynamically more stable ***ent-trans*-21**. An additional observation was the formation of isoflavone **22**, albeit in low yield, which was attributed to a hypothetical intermediate **E** generated from **D**, or which might alternatively have proceeded via the direct 1,2-migration of the electron-rich aryl group as in **F**.

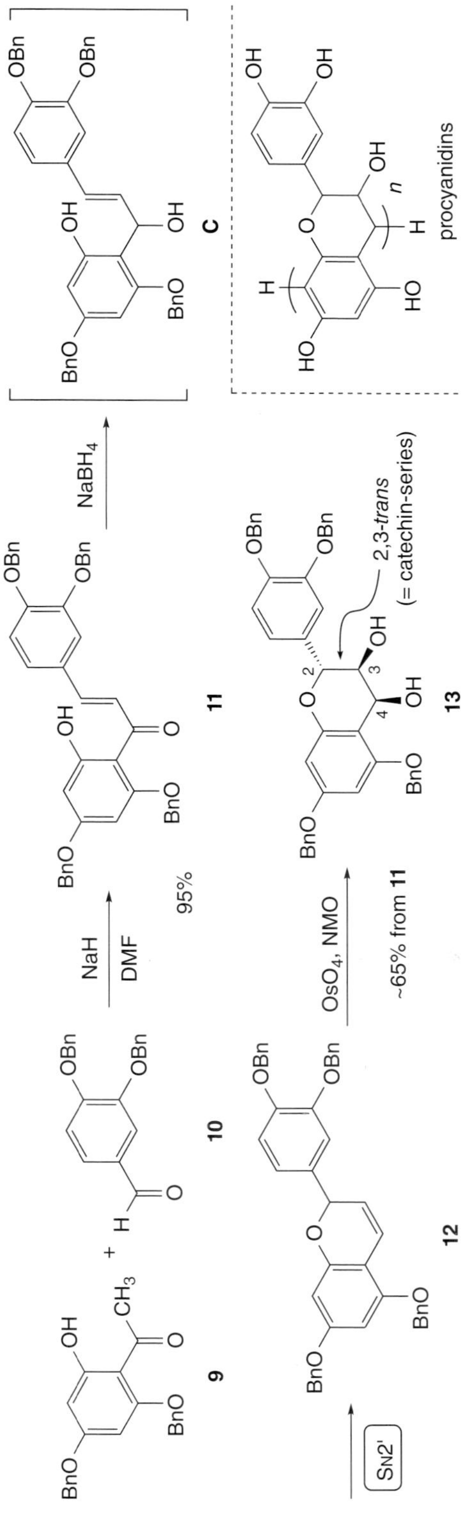


**Fig. 12.6** Kawamoto's flavan-3,4-diol synthesis.

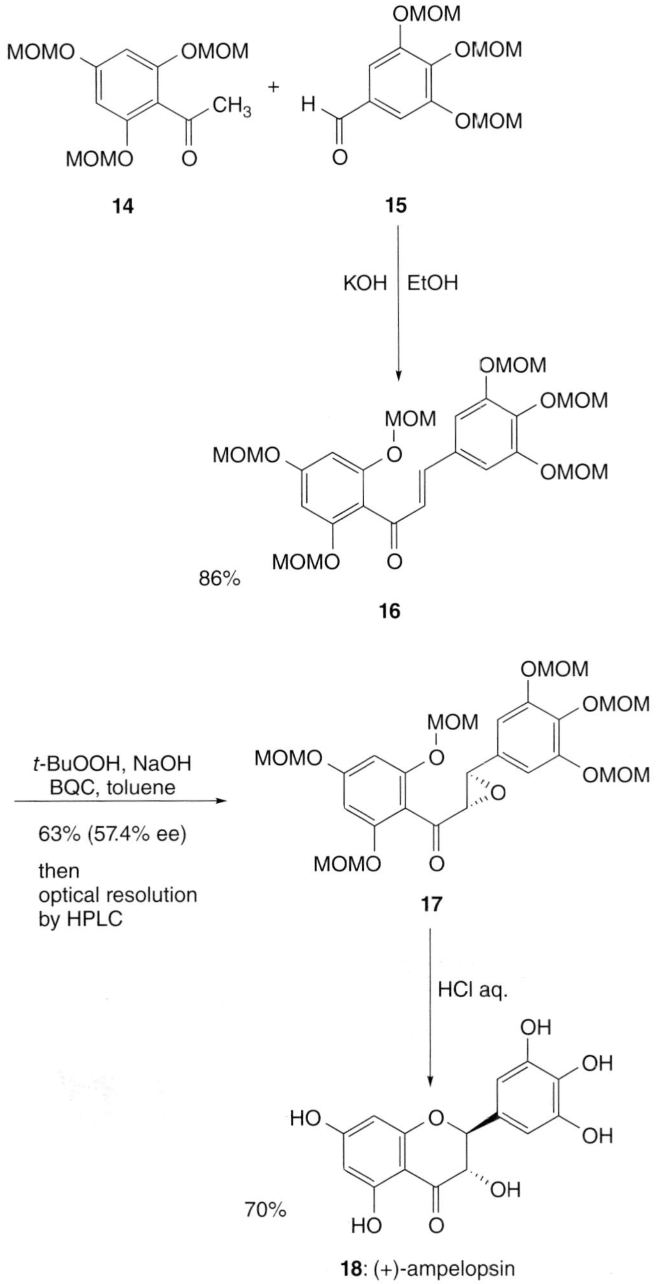


**Fig. 12.7** Onda's chalcone–epoxy ketone route to (+)-ampelopsin (**18**).

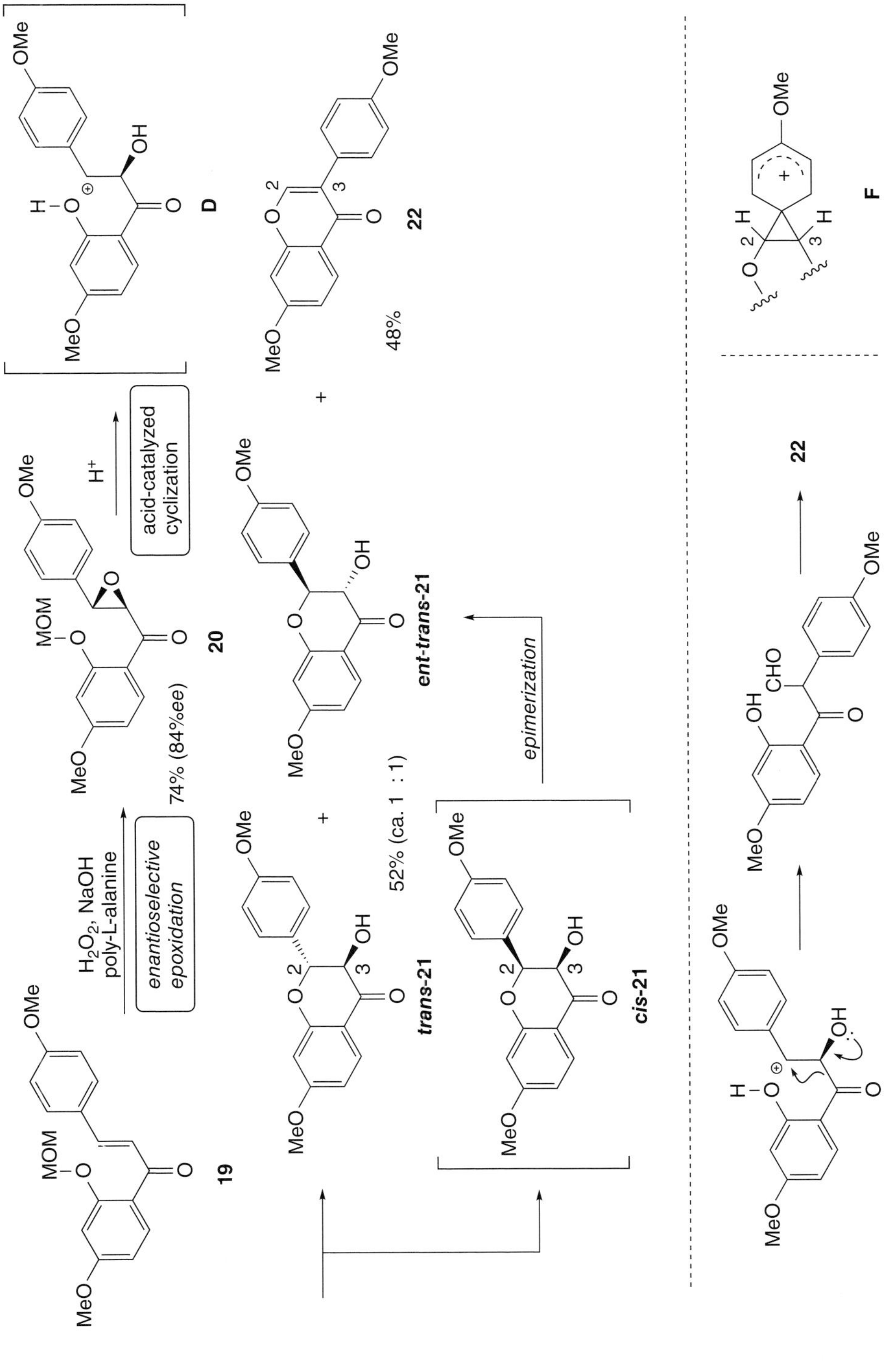


**Fig. 12.8** Potential mechanisms for the loss of stereochemical integrity and skeletal rearrangement in the chalcone–epoxy ketone route.

In order to enhance the $S_N2$ nature of the pyran-ring closure, the intramolecular Mitsunobu reaction was studied (Jew *et al.*, 2000) (Fig. 12.9). The cyclization precursor **27** was prepared in highly enantio-enriched form via Sharpless dihydroxylation of the caffeic acid derivative **23**. Conversion of **23** into aldehyde **24**, coupling with aryllithium **25**, oxidation of the benzylic alcohol **26** ($n$-$Pr_4NRuO_4$, NMO, MS4A), and deprotection gave hydroxy ketone **27**. Treatment of **27** with $Ph_3P$ and diethyl azodicarboxylate (DEAD) resulted in the smooth formation of the *trans*-3-hydroxy flavanone **28** in 51% yield, where the cyclization proceeded stereospecifically with inversion at C(2). This method proved applicable for the preparation of other 3-hydroxy derivatives, and was indeed employed in the synthesis of free (+)-catechin (**1**) in enantiopure form (see Section 12.3.1).

## 12.3 Stereoselective approaches to flavan-3-ols

### *12.3.1 Synthesis of catechin-series (= 2,3-*trans*) derivatives*

In 1997, a stereoselective approach to catechin skeleton possessing the natural oxygenation pattern was reported. The synthesis relied on asymmetric dihydroxylation of (*E*)-1,3-diarylpropenes (deoxodihydrochalcones) followed by acid-promoted cyclization to form the pyran ring (Van Rensburg *et al.*, 1997) (Fig. 12.10). The key substrate, (*E*)-1,3-diarylpropene **31**, was prepared from the chalcone **29** by two consecutive reduction steps ($H_2$, Pd/C; $NaBH_4$) followed by water elimination from the resulting benzyl alcohol. The olefinic intermediate **30** was then subjected to Sharpless dihydroxylation, providing 1,2-diol **31** in high yield with 99% *ee*. Treatment of **31** with acid (3 M HCl aq.) to form the pyran skeleton followed by acetylation afforded the cyclized product **32** as a *trans*/*cis* mixture (3 : 1 ratio). Although the potential utility of this approach was verified by the synthesis of various flavan-3-ol derivatives, there remained the problems of poor stereospecificity in the pyran cyclization step, as well as difficulty in the removal of the methyl-protecting groups.

To support biological studies on the metabolism of flavans, $^{13}C$-labelled catechin was prepared on a practical scale (Nay *et al.*, 2000) (Fig. 12.11). Friedel–Crafts reaction of phloroglucinol tribenzyl ether **33** with cinnamic acid **34** proceeded by use of trifluoroacetic anhydride (TFAA) at 0 °C, providing the pentabenzyloxychalcone **35** (58% yield), in which a $^{13}C$ atom was incorporated at the C(4)-position (flavan numbering; the location of the $^{13}C$-labelling is indicated in Fig. 12.11). Selective detachment of one benzyl group in **35** was effected using $TiCl_4$ to give the requisite hydroxychalcone **36** (63% yield), although an unexpected vicinal *C*-benzylation occurred as a side reaction (8% yield).

The key pyran-ring formation was effected via the Clark-Lewis protocol (see Section 12.2): 1,2-reduction of enone **36** by $NaBH_4$ followed by $S_N2'$ cyclization in the presence of $BF_3{\cdot}OEt_2$ gave the unstable flavene **37** in racemic form, which was employed in the next step without purification. Dihydroxylation of **37** employing $OsO_4$ and *N*-methylmorpholine *N*-oxide (NMO) yielded the flavan-3,4-diol **38** as a single diastereomer with 2,3-*trans* and 3,4-*cis* relative stereochemistries. Compound **38** has potential utility as a monomer unit for preparing oligomeric catechins, which has been intensively

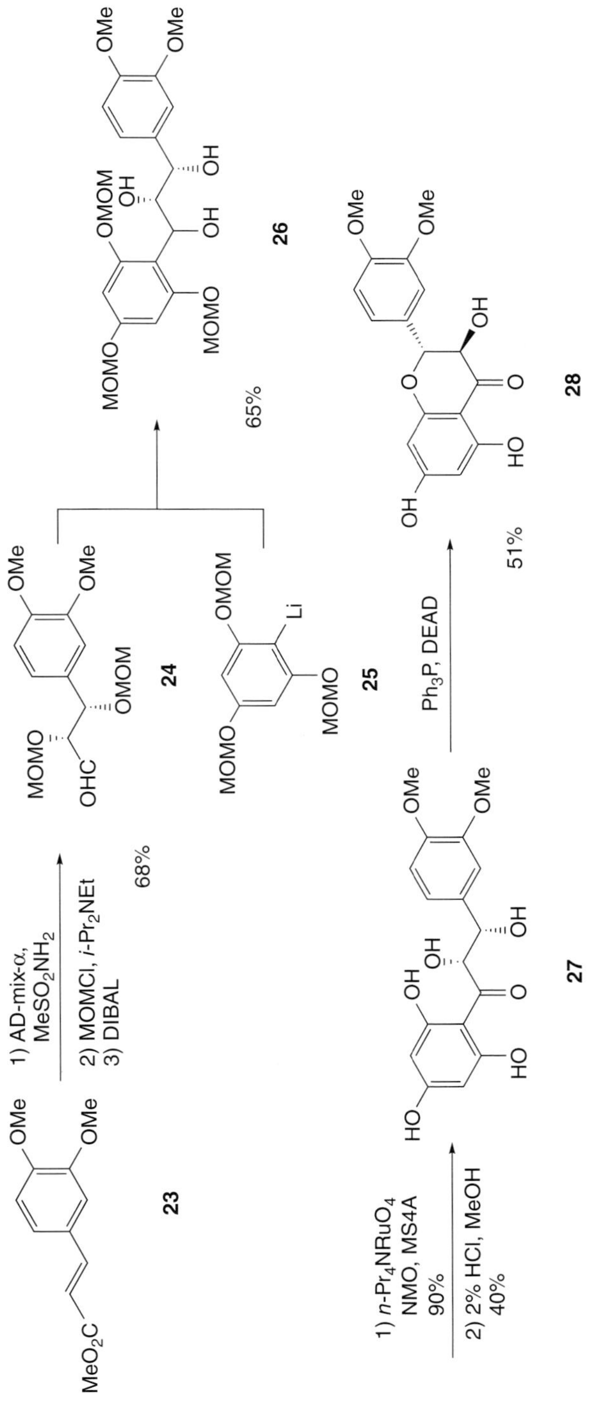


**Fig. 12.9** Pyran formation by intramolecular Mitsunobu reaction.

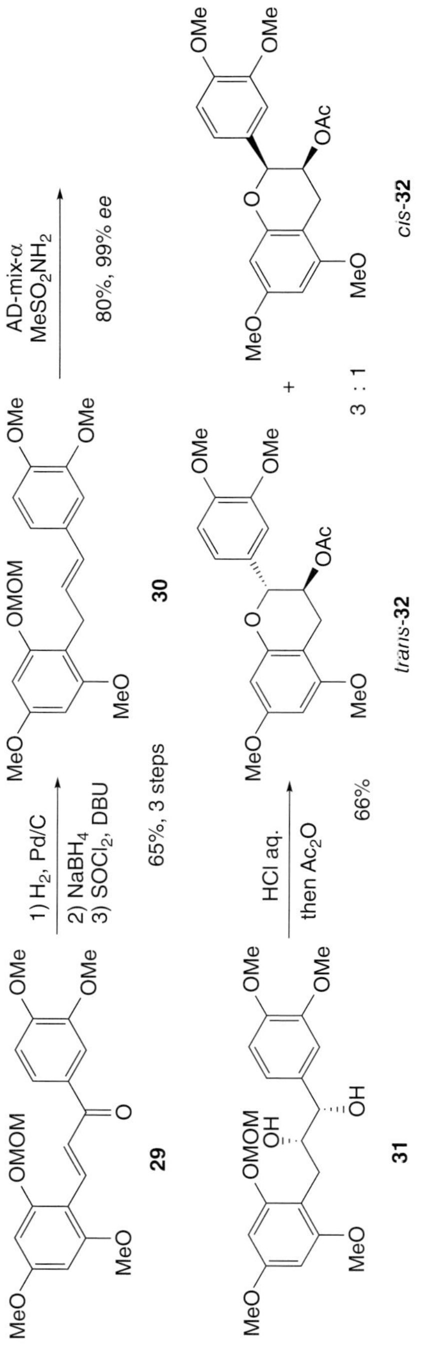


**Fig. 12.10** Acid-promoted cyclization of (*E*)-1,3-diarylpropenes.

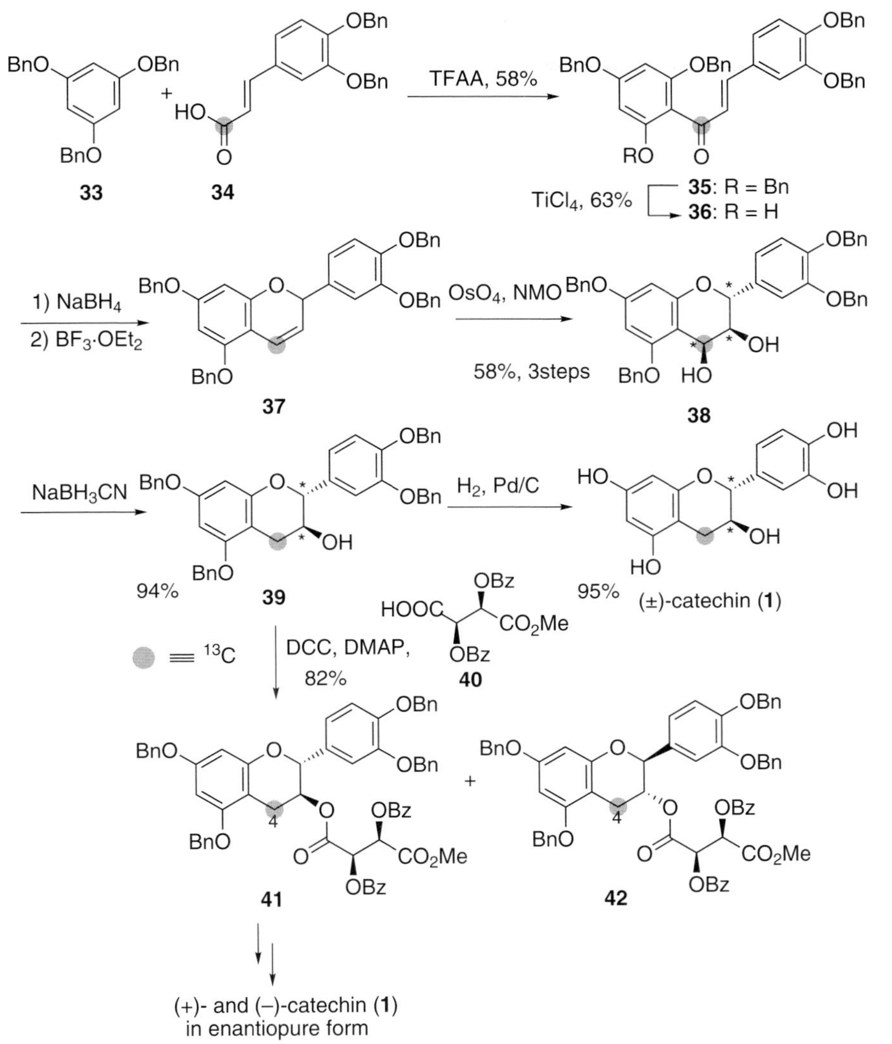


**Fig. 12.11** Synthesis and optical resolution of (+)- and (−)-catechin.

investigated by other research groups (for reviews, see Kozikowski & Tückmantel, 2008; Ferreira & Coleman, 2011; Ohmori & Suzuki, 2012). Selective deoxygenation at C(4) using $NaBH_3CN$ in the presence of acetic acid followed by hydrogenolysis of the benzylic protecting groups yielded racemic C(4)-labeled ($^{13}C$) catechin (**1**) in quantitative yield. Furthermore, optical resolution was carried out after condensation with the D-tartaric acid derivative **40** to form the diastereomeric esters **41** and **42**. Saponification and hydrogenolytic removal of all benzyl groups afforded enantiomerically pure (+)- and (−)-catechin (**1**).

Fig. 12.12 illustrates another excellent application of the Clark-Lewis method for the synthesis of gallocatechin-3-*O*-gallate derivative **47** (Zaveri, 2001). Condensation of aldehyde

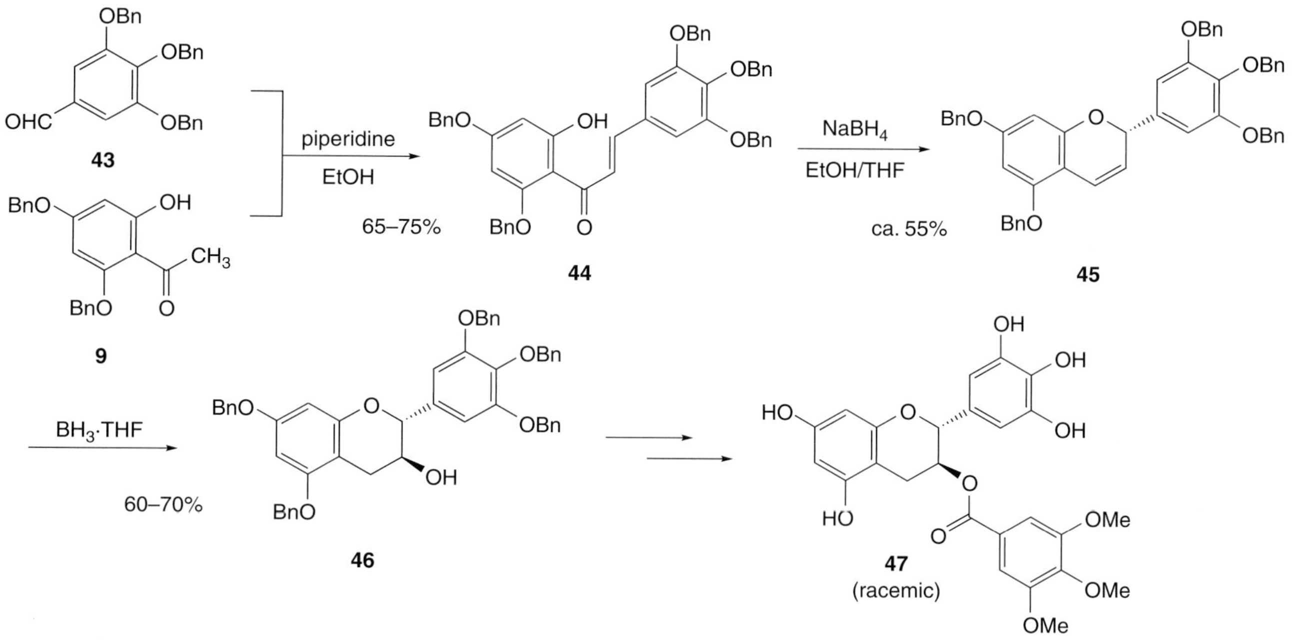


**Fig. 12.12** Stereoselective hydroboration route to the gallocatechin 3-*O*-gallate derivative **48**.

**43** and acetophenone **9** followed by $S_N2'$ cyclization gave chromene **45**, which was directly converted to flavan-3-ol **46** (cf. Clark-Lewis's approach) upon treatment with $BH_3$·THF followed by oxidative workup. The reaction proceeded with complete *trans*-selectivity.

Li & Chan (2001) developed a reliable synthetic method by which to access catechin-series derivatives (Fig. 12.13). Union of phloroglucinol derivative **48** with cinnamyl alcohol **49** was effected by Friedel–Crafts reaction, affording (*E*)-1,3-diarylpropene **50**, which was then subjected to Sharpless asymmetric dihydroxylation (**50** → **51**) followed by desilylation to give optically active diol **52**. Subsequent treatment with triethyl orthoformate (PPTS in $CH_2Cl_2$, r.t. → 60 °C) directly converted this intermediate, via orthoester **53**, to pyran **54** in high yield. The reaction stereospecifically proceeded via an $S_N2$ mechanism, securing the 2,3-*trans* relationship in **54**. Saponification by $K_2CO_3$/MeOH gave gallocatechin derivative **47**.

Jew *et al.* (2002) modified their synthetic strategy (cf. Fig. 12.9) to develop a concise synthetic route to free, optically pure catechin (**1**) (Fig. 12.14). Intermediate **58** was assembled from aldehyde **56** and aryllithium **57**. Barton deoxygenation followed by detachment of all MOM groups produced the cyclization precursor **59**. Treatment of **59** with $Ph_3P$ and DEAD gave free (+)-catechin (**1**) in 50% yield.

Krohn *et al.* (2009) also took advantage of the Sharpless/Mitsunobu sequence for the enantioselective synthesis of 2,3-*trans*-flavan-3-ols (Fig. 12.15). Asymmetric dihydroxylation of *trans*-alkene **60** using AD-mix-$\alpha$ afforded 1,2-diol **61** in excellent enantiomeric purity. Treatment of **61** with DEAD and $Ph_3P$ effected regioselective and stereoselective/specific cyclization at the benzylic position, giving 4′,7-di-*O*-benzyl guibourtinidol (**62**).

Hirooka *et al.* (2008) reported the formation of the (−)-5,7-dideoxy gallocatechin 3-*O*-gallate (**69**) in enantiopure form (Fig. 12.16). The Julia–Kocienski reaction of aldehyde **63** with sulfone **64** unexpectedly afforded *cis*-olefin **65**, although the origin of the *Z*-selectivity was not clear. The product, however, was easily isomerized to its *trans*-counterpart **66** by treatment with catalytic iodine. Subsequent Shi's asymmetric epoxidation (Wu *et al.*, 2002) of **66** gave epoxide **67** in 92% *ee*. After desilylation, exposure of the resulting epoxy alcohol to acidic conditions led to a 6-*endo* cyclization to give the requisite *trans*-dihydrobenzopyran **68** in 61% yield (two steps). In stark contrast, the *cis*-counterpart of **67** gave only a 1 : 1 mixture of *trans*/*cis* cyclization products. *Trans*-pyran **68** was further converted into (−)-5,7-dideoxy-gallocatechin 3-*O*-gallate (**69**).

A different approach to the construction of a catechin-series flavan skeleton was described by Ohmori and Suzuki (Higuchi *et al.*, 2006) (Fig. 12.17), which involved three critical steps: (i) union of two building blocks, iodophenol **I** and epoxy alcohol **II**, to produce the ether **III**, thus establishing the C–O bond prior to formation of the C–C bond with inversion of the C(2) stereogenic center; (ii) ring-opening of the oxirane to form the halohydrin **IV** in order to avoid an unfavorable 6-*endo* ring closure (Baldwin's rule) in the following step – pyran ring closure through halohydrin **IV** constitutes a more facile 6-*exo* cyclization; and (iii) intramolecular C–C bond formation to construct the pyran ring as in structure **VI**. The cyclization sequence comprises (i) iodine–metal exchange (I → *Met*) to generate the aryl anion species **V**, which takes place chemoselectively in the presence of the second, aliphatic halogen X and (ii) intramolecular $S_N2$ displacement to form the pyran ring.

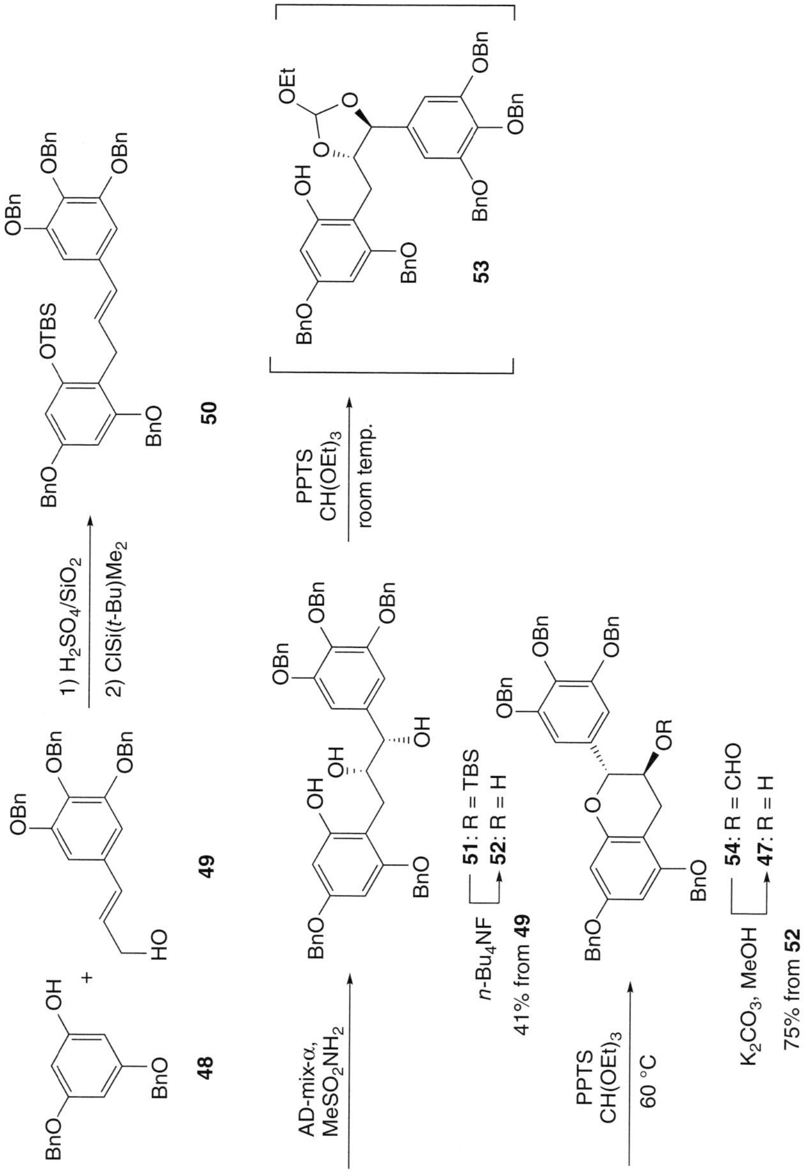


**Fig. 12.13** Synthesis of gallocatechin derivative **47** via the orthoester route.

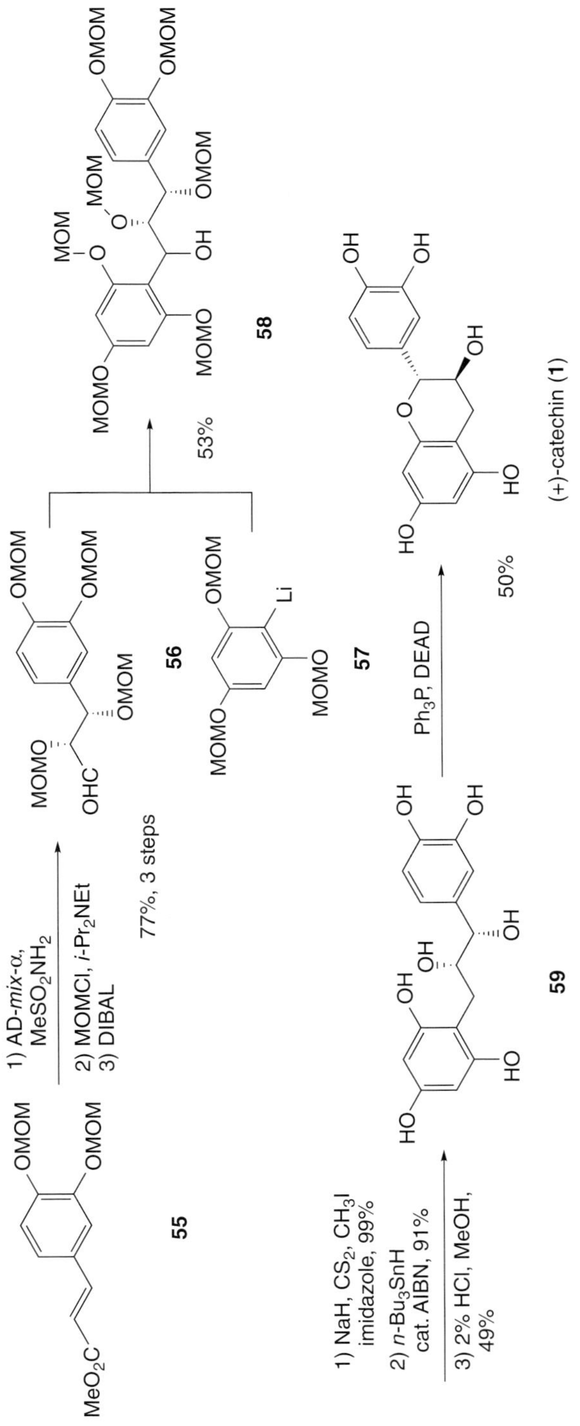


**Fig. 12.14** Intramolecular Mitsunobu strategy for catechin synthesis.

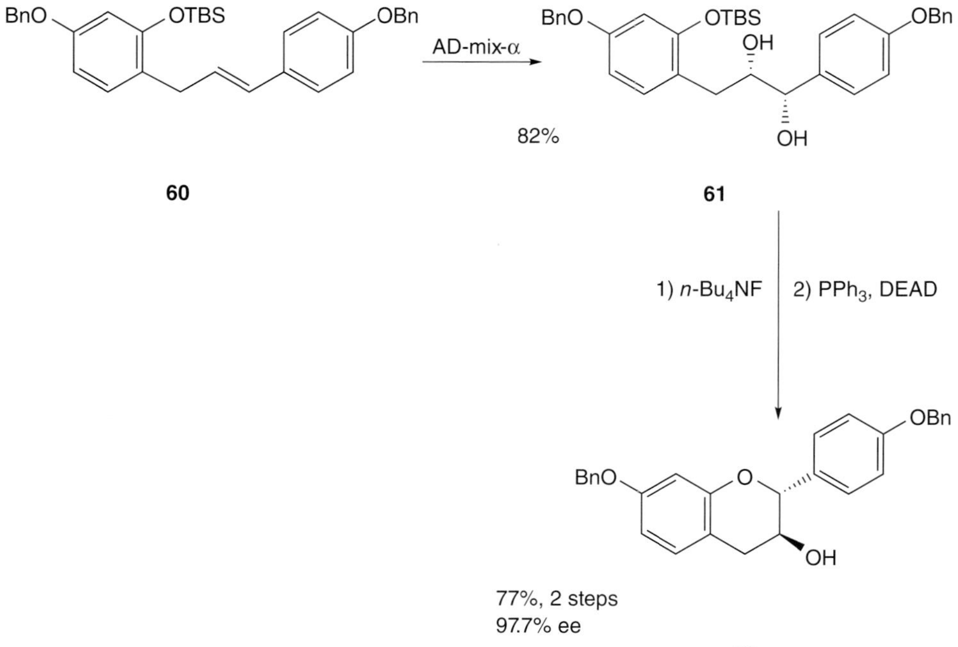


60 61 62

**Fig. 12.15** Concise synthesis of a guibourtinidol derivative via a Sharpless/Mitsunobu reaction sequence.

To implement this strategy, modified Mitsunobu reaction conditions (Tsunoda *et al.*, 1994) (Fig. 12.18) employing *n*-$Bu_3P$ and *N,N,N′,N′*-tetramethylazodicarboxamide (TMAD) were used to join epoxy alcohol **72** and phenol **71** (for preparation of these fragments, see Higuchi *et al.*, 2006) in order to obtain the aryl ether **73** in high yield with complete inversion at the C(2) benzylic position. Upon treatment of **73** with $Ph_3MgLi$ in the presence of HMPA, chemoselective iodine–lithium exchange and subsequent intramolecular $S_N2$ reaction smoothly proceeded to give the *trans*-dihydrobenzopyran **75**. Finally, stepwise removal of the silyl and all benzyl-protecting groups in **75** afforded (−)-gallocatechin (**76**) in good yield.

This synthetic approach serves as a reliable and facile method, enabling access to various catechin derivatives with 2,3-*trans* stereochemistry. However, the stereospecificity of the Mitsunobu reaction step (see Section 12.2) was a concern, as the stereochemical integrity could be lost if a competing $S_N1$ ionization at the C(2)-benzylic position took place. The stereospecificity of the reaction was investigated for various substrates (Ohmori *et al.*, 2009) (Fig. 12.19) and it was shown that the inversion/retention ratio was not linearly related to the number of B-ring oxy-functions. Thus, the extent of the competing $S_N1$ pathway relative to the number of oxy-group increases in the sequence $1 > 2 \gg 3 \simeq 0$.

Apparently, one needs to separate the inductive and the mesomeric effects. While the *para*-oxygen works to stabilize the benzylic cation, the *meta*-oxy groups are out of conjugation, leading to destabilization (−I effect). This accounts for the occurrence of the $S_N2$ pathway as the number of *meta*-oxygens increases. An additional factor, particularly

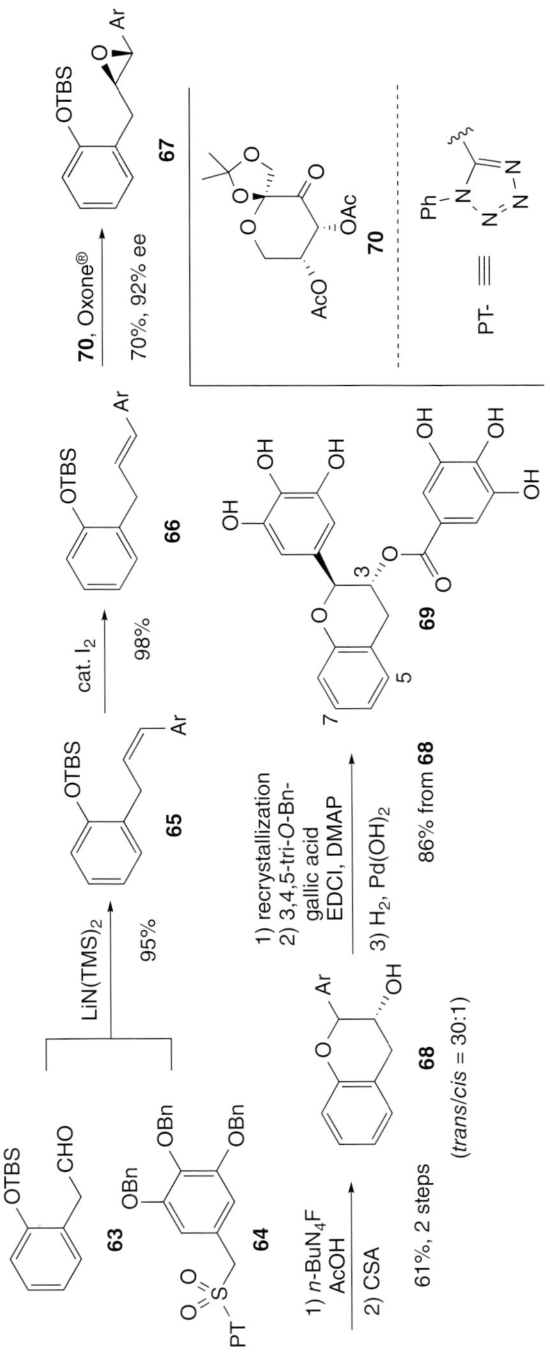


**Fig. 12.16** Construction of a *trans*-flavan-3-ol by Shi's asymmetric epoxidation and stereospecific pyran-ring closure.

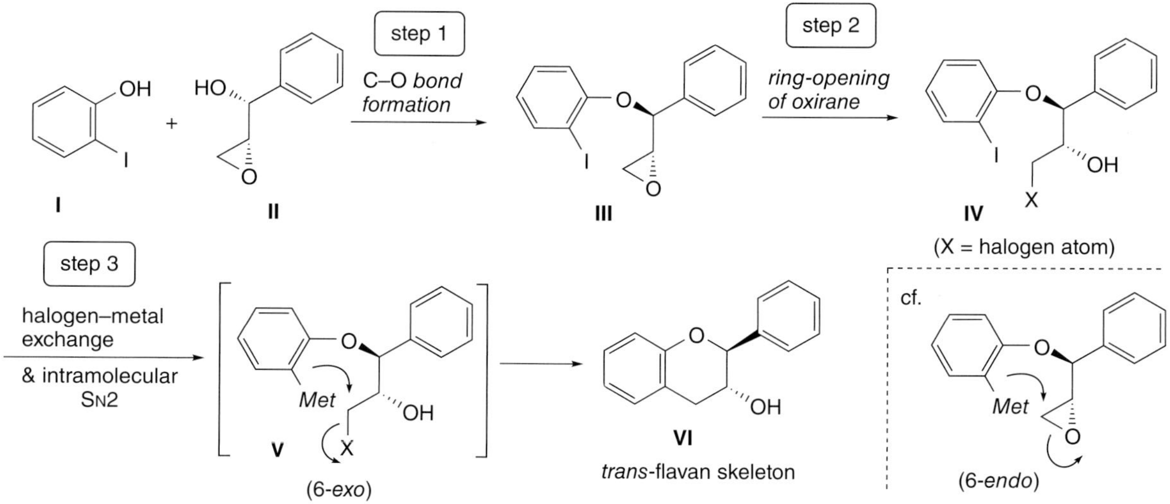


**Fig. 12.17** The Ohmori–Suzuki *C–O* tethering strategy for the construction of flavans with 2,3-*trans* stereochemistry.

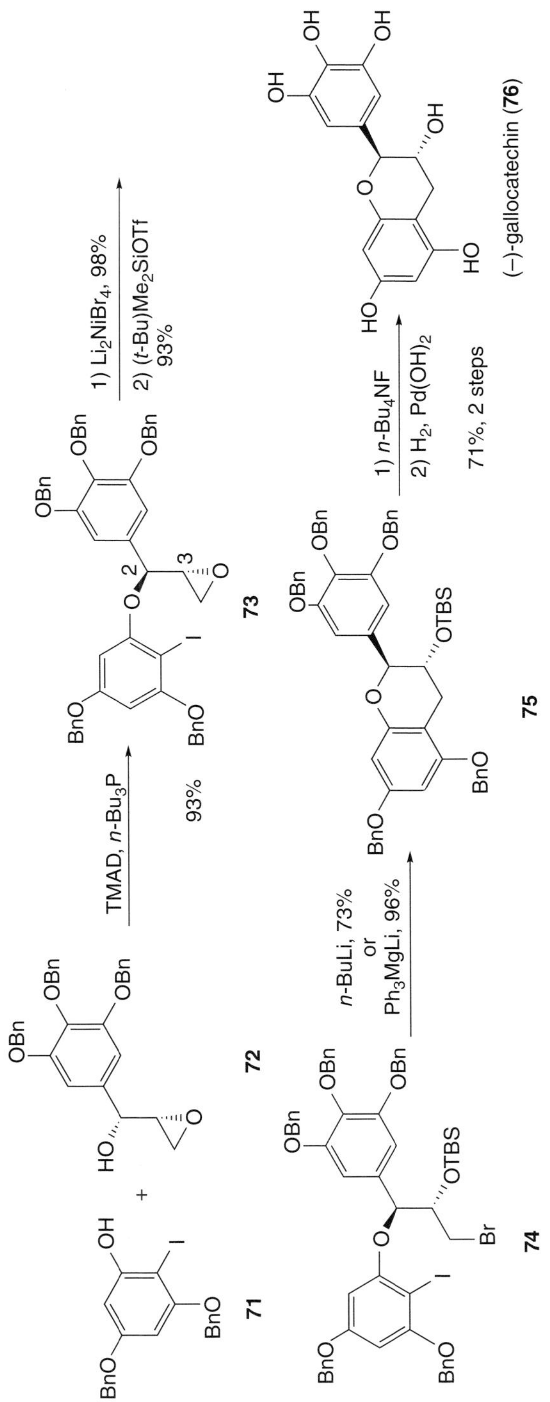


**Fig. 12.18** Application of the *C–O* tethering strategy to the synthesis of (−)-gallocatechin (**76**).

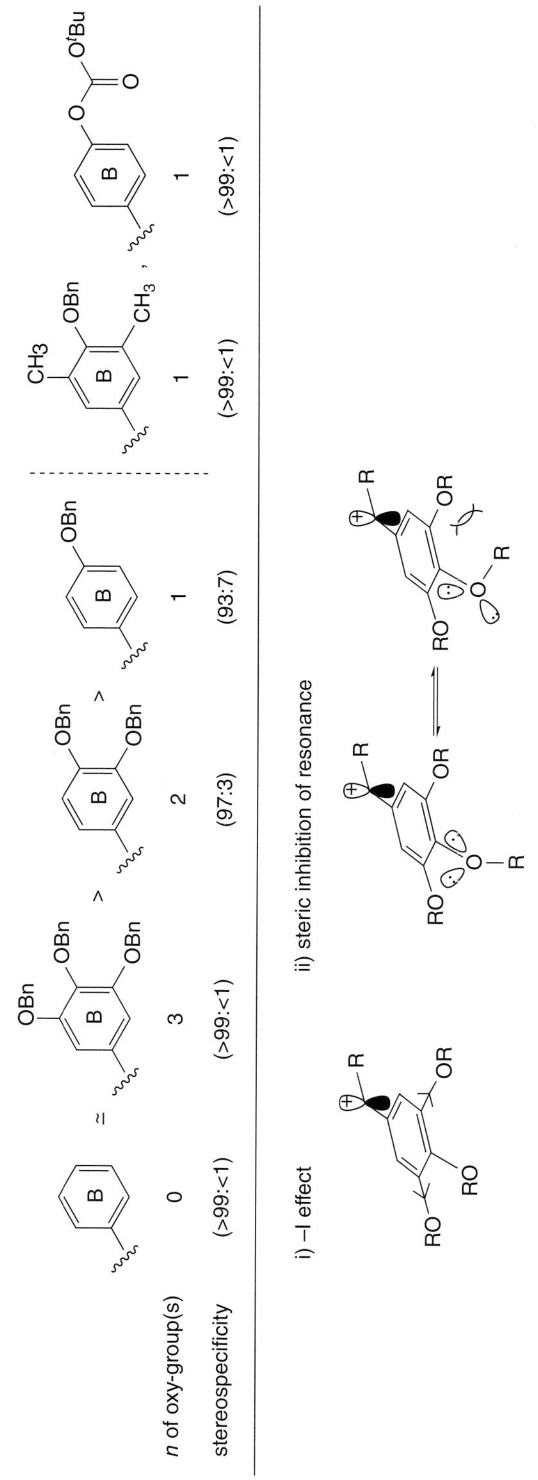


**Fig. 12.19** Stereosepecificity of the Mitsunobu condensation and its rationale.

for the 3,4,5-trioxy substrate, is the *steric inhibition of the resonance*. As experimental support for this idea, the 4-mono-oxy substrate bearing 3,5-dimethyl substituents underwent a clean $S_N2$ reaction. Furthermore, an electron-withdrawing protecting group at the 4-hydroxy group engendered complete inversion even at room temperature. In a further application of the C–O tethering synthetic strategy, a novel and promising protocol for pyran-ring formation was reported by Liu *et al.* (2008) (Fig. 12.20). The authors developed the thiourea-catalyzed cyclization reaction of aryl epoxides under Lewis-acidic conditions, thereby converting the epoxide precursor **III** (cf. Fig. 12.17) directly to the corresponding flavan derivative. Thus, reaction of phenol **77** with epoxyalcohol **78** under basic conditions followed by a two-step protocol (TsCl; $K_2CO_3$) gave the key cyclization precursor, epoxide **80**. Treatment of this compound with a catalytic amount of $AuCl_3$ (5 mol%), AgOTf (15 mol%), and thiourea **82** (6 mol%) allowed the smooth formation of the flavan skeleton via an intramolecular Friedel–Crafts reaction, affording **81** in good yield. As for the mechanistic aspect, C–H activation by the Au(III) catalyst to generate an arylgold (III) species was suggested (Shi & He, 2004). A cooperative effect of the Lewis acid and thiourea could be conceived to activate the oxirane ring, forming a betain species **G** and thereby allowing the cyclization to proceed in a 6-*exo-tet* fashion.

### *12.3.2 Synthesis of epi-series (= 2,3-*cis*) catechins*

Epi-series catechins, such as EC and EGCg, which are major ingredients in green tea, are attracting much attention for their significant bioactivities, including antioxidant, anticancer, and antiviral effects. However, the synthesis of these catechins remains a challenge, due to the difficulty in establishing the 2,3-*cis* stereochemistry, and because the C(2)-chiral center undergoes easy epimerization under acidic or basic conditions (see Section 12.2).

In addressing this stereochemical issue, a remarkable contribution was reported by Tückmantel *et al.* (1999), who developed a facile protocol to convert a 2,3-*trans*-flavan into its 2,3-*cis*-counterpart via successive oxidation/reduction at the C(3)-position (Fig. 12.21). This method proved applicable for the preparation of various epi-series catechins (Li & Chan, 2001; Zaveri, 2001; Wan *et al.*, 2006; Furuta *et al.*, 2007; see also Sharma *et al.*, 2007). However, the synthesis necessarily involves additional steps, calling for a more direct method.

Aiming at a direct inversion at the C(3) stereogenic center, the Mitsunobu reaction of compound **83** with *p*-nitrobenzoic acid was attempted, but, unfortunately, led to a facile E2 elimination, giving olefin **84** in 87% yield (Tückmantel *et al.*, 1999) (Fig. 12.22). However, further research by Ding *et al.* (2006) revealed that the Mitsunobu reaction of **86** with 3,4,5-tri-*O*-benzylgallic acid proceeds well to give the corresponding 3-*O*-gallate **87** in moderate yield with complete inversion at the C(3). This method may be of merit in the synthesis of the 3-*O*-gallates of epi-series congeners, although further optimization and clarification of its scope and limitations appear necessary.

Li and coworkers returned to their orthoester approach (see Fig. 12.13) to develop a novel pathway to epi-series catechins, based on a double-inversion strategy (Li & Chan, 2001) (Fig. 12.23). When orthoformate **53** (Ar = phenyl), prepared as described before, was treated with acetyl bromide, the bromine atom was incorporated at the C(2)-position

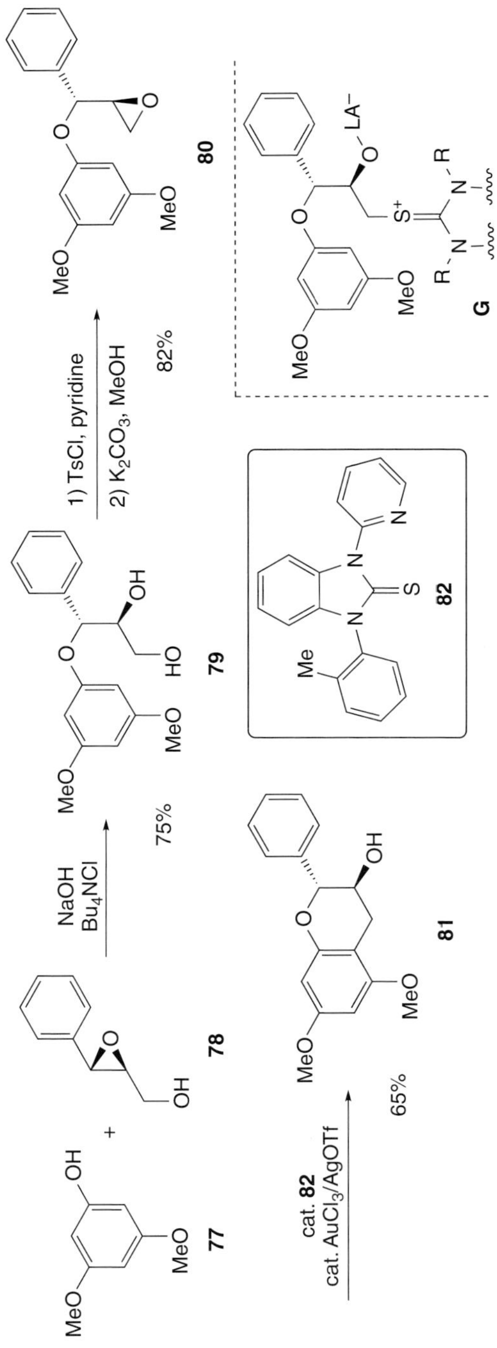


**Fig. 12.20** Gold/silver/urea co-catalyzed pyran-ring formation from a tethered epoxide.

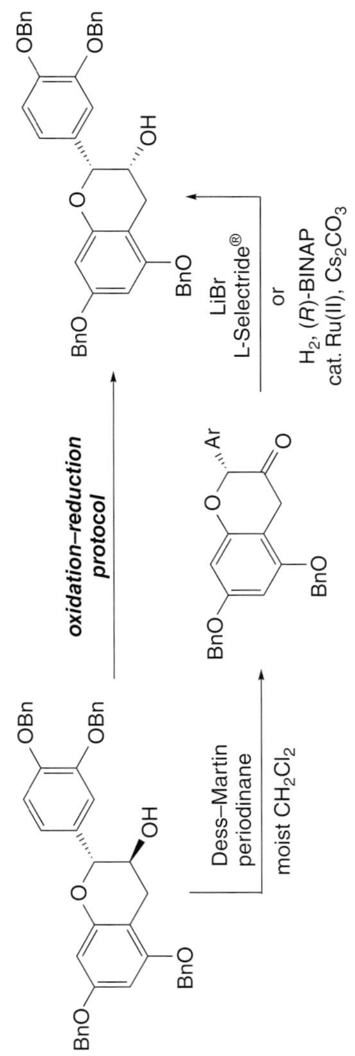


**Fig. 12.21** Kozikowski–Tückmantel oxidation/reduction protocol for the conversion of 2,3-*trans*- to 2,3-*cis*-flavans.

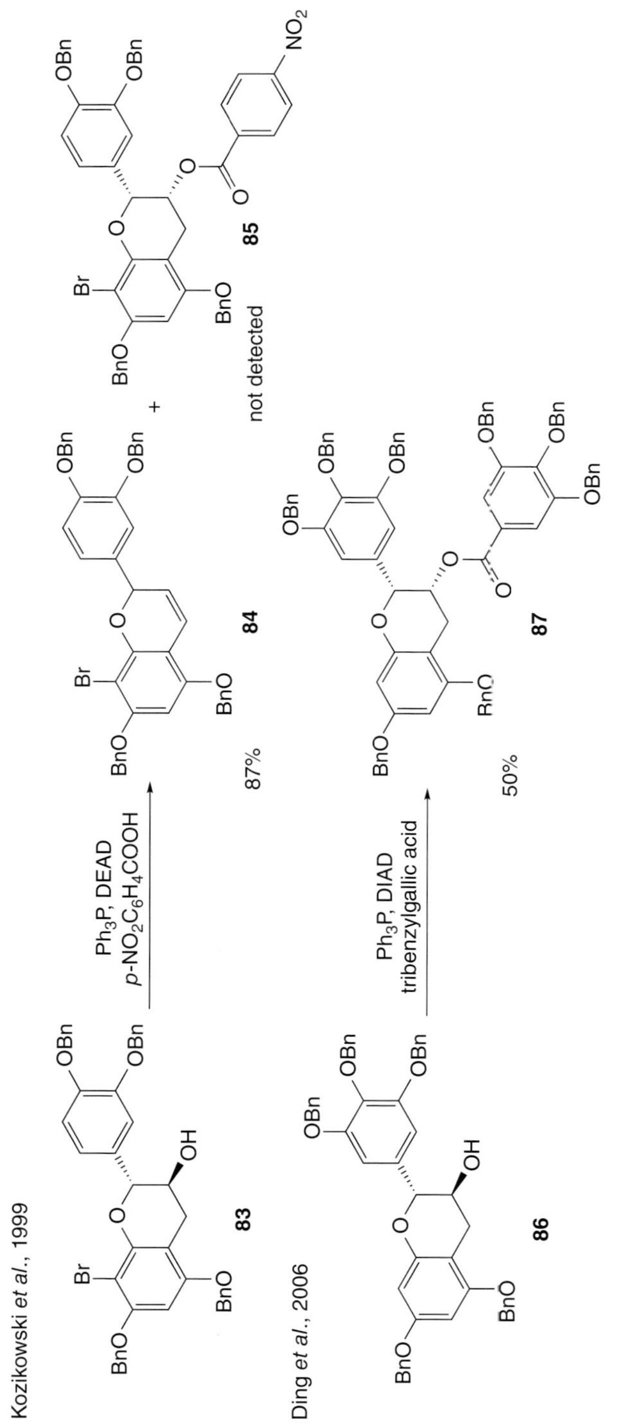


**Fig. 12.22** Mitsunobu reaction of 2,3-*trans*-flavan-3-ols. (Adapted from Tückmantel *et al.*, 1999 and Ding *et al.*, 2006.)

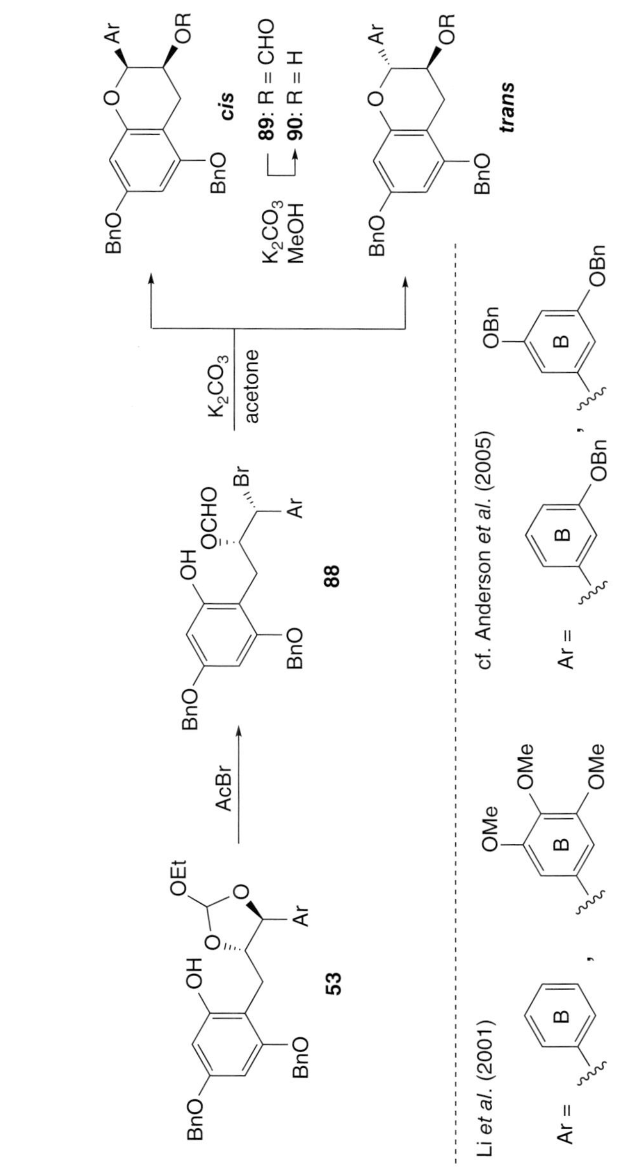


**Fig. 12.23** Orthoester route to 2,3-*cis*-flavans. (Adapted from Li and Chan, 2001 and Anderson *et al.*, 2005.)

(flavan numbering) with complete inversion, giving the benzylic bromide **88** together with a small amount of the regioisomeric homobenzylic bromide. Without purification, the crude compound **88** was cyclized by treatment with potassium carbonate in acetone to afford the benzopyrans *cis*-**89** and *trans*-**89**. After removal of the formyl group, the flavan-3-ols *cis*-**90** and *trans*-**90** were obtained together in a ratio of 9 : 1. When the same reaction sequence was repeated with the 3,4,5-trimethoxyphenyl derivative, cyclization followed by removal of the formyl group gave a diastereomeric mixture in a ratio of 1 : 4 in favor of the *trans* isomer. These results suggest that the *p*-oxy-group on the B ring is critical in the sense that *para*-oxygenated substrates, such as the 3,4,5-trimethoxy derivative, lose their stereochemical integrity, while the *meta*-oxy function on the B ring does not diminish the stereospecificity (Anderson *et al.*, 2005). On the other hand, a *meta*-oxy substituent is rather effective in lowering the $S_N1$ reactivity at the C(2)-benzylic position, presumably due to its −I effect. Thus, the 3,5-dioxy congener, lacking the 4-oxy function, showed complete stereospecificity to give the corresponding 2,3-*cis*-flavan in high yield (Anderson *et al.*, 2005). Despite this progress, a more general methodology for the synthesis of the epi-series catechins was still needed.

In 2006, Kitade *et al.* (2006) reported an efficient way to establish the 2,3-*cis* stereochemistry in a flavan skeleton (Fig. 12.24). Hydrosilylation of enone **94**, which was prepared by aldol condensation of aldehyde **92** and methyl ketone **93**, provided enol silyl ether **95**, which was subsequently oxidized by dimethyldioxirane to give $\alpha$-hydroxy ketone **96**. After esterification with the gallic acid derivative **97**, the resulting ester **98** was treated with $Et_3SiH$ in the presence of TFA to effect removal of the 2-(trimethylsilyl)ethoxymethyl (SEM) group, followed by a stereoselective reductive cyclization, providing *cis*-dihydrobenzopyran **99**. A plausible reaction mechanism involves intermediate **H** through neighboring-group participation of the gallate moiety, which shields the $\alpha$-face of the cationic center in **H**.

This study was extended to a solid-phase synthesis (Tanaka *et al.*, 2007) (Fig. 12.25). The aldol condensation of polymer-supported aldehyde **101** with 3,4,5-tris-(benzyloxy)phenyl methyl ketone followed by oxidation gave epoxide **102**. $Zn(OTf)_2$-mediated addition of a thiol followed by esterification with protected gallic acid gave $\alpha$-acyloxy ketone **103**. In the next step, direct conversion of **103** to **104** with trifluoroacetic acid and $Et_3SiH$ was noteworthy for the multiple events that successively occurred: (i) liberation of the substrate from the resin, where the bromo-protection at C(8) prevented Friedel–Crafts alkylation of the cleaved *para*-alkyloxybenzyl linker; (ii) departure of the benzyl thiolate and reduction; (iii) removal of bromide; and (iv) reductive etherification, which resulted in the formation of the flavan skeleton with a 2,3-*cis* relative stereochemistry. This method proved highly efficacious for the preparation of a chemical library of epi-series catechins, albeit in racemic form (Tanaka *et al.*, 2010).

Recently, the same research group reported a divergent approach to *cis*- and *trans*-flavans (Tanaka *et al.*, 2012). Union of phenol **105** and oxirane **106** was a branching point for stereoselective access to each catechin series (Fig. 12.26). The ring-opening reaction of oxirane **106** (racemic and diastereomer mixture) with phenol **105** under the basic conditions ($K_2CO_3$, $CH_2Cl_2$, 45 °C) proceeded via $S_N2$ reaction with complete inversion at C(2), forming the *anti*-product **107**. In contrast, heating (reflux) of the mixture of **105** and **106** in $CH_2Cl_2$ *without* $K_2CO_3$ cleanly provided the *syn*-product **108**. This stereochemical

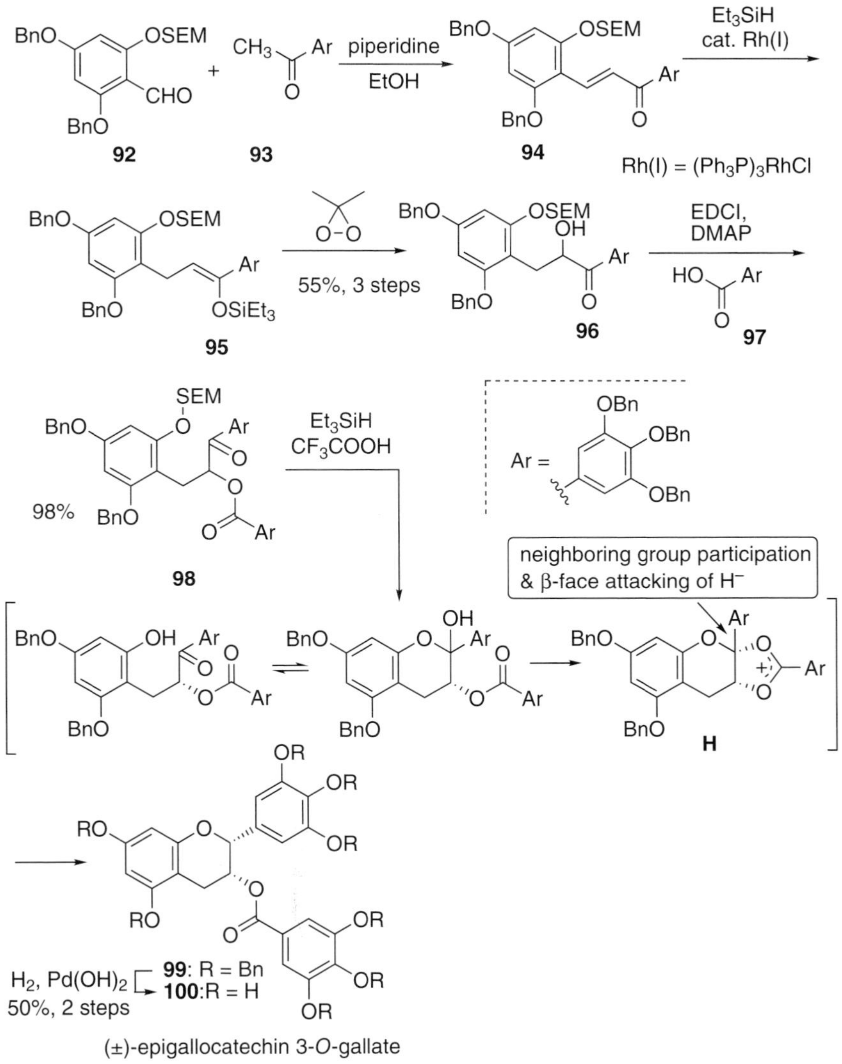


**Fig. 12.24** Stereoselective formation of *cis*-flavans via neighboring-group participation of a C(3) galloyl group.

changeover might be rationalized by the activation of the oxirane ring by the proton source, alcohol **105**, resulting in the internal delivery of the phenoxy group of **105** into the C(2)-benzylic position with retentive manner. After esterification, the Pummerer-type activation of the *S*,*O*-acetal moiety in **107** or **108** by trifluoroacetic anhydride (TFAA) in the presence of $Et_3SiH$ allowed the intramolecular electrophilic aromatic substitution to form the corresponding flavan skeleton with *cis*- (from **107**) and *trans*- (from **108**) stereochemistry.

Yoshida *et al.* (2011) concisely synthesized an EGCg derivative (Fig. 12.27), which served as a molecular probe. After condensation of A-ring sulfone **109** with the B-ring aldehyde by a Julia–Kocienski reaction, Suzuki–Miyaura coupling of the resulting

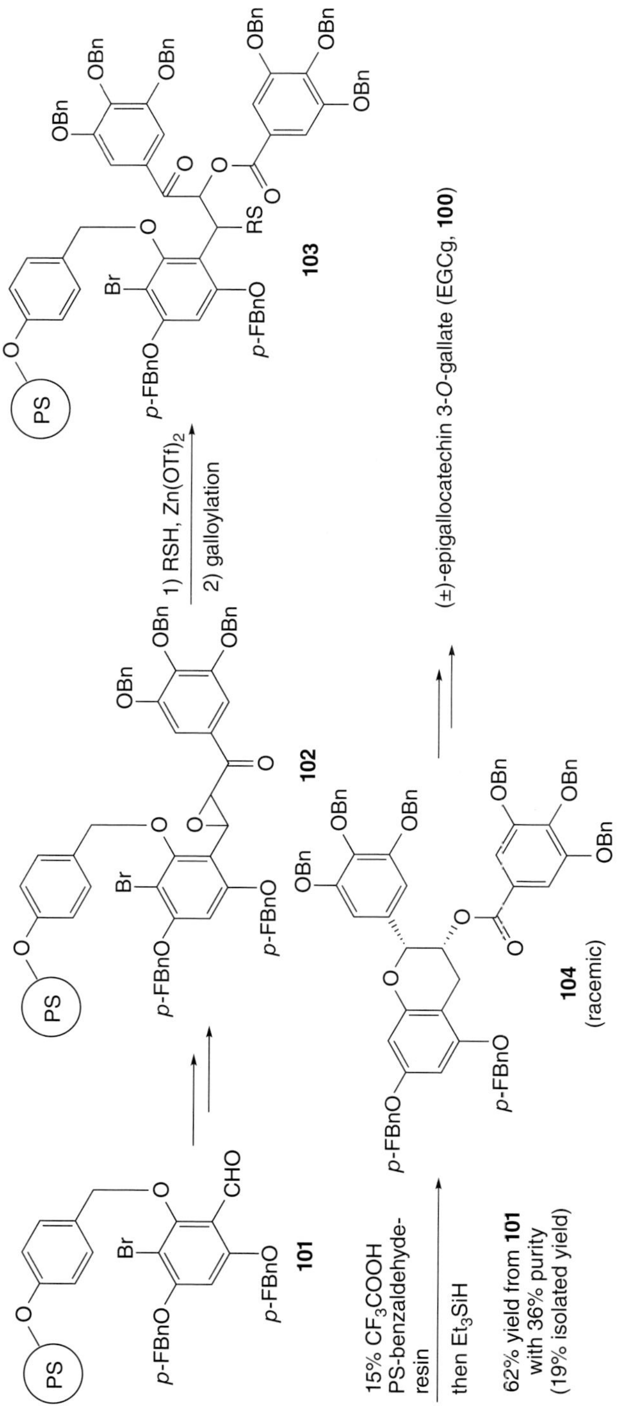


**Fig. 12.25** Polymer-supported synthesis of epi-series catechins.

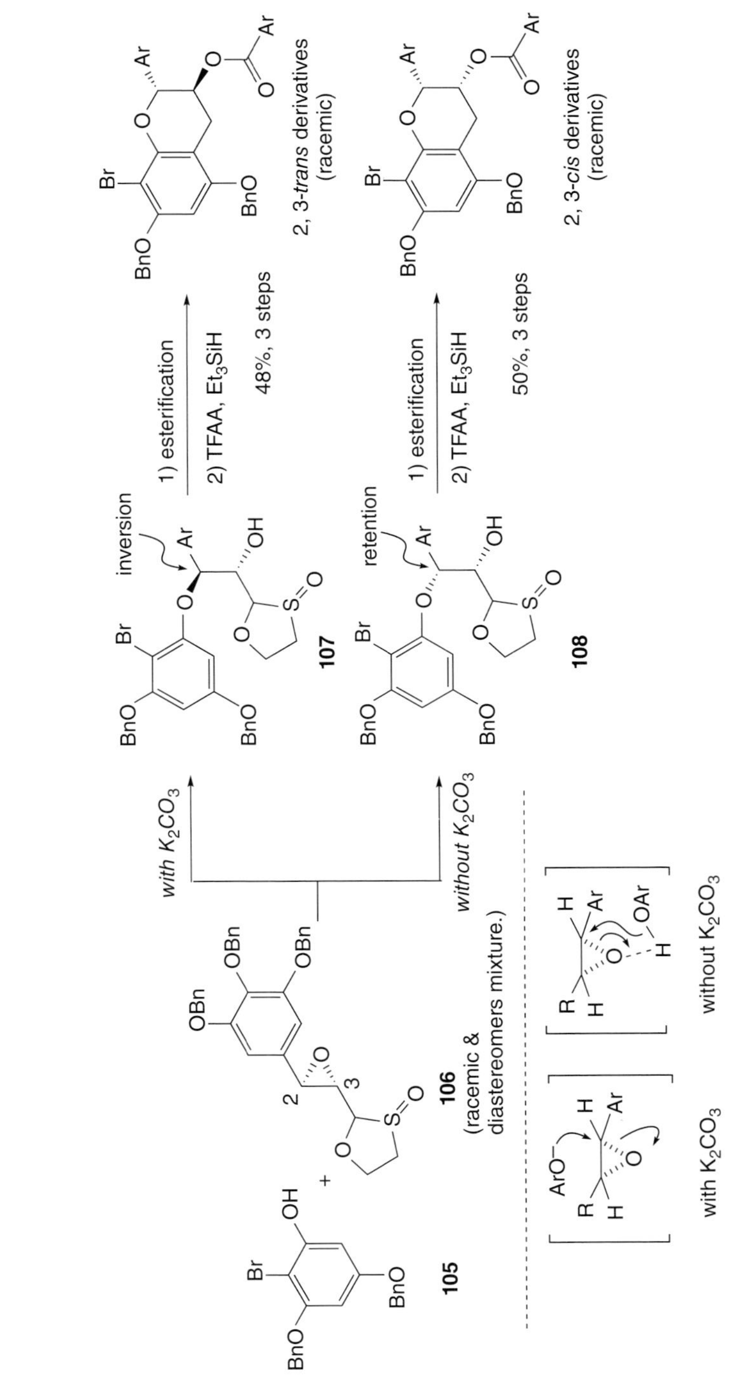


**Fig. 12.26** Change in the stereochemical course of oxirane ring-opening *en route* to 2,3-*trans*/*cis* flavans.

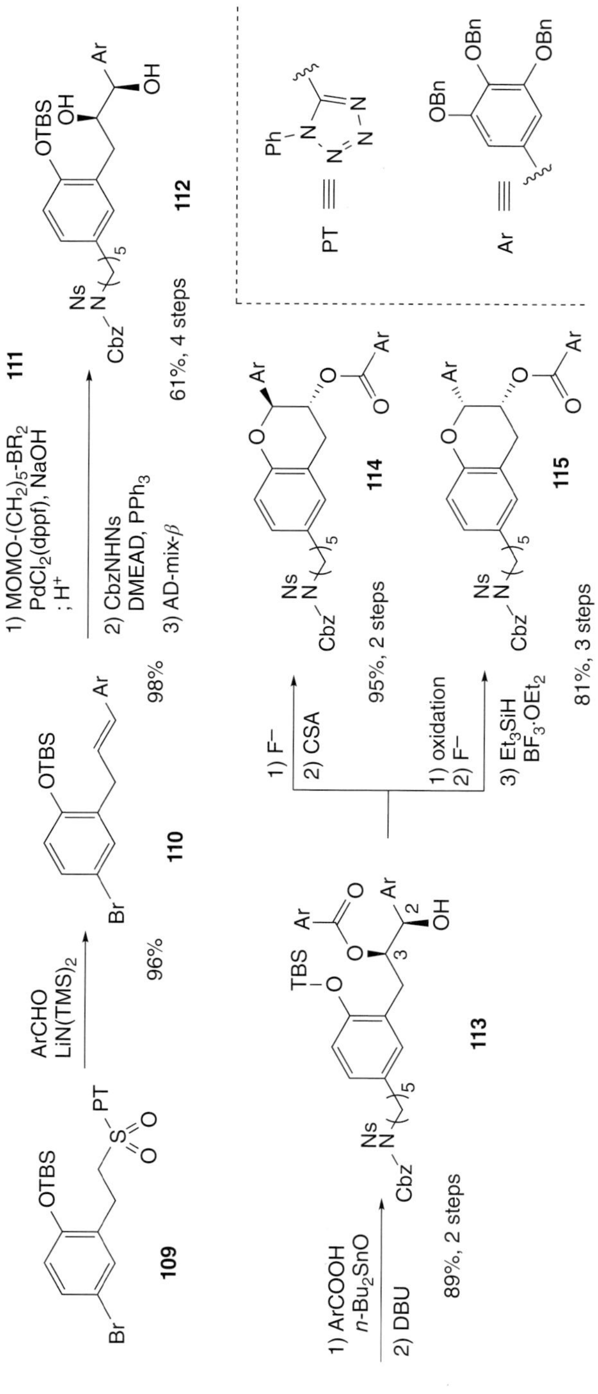


**Fig. 12.27** Synthesis of EGCg derivatives for use as molecular probes.

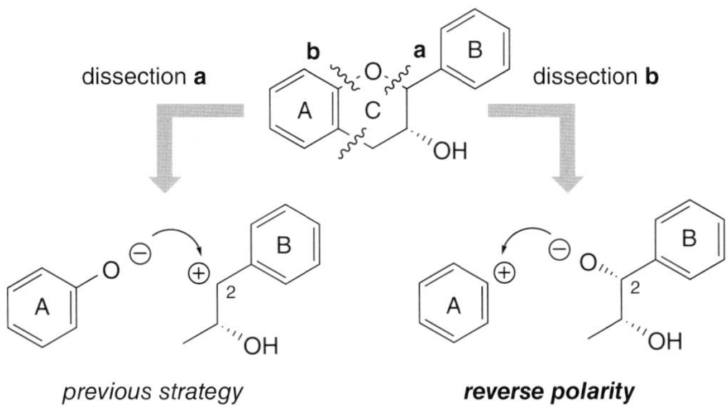


**Fig. 12.28** Dissection of the flavan skeleton and analysis of fragment polarities.

*trans*-olefin **110** with borane **111** attached a side chain to the A ring. Subsequent Mitsunobu reaction with the sulfonamide CbzNHNs introduced a nitrogen function at the terminal position of the side chain. Sharpless asymmetric dihydroxylation led to diol **112**, whose substituent treatment with protected gallic acid in the presence of *n*-$Bu_2SnO$ and DBU gave ester **113**. The galloyl group was initially introduced on to the C(2) hydroxy group and then migrated to the sterically less hindered C(3) hydroxy group. The resulting benzyl alcohol **113** cyclized upon treatment with CSA to provide 2,3-*trans*-flavan **114**, which is the thermodynamically stable diastereomer. For the synthesis of the *cis*-isomer **115**, the alcohol **113** was oxidized to the corresponding ketone, which was then treated with $Et_3SiH$ and $BF_3{\cdot}OEt_2$ to effect the reductive cyclization via an intermediate lactol to form the 2,3-*cis*-flavan **115** (cf. Fig. 12.24).

Recently, the author's group reported a new C–O bond-tethering approach (cf. Fig. 12.18), based on the reverse-polarity strategy, that enables access to epi-series derivatives (Ohmori *et al.*, 2010). The key point is the change in the connection mode of the A- and B-ring fragments (Fig. 12.28): instead of dissecting the C ring via path **a**, disconnection at the Ar–O and Ar–C bonds via path **b** provides a new fragment pair, which can be combined in a reversed polarity (umpolung) fashion without affecting the C(2)-stereochemistry.

To realize this scenario, an aromatic nucleophilic substitution ($S_NAr$) was envisaged to form *syn*-**III** by employing electrophilic unit **I** and the alkoxide derived from *syn*-**II** (Fig. 12.29). The choice of the two substituents X and Y was critical. The Y group had to be a strongly electronegative group such as fluorine, capable of attracting the incoming nucleophile to form the Meisenheimer complex **G**. On the other hand, the X group needed to play two roles: (i) that of a $\pi$-electron acceptor, to facilitate the alkoxide's conjugate attack; and (ii) that of a carbanion precursor at the stage of the pyran-ring closure later in the synthetic scheme. With these requirements in mind, a sulfinyl group was chosen as X, with the expectation that a sulfinyl–metal exchange would serve for the latter purpose.

Fig. 12.30 illustrates the execution of this concept. Lithiation of 1,3,5-trifluorobenzene (**116**) with *n*-BuLi followed by trapping of the resulting anion with $PhSSO_2Ph$ gave the aryl sulfide in 87% yield. Upon careful treatment of this sulfide with sodium benzyloxide,

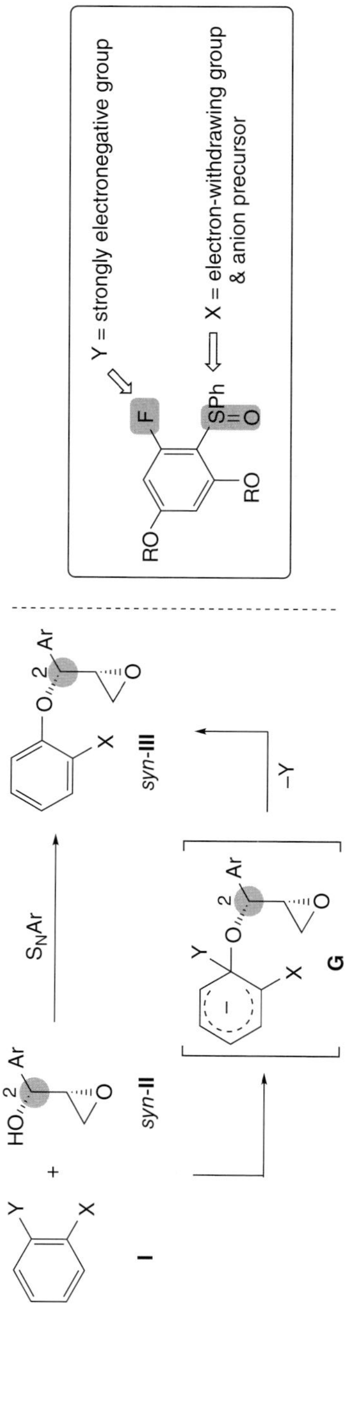


**Fig. 12.29** Realization of the reverse polarity strategy by $S_NAr$ reaction.

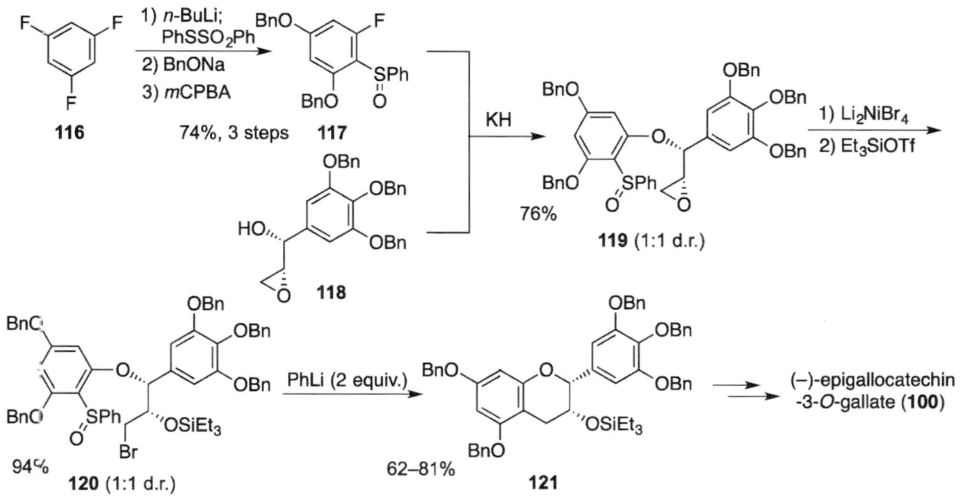


**Fig. 12.30** Synthesis of (−)-epigallocatechin 3-*O*-gallate (**100**) based on the reverse-polarity strategy.

two fluorine atoms were displaced with high regioselectivity, to obtain the bisbenzyl ether **117**. The $S_N$Ar reaction proceeded upon treatment of **117** with the sodium alkoxide of **118** in a mixed solvent (toluene/*N,N'*-dimethylpropylene urea = 4 : 1) at 0 °C. Notably, the Payne rearrangement of **118** was not observed under these conditions, giving the desired $S_N$Ar-product **119** in high yield as a mixture of diastereomers. Separation by column chromatography on silica gel gave the diastereomeric epoxy ethers **119** (ca. 1 : 1 d.r.) in 76% yield, whose relative configuration at the sulfoxide moiety was not established. Oxirane cleavage of **119** (diastereomeric mixture) with $Li_2NiBr_4$ followed by silylation with triethylsilyl triflate afforded the bromide **120** as a mixture of diastereomers (ca. 1 : 1 d.r., 94% yield).

The next step was to construct the flavan skeleton by cyclization of **120** via sulfinyl–metal exchange followed by an intramolecular nucleophilic substitution. Upon treatment of sulfoxide **120** (diastereomeric mixture) with 2.0 equiv. of PhLi, this reaction took place smoothly, giving the single cyclized product **121** (62–81%). It is noteworthy that the aryllithium species generated by the sulfinyl–metal exchange could be exploited in such an efficient C–C bond-forming process, a finding that should be of sizable potential for other purposes. The cyclized product **121** was uneventfully converted into (−)-epigallocatechin 3-*O*-gallate (**100**) in two steps.

Recently, a more concise synthetic route to epi-series catechins was developed by the authors (Stadlbauer *et al.*, 2012) (Fig. 12.31). This synthesis commenced with the aryl fluoride **122**, derived from 1,3,5-trifluorobenzene (**116**). Regioselective lithiation was accomplished by treatment of **122** with *n*-BuLi, and subsequent reaction with epoxide **123** in the presence of $BF_3 \cdot OEt_2$ gave the adduct **124** in 88% yield. After protection of the free hydroxy group as a methoxyethoxymethyl (MEM) ether and removal of the TBS group, treatment of the resulting alcohol with KH resulted in the desired $S_N$Ar reaction, and the 2,3-*cis* flavan derivative **126** was obtained in high yield (88%, three steps). Finally, stepwise deprotection of the MEM and Bn groups provided (−)-epicatechin (**2**) in 87% yield (two steps).

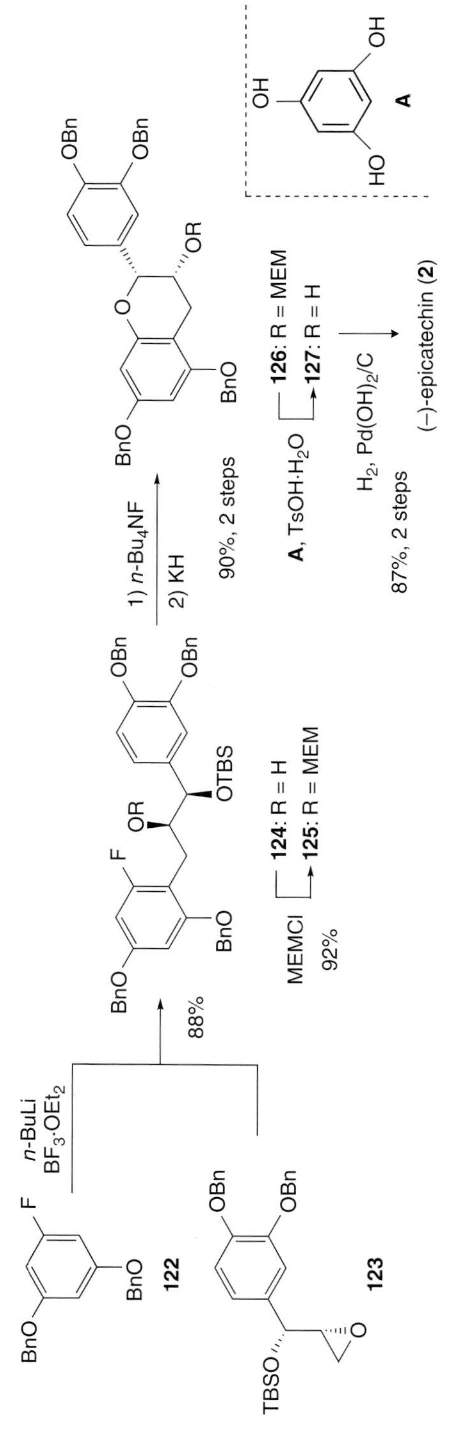


**Fig. 12.31** Concise synthesis of (−)-epicatechin (**2**) via the intramolecular $S_NAr$ approach.

## 12.4 Conclusions

We have reviewed some methods for the stereoselective synthesis of catechin and related flavonoids, which we hope will provide ideas for further developments in this area, and might also stimulate further interest in the synthesis of more complex derivatives, such as oligomeric and polymeric polyphenols. Apparently, such investigations have been limited by the scarce availability of their monomeric constituents, except for (+)-catechin. Moreover, progress in the area needs a reliable synthetic method capable of providing oligomeric catechins with rigorous control of their stereo- and regiochemistry, as well as their degree of oligomerization.

## Abbreviations

| | |
|---|---|
| Ac | acetyl |
| AIBN | 2,2′-azobisisobutyronitrile |
| BINAP | 2,2′-bis(diphenylphosphino)-1,1′-binahptyl |
| Bn | benzyl |
| Bz | benzoyl |
| Cbz | benzyloxycarbonyl |
| *m*CPBA | 3-chloroperoxybenzoic acid |
| CSA | 10-camphorsulfonic acid |
| DBU | 1,8-diazabicyclo[5.4.0]undec-7-ene |
| DCC | *N,N*′-dicyclohexylcarbodiimide |
| DIAD | diisopropyl azodicarboxylate |
| DIBAL | diisobutylaluminum hydride |
| DEAD | diethyl azodicarboxylate |
| DMAP | 4-(dimethylamino)pyridine |
| DMEAD | di-2-methoxyethyl azodicarboxylate |
| DMF | *N,N*-dimethylformamide |
| dppf | 1,1′-bis(diphenylphosphino)ferrocene |
| EDCI | 1-ethyl-3-(3-dimethylaminopropyl)carbodiimide hydrochloride |
| *p*-FBn | 4-fluorobenzyl |
| HMPA | hexamethylphosphoramide |
| L-Selectride | lithium tri-*sec*-butylborohydride |
| MEM | 2-methoxyethoxymethyl |
| MOM | methoxymethyl |
| MS4A | molecular sieves 4Å |
| NMO | *N*-methylmorpholine *N*-oxide |
| Ns | 4-nitrophenylsulfonyl |
| Oxone | potassium peroxomonosulfate |
| PPTS | pyridinium *p*-toluenesulfonate |
| pyr | pyridine |
| SEM | 2-(trimethylsilyl)ethoxymethyl |

| | |
|---|---|
| Tf | trifluoromethanesulfonyl |
| TBS | *tert*-butyldimethylsilyl |
| TFAA | trifluoroacetic anhydride |
| TMAD | *N,N,N′,N′*-tetramethylazodicarboxamide |
| TMS | trimethylsilyl |
| Ts | *p*-toluenesulfonyl |
| THF | tetrahydrofuran |

## Acknowledgments

Some of the authors' works listed in this review were supported by *Grant-in-Aid for Specially Promoted Research* (No. 23000006) from JSPS, which is gratefully acknowledged.

## References

Anderson, J.C., Headley, C., Stapleton, P.D. & Taylor, P. W. (2005) Asymmetric total synthesis of B-ring modified (−)-epicatechin gallate analogues and their modulation of $\beta$-lactam resistance in Staphylococcus aureus. *Tetrahedron*, 61, 7703–7711.

Augustyn, J.A.N., Bezuidenhoudt, B.C.B. & Ferreira, D. (1990) Enantioselective synthesis of flavonoids. Part I. Polyoxygenated chalcone epoxides. *Tetrahedron*, 46, 2651–2660.

Clark-Lewis, J.W. & Skingle, D.C. (1967) Flavan derivatives XVIII. Synthesis of hemiketals containing the peltogynol ring system: 3(3′)-hydroxyisochromano(4′,3′:2,3)chromans. Conversion of 2′-hydroxychalcones into flav-3-enes and its biosynthetic implications. *Australian Journal of Chemistry*, 20, 2169–2190.

Ding, T.-J., Wang, X.-L. & Cao, X.-P. (2006) A fast assembly of (−)-epigallocatechin-3-gallate [(−)-EGCG] via intra- and inter-molecular Mitsunobu reaction. *Chinese Journal of Chemistry*, 24, 1618–1624.

Ferreira, D. & Coleman, C.M. (2011) Towards the synthesis of proanthocyanidins: half a century of innovation. *Planta Medica*, 77, 1071–1085.

Freudenberg, K. & Weinges, K. (1958) Leuco- und pseudoverbindungen der anthocyanidine. *Justus Liebigs Annalen der Chemie*, 613, 61–75.

Furuta, T., Hirooka, Y., Abe, A., *et al.* (2007) Concise synthesis of dideoxy-epigallocatechin gallate (DO-EGCG) and evaluation of its anti-influenza virus activity. *Bioorganic & Medicinal Chemistry Letters*, 17, 3095–3098.

Higuchi, T., Ohmori, K. & Suzuki, K. (2006) General and convenient approach to flavan-3-ols: Stereoselective synthesis of (−)-gallocatechin. *Chemistry Letters*, 35, 1006–1007.

Hirooka, Y., Nitta, M., Furuta, T. & Kan, T. (2008) Efficient synthesis of optically active gallocatechin-3-gallate derivatives via 6-endo-cyclization. *Synlett*, 3234–3238.

Jew, S.-S., Kim, H.-A., Bae, S.-Y., *et al.* (2000) Enantioselective synthetic method for 3-hydroxyflavanones: an approach to (2*R*,3*R*)-3′,4′-O-dimethyltaxifolin. *Tetrahedron Letters*, 41, 7925–7928.

Jew, S.-S., Lim, D.-Y., Bae, S.-Y., *et al.* (2002) Enantioselective synthesis of (2*R*,3*S*)-(+)-catechin. *Tetrahedron Asymmetry*, 13, 715–720.

Kawamoto, H., Nakatsubo, F. & Murakami, K. (1989) Synthesis of a condensed tannin model compound, 4-(2,4,6-trihydroxyphenyl)flavan-3,3′,4′,5,7-pentaol. *Journal of Wood Chemistry and Technology*, 9, 35–52.

Kitade, M., Ohno, Y., Tanaka, H. & Takahashi, T. (2006) An efficient synthesis of (±)-epigallocatechin gallate by reductive intramolecular etherification. *Synlett*, 2827–2829.

Kozikowski, A. P. & Tückmantel, W. (2008) Methods for synthesizing the cocoa-derived oligomeric epi-catechins – Observations on the anticancer activity of the cocoa polyphenols In: *Recent Advances in Polyphenol Research* (eds F. Daayf & V. Lattanzio), Vol. 1, pp. 88–112. Wiley-Blackwell, Oxford.

Krohn, K., Ahmed, I. & John, M. (2009) Enantioselective synthesis of flavan-3-ols using a Mitsunobu cyclization. *Synthesis*, 5, 779–786.

Li, L. & Chan, T.H. (2001) Enantioselective synthesis of epigallocatechin-3-gallate (EGCG), the active polyphenol component from green tea. *Organic Letters*, 3, 739–741.

Liu, Y., Li, X., Lin, G., *et al.* (2008) Synthesis of catechins via thiourea/AuCl(III)-catalyzed cycloalkylation of aryl epoxides. *Journal of Organic Chemistry*, 73, 4625–4629.

Marais, J.P.J., Ferreira, D. & Slade, D. (2005) Stereoselective synthesis of monomeric flavonoids. *Phytochemistry*, 66, 2145–2176.

Nay, B., Arnaudinaud, V., Peyrat, J.-F., *et al.* (2000) Total synthesis of isotopically labelled flavonoids, 2. $^{13}$C-Labelled (±)-catechin from potassium [$^{13}$C]cyanide. *European Journal of Organic Chemistry*, 2000, 1279–1283.

Ohmori, K. & Suzuki, K. (2012) Synthetic strategies and tactics for oligomeric proanthocyanidins. *Current Organic Chemistry*, 16, 566–577.

Ohmori, K., Takeda, M., Higuchi, T., *et al.* (2009) Stereocontrolled synthesis of (−)-afzelechin: general route to catechin-class polyphenols by solving an $S_N2$ vs. $S_N1$ Problem. *Chemistry Letters*, 38, 934–935.

Ohmori, K., Yano, T. & Suzuki, K. (2010) General synthesis of epi-series catechins and their 3-gallates: reverse polarity strategy. *Organic & Biomolecular Chemistry*, 8, 2693–2696.

Onda, M., Li, S., Li, X., *et al.* (1989) Heterocycles, XXIV. Synthesis of optically pure 2,3-*trans*-5,7,3,4,5-pentahydroxyflavan-3,4-diols and comparison with naturally occurring leucodelphinidis. *Journal of Natural Products*, 52, 1100–1106.

Sharma, P. K., Kolchinski, A., Shea, H.A., *et al.* (2007) Scale-up synthesis of two naturally occurring procyanidins: (−)-epicatechin-(4*β*, 8)-(+)-catechin and (−)-epicatechin-3-*O*-galloyl-(4*β*, 8)-(−)-epicatechin-3-*O*-gallate. *Organic Process Research & Development*, 11, 422–430.

Shi, Z. & He, C. (2004) An Au-catalyzed cyclialkylation of electron-rich arenes with epoxides to prepare 3-chromanols. *Journal of the American Chemical Society*, 126, 5964–5965.

Stadlbauer, S., Ohmori, K., Hattori, F. & Suzuki, K. (2012) A new synthetic strategy for catechin-class polyphenols: concise synthesis of (−)-epicatechin and its 3-*O*-gallate. *Chemical Communications*, 48, 8425–8427.

Tanaka, H., Miyoshi, H., Chuang, Y.C., *et al.* (2007) Solid-phase synthesis of epigallocatechin gallate derivatives. *Angewandte Chemie International Edition*, 46, 5934–5937.

Tanaka, H., Yamanouchi, M., Miyoshi, H., *et al.* (2010) Solid-phase synthesis of a combinatorial methylated (±)-epigallocatechin gallate library and the growth-inhibitory effects of these compounds on melanoma B16 cells. *Chemistry an Asian Journal*, 5, 2231–2248.

Tanaka, H., Chino, A. & Takahashi, T. (2012) Reagent-controlled stereoselective synthesis of (±)-gallo- and (±)-epigallo-catechin gallates. *Tetrahedron Letters*, 53, 2493–2495.

Tückmantel, W., Kozikowski, A.P. & Romanczyk, L.J. Jr., (1999) Studies in polyphenol chemistry and bioactivity. 1. Preparation of building blocks from (+)-catechin. Procyanidin formation. Synthesis of the cancer cell growth inhibitor, 3-*O*-Galloyl-(2*R*,3*R*)-epicatechin-4*β*, 8-[3-*O*-galloyl-(2*R*,3*R*)-epicatechin]. *Journal of the American Chemical Society*, 121, 12073–12081.

Tsunoda, T., Otsuka, J., Yamamiya, Y. & Itô, S. (1994) *N,N,N′,N′*-Tetramethylazodicarboxamide (TMAD), a new versatile reagent for Mitsunobu reaction. Its application to synthesis of secondary amines. *Chemistry Letters*, 539–542.

Van Rensburg, H., Van Heerden, P.S. & Ferreira, D. (1997) Enantioselective synthesis of flavonoids. Part 3. *Trans-* and *cis*-flavan-3-ol methyl ether acetates. *Journal of Chemical Society, Perkin Transactions* 1, 3415–3422.

Wan, S.B., Ping Dou, Q. & Chan, T.H. (2006) Regiospecific and enantioselective synthesis of methylated metabolites of tea catechins. *Tetrahedron*, 62, 5897–5904.

Weinges, K. (1958) Über Catechine und ihre Herstellung aus Leukoanthocyanidin-hydraten. *Justus Liebigs Annalen der Chemie*, 615, 203–209.

Wu, X.-Y., She, X. & Shi, Y. (2002) Highly enantioselective epoxidation of $\alpha,\beta$-unsaturated esters by chiral dioxirane. *Journal of the American Chemical Society*, 124, 8972–8973.

Yoshida, A., Hirooka, Y., Sugata, Y., *et al.* (2011) Concise synthesis of catechin probes enabling analysis and imaging of EGCg. *Chemical Communications*, 47, 1794–1796.

Zaveri, N.T. (2001) Synthesis of a 3,4,5-trimethoxybenzoyl ester analogue of epigallocatechin-3-gallate (EGCG): a potential route to the natural product green tea catechin, EGCG. *Organic Letters*, 3, 843–846.

# Index



*Recent Advances in Polyphenol Research, Volume 4*, First Edition.
Edited by Annalisa Romani, Vincenzo Lattanzio, and Stéphane Quideau.